DIGITALIZATION AND ANALYTICS FOR SMART PLANT PERFORMANCE

DIGITALIZATION AND ANALYTICS FOR SMART PLANT PERFORMANCE

Theory and Applications

FRANK (XIN X.) ZHU
Honeywell UOP, Des Plaines, IL, USA

WILEY

AIChE
The Global Home of Chemical Engineers

A Joint Publication of the American Institute of Chemical Engineers and John Wiley & Sons, Inc.

Published by John Wiley & Sons, Inc., Hoboken, New Jersey.
Published simultaneously in Canada.

Library of Congress Cataloging-in-Publication Data:

Names: Zhu, Frank, 1957– author.
Title: Digitalization and analytics for smart plant performance : theory and applications / Frank (Xin X.) Zhu.
Description: First edition. | Hoboken, New Jersey : Wiley-AIChE, [2021] | Includes bibliographical references and index.
Identifiers: LCCN 2020040694 (print) | LCCN 2020040695 (ebook) | ISBN 9781119634034 (cloth) | ISBN 9781119634119 (adobe pdf) | ISBN 9781119634102 (epub)
Subjects: LCSH: Chemical plants–Management–Data processing. | Chemical engineering–Data processing. | Industrial productivity–Data processing. | Information storage and retrieval systems–Chemical engineering.
Classification: LCC TP155.6 .Z49 2021 (print) | LCC TP155.6 (ebook) | DDC 660.0285–dc23
LC record available at https://lccn.loc.gov/2020040694
LC ebook record available at https://lccn.loc.gov/2020040695

Cover Design: Wiley
Cover Image: © Yagi Studio/Getty Images
© Thomas M. Scheer/EyeEm/Getty Images
© MR.Cole_Photographer/Getty Images
© Yuichiro Chino/Getty Images

Set in 10/12pt TimesLTStd by SPi Global, Pondicherry, India

10 9 8 7 6 5 4 3 2 1

To Jane, Kathy, and Joshua
These three remain: faith, hope, and love.
The greatest of these is love.
1 Corinthians 13:13

CONTENTS

PREFACE xiii

ACKNOWLEDGMENTS xvii

PART 1 CHALLENGES AND OPPORTUNITIES FOR DIGITALIZATION 1

1 Challenges for Operation Excellence 3

1.1 Introduction, 3
1.2 Operation Activities in a Process Plant, 4
1.3 The Major Challenges Facing the Industries, 5
1.4 The Methodology of Connected Plant, 11
1.5 Digitalization Enabling Connected Plant, 12
1.6 What is the Digitalization Journey?, 18
1.7 Overview of the Book Structure, 19
References, 21

2 Mission of Connected Plant 23

2.1 What is Connected Plant?, 23
2.2 Major Functions of Connected Plant, 24
2.3 Digital Twins: The Core of Connected Plant, 27
2.4 Conclusions, 32
References, 33

3 Data Analytics for Operation Excellence **35**

3.1 Introduction, 35
3.2 Process Data Overview: Characteristics and Attributes, 37
3.3 Unique Attributes of Process Data Analytics, 39
3.4 Model Types and Characteristics, 40
3.5 First Principle Modeling and its Characteristics, 42
3.6 Statistic Modeling and its Characteristics, 45
3.7 Optimization Models, 47
3.8 Artificial Intelligence (AI) and Machine Learning (ML) Models, 50
3.9 Put All Together: Digital Twin as a Data Science Platform, 55
References, 59

PART 2 MODEL THINKING FOR SMART OPERATIONS **63**

4 Statistics Basics **65**

4.1 Introduction, 65
4.2 Normal Distribution, 65
4.3 Conditional Probability, 72
4.4 Bayes' Probability, 73
4.5 Statistic Tests, 75
References, 84

5 Advanced Statistic Modeling **85**

5.1 Introduction, 85
5.2 Distribution Models, 85
5.3 Correlation Models, 94
5.4 Advanced Modeling Techniques, 101
5.5 Data Mining 106
5.6 Summary, 107
References, 107

6 Rigorous Process Modeling **109**

6.1 Introduction, 109
6.2 Reaction Kinetic Modeling, 110
6.3 Reactor Types and Modeling, 126
6.4 Integrated Kinetics and Reactor Modeling, 131
6.5 Catalyst Deactivation Root Causes and Modeling, 135
6.6 Distillation Modeling, 136

6.7 Process System Modeling and Simulation, 138
6.8 Separation Technology Overview, 142
References, 144

7 Linear Optimization Modeling **147**

7.1 Introduction, 147
7.2 Linear Optimization for Planning, 148
7.3 How to Deal with Nonlinear Terms?, 151
7.4 Delta Vector as Linear Approximation of Nonlinear Yield Models, 154
7.5 Successive Linear Programing (SLP) Approach, 159
References, 160

8 Nonlinear Optimization Modeling **161**

8.1 Introduction, 161
8.2 Successive Quadratic Programming (SQP) Approach, 162
8.3 Local Versus Global Optimum, 162
8.4 Optimality Conditions, 166
8.5 Nonlinear Process Optimization Model, 167
8.6 Stochastic Programming, 171
8.7 Simulation-Based Optimization, 178
8.8 A Case Study for Process Optimization, 180
8.9 Concluding Remarks, 188
References, 190

9 Process Control and APC Modeling **193**

9.1 Introduction, 193
9.2 Process Modeling in Control, 194
9.3 Regulatory Control: Managing Individual Variables, 207
9.4 PID Controller Modeling, 211
9.5 Advanced Process Control (APC), 221
References, 233

10 AI And Machine Learning Modeling **235**
Amit Gupta and Frank (Xin X.) Zhu

10.1 Introduction, 235
10.2 Artificial Neural Networks, 235
10.3 Key Concept in ML: Perceptron, 238
10.4 Machine Learning, 242
10.5 Ml Applications in the Process Industry, 246
References, 248

PART 3 CONNECTED PLANT FOR SMART OPERATIONS **251**

11 Connected Metering and Measurements **253**
Martin Bragg

11.1 Introduction, 253
11.2 Review of Metering Devices, 254
11.3 Connected Metering, 258
11.4 Positive-Unexpected Consequences of the Digital Economy, 267
11.5 The Outlook for Connected Metering, 269
11.6 Conclusions, 273
References, 274

12 Connected Asset and Safety Management **275**
Frank (Xin X.) Zhu and Tony Downes

12.1 Introduction, 275
12.2 Review of Different Maintenance Strategies, 276
12.3 The Concept of Operating Windows, 280
12.4 The Major Gaps in Current Asset Management, 283
12.5 Digitalized Asset Management, 284
12.6 Process Safety Management, 290
12.7 Case Study: APM Drives Capacity Improvement, 299
Reference, 301

13 Integrated Production Planning and Process Control **303**

13.1 Introduction, 303
13.2 Current Practice in Site-Wide Optimization and Control, 304
13.3 Simultaneous Approach for Site-Wide Optimization and Control, 304
13.4 General Decomposition Strategy, 309
13.5 MPC-Based Integration Approach, 314
13.6 Rigorous Model-Based Integration Approach, 322
13.7 Comparison Between the MPC and Rigorous Model-Based Approaches, 324
References, 325

14 Digitalizing the Energy Management **327**

14.1 Introduction, 327
14.2 The Concept of Energy Intensity, 328
14.3 Energy Benchmarking for Processes, 337
14.4 The Concept of Key Indicators, 340
14.5 Set Up Targets for Key Indicators, 346

14.6 Economic Evaluation for Key Indicators, 350
14.7 Site-Wide Energy Management Strategy, 354
14.8 Digital Twin for Energy Management, 360
14.9 Establishing Energy Management System, 361
References, 365

15 Integrating the Workflows **367**
Frank (Xin X.) Zhu and Joe Ritchie

15.1 Introduction, 367
15.2 Key Elements of Industrial Supply Chain, 368
15.3 Little Integration of Supply Chain Work Processes, 381
15.4 Gaps Existing in Current Supply Chain Management, 382
15.5 Integrated Work Process for Supply Chain Management, 383
15.6 Supply Chain Digital Twin: One Platform for Workflow Integration and Automation, 385
15.7 Integration of Engineering Models with Supply Chain Digital Twin, 387
References, 388

16 Digitalizing the Workforce **389**
Rohan McAdam

16.1 Introduction, 389
16.2 Enabling the Workforce, 390
16.3 Empowering the Workforce, 398
16.4 Digitalization Challenges, 410
16.5 Summary, 416
References, 416

PART 4 DIGITAL SOLUTIONS FOR SMART OPERATIONS **419**

17 Honeywell Forge: The Platform for Connected Plant **421**
Matt Burd and Frank (Xin X.) Zhu

17.1 Honeywell Forge: A Digital Platform for Connected Plant, 421
17.2 IIoT for Data Infrastructure, 421
17.3 How It Works?, 423
17.4 Intelligent Models Behind Digital Twins in Honeywell Forge, 429
17.5 Cybersecurity, 434
Reference, 436

18 Digital Reediness Assessment and Six-Step Digitalization Journey 437

18.1 Introduction, 437
18.2 Digital Readiness Assessment, 438
18.3 The Six-Step Digitalization Journey, 449
18.4 Recommendations: A Digital Transformation Management System, 454
18.5 Establishing a Digital Transformation Management System, 455
References, 457

19 Digital Project Evaluation and Development 459

19.1 Introduction, 459
19.2 Business Case Evaluation, 459
19.3 Digital Project Development Steps, 461
19.4 Remarks on Digital Project Development, 465
19.5 S-Curve for Project Review and Management, 469
19.6 Basics of Economic Analysis, 472
Reference, 475

20 Application Case Studies 477

20.1 Introduction, 477
20.2 Application Cases from Digital Twins, 477
20.3 Applications from Other Digital Projects, 481
References, 506

INDEX 507

PREFACE

WHY DIGITALIZATION FOR INDUSTRIES?

In today's competitive environment, process plants are under more pressure than ever to deliver improved operating performance. Making the things even more challenging, the workforce is evolving, as an information technology savvy, but less experienced generation moves into roles once held by industry veterans, who have retired or are looking to retire. These trends are making the industrial norms to be rewritten, affecting profitability, reliability, and safety.

Thus, the industry is at a turning point. The needs for capturing deep knowledge from veterans, improving operation performance from data and models, and automating workflows via digital solutions must be addressed. Thus, the answer lies in the digital transformation, namely digitalization. To overcome the challenges, pacesetting companies have already applied digitalization solutions. The experience demonstrates that digitalization can turn a distracted organization—that is bogged down in reacting to day-to-day issues—into an agile, well-oiled machine that proactively anticipates issues and organizes to prioritize and solve them before they occur and escalate.

To achieve the best digitalization outcome, it requires optimal integration of information technology (IT), operation technology (OT), and engineering technology (ET) and data analytics technology (DT). IT consists of computers, networking devices, and other devices to generate, process, store, secure, and exchange all forms of electronic data. OT mainly includes industrial sensors and control systems, while ET comprises process simulation, models, and best practices for operation. DT consists of data analytics and machine learning technology.

Thus, industrial companies can rely on digital solutions to integrate different data sources with advanced analytics to make decisions in real time to run their production processes more efficiently. Done right, it can empower personnel and makes the decision and implementation process more effective and can solve problems that could not be solved before.

However, there are some misconceptions for digitalization in the process industries. Many people thought they had been doing digital solutions for measurements, DCS, control, and automation for decades. This kind of digital efforts is largely conducted though IT department, and it is called digitization. In this book, we will focus on digitalization, which aims to create new value beyond digitization by enhancing companies' business capabilities, workflow management, and better use of date to drive new insights across the entire operation value chains. Experience demonstrated that mere digitization from IT deployment is no longer a competitive advantage without digitalization of business and technology integration, software and hardware integration as well as work processes integration. These integrations create significant value for companies and plants and will lead to new business opportunities and growth.

MOTIVATION OF WRITING THIS BOOK

There are no dedicated books available to discuss digitalization for the process industries. To fill this gap, the book will explain the principles and practices that lead to successful transformation to Connected Plant performance: what are the gaps associated with traditional work process and how can the gaps be closed by applying Connected Plant concept. From this book, one can learn:

- How to create a digital vision and develop digital tasks?
- What takes to establish Connected Plant performance?
- How to digitally enhance profit, reliability, operation, and control?
- How to integrate work processes digitally?
- How to digitally connect people and enhance workforce skills?

While many companies have looked at various ways for implementing digitalization, very few have taken the holistic view of applying digital transformation to close gaps in current business and technical processes. Digital solutions should be developed centered on the pain points identified with a clear perspective of prioritized opportunities and with the outcome-based approach. The book also share some successful applications to show the evidence of what some companies are doing to take advantage of industrial digitalization and to thrive for now and in coming years.

OVERVIEW OF THIS BOOK

The intent for this book is to break down the myth surrounding digitalization and data science. This book describes how a new breed of real-time analytic solution enables continuous monitoring, optimization, and control of key plant operating conditions and allows for the early indication of performance degradation and the appropriate prioritization to address the situation. By taking advantage of advanced "digital twin" technology enabling the most advanced monitoring, analytical, and predictive capabilities, process engineers can implement around-the-clock monitoring of plant data and ongoing operational health checks and take informed actions to close performance gaps.

As the book covers a wide range of topics, I have attempted to organize the materials in a contextual manner that aids the reader to locate the relevant materials quickly, to be able to understand them readily, and to apply them in the right context. Furthermore, the structure of the book is carefully designed to help readers to avoid losing sight of the forest for the trees. The book starts with provision of an overall context of operation excellence, followed with concepts and models to gain basic understanding of data science and machine learning, gradually transitions to practical needs and methods for instrument, asset, and safety management, from production planning to process control, from workflow integration to workforce digitalization, and culminates in developing relevant digital solutions and infrastructure to sustain the benefits.

To accomplish the above goals, four parts with 20 chapters in total are provided in this book. Three chapters in Part 1 describe the major challenges that industries are facing and the need of digital solutions for Connected Plant. In addition, it provides an overview of data science for industrial applications. Seven chapters in Part 2 discuss individual modeling techniques including first principles, statistics, optimization, artificial intelligence, and machine learning, as many-model thinking and applications can empower digital solutions to convert data to new insights and actions. Six chapters in Part 3 focus on individual components of Connected Plant, which aim to achieve operation excellence for economical, efficient, reliable, and safe operation. The last four chapters in Part 4 explain the data infrastructure and open architect for the digital solutions for operation excellence. Furthermore, digital readiness assessment for industrial companies is explained as well as an effectivity digital journey is described with a roadmap showing signposts of milestone progress. Finally, application case studies in Chapter 20 are provided to share the success stories in applying digital solutions for operation excellence.

THE AUDIENCE

This book is written with the following people in mind: company executives and plant managers, engineers, and operators as well as data scientists working in the process industries. The book also targets engineering and data analytics firms who face challenges and wish to find opportunities for helping operating companies to improved process operation and design.

In addition, this book can be valuable to university students, and to those who want to equip themselves with real-world problems and practical methods that will allow them to become more employable after graduation. The contents of this book are based on the knowledge and experience the authors gained over many years in research, engineering, consulting, and service support for chemical processes.

The book is aimed to provide understanding of basic theory, practical tools, and applications to people who face challenges and wish to find opportunities for improved operation. I hope that this book can demystify digitalization and data science with clearly explained concepts, theories, methods, and tools in the digital field and hope that this book is able to convey them in a straightforward and practical manner.

It is my sincere hope that readers are able to take away concepts, methods, and techniques, which are required to identify, assess, and implement digital opportunities for operation excellence. More importantly, I hope that this book can help readers to build mental models for critical thinking to enhance your capability to reason, explain, predict, communicate, and explore.

ACKNOWLEDGMENTS

Clearly, it was not a small effort to write this book; but it was the strong need to help people understand the practical methods for industrial digitalization, which spurred me on. In this endeavor, I owe an enormous debt of gratitude to many colleagues at Honeywell for their generous support in this effort.

First of all, I am grateful to Honeywell leaders including Scott Zhang, William Yu, Lubo Zhou, Kelly Seibert, Carrie Eppelheimer, John Simley, and Christophe Romatier for identifying the market needs for this book and showed their strong support. Furthermore, I wish to recognize Que Dallara, president of Honeywell Connected Enterprise (HCE), whose vision enlightened the writing of this book.

Several subject matter specialists made good writing contribution to this book. My special gratitude goes to Martin Bragg for Chapter 11, Rohan McAdam for Chapter 16, and Matt Burd for Chapter 17 as well as Amit Gupta, Tony Downes and Joe Ritchie who contributed to Chapters 10, 12 and 15, respectively.

I am very grateful to many colleagues for their constructive suggestions and comments on the materials contained in this book. My special thanks go to Paul McLaughlin, Ginger Shao, and Yanling Wu for their critic review of the book with constructive comments, as well as Mark James, and Jason Skinder for their IP and legal review. I would also like to thank the following people for their valuable comments: Jayson Keels for Chapters 4 and 5, Robert Tsai for Chapters 6 through 8, Selman Erisken for Chapter 7, Grzegorz Oleszczuk for Chapter 9, Joseph Lu for Chapter 13, Joe Montalbano for Chapter 15, and Paul McLaughlin for Chapters 11, 12, and 16, respectively. I would like to express my sincere gratitude to all my colleagues for their help with the book and my debt to them is very great, and I apologize if any names are unmentioned. I would like to stress that any deficiencies are my responsibility as this book reflects my own opinions but not that of UOP and Honeywell.

I would also like to thank my copublishers, AIChE and John Wiley, for their help. Special thanks go to Cindy (Cynthia) Mascone and Kate McKay at AIChE, Michael Leventhal, Beryl Mesiadhas, and Elke Morice-Atkinson at John Wiley for their guidance as well as Vishnu Priya and Grace Paulin Jeeva for managing the copyediting and typesetting.

FRANK (XIN X.) ZHU

Des Plaines, Illinois, USA
May 30, 2020

PART 1

CHALLENGES AND OPPORTUNITIES FOR DIGITALIZATION

This part consists of three chapters. Chapter 1 provides the challenges that industries are facing, which defines the needs for digitalization, while Chapter 2 introduces the concept of Connected Plant, which can overcome the challenges. Furthermore, Chapter 3 provides an overview of big data and data analytics for industrial applications.

The industry is at a turning point. On one side, environmental, quality, and safety regulations are becoming ever stricter, and meeting the regulations is the license to operate manufacturing plants. On the other hand, the workforce is evolving, as an information technology (IT) savvy, but less experienced generation moves into roles once held by industry veterans, who have retired or are looking to retire. These trends are making the industrial norms to be rewritten, negatively affecting profitability, reliability, and safety.

The answer lies in the digital transformation. To overcome the challenges, pacesetting companies have already applied digitalization. The experience demonstrates that digitalization can turn a distracted organization—that is bogged down in reacting to day-to-day issues—into an agile, well-oiled machine that proactively anticipates issues and organizes to prioritize and solve them before they occur and escalate. Thus, industrial companies need digital solutions to integrate different data sources and advanced analytics incorporating with organization culture and day-to-day processes to make more informed decisions in real time to run their production processes more efficiently. Done right, it can empower personnel and makes the decision and implementation process more effective and can solve problems that could not be solved before. It can build a culture with digital

Digitalization and Analytics for Smart Plant Performance: Theory and Applications,
First Edition. Frank (Xin X.) Zhu.

information at the core. In short, digitalization seeks to achieve all that we sought from digitization in the past, albeit more effectively and efficiently. More importantly, digitalization aims to create new value beyond digitization by enhancing companies' business capabilities and achieving operation excellence across the entire operation value chains.

The best outcome for digitalization is achieved by the proper integration of IT (data centric), OT (Operation Technology; control centric), and ET (Engineering Technology; model centric), which is designed to monitor current events and predict new events, detect early warning signs of failure, and make changes to process operations and enterprise business for better profit, efficiency, reliability, and safety. The most representative of digitalization is digital twin. By taking advantage of advanced "digital twin" technology, process engineers can implement around-the-clock monitoring of plant performance, and ongoing operational health checks and optimization to maintain the best plant performance every day.

1

CHALLENGES FOR OPERATION EXCELLENCE

1.1 INTRODUCTION

Environmental, quality, and safety regulations are becoming ever stricter, and meeting the regulations is the license to operate manufacturing plants. In today's competitive environment, process plants are under more pressure than ever to deliver improved operating performance. Making the things even more challenging, the workforce is evolving, as an information technology (IT) savvy, but less experienced generation moves into roles once held by industry veterans, who have retired or are looking to retire. These trends are making the industrial norms to be rewritten, affecting profitability, reliability, and safety.

Thus, the industry is at a turning point. The answer lies in the digital transformation. To overcome the challenges, pace-setting companies have already applied digital transformation. The experience demonstrates that digitalization can turn a distracted organization—which is bogged down in reacting to day-to-day issues—into an agile, well-oiled machine that proactively anticipates issues and organizes to prioritize and solve them before they escalate. Thus, industrial companies need digital solutions to integrate different data sources with advanced analytics to make decisions in near real time to run their production processes more efficiently. Done right, it can empower personnel and makes the decision and implementation process more effective and can solve problems that could not be solved before. It can build a culture with digital information at the core. In short, digitalization seeks to

Digitalization and Analytics for Smart Plant Performance: Theory and Applications,
First Edition. Frank (Xin X.) Zhu.

achieve all that we sought from digitization in the past, albeit more effectively and efficiently. More importantly, digitalization aims to create new value beyond digitization by enhancing companies' business capabilities and achieving operation excellence across the entire operation value chains.

For most industrial plants, mere Industrial Internet of Things (IIoT) deployment is no longer a competitive advantage without understanding the implications of these insights. These advantages could be anything from improving safety to productivity to operational efficiency to workflow integration and automation. This is the part where true business outcome is delivered. To achieve the best digitalization outcome, it requires optimal integration of IT, operation technology (OT), engineering technology (ET), and data technology (DT). IT consists of computers, networking devices, and other devices to generate, process, store, secure, and exchange all forms of electronic data. OT mainly includes industrial sensors and control systems, while ET comprises process models simulation and optimization and best practices for operation. DT consists of data analytics and machine learning (ML) technology.

This integration of IT/OT/ET/DT creates significant value for companies and will lead to new business opportunities and growth. A properly designed digital twin is the best representation of this integration. The book will focus on the methodology of Connected Plant, which is designed to achieve the optimal integration of IT, OT, ET, and DT for operation excellence in the form of digital twins.

1.2 OPERATION ACTIVITIES IN A PROCESS PLANT

Process operation activities can be represented as a hierarchy (Figure 1.1). The foundation of the structure is composed of field devices (Level 1) such as sensors, transmitters, actuators, and valves that measure process variables such as flows,

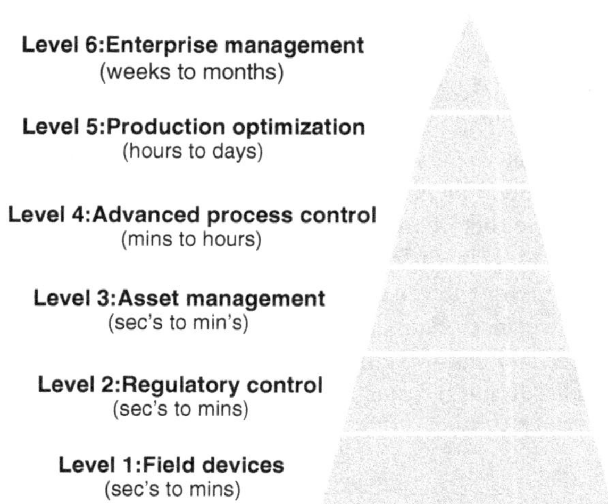

FIGURE 1.1. Six-level hierarchy for process control and optimization.

compositions, temperatures, and pressures, which are required for any industrial control system.

The next level (Level 2) consists of regulatory controllers, which maintain measured values within specified limits. Process control plays a key role in ensuring process safety and protecting personnel, equipment, and the environment (Level 3).

While regulatory controllers tackle each variable individually, the advanced process control such as multivariable and constraint controllers (Level 4) evaluates multiple variables simultaneously and enables an entire process unit close to a set of limiting constraints. To determine the optimum set points for a set of controllers, real-time optimization (Level 5) is employed, which also consider advanced control strategies beyond control modules, process supervision and coordination, shutdowns and start-ups, reliability and safety management, maintenance scheduling, etc.

At Level 6, enterprise management involves organizing and coordinating the operations of multiple plants. This level is the highest in the plant hierarchy and handles various functions including feed purchasing, production planning, scheduling, optimization procedures relevant to general production planning, coordination between different plants, feed movement and product delivery, etc.

The successful implementation of these six-level activities is a critical factor in making plant operation as profitable and reliable as possible. Levels 1, 2, and 3 are required for all process plants while the activities in Levels 4, 5, and 6 are designed for maximizing profit. Hence, the decision to implement the latter activities is driven by the company's financial priorities and economic considerations. The frequency of execution is much lower for higher-level activities, and the time scale of each activity increases up the hierarchy from less than a second to days and months. This is due to the need of production change cycle as well as the increase in computational requirements and analysis time from the lowest level to the highest. These activities are related and should be carefully coordinated. Levels 4 and 5, which focus on the control functions of advanced process control and real-time optimization for a process plant, are especially beneficial for a plant's operations.

In the above six levels, each layer depends on the previous layer in the hierarchy, and the control performance of an individual layer directly affects the stability of the process, the quality of the product, and the costs associated with making the product. However, running a single unit at its local optimum is not necessarily the best strategy for achieving overall profitability of the facility. This is where the production management layer steps in to manage the individual units' APCs in concert to accomplish plant-wide optimization. Furthermore, enterprise management coordinates multiple plant operations via optimizing feed selection and product generation and delivery to the market. That is where global optimum can be achieved for a company as a whole.

1.3 THE MAJOR CHALLENGES FACING THE INDUSTRIES

Changing market conditions put pressure on industrial companies. Increased competition and consolidation force businesses to speed up construction, engineering, and design cycles. Environmental, quality, and safety regulations are becoming ever stricter, and meeting the regulations is the license to operate manufacturing plants. Making the things even more challenging, the workforce is evolving, as an IT savvy, but less experienced generation moves into roles once held by industry veterans, who have retired or are looking to retire. These trends are making the industrial norms to be rewritten, affecting profitability, reliability, and safety.

1.3.1 Challenge 1: Supply Chain Management

In today's business environment, a lack of agility can hinder an organization's ability to be competitive. With high capital costs and ever-increasing energy expenditures, an organization can become locked into patterns where it is implementing shortsighted, cost-cutting measures that can adversely affect operational productivity—decreasing overall market responsiveness.

The bottom line is: maintaining a big picture view of key metrics—such as selecting and purchasing cheap raw materials and increasing manufacturing throughput, as well as transportations and logistics accuracy—helps minimize overall costs. This is done by optimizing the whole supply chain including business and production planning and operation scheduling of production processes, which plays key roles in the search for better profits.

Consider the oil refining sectors as an example. The supply chain of an integrated oil company stretches from the production and purchasing of crude oil to customers buying petrochemical products or fuel for heating or transport. Industrial supply chain consists of feed purchasing, delivery, feed mixing and scheduling, operation, and control (Figure 1.2). The management of such enterprise involves making decisions on what feed to buy and where and when? What processes to operate and how? What products to make? When each product is blended to make on-spec product to sell? What maintenance schedule for key equipment? How to integrate refining and petrochemical production? These decisions will be made repeatedly when demands and supplies change, production upsets occur, and transportation logistics change. A typical medium refinery may make products worth around \$10 million/day. The difference in profit margin between best and worst supply chain management could be at least around \$1/barrel, which is amount to \$100,000/day for a refinery of 100,000 barrel/day capacity. The value for the gap is very significant. Thus, refineries strive to optimize operation effectively via digital systems in order to close the gap and capture the opportunity.

Digital solutions for supply chain management can take advantage of cheap shale gas, renewable feedstocks and a wide variety of crudes that are available today while meeting product demand. In addition, many refiners have revamped their technology and expand petrochemical capacity so that they can achieve flexible operation between making transportation fuels and petrochemical products. These two trends have significantly increased opportunities for many refineries, helping them maximize and sustain margins in a competitive global environment. To take advantage of these options, agility in feedstock selection and accuracy in planning and scheduling as well as optimizing process operation and equipment maintenance to increase utilization is paramount.

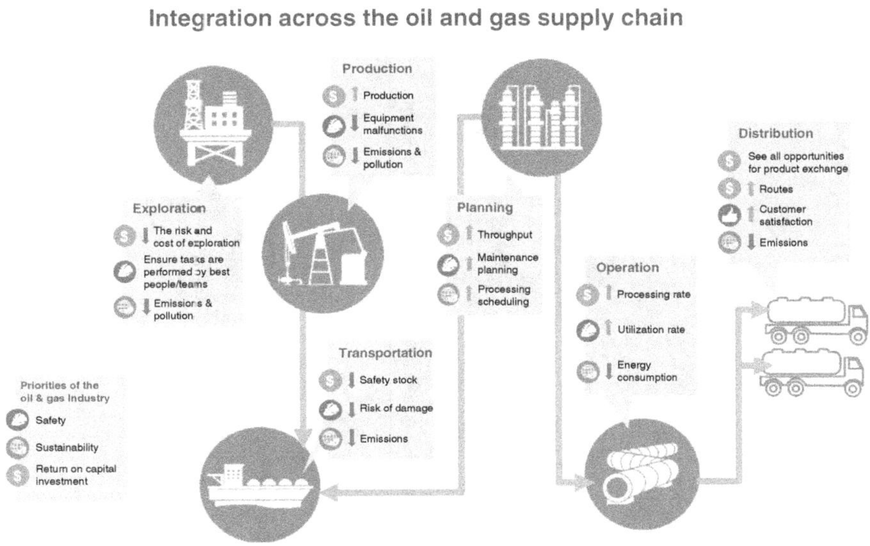

FIGURE 1.2. Manufacturing supply chain.

1.3.2 Challenge 2: Process Efficiency Improvement

The best process efficiency consists of three components: chemical efficiency, design efficiency, and operation efficiency. Chemical efficiency is the efficient use of raw materials. To get the best chemical efficiency, the key is to select the most cost-effective raw materials and select the best fit catalyst. A molecular modeling platform integrating reaction chemistry, kinetics modeling, and catalyst development is the key enabler to achieve high chemical efficiency. Design efficiency is about achieving the best technology selection and process configurations to obtain optimal system designs featuring low capital and operating costs. Process integration and intensification is the key enabler to achieve high design efficiency. Operation efficiency is how the teams operate the processes with high economic profit and minimum energy consumption via optimizing process conditions and better control. Integrated process optimization and control is the key enabler to achieve high operation efficiency.

1.3.2.1 Variability It is common that operation variability is a major cause of operation inefficiency. In general, there are two kinds of variability that can be observed in reality: inconsistent operation and consistent but nonoptimal operation. Figure 1.3 represents the operating data of a reboiler heat input as measured by condensate rate in the deheptanizer in a xylene isomerization unit. In Figure 1.3a, the heat input data points appear to be randomly scattered showing an example of inconsistent operation. This is usually caused by either poor control strategy or different operating philosophies used by operators for running the column. In contrast, Figure 1.3b shows a consistent operation, but one which is nonoptimal. In

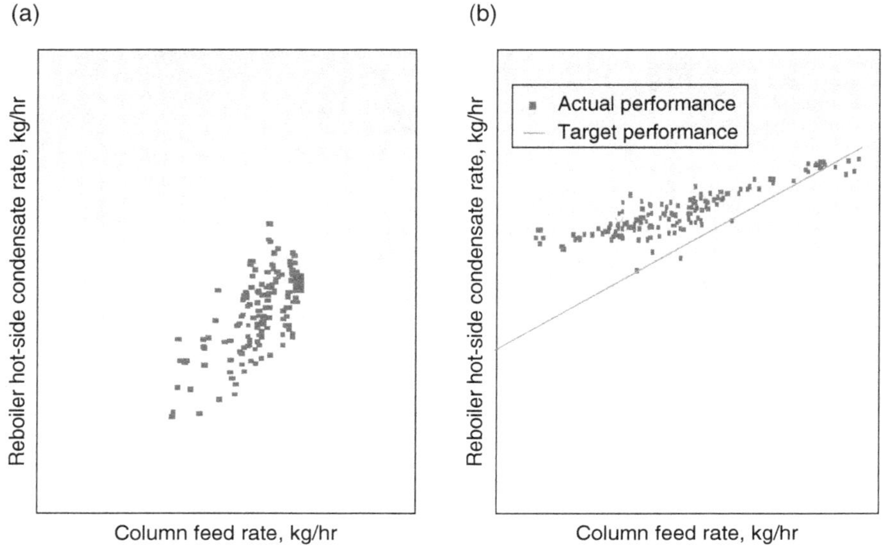

FIGURE 1.3. Two common operating patterns: (a) inconsistent operation and (b) consistent operation but nonoptimal. *Source*: From Zhu (2014). © 2014, American Institute of Chemical Engineers, Inc. Reprinted with permissions of American Institute of Chemical Engineers.

this case, a consistent operating strategy was adopted, but it was far away from the target for adjusting reboiling duty against column feed rate. The target operation represents the minimum reboiling duty required to achieve product specification.

1.3.2.2 Optimization If the operation practice or control strategy improves, the variability could be minimized to achieve more consistent operation (Figure 1.4b) than the base operation in Figure 1.4a, but the performance in Figure 1.4b still remains far away from the specification limit. The limit is usually set by a physical limit such as product purity specification, maximum temperature or pressure, maximum valve opening, maximum vapor loading in a separation column, maximum space velocity in a reactor, and so on.

Digital operation efficiency tools can help operators to improve further (Figure 1.4c), moving the average closer to the limit by adopting an integrated optimization and control strategy.

Digital solutions for process optimization and control can determine optimal reaction and separation operation as well as optimal control set points to maintain higher production efficiency under varying feed and product prices.

1.3.3 Challenge 3: Rigorous Model-Based Control

The fierce competition in operating companies has forced most companies to find ways to optimize plant operation. At the same time, increased computing power has provided opportunity for using comprehensive models to guide APC. In the past,

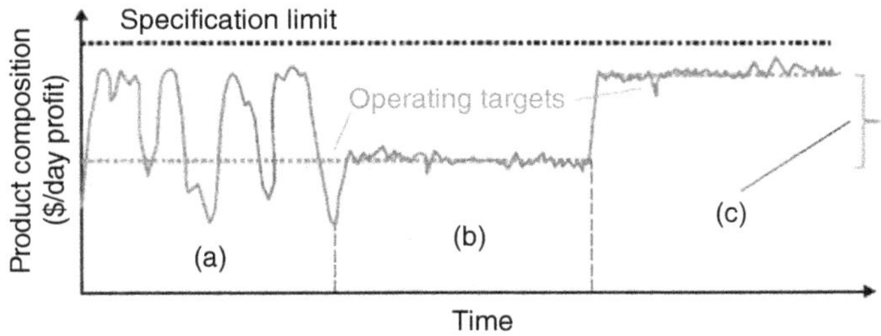

FIGURE 1.4. Operation performance: (a) current operation, (b) better control and reduced variability, and (c) increased profit by optimizing the set point. *Source*: From Zhu (2014). © 2014, American Institute of Chemical Engineers, Inc. Reprinted with permissions of American Institute of Chemical Engineers.

several modeling approaches are applied in the context of APC and regulatory control. These approaches range from empirical correlations to neural network (black box) and to shortcut models to rigorous models.

In recent years, the use of first principle based with autocalibrated models gains much attraction due to its ability in accurate representation of molecular information and nonlinear behavior toward more informed actions and implementation. This fact is even more compounded by the ever-increasing computing power at low cost and capability of data handling and nonlinear solvers for large problems. But it is a double-edged situation. In one hand, rigorous models can capture highly nonlinear behavior at fundamental level and thus can predict effects of changing feed types and process conditions more accurately. On the other hand, it is harder for process engineers to understand the prediction results than traditionally linear equation-based APC. In addition, it could become very difficult for troubleshooting of the root causes in the case of wrong prediction. Aggressive model turning can only shift error from one place to the other. Thus, the challenge is how to capture the benefit of rigorous models but avoid the difficulty of applying it for optimization and control purpose.

1.3.4 Challenge 4: Asset Management

In manufacturing, what is more crucial than the equipment itself? It is downtime, which damages business and costly for manufacturers. That is the reason why asset managers try their best to keep their equipment performing, despite constrained budgets and aging infrastructure. That is why equipment availability is one of the most important metrics for asset management.

Yet many organizations lack the insight and visibility they need to proactively refine their operations in a cost-effective way. In fact many organizations struggle to establish a benchmark from which to build an asset management strategy based on the basic information. Such information might include:

- Do we have a register of assets down to a significant level?
- Do we know the physical location of these assets?

- Do we know how many?
- Do we know the condition of the assets?
- Do we know, or are we able to report in the future, the life cycle costs?
- Have we assessed and determined a risk profile for these assets?

Digital solutions for asset performance management (APM) can reduce costs even as they reduce risk. The APM can analyze patterns, provide early warning signs for equipment failure, and respond with greater agility. The APM can also perform at higher levels of reliability, quality, and safety without needing more resources.

1.3.5 Challenge 5: Safety Management

Chemical process industry is a high-risk industry that can result in catastrophic casualties, loss of properties, and environment damage. Process safety should always come first in the chemical process industry. Effective process safety management is the guarantee for process safety in operation.

For many years, the approach to process safety management was more reactive. When some incident happened, the plant will review and revise its approach and then carry on until the next incident. Although the reactive safety management approach is useful when dealing with technological failures, or unusual events, its focus is on examination of actual failures only and compliance with the minimum safety requirements. This is a bit like driving a car by looking through the rear-view mirror. It is possible to stay in your lane, so long as there are no "curves." This is an expensive way to learn and manage safety. The chance of damaging the corporation's reputation is very real.

All major companies in the process industries have moved beyond this purely reactive mode toward more proactive approaches where major hazards are identified proactively and their risk is estimated and then managed. The team-based reviews are revalidated often in every five years. Over time, process industry companies develop a wide variety of safeguards to protect them against the potential incidents that the hazard identification teams have identified. But no safeguards are perfectly reliable—they tend to deteriorate over time.

This brings us to the newest developments in managing process safety risks— "digital twin for safety." With the real-time safety management (RSM) approach enabled by digital twin solutions, it is an even more proactive approach.

1.3.6 Challenge 6: Competency Management

Competencies are what people need to be successful in performing their jobs including the leaders and workforce. Competencies include all the related knowledge, skills, abilities, and attributes that define the competency of a person's job. This set of context-specific qualities is correlated with superior job performance.

However, the aging of the workforce worldwide has been called the "silver tsunami." With experienced personnel are set to retire, the workplace needs to prepare for a veritable tidal wave of turnover. The challenge is training young staffs to acquire the desired skill sets in timely manner. Digital solutions for competency management could provide effective training tools such as dynamic simulation, virtue reality (VR), augmented reality (AR), and so forth for training young professional to acquire and apply necessary skills and knowledge effectively.

1.3.7 Challenge 7: Workflows Integration and Automation

Technology on its own does not guarantee outcomes. The missing ingredient in order to achieve business outcomes is an organizational culture of excellence implying operating workflows and best practices—the more that these tasks are automated, the more repeatable the outcome. If a start-up or shutdown procedure can be automated, experience shows that the number of unsafe and costly incidents is reduced. If it pays to drive a plant up against its operating constraints, then automating that strategy so it can chase fleeting opportunities for improvement has been shown to generate huge returns.

Automation comes at a cost, however, and with some risk. The cost is the number of sensors required and the need for them to be reliable. The risk is how it behaves in the face of circumstances that was not designed to encounter. Our digitalization philosophy, therefore, is for our customers to invest heavily but wisely in automation and to mitigate the risks through good digital tool design practices and supervision.

Replicating existing business and technical processes in a digital environment does not assure delivery of superior results. Digitalization is more about the improvement of workflows and processes, organizational culture, and the application of knowledge and information in more effective manners rather than "just data." The focus should be on improving the existing best practices and work processes and thereafter finding the digital means to institutionalize automation of as much of these improved workflows as possible.

1.4 THE METHODOLOGY OF CONNECTED PLANT

To overcome the above challenges, the Connected Plant methodology was developed (Edwards, 2016). This methodology aims to connect **people**, **processes**, and **assets** within the organization much better than what has been doing with traditional digital solutions. The Connected Plant wants to achieve the following outcome for customers:

- Every day is your best day of production.
- Everyone in your organization is a leading expert.

- Can maximize the overall profit based on the market conditions and asset capability.
- Can maximize asset utilization via predicting and correcting abnormal situations.
- Can make the company be the best in the industry.

The Connected Plant is more than just data collection and software tools. It is about applying intelligent models to reveal insights. It is about integrating new digital solutions with existing data infrastructure. It is about better management of asset, workflows, and knowledge via digital solutions.

Industrial plants can benefit from being connected since they typically operate below their optimum output capacity due to process variations, unplanned downtime, and inconsistent workforce competency. Reaching and sustaining optimum levels of plant performance requires the transformation of real-time data into actionable insights. This will allow processes to operate at optimum levels for longer periods of time, identify and resolve asset issues before unplanned outages, and help every worker to operate as an industry leading expert. A Connected Plant is a single pane of glass software solution that collects real-time data and runs it through a digital plant model to predict and control deviations leading to optimized plant performance.

1.5 DIGITALIZATION ENABLING CONNECTED PLANT

The Connected Plant creates value through real-time data analytics, models, and optimization to reveal new insights for opportunities leading to turn-key solutions. It also applies advanced analytics and ML techniques to develop new heuristics and enable the solving of issues that we were previously unable to. It leverages the long-standing experience, paired with new digital technologies to resolve customers' issues better, faster consistently. It uses the cloud and edge capabilities to connect experts and project that knowledge back to operating companies. Use of the Connected Plant method can result in faster detection and resolution of operating problems as well as more effective best practice sharing and knowledge management.

Some might say, a digital system is about sensors, computers, software, and IIoT. The answer is yes and no. It is true that a digital system consists of these components; but that is not the point. It is what they do for value creation that matters. Digitalization solutions enable companies to manage day-to-day performance and achieve operation excellence. They can also help companies to response to market price swings quickly, operate the plant at a true optimum in order to squeeze the gap between potential and realized margin, and outmaneuver the competition. Let us dig deeper a little more.

1.5.1 Four Key Components of a Digitalization System

A digital system is an information gathering and processing system that converts data to insight via digital twin models, which helps plant staffs to make more intelligent decisions. There are four major components in a digital system, namely,

IT technology, OT, ET, and data technology (DT). IT consists of computers, networking devices, and other devices to generate, process, store, secure, and exchange all forms of electronic data. OT mainly includes industrial sensors and control systems, while ET various process models simulation and optimization and best practices for operation. DT consists of data analytics and ML technology.

1.5.2 What Is New That Digitalization Can Offer?

Industrial companies have been using traditional IT, OT, and ET in combination for many years. DT is the new component that will be added. So, what is new that the digitalization can offer?

Digitalization is mainly originated from several major breakthroughs in IT and data science in recent years. First is that the computing power has increased exponentially together with cloud technology, both of which make processing and storing of big data possible.

Second, with the advent of Internet of Things (IoT), it connects data from different sources (on-premise, edge, and cloud) and makes the internal and external connectivity possible via smart sensing/wireless sensing technology. The smart sensing and wireless sensing technology can measure equipment performance from remote locations so that the measurements in various locations can be connected to the data center in a cost-effective manner.

Third, digital twin creates a virtual plant based on rigorous and analytics models and data available in plant historian to mimic the real plant. Digital twin can analyze "what-if" scenarios, predicting "what's wrong" events and determining "what's best" conditions in an automated manner. Due to cloud, digital twin can be provided and maintained by third parties, and the predicted results are displayed in cloud for visualization. External specialists can provide technical supported if needed.

Fourth, the advancement in data science technology (DT) and artificial intelligence (AI) introduces new capability of prediction based on statistical and ML algorithms, which are mainly applied for predicting equipment failure mode in manufacturing plants.

Therefore, digitalization, based on new IT, smart sensing, rigorous models, and data science technologies, can connect people, processes, assets, and products in both the physical and digital worlds and enhance operation excellence to a leapfrog new level. It is natural for companies to consider to take advantage of the digitalization for the goodness of the entire organization via improving connectivity among people, assets, process, workflows, and tools under stronger competition in the market and a growing scarcity of skilled staff.

1.5.3 Collective Excellence: The Objectives of Digitalization

Digitalization should be developed for value creation via IIoT, cloud and edge, and models for deriving insights and integration for automated workflows to address end-to-end enterprise touchpoints (Figure 1.5). The overall goal that we wish to accomplish via digitalization is operational excellence, which is defined as "the execution of the business strategy more consistently and reliably than the

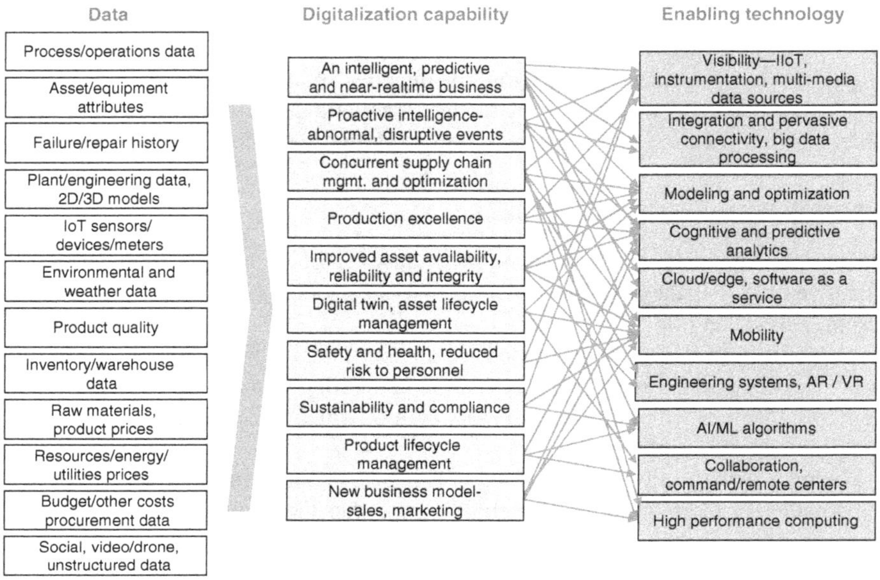

FIGURE 1.5. Overview of industrial digital system.

competition. Operational Excellence is evidenced by results." Given two companies with the same strategy, the operationally excellent company will have lower operational risk, lower operating costs, and increased revenues relative to its competitors, creating value for customers and shareholders (Francisco, 2017).

Specifically, operation excellence includes the following elements:

- Identify and sustain the value from continuous changes.
- Discover new revenue from unexploited feed selection and product gains.
- Maximize process throughout and equipment utilization.
- Predict failure mode with much early warning signs.
- Preventive maintenance.

While many companies have looked at various ways for implementing digitalization, very few have taken the holistic view of applying digital transformation to close gaps in current business and technical processes with a clear perspective of prioritized opportunities and adopt the outcome-based approach. Digitalization transformation should be centered around value creation instead of doing it for the sake of IT technology under peers' pressure.

Companies can see the potential of the digitalization in helping them to improve connectivity among people, assets, tools, and workflows to make the whole company/organization more *collectively intelligent*, agile, and efficient via integrating business, technical, and operation activities.

1.5.4 What Is Collective Intelligence?

Broadly speaking, intelligence includes the capability of problem solving using combination of logic, understanding, creativity, experience, wisdom, and learning as well as emotional knowledge and self-awareness. In the context of this book focusing on industrial topics, we adopt the definition by Tegmark (2017):

$$\text{Intelligence} = \text{ability to accomplish complex goals} \tag{1.1}$$

Collective intelligence is the combination of different expertise, methods, and tools toward the same goal that is operational excellence in the context of outcome-based digitalization. Thus, collective intelligence can be expressed as

$$\text{Collective intelligence} = \text{goal} + \text{agency} + \text{smart} = \text{value fully realized} \tag{1.2}$$

The operation excellence companies are collective intelligent featuring three key qualities: a sense of purpose, agency, and smartness.

Thinking of purpose here as a lofty goal of operational excellence that provides a clear direction for all personnel within the company and everyone plays a part in it. But it takes extraordinary initiative involving effort, knowledge, experience, and wisdom to accomplish the lofty goal. This initiative is called agency. People with strong agency will take the initiative, "let us do this!" and then follow through with actions for implementation.

Regarding smart, it is similar to intelligence as defined in Equation (1.1), which can be complemented by the Oxford Dictionary definition "the ability to acquire and apply knowledge and skills to solve problems." Thus, we think "company smarts" as: the ability for the company to face challenging tasks with collective intelligence to learn and figure out and take informed strategic decision to solve the problems.

So, the million-dollar question is: how can the digitalization help improving company's collective intelligence? The answer is very positive. The pace-setting companies are already using digitalization to holistically integrate business and technical data and activities via integrating data and workflows to determine:

- Convergence of planning, scheduling, and operation optimization:
  - Planning optimization (what): what feedstocks to buy and what products to make, and in what quantity?
  - Scheduling optimization (when and where): when to receive what feeds in which tanks, when to blend them (if needed), and when to blend and deliver finished products, in what quantity?
  - Operation optimization (how): how to run individual processes via setting operating targets to guide APC and regulatory control systems?

- Asset management (what-if and what's next): what failure mode could occur to which equipment and when?
- People and skill management (who and what): who to be assigned to what tasks based on the skill requirements?

Furthermore, these companies use digital twin models for operation excellence. The digital twin located in cloud or edge gathers data from different sources including sensors and soft sensors, DCS, historians, and labs as well as feedstock and energy pricing data. The digital twin can not only determine optimal operating conditions but also assess what-if scenarios and predict future failure events.

Therefore, digitalization can enhance the capability of accomplishing complex goals. In addition, use of data analytics and machines learning enhances the ability of learning from existing data and deriving new insights. Now we can conclude that digitalization can improve company's collective intelligence for problem solving proactively and creating value for the organization via converting the data and models to insights that helps human to make more informed decisions.

The best outcome for digitalization is achieved by the proper integration of IT (big data and connectivity centric), OT (procedure and control centric), and ET (model and optimization centric) together with DT (predictive analytics centric). The outcome is to have effective response to market environment and achieve company's objectives via making recommendations for business changes, deriving new insights for operation, predicting new events and opportunities, detecting early warning signs of failure, optimizing workflows, and enabling more intelligent actions for better profit, efficiency, reliability, and safety (Figure 1.6).

1.5.5 What Is the Human's Role in Digitalization?

No matter how intelligent the digital solutions are, human must act upon. Otherwise, nothing would happen. It takes deep know-how, experience and courage to take intelligent

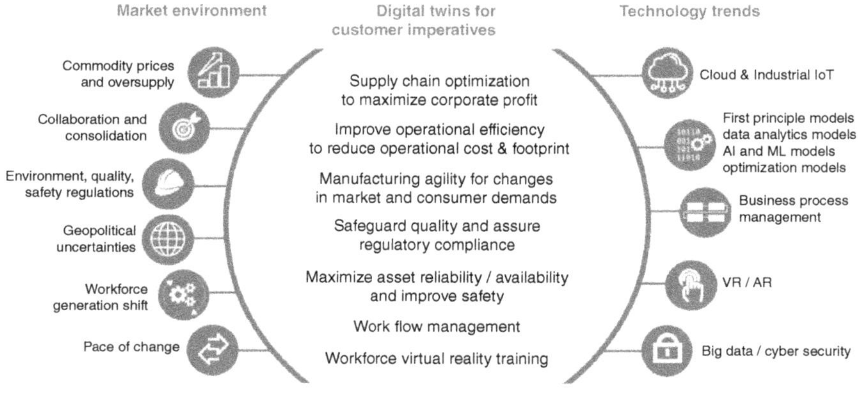

FIGURE 1.6. Integration of multiple digital twins: outcome-based digitalization.

actions with confidence to make successful implementation to capture value. The simple truth is that digital solutions are not taking over the world nor are they obviating the need for human in the workplace. They are simply amplifying people's skills and collaborating with plant personnel to achieve the true potentials of digitalization.

Digitalization is not just a bunch of technologies and people may think the job is done when all the necessary IT technologies are installed. No! Digitalization is just an enabler for achieving operation excellence more effectively. Digitalization is only the starting of the journey for achieving operation excellence. A continuous system with effective work processes in place for operation excellence must be developed with strong leaders driving it. It is a never-ending journey.

With the digitalization, companies need special skilled people in the IIoT environment. Simple and monotonous processes are being automated via digitalization, but other processes become more complex and intertwined.

There are three major components of human interactions with the digitalization process, which could miss in the middle between human actions and digital solutions (Figure 1.7). We call this the "missing middle"—"missing" because virtually no technology vendors talk about it; only some pace-setting companies are taking dedicated effort to fill this crucial gap. These three missing-in-the-middle components are briefly defined below:

(a) Trainer: Validate the data quality and digital twin models when feed type and catalyst type change significantly; digital twin models need to be calibrated manually prior to deploying online. For machine leaning models, it requires extensive training prior to applications. The model turning and training are conducted by subject matter experts from technology vendor. Once the models are running online in nominal situations, data validation and model turning are conducted automatically within digital twin. If digital twin is located in cloud and supported by technology vendors, the role of trainer is performed by vendors.

(b) Explainer: Interpret the model results with the relationships between parameters of interest and their implications and value to the company. The responsibility of the explainer's job is to bridge the gap between technical

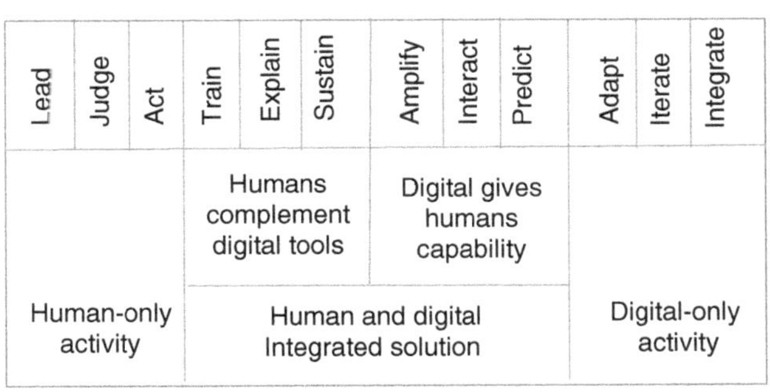

FIGURE 1.7. Human's roles in digitalization.

personnel and business leaders. As companies rely on the digital twin to determine actions, some of the actions that push operating limits may affect profit, reliability, and safety. Explainers need to able to explain and justify those decisions. They must understand the inner working mechanism of complex models as well as practical operating limits and able to explain the reasoning of digital twin and rationale of the recommendations. In addition, they may also assist the implementation team to provide support for conducting what-if analysis using digital twin during the implementation process and estimate the value capture during and after the implementation.

(c) Sustainer: Subject matter experts may act (like a "babysitter") as sustainers of digital twin. They must work continually to ensure that digital twin systems are functioning properly to avoid any misleading predictions, which could lead to operation trips or disaster in the worst case. They oversee the digital twin performance, remove bad data, impose new limits (reliability, economic, legal or ethical), maintain models and make exceptions where necessary.

1.6 WHAT IS THE DIGITALIZATION JOURNEY?

To decide what digital applications and capabilities are required, one must identify the problems or pain points that negatively impact operational excellence. Once these problems are known and understood, the next task is to develop value propositions or business cases for solutions to overcome these problems and prioritize them and define use cases. Then the team needs to develop a holistic digitalization approach for achieving operational excellence that addresses the selected use cases simultaneously. Not only will this eliminate duplicate or conflicting effort; it assures an optimum outcome. Furthermore, the team must determine the digitalization readiness as to where the company stands in the journey and what takes to achieve digitalization fulfillment for achieving operation excellence. A typical digitalization journey takes several steps as shown in Figure 1.8.

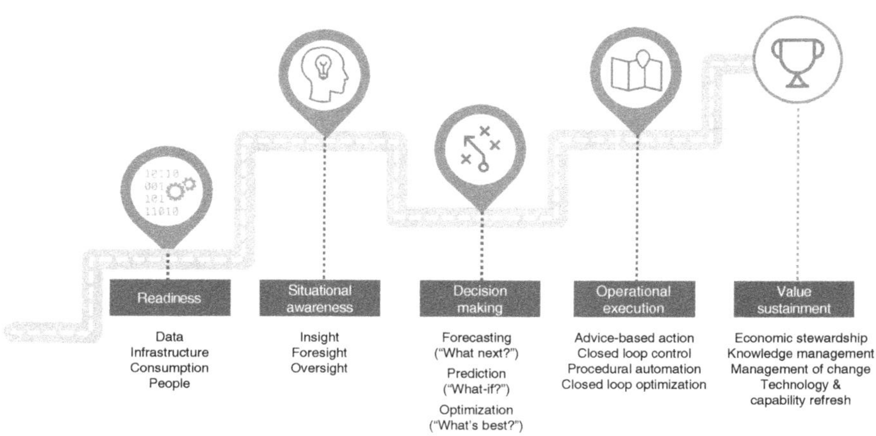

FIGURE 1.8. A typical digitalization journey.

1.7 OVERVIEW OF THE BOOK STRUCTURE

The challenges presented for industrial digitalization can also be viewed as opportunities—to grow, connect, streamline, and build chemical manufacturing value chain. However, the path to digitalization can be fearful to many people because of the changes that may involve and can be confusing at the same time for what steps that digitalization could take to implement and what values that digitalization could generate. This book will explain recent emerging technologies in the field of industrial digitalization and their implications. More importantly, the book will focus on the key enabler so-called Connected Plant and its applications.

What does the Connected Plant Performance (CPP) look like? In overall, the CPP adopts first principle models as the brain supported by data analytics and ML algorithms, digital twin as the foundation, IIoT as the data infrastructure, and cloud or edge as the platform. CPP connects processes, assets, people, and work processes. CPP allows effective cyber–human interactions where digital solutions provide intelligent indications and recommendations that enable plant personnel to act with agility and confidence to respond to threats and opportunities. A Connected Plant can see operations with a reduced number of unexpected events: excellent levels of reliability, safety, performance and environmental compliance, and assets running with extraordinary efficiency in terms of energy and process yield.

The Connected Plant is more than just data collection and software tools. It is about applying effective sensing, cloud computing, IIoT, and data analysis for revealing insights, process, and asset optimization and automation to address end-to-end customer touchpoints and desired outcomes. It creates value through real-time data analytics, models, and optimization, via revealing new insights for opportunities leading to turn-key solutions. It is about horizontal and vertical integration of different levels of activities, from business planning for feed and product selection at corporate level to operation planning and scheduling at the plant level and to process optimization and control at the DCS level. At the same time, applying Connected Plant results in faster detection and resolution of operating problems as well as more effective best practice sharing and knowledge management. For the first time, these aspects will be discussed and brought together in one single book, which covers the end-to-end execution of digital transformation—from leadership-level strategy to sensors, data analytics, process models, optimization models, data infrastructure, and integrated workflows and to on-the-ground team implementation for industrial companies. This is the main feature of this book.

The book will explain the principles and practices that lead to successful transformation to CPP: what are the gaps associated with traditional work process and how can the gaps be closed by applying Connected Plant concept. From this book, one can learn the following:

- How to create a digital road map and develop digital tasks?
- What takes to establish CPP?
- How to digitally enhance profit, reliability, operation, and control?
- How to integrate different levels of activities and then automate integrated processes digitally?

While many companies have looked at various ways for implementing digitalization, very few have taken the holistic view of applying digital transformation to close gaps in current business and technical processes. Digital solutions should be developed centered on the pain points or gaps identified with a clear perspective of prioritized opportunities and with the outcome-based approach. The book also share some successful applications to show the evidence of what some companies are doing to take advantage of industrial digitalization and to survive and thrive for now and in coming years.

The intent for this book is to break down the myth surrounding digitalization and data science.

This book describes how a new breed of real-time analytic solution enables continuous monitoring, optimization, and control of key plant operating conditions and allows for the earliest possible indication of performance degradation and the appropriate prioritization to address the situation. By taking advantage of advanced "digital twin" technology enabling the most advanced monitoring, analytical, and predictive capabilities, process engineers can implement around-the-clock monitoring of plant data and ongoing operational health checks and recommendations to close performance gaps.

To accomplish the above goals, the book comprises four parts with 20 chapters in total, and the structure of the books is explained as below:

- Part 1—Challenges and Opportunities, which consists of three chapters. An overview of industrial challenges, which defines the needs for digitalization, is provided in Chapter 1. The concept of Connected Plant to overcome the challenges is described in Chapter 2. Furthermore, Chapter 3 provides an overview of data science for industrial applications.
- Part 2—Data Analytics for Smart Operations, which consists of seven chapters. In this part, individual modeling techniques including first principles, statistics, optimization, AI and ML are discussed in details. Intelligent models can empower digital solutions to convert data to new insights and actions.
- Part 3—Connected Plant for Smart Operations, which includes six chapters, which focus on individual components of Connected Plant ranging from asset and safety management to integrated planning and control, from workforce enhancement and workflow management.
- Part 4—Digital Solution for Smart Operations, which includes four chapters. In Part 4, the data infrastructure and open architect for the digital solutions is discussed. Furthermore, digital readiness assessment for industrial companies is explained, and an effectivity digital journey is described with a roadmap showing signposts of milestone progress. Finally, application case studies are provided to share the success stories in applying digital solutions for operation excellence.

REFERENCES

Edwards D (2016), Honeywell launches industrial internet of things service to improve plant performance, *Robotics & Automation News*, September.

Francisco S (2017), A Better Definition of Operational Excellence. Retrieved 2017-12-07. Wikipedia.

Tegmark M (2017), Life 3.0 – Being Human in the Age of Artificial Intelligence, Alfred A. Knopf, New York

Zhu XX (2014), Energy and Process Optimization for the Process Industries, Wiley/AIChE.

2

MISSION OF CONNECTED PLANT

2.1 WHAT IS CONNECTED PLANT?

The Connected Plant is a suite of modern digital applications that deliver solutions for safety, reliability, efficiency, and profitability. These proven industry solutions are based on decades of domain knowledge, optimization and and controls experience (Chandran, 2018). The Connected Plant method aims to connect **people**, **processes**, and **assets** within the organization much better than what has been done with traditional digital solutions (Edwards, 2016). The Connected Plant becomes a reality because of advent of new technology available, which include IIoT, cloud/edge computing, data analytics, virtual reality, and digitalization for integrated workflows (Figure 2.1).

The Connected Plant intends to address a critical issue with existing practice: key activities and data are in silos and thus unable to convert data into actions in a systematic and timely manner. The key activities include supply chain management, production monitoring and optimization, DCS and control, and asset management and maintenance. In other words, all of these systems generate a lot of data, making it practically impossible for site engineers to gather, correlate, and analyze data in a timely manner to predict operating issues and equipment failures to come up effective actions for prevention and optimization.

The Connected Plant is more than just data collection and software tools. It is about applying model-based digital twin to convert data to insights. It is about integrating new digital solutions with existing data infrastructure. It is

Digitalization and Analytics for Smart Plant Performance: Theory and Applications,
First Edition. Frank (Xin X.) Zhu.

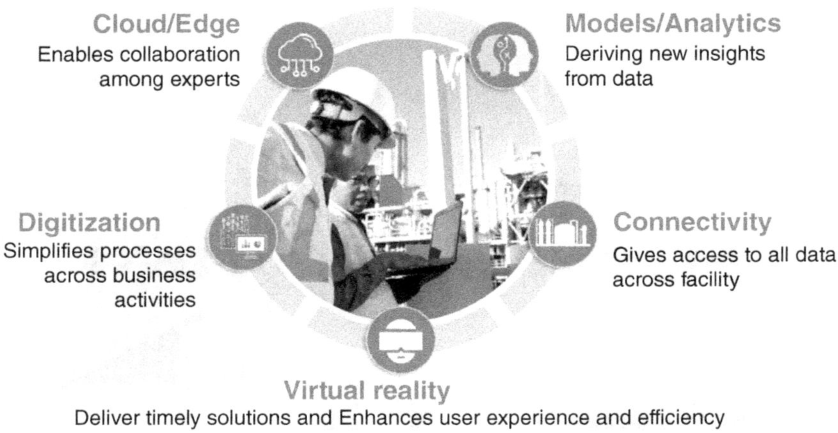

FIGURE 2.1. Technology makes Connected Plant a reality.

about better management of asset, workflows, and knowledge. The Connected Plant creates value through real-time data analytics, models, and optimization to reveal new insights for opportunities, leading to turn-key solutions. It also applies advanced analytics and machine learning (ML) techniques to develop new heuristics and enable the solving of issues that we were previously unable to. It leverages the long-standing experience, paired with new digital technologies to resolve operating issues better, faster consistently. Use of the Connected Plant method can result in faster detection and resolution of operating problems as well as more effective best knowledge sharing and management.

2.2 MAJOR FUNCTIONS OF CONNECTED PLANT

Digitalization implementation should be approached with the objective of maximizing value by identifying key value drivers that will significantly impact the performance of typical manufacturing companies. In order to identify and prioritize opportunities along the digital thread, the Connected Plant method focuses on optimality, reliability, and human capital. For each of these value drivers, the Connected Plant connects the most important levers to the key value drivers, resulting in a framework that companies can leverage to systematically identify potentials and take actions to create value.

2.2.1 Function 1: Reliability Enhancement

The Connected Plant offers reliability monitoring that includes process models and fault models. This solution provides early detection of process and equipment issues and includes troubleshooting recommendations to resolve these issues before they become costly. Through the dashboard, recommendations are shown in a clear and timely manner to deliver improved learning to customers' organizations.

The Connected Plant consists of a Reliability Advisor in the form of digital twin for fault detection and predictive maintenance, which can drive value by decreasing planned machine downtime, unplanned machine downtime, or changeover times. Typically, predictive maintenance decreases the total equipment downtime by 30–50% and increases machine life by 20–40%.

Non-reliable operation can be eliminated if detected and documented; thus the physical production process needs to be mapped along the digital thread—based on the collection of real-time data in an automated way and with historical data points. This requires moving from selected, sampling-based measurements for quality control purposes to a full coverage of the production process using in-line sensors and measurement devices to collect information for every single workpiece.

As an example, Reliability Advisor for a fluid catalytic cracking (FCC) unit was installed for a key US Gulf Coast refiner in 2017. The customer benefited in two key areas:

- Identification of debottlenecking opportunities around the caustic system operations to mitigate H_2S contamination that would enable the customer to increase throughput by 10%.
- The customer also benefited from the early event detection when over the weekend a portable air compressor was lost; the service communicated the resulting high CO & SOx to the customer before the end of the weekend and recommended operating adjustments to mitigate effects so as not to affect overall production

2.2.2 Function 2: Integrated Energy and Process Optimization

The process optimization service, as an important part of the Connected Plant, rigorously simulates an entire complex of process units to determine the maximum profitability of the complex, given current constraints and market conditions. Process optimization is about improving a process in terms of material consumption, energy use, and yield optimization. The Connected Plant consists of a Process Optimization Advisor in the form of digital twin to achieve these goals. This solution embeds first principle models integrated with nonlinear solver to provide optimal conditions consistently to customers. Typically, real-time process optimization yields an improvement in throughput of up to 5%.

As an example, the Process Optimization Advisor was installed for a naphtha complex located in Southeast Asia that includes a naphtha hydrotreater with the overhead going to a UOP Penex™ and the bottoms to the UOP CCR Platforming™ unit with the appropriate fractionation columns. Based on the actual measures, it optimizes the operation conditions for the overall complex and alerts operators with the optimal conditions as operating targets. The implementation of the Process Optimization Advisor has enabled the customer to increase gasoline production over 5%. Based on the success of this engagement, the customer expanded the scope to include Reliability Advisor for the naphtha complex.

2.2.3 Function 3: Asset Performance Management

The Connected Plant includes asset performance management (APM), which goes beyond traditional machine monitoring and data gathering. By merging together decades of machine modeling experience with modern analytics, the APM digital twins predict machinery availability, drill to the root causes of inefficient machine operation, and bring order to reliability and maintenance planning.

Cloud-based digital twins drive real-time equipment performance modeling and predictive analytics. Integrated process and asset models reveal root causes of bottlenecks, identify new levels of untapped productivity, and uncover potential problems. APM deploys rapidly through a wealth of standard asset libraries and tools to integrate data sources quickly. On average, installation costs are recouped in less than 6 months.

One example was installing APM to UOP Oleflex™ process in a large Middle East petrochemical plant in 2017. Within the first month of the engagement implemented, the customer was able to make operational adjustments that improved unit performance and efficiency to increase profits by over 1.2 million/year. Other benefits include:

- Relaxing constraints—Used the rigorous flowsheet (plus the process model) that matches the unit configuration and as-built operation to provide constraint awareness. Here the customer was aware of a key operating constraint around the cold box (key equipment section of the Oleflex), but with the flowsheet the Reliability Advisor could provide operational guidance to drive closer to the constraint and improve overall production.
- Improve process effectiveness—The process model with the fault model immediately identified that the reactors were not operating at ideal pressure. Using the model Reliability Advisor provided justification for lowering the pressure and provide recommendations to meet the best practice.
- Reduce downtime—The equipment analytics model identified unique "shift" patterns in the machines' bearing vibration and temperature signatures. Interestingly, the shifts revealed the ability to predict failure of a nearby packing gland (i.e., seal) rather than the bearing itself. Previously the gland was a preventative maintenance item and replaced early, but with the new insight the plant engineers realized they could confidently prolong the change-out interval. By doing so the plant avoided 3 days of unnecessary downtime. Instead, the plant was able to run uninterrupted through the end of the year and plan the packing replacement in conjunction with the plant's planned turnaround.

- Beyond the fault and process models, the catalyst health model can monitor catalyst transfer anomalies to alert the user of potential transfer disruptions that could lead to shutdown.

2.2.4 Function 4: Profit Performance Optimization for the Total Site

The Connected Plant includes site-wide performance monitoring and optimization, which enables collaboration to understand what is limiting APC from driving the individual units to its best potential. It combines the power of multiple digital twins with cloud-based collaboration. It provides visibility into the economic consequences of process performance as cost of lost opportunity and ensures sustainable benefits over varying operating conditions, visualizing real-time unit profitability. Recommendations are provided by multiple digital twins through an online, cyber-secured collaboration portal. Real-time discussion and exchange of ideas can take place among subject matter experts (SMEs), plant-based engineering, operations, maintenance, and planning groups.

2.2.5 Function 5: Workforce Competency Management

Digitalization enables the workforce in two broad areas—enabling the workforce through competency development and knowledge sharing and empowering the workforce through the application of digital twins to transform work practices. Through a well-planned digitalization strategy, a range of solutions implemented allow workers to do more, do it more efficiently, and do it more safely. In many cases, these solutions are becoming essential to the ongoing viability of industrial enterprises as they ride out demographic change, changing market expectations, and a changing regulatory environment. By carefully considering the application of digitalization to areas of greatest value along with strategies for mitigating the challenges that can arise, a significant step change is observed in how the workforce is equipped to do the job and how they go about doing it.

2.3 DIGITAL TWINS: THE CORE OF CONNECTED PLANT

A data analytics platform for Connected Plant, in the form of digital twins, is developed (Dignan, 2019) that provides industrial companies with real-time data, first principles, analytics, ML, visual intelligence, and actionable recommendations to help them optimize production, improve asset reliability and safety, and increase profitability.

A digital twin is a virtue asset that mimics the physical asset. A digital twin aims to be an accurate representation of an asset over its full range of operation and its full life cycle. It is created during the initial study to evaluate the feasibility of the asset. It works in the present, mirroring the actual plant in simulated mode, but with full knowledge of its historical performance and accurate understanding of its future potential.

Digital twin contains intelligence via built-in predictive analytics to predict what is going to happen and provide recommendations for what actions to take and what outcomes may result—this becomes the basis for human operators to make informed and intelligent decisions. The digital twin gathers data from different sources including DCS, historians, and labs as well as feedstock and energy pricing. It can not only determine optimal operating conditions but also evaluate what-if scenarios and predict future failure events.

In this manner, the digital twin system provides a single source of the truth for what is going on inside the plant. It enables everyone to see inside and perceive things that are not being directly measured. It is wired up so that the insights are instantly available without data and model wrangling by end users and run in a consistent way that everyone can understand and agree on. It drives convergence in comprehension and unity in decision making and action.

A digital twin can be located in cloud or at edge. The former is used for trending analysis, supply chain management, production supervision, communications, and so on. The latter is applied for time-critical applications such as reliability and safety for critical equipment, inferential calculations (or soft sensors) for control, and so on. By means of virtue representation and model prediction, the digital twin can be used for training of the staff who will operate the asset.

The foundation for the digital twin system is the process and equipment simulation based on first principle models, which are enhanced by the embedded data analytics and ML algorithms for improving prediction accuracy. The first principle models capture known part of mechanisms, while the analytics and ML algorithms cover unknown part of disturbances. The plant models inside the simulator synchronizes with the plant models inside the enterprise process historian (EPH) to ensure their separate representations of the plant remain aligned. In this manner, the digital twin can allow "What if?" and "What's best?" scenarios to be run automatically to determine available strategies that maximize profitability. Experts can review the recommended strategies to assess the impact of each recommended approach. Planning and scheduling linear submodels can be updated to reflect the proposed changes. Only in this way digital twins have teeth—a digital twin is not just a technological nicety but is a powerful and effective system that you can trust and rely on for the quality and accuracy of its predictions (KBC, 2019).

The benefits of digital twins can be summarized as follows:

- Unifies existing data silos into a virtual entity.
- Federates data across different applications to drive end-to-end integration.
- Understand how a plant responds to changes to drive increased reliability and optimization and workflow management using data analytics, process simulation, optimization, and control technology.

- Enables those with strong domain expertise to speed up diagnosis and troubleshooting.
- Quickly spot opportunities and provide recommendations to improve processes, equipment performance, and automation.
- Real-time data and data analytics to predict operating issues, optimize production, and increase profitability.

There are several key features for the digital twin software for Connected Plant, which are discussed as below.

2.3.1 Feature 1: IIoT as the Data Infrastructure

Given the importance of data and information, it is important to treat and develop data as a central business asset. This means data is actively managed at all stages of the data life cycle (i.e., collected, stored, analyzed, shared, and archived) through defined data practices, standards, and policies. Figure 2.2 shows the data life cycle for the Connected Plant services. The process and lab data are collected directly from the site historian (see bottom left) and then enters the data platform that connects different sources of data. Cloud offers several advantages including fast computation, large storage, and transparent data overview.

2.3.2 Feature 2: Model Driven

The cornerstone of digitalization is the power to convert data to insights. When connecting equipment, workflows, feeds and products, and personnel, huge amounts of data will be generated through sensors, monitoring, operation, control systems, etc. But data is not useful unless it is processed in a way that provides context, meaning, and intent that can be understood by the right personnel. Just connecting sensors to a machine or connecting a machine to another machine will

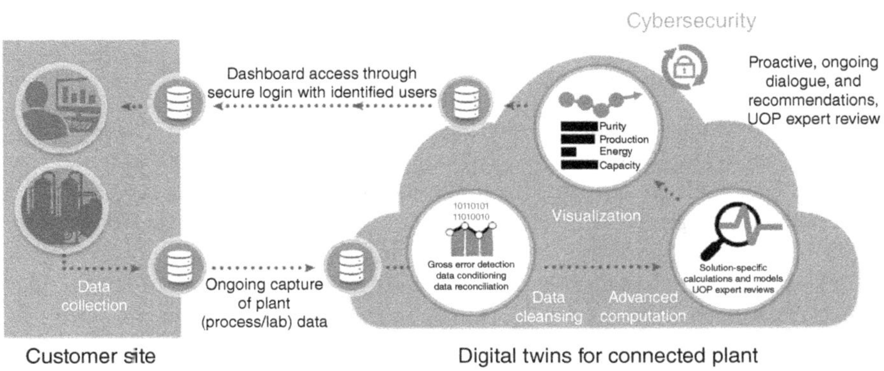

FIGURE 2.2. Digital platform for Connected Plant.

not give users the insights needed to make better decisions. To transform data into actionable information requires both the right technologies and the right people. The right technologies are analytical technologies, which consist of first principle models, statistical models, optimization models, and ML models. Good models can generate optimal conditions and set points for control as well as provide failure prediction and recommendations for actions to avoid abnormal operation and hazards.

Let us conduct a thought experiment to see how different data analysis methods can be integrated to help derive cross-functional insights. Delivering new insight of the behavior of a chemical or physical property in the context of operation scenario is a good place to start. Let us consider the declining catalyst activity (property) due to catalyst deterioration (scenario) in a hydrocracker. Using a statistic model analyzing the online data, the key variables that affect the activity can be identified, and thus the correlation relationship can be determined. However, the correlation only provides a summary of the past operation. How can this correlation be used to predict the catalyst performance in the future? ML techniques could be employed in this case. The historical data can be used to initially train the ML algorithm so that it fits the current operation behavior. Then the ML algorithm can be extended for future prediction via adopting a rewarding/penalty function in the reinforcement learning technique. The weighting factors in the ML algorithm is adjusted automatically via this function whenever making a single prediction compared with the reality. This is like self-training. As a result, the improved ML algorithm can predict the catalyst performance for the future with high accuracy. The prediction can be used for guiding operation via changing some of the key operating parameters or determining the schedule for reloading catalyst in the future.

This simple example answers some of the questions for data science:

- Can the analytics model predict correctly the events that are forthcoming?
- Does the model derive new insights?
- Does the insight drive actions in operation?

2.3.3 Feature 3: Integrate Simulation and Analytics

The digital twin enables end users to fully leverage current installed technology—such as process simulation tools—in a manner that serves both its traditional role and integrates with modern analytics techniques. Process simulation can be used to mimic the design, conditions, and limits of actual processes and equipment based on real-time operation and lab data using proprietary rigorous reaction models, product fractionation and impurity separation, rotating equipment, fired heaters, heat exchangers, operating limits, and many more (Zhu et al., 2019). In this manner, process design is integrated with varying feed conditions and production requirement. Once the known part of the process and equipment is described by simulation, fault detection and diagnostics can be conducted by applying data analytics and ML techniques.

2.3.4 Feature 4: Leverage Domain Knowledge into Analytics Models

Digital twins adequately leverage the proven diagnostic methods that SMEs historically utilize to perform machinery and process diagnostics. Any successful digital twin is to accentuate the imperative role that SMEs play now and will forever play. The goal of emerging simulation with analytics and AI/ML techniques must align with what experts already do well in a manner they can trust.

The advancement of data analytics can automatically process complex data sets and make prediction. However, it has been observed that many examples of digital solutions failed to deliver their envisioned added value. Often the root cause of unrealized outcomes rests largely in the disconnect between the analytics models and domain knowledge in troubleshooting and control and process optimization.

2.3.5 Feature 5: Visualization and Collaboration

Key plant performance can be visualized clearly, which can guide users to focus on key issues that require attention. As example, dashboards (Figure 2.3) provide a high-level overview with drill-downs to support deeper investigations.

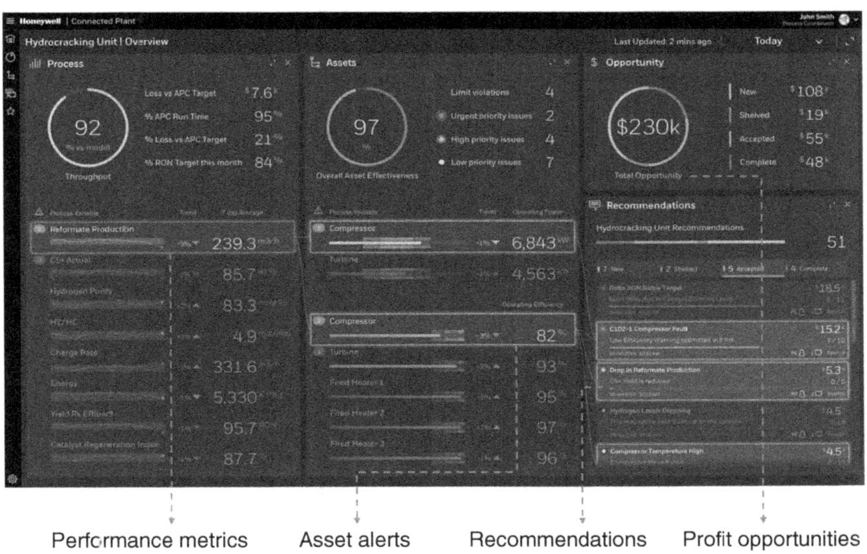

FIGURE 2.3. Visual display of Honeywell Forge as an example (Honeywellprocess.com, 2020).

Recommendation summaries give users prioritized actions to address current or potential issues and opportunities. Data analysis tools support ad hoc data exploration to assist in troubleshooting and process improvement.

2.3.6 Feature 6: Cybersecurity

As businesses move to digitally connect their operations, organizations leading digital transformation are developing enterprise-wide cybersecurity solutions that drive greater consistency and protection across operational technology environments. Solutions should meet industry standards and regulatory compliance requirements while remaining easy to use by teams stretched with increased responsibilities.

The application experiences indicate the need of integrating cybersecurity into core processes and establish cyber-resilience as part of the enterprise-wide risk management process.

- Secure the core production processes by integrating cyber defense.
- Have regular trainings/war games examining the possibility of cyberattacks, focusing on short reaction times.
- Develop deep integration of security into the technology for all connected technologies.
- Automate defense as far as possible to enable cybersecurity staff to focus on safeguarding the technology against new threats.

2.4 CONCLUSIONS

The biggest drivers of Connected Plant for operation excellence are:

- Inefficient supply chain management due to isolated work-process and tools;
- Inefficient operation of processes;
- Underutilization of assets;
- Unplanned downtime;
- Safety and compliance;
- Human capital management;
- Foot print and climate change impact.

The Connected Plant is about human–digital interactions. The method provides insights, and it requires people to make informed decisions and take smart actions for things to happen. One must understand the fact that new technology is the enabling tool; it is not the solution. If just expecting things will happen naturally after digital analytics are installed, it is doomed to disillusion and failure.

REFERENCES

Chandran S (2018), The honeywell connected plant. In: Honeywell Users Group (HUG) EMEA Conference, October 2.

Dignan L (2019), Honeywell Launches Industrial IoT Platform Called Honeywell Forge, Special Report, Between the Lines, June 6.

Edwards D (2016), Honeywell launches industrial internet of things service to improve plant performance, *Robotics & Automation News*, September.

Honyewllprocess.com (2020), Honeywell Forge for Industrial.

KBC (2019), Digitalization Manifesto – Now Is the Time, https://www.digitalrefining.com/literature/.

Zhu FX, Johnson JA, Ablin DW, Ernst GA (2019), Efficient Petrochemical Processes – Technology, Design and Operation, (ISBN-13: 978-1119487869), Wiley, Hoboken, NJ.

3

DATA ANALYTICS FOR OPERATION EXCELLENCE

3.1 INTRODUCTION

Readers may ask this question to start with: what is the distinguished feature that predictive analytics can offer compared with other traditional analytics? A straightforward analogy would be running a company or plant without intelligent data-centric analytic models. It'd be like taking a road trip via driving toward a remote destination using a large map book versus a predictive analytics built-in GPS navigator. The later could guide you to the destination without headache as the GPS can automatically adjusts to traffic patterns, forecasts delays, and receives the real-time data from satellite so that you can make informed decisions on your route. Predict what is going to happen and provide options and recommend the effective actions—that is the value what data science can provide for your organization (Bardin, 2017).

3.1.1 What Is Big Data?

In the age of digitalization, the Internet and the much increased computing power are leveraged in order to achieve strong connectivity among people, asset, and workflows. The amount of data that exists in unprecedented dimensionality and granularity results in so-called big data. A more precise definition of big data is data volume, variety, velocity, and veracity, which are the 4-V's representation of big data. At the same time, people apply data science to convert data to intelligent

Digitalization and Analytics for Smart Plant Performance: Theory and Applications,
First Edition. Frank (Xin X.) Zhu.

models for prediction and taking actions. Today with data science, the modeling techniques include statistics, first principle models, correlations, data mining, and machine learning (ML) and combination of these with the aim to derive insights and actionable decisions for value creation. In this sense, "value" should be added to the 4-V's to make the 5-V's representation of big data.

3.1.2 What Is Data Science?

Data science is a platform to apply fundamentals (first principle based), statistics (correlations based on historical data), and ML (algorithms based on past data and newly generated data) and their related methods in order to understand and analyze actual phenomena with data. It employs techniques and theories drawn from many fields such as science, mathematics, statistics, computer science, and information science. At its core, data science intends to answer the following question: "What is more likely to happen based on the current operation, and what actions should be taken to influence the outcome?" It selects appropriate models to predict the future based on data.

There are fundamental differences between data science and conventional data analytics techniques. The conventional methods are limited in the amount of data to be processed and the range of the applications due to the limited amount of data and computation power. However, there is much value to be acquired with advent of Internet of Things (IoT), artificial intelligence (AI), and ML methods as well as more advanced predictive data analytics capability in terms of the scale and sophistication. That is the reason why advanced predictive analytic technologies are developed to meet the challenges.

The ever-increasing availability of data requires the use of models that can convert data to insights and recommendations for actions for operation excellence. In the most simplified perspective, one may broadly classify models into two categories, namely, mechanism-driven models based on first principle and data-driven models based on analytics. Combinations of these classes lead to hybrid models. With the advent of smart sensing and the IoT, data science offers powerful tools for operation monitoring, fault diagnosis and prediction, asset and safety management, abnormal operation management, and integrated data and workflow management.

White (2016) reported an application case in which GE uses predictive analytics to monitor performance of over 1,500 turbines and generators in 58 countries. Each turbine typically has more than 100 physical sensors and over 300 virtual sensors that monitor temperature, pressure, vibration, speed, and so on. Data are routed over the Internet to a data center—central monitoring location. At the data center, predictive analytics algorithms check current and recent operating conditions against models of more than 150 failure scenarios to detect sign of impeding failures.

Routing data into central storage makes efficient use of highly skilled workers, allowing specialists to monitor all of the turbines and generator installations. This single-database method also helps to ensure that predictive models are continually refined and improved and enables best practices to be shared with customers simultaneously rather than on an incremental basis during system upgrades. GE estimated that this approach collectively saved US $70MM in the last a couple of years for the GE customers.

Predictive analytic models perform more effective and faster when implemented near the data source as they can intercept data and perform time-critical analysis almost immediately after the data are generated. The prediction occurs before the data are written to data storage. This is possible because processing and communication techniques have become smaller, cheaper, and yet more power ful. Microprocessors can be embedded in turbines, and predictive analytics software can execute on the device itself. Time-critical prediction of failure alerts can be performed using the software installed on the device. For trend analysis that are not time critical, the data can be aggregated and sent to long-term data storage.

3.2 PROCESS DATA OVERVIEW: CHARACTERISTICS AND ATTRIBUTES

Process historical databases are massive due to the use of process control computers and manufacturing execution systems. The data come from the six-level plant activities as shown in Figure 1.1. The variety of process measurement technology ranges from conventional process sensors to images, videos, and other indirect methods such as spectroscopy and vibration measurements.

Process data often include process variables, quality data, and measured disturbances. These variables are often connected by the process flows and controller feedback, making them highly interactive and dynamic. Process data are measured at rather fast sampling rates, which can go from seconds to minutes for online measurement and control-related variables. On the other hand, property data from laboratories are often collected at a much lower sampling rate in the order of days.

3.2.1 Numerical Data

This is the most common data type. Basically, it represents some quantifiable thing that can be measured. Four major measurements are flow rate, temperature, pressure, and level:

- **Temperature measurement devices** such as thermocouples and resistant temperature detectors are the most common instruments found in industrial plans.
- **Pressure transmitters** are another common instrument in the form of differential pressure (DP) between one process connection and a reference pressure (e.g., atmosphere) or another process connection. DP transmitters are used to infer other process variables, such as flow and level.
- **Flow measurements.** Flow rates can be calculated from DP measurement across an orifice plate. Volumetric flow can be calculated from the velocity, which is proportional to the square root of DP. Mass flow rate can be calculated from the density of the material. Accurate flow measurements are essential for regulatory and APC applications.

- **Level measurements.** The level or height of materials in a vessel can be measured directly in some cases, but DP transmitters are commonly used to infer level as a function of hydrostatic pressure and density of the material.

Beyond single-point measurements, time series measurement data is another common data type in process industry. Usually, this kind of data is acquired on a regular time interval. The time intervals may need to be adjusted to meet the requirement for a desired application—for example, converting current data to 1 min based intervals for control purposes, to hourly based intervals for reliability evaluation, or to daily based intervals for benefit evaluation.

3.2.2 Analyzer Data

Process analyzers are integral parts of many APC applications. They come in a variety of types and process specific configurations. For example, an analyzer may provide continuous results or update intermittently and measure a property or composition of a stream.

3.2.3 Engineering Data

In chemical process problems, inferential or soft sensors are used to calculate, based on other easily measured ones, those critical but hard-to-measure variables including feed properties and product quality and so on.

Once the data is converted to desired format, soft sensors, engineering calculations, and process simulation become analytical building blocks for assisting plant operation, which helps to identify patterns of operation and predict optimal conditions and failure operating modes.

In many cases, it is the analytical features rather than the raw data itself that become the starting point for statistic modeling and/or ML algorithms. If the subject matter expert has validated that the features and statistical metrics contain sufficient information about the data, then there is a much higher likelihood of running successful ML algorithms on the data at scale. In many cases, well-crafted and well-engineered features may well become the data quality benchmarking basis for data analytics applications.

3.2.4 The Needs of Data Analytics

With the large amount of data available in a manufacturing plant, one can apply principal component analysis (PCA) to analyze the process data to visualize variation and strong patterns. To reveal the relationship between quality data and process operating variables, quality data from laboratory and quality prediction with inferential sensors can be analyzed jointly with process data via supervised learning techniques such as partial least squares (PLS) and neural networks.

For condition-based maintenance, PCA can be used to identify the most important contributor to characterize the conditions of equipment, while PLS is used to predict the system state and detect abnormal behavior, which helps to determine maintenance tasks. For more complex fault detection, advanced ML techniques such as Q-learning and convolutional neural networks can be applied, and these techniques can furthermore learn on their own. The detailed discussions for PCA, PLS, neural networks, and Q-learning will be given in Part 2 of this book. It is believed that process data analytics will be a cornerstone for Industry 4.0 where proactive data analytics are central for process operations and decision making (Qin and Chiang, 2019), which are discussed below.

3.3 UNIQUE ATTRIBUTES OF PROCESS DATA ANALYTICS

What set process data analytics apart from other big data analytics, such as business intelligence? This question must be answered to make an impact in practice. There are four important and unique attributes of process data attributes discussed by Qin and Chiang (2019):

- Working compatibly with first principles or process knowledge;
- Applying mathematical programming models for profit optimization;
- Dealing effectively with uncertainties that are usually time varying;
- Generating interpretable solutions.

Chemical processes are designed via applying first principles based on chemistry and physics. First principle models are largely represented by differential and algebraic equations (DAEs). These models become the basis for process simulations, optimization, and control. They are the foundation for how processes are designed and operated with confidence.

To achieve economic targets while observing reliable and safe operation, mathematical programming methods such as linear (LP), nonlinear (NLP), and mixed integer linear programing (MILP) have been applied widely. These methods use an objective function for profit maximization and constraints for reliability and safety. The relationship between operating variables can be approximated using piecewise linear equations or described by nonlinear equations. Discrete events such as feed or equipment selection can be represented by integer or binary variables.

Industrial process operations are like a coin with two sides. On one side are the first principle and optimization models, which are so-called white models because the modeling equations used can clearly describe technical and economic relationships. Uncertainties are the other side of the coin; they are everywhere in industrial operations and control and are time varying and dynamically changing in mean and variance. Data-driven models based on statistics and ML can be applied to model uncertainty. Since no fundamental relationships are employed in these models, they are termed as black-box models. With sufficient learning and tracking of unknown disturbances, it is possible to characterize and then reduce uncertainties for better prediction to achieve high reliability and safety.

The fourth attribute to make ML and analytics applicable in process industries is that they need to be interpretable or understandable by operation personnel. Even if certain operations can be automated, human operators and decision makers must be in the loop and understand the prediction results for more confident implementation. If the people do not understand what the ML is doing, they will not take the risk to act upon it. In a modern manufacturing environment, there are usually a hierarchy of people with clear responsibilities at each level. They must be convinced that the new solutions offer more values than risks to be adopted.

Comparing first principle models with ML models, industrial personnel tend to trust first principle models more than data-driven models because they are more interpretable.

Therefore, sophisticated ML methods including deep learning (DL) and deep reinforcement learning (RL) have to pass the interpretability test before they can be widely adopted in industrial practice. This was also the case when model predictive control (MPC) was initially applied in process industries, where model predictions have to be observed and trusted by the operation personnel for a period of time before the controller is commissioned to closed loop (Qin and Badgwell, 2003).

In summary, process data analytics should have the following desirable features (Qin, 2014):

- Interpretable and scalable (up to thousands of variables);
- Making use of all kinds of data (e.g., process data, spectra, vibration, and image data);
- Relatively easy to apply to real processes comparing to other methods;
- Online use for real-time monitoring and decision making;
- Offline troubleshooting as a valuable tool for continuous improvement.

3.4 MODEL TYPES AND CHARACTERISTICS

In general, models can be classified into five categories:

1. **First principles** are applied to model chemical processes based on chemistry and physics. First principle models are largely represented by DAEs, and thus this kind of models are mechanism driven. These models become the foundation for process simulations and control.
2. **Statistics** is the analysis and interpretation of data to reveal data patterns (e.g., normal distribution) and relationships (e.g., linear regression) via data fitting without relying on first principles. Thus, this kind of models are data driven.
3. **AI** is an area of computer science to enable computers to perform tasks and functions that would require human intelligence, such as speech recognition, image recognition, natural language processing, and decision making. The objective of AI research is to make computers to learn, reason, and respond with

intelligence similar to and better than human. **ML** is a major part of AI, and it is used for learning and prediction based on big data. It is developed in the form of neural network algorithms. In ML, the breakthrough is the RL (reinforcement learning) that enables self-learning and self-adjustment to the prediction based on combination of the feedforward and feedback mechanisms.

4. **Mathematical programming**, also known as optimization modeling, consists of a set of equations including an objective function and several constraints. The purpose is to select optimal independent variables that can achieve the objective function.
5. **Hybrid models** is combination of any of the above models.

People also classify the model types based on the transparency that a model provides. Since first principles can better describe the known part of the processes, this kind of models are called white models. In contrast, although data-driven models are more efficient at using data to reveal the unknown part of the processes, it is treated as a black box without revealing the fundamentals; this kind of models are called black models. Hybrids of the white and black models are termed as gray models.

Many-model thinking is about developing a better understanding of complex physical and chemical phenomena. Selecting the right model for a certain application helps us to make sense of reality and search for the truth. Mastery of model thinking improves our ability to reason, design, predict, communicate, and act.

Statistics was originally understood as the sole method for analysis and interpretation of data. Due to the search for truth in the 1950s, scientists and engineers represented by Amundson applied fundamentals to describe physical and chemical phenomena and relied on fundamentals in the form of DAEs to tackle problems associated with chemistry, thermodynamics, transport phenomena, reaction engineering, etc. DAE models the process units with the relationship among process variables based on first principles.

Starting from the 1960s and beyond, mathematical programming led by Sargent, Grossman, and others gained attraction who intend to search for economically optimal solutions to problems in production planning, scheduling, product blending, process design, control and so on. Basically, mathematical programming methods model the relationship between an objective function for economic performance and constraints for technical feasibility.

The next significant development in this long arc of modeling paradigms has been the introduction of knowledge representing and reasoning techniques with symbolic relationships from AI since the early 1980s led by Westerberg, Stephanopoulos, Venkatasubramanian, and others. AI and ML are good at causal reasoning and are thus most suitable for process monitoring, fault diagnosis, and process hazards analysis, which are important for achieving equipment reliability and safety in plant operation.

Conceptually, DAE's model relationships between process variables based on first principles, statistics model data distribution patterns and relationships based on statistic regression, mathematical programming models constraints, and AI and ML models relationships between symbolic variables and symbolic structures. In the early years, logic was considered the discipline best suited to provide the formal foundations of AI, but recent developments seem to suggest that probability, statistics, and neural network science are perhaps better suited. The truth might lie in a hybrid approach, depending on the application (Venkatasubramanian, 2019).

Many-model thinking leads to advent of **predictive analytics**, which mainly relies on first principles and AI/ML techniques as well as optimization methods and advanced statistics methods such as Bayesian probability method, PCA, PLS, and time series models, to predict future trends based on historical data.

For example, for condition-based maintenance, PCA methods can be applied first to identify the most important contributor to characterize the conditions of equipment, while PLS methods can be used to predict the system state and detect abnormal behavior, which triggers certain maintenance tasks. Furthermore, first principle-based models can fundamentally describe the effects of the critical operating parameters, while neural network methods can be employed to describe the hidden relationships of other key operating parameters in statistic terms. ML algorithms implementing the knowledge and patterns learned from the above can predict the impeding failure by comparing current and past performance of a device.

Predictive analytics can incorporate first principles describing the structure of the processes in data-driven modeling, which is a powerful technique for applications including fault diagnosis and predictive maintenance. Fault diagnosis requires information about the causal structure of the systems, for which conventional data-driven methods cannot provide. Predictive modeling also requires this kind of causal knowledge for process control and optimization applications. Bayesian approaches together with hybrid models are promising to introduce a priori knowledge into data-driven modeling.

The following discussions explain each individual modeling methods and their applications.

3.5 FIRST PRINCIPLE MODELING AND ITS CHARACTERISTICS

First principle models should be considered first if possible because they describe the laws of nature that govern chemistry, physics, transport phenomena, reaction kinetics, thermodynamics, fluid dynamics, process conditions and so on. A first principle model is not like a statistical or pure empirical model based on data fitting and thus provides better extrapolation than the interpolation by statistical models.

Therefore, a first principle models strips away "confusion factors" of the problem and gets to the core. These types of models are appealing because they can be extrapolated to a wide variety of situations with few assumptions. First principle models are also called "theoretical models," "fundamental models," or "mechanistic models," which mean the same thing.

The most typical of first principle models are the DAE models, which are suitable for a certain class of problems—chemical engineering has abundant examples in thermodynamics, transport phenomena, and reaction engineering. For example, DAE models can be developed for describing reaction rate as a function of kinetic constant and reactant concentrations, reactor size, and operating conditions (temperature, pressure, catalyst, flow distribution, etc.). Other DAE models can be developed for separation, heat transfer, hydraulics, fluid mechanics, etc.

Some examples of DAE equations are listed below.

Total mass balance in a process system:

$$\frac{d(\rho V)}{dt} = \sum_{i=1}^{N} \rho_i F_i - \sum_{j=1}^{M} \rho_j F_j \tag{3.1}$$

Mass balance on individual component A:

$$\frac{dn_A}{dt} = \frac{d(c_A V)}{dt} = \sum_{i=1}^{N} c_{Ai} F_i - \sum_{j=1}^{M} c_{Aj} F_j \pm r_A V \tag{3.2}$$

Total energy balance:

$$\frac{dE}{dt} = \frac{d(U + K + P)}{dt} = \sum_{i=1}^{N} \rho_i F_i h_i - \sum_{j=1}^{M} \rho_j F_j h_j + Q + W \tag{3.3}$$

where

V = the volume of the system; N = the number of feeds; M = the number of products; ρ = the density of material; F = the volume flow of material; n_A = number of moles in component A; c_A = molar concentration of component A; c_{Ai}, c_{Aj} = molar concentration of component A in the ith feed and jth product, respectively; r_A = reaction rate for component A; a positive (negative) r_A indicates A is produced (consumed); U, K, P = internal, kinetic, and potential energy of the system, respectively; Q = amount of heat into the system from the surroundings; and W = shaft work done by the surroundings on the system.

Example 3.1 Develop a Process Model for an Isothermal Reactor (Erickson and Hedrick, 1999)

Consider the continuous flow stirred tank reactor (CSTR) of Figure 3.1. It consists of a tank with an inlet and outlet flow. This process is a very simple reactor where chemical A reacts to form chemical B. This reaction generates no heat, and the volume of the liquid is constant due to the nature of both chemicals.

Solution

There is no energy transfer in this system since it is isothermal, and so the differential equations are developed based on the mass balances:

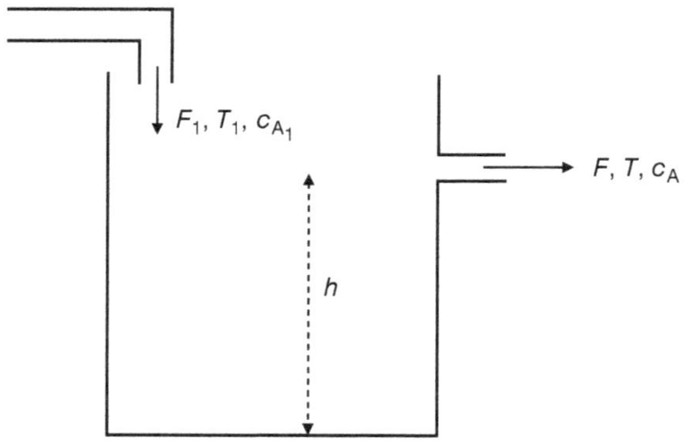

FIGURE 3.1. CSTR reactor for Example 3.1.

Overall mass balance:

$$\frac{d(\rho V)}{dt} = \frac{d(\rho Ah)}{dt} = \rho F_1 - \rho F \tag{3.4}$$

Component mass balance:

$$\begin{aligned} \frac{d(n_A)}{dt} &= \frac{d(c_A V)}{dt} = c_{A_1} F_1 - c_A F - r_A V \\ \frac{d(c_A V)}{dt} &= c_A \frac{dV}{dt} + V \frac{dc_A}{dt} = c_{A_1} F_1 - c_A F - k_o e^{-E/RT} c_A V \end{aligned} \tag{3.5}$$

where V is the reactor volume. Applying Equation 3.4, Equation 3.5 becomes

$$\frac{dc_A}{dt} = \frac{F}{V}\left(c_{A_1} - c_A\right) - k_o e^{-E/RT} c_A \tag{3.6}$$

Therefore, this system consists of variables c_A and F. c_{A_1} is an external stimulus, and other symbols represent specified constants and depend on the chemicals involved and reactor size. If T is considered as a variable as well, Equation 3.6 is DAE with a highly nonlinear (exponential) term as a function of temperature.

Example 3.2 Develop a Kinetic Model for a Reaction
The general form of a reaction rate model can be expressed as

$$r = \frac{dC_i}{dt} = k(T) \cdot f\left(c_1, \ldots, c_i, \ldots, c_{nc}\right) \tag{3.7}$$

For the reaction with the stoichiometric relationship as $a \cdot A + b \cdot B \rightarrow p \cdot P + q \cdot Q$, the rate equation is expressed as

$$r = k(T)[\mathrm{A}]^n [\mathrm{B}]^m \tag{3.8}$$

where k is the reaction constant and [A] and [B] are the concentrations of each reactant. For gas-phase reactions, the rate is often alternatively expressed by partial pressures. Reaction constant k and reaction order n and m are determined from experiments.

k is the reaction constant although it is not really a constant, because it includes all the parameters that affect reaction rate, except for concentration, which is explicitly considered. Of all the parameters influencing reaction rates, temperature is normally the most important one and is accounted for by the Arrhenius equation as

$$k = k_0 e^{-E_a/RT} \tag{3.9}$$

The rate constant k is a function of activation energy (E_a) and temperature (T), where R is the universal gas constant. As denoted in Equation 3.9, the higher the temperature, the larger the k value and thus the higher rate. This is because with a high temperature, the kinetic energy of the reactants increases, allowing far more collisions between the molecules. This, therefore, allows for products to be formed faster. On the other hand, use of catalyst in a reaction reduces E_a and thus increases k for speeding up a reaction as it enables a reaction starting at lower temperature.

Detailed discussions of theoretical modeling methods are given in Chapter 6.

3.6 STATISTIC MODELING AND ITS CHARACTERISTICS

Statistics plays a central role in data science. The central dogma of statistical inference is that there is a certain randomness in data, and the focus is to reveal the degree of certainty or uncertainty. A statistical framework allows us to distinguish between causation and correlation and thus to identify interventions that will cause changes in outcomes. Simply put, statistical methods allow us to reveal some hidden knowledge and discover a pattern of behavior.

Statistics are applied to answer the following questions regarding a data set:

- Sample mean and median;
- Sample variance and standard deviation;
- Sample quantiles besides the median, for example, quartiles and percentiles;
- Regression between variables;
- Test statistics, such as t statistics, chi-squared statistics, and f statistics.

Example 3.3 Statistic Control Chart for Minimizing Operation Variability
The control chart (Figure 3.2) is a statistical process control tool developed by Sheward (1986) to minimize operation variability, which is a major cause of operation inefficiency. The essential idea is to determine if a manufacturing or business process is in a state of stable control. Control charts take advantage of the normal distribution curve. As such, data should be normally distributed (or transformed) when using control charts, or the chart may signal an unexpectedly high rate of false alarms.

Mathematically, obtaining a control chart is to establish the two limits (X_L, X_U) between which we are 90% confident that 95% of values lie within. The data outside of the limits are recognized as abnormal operation. X_L and X_U can be calculated as

$$\begin{aligned} X_U &= \bar{X} + K \cdot S \\ X_L &= \bar{X} - K \cdot S \end{aligned} \tag{3.10}$$

where $\bar{X}$ is the average and S the standard deviation in the data set. K is the statistical factor dependent on the sample size N, the desired probability (P) of the data belonging to the upper and lower limits defined by Equation 3.10. For a large sample size, $S = \sigma$ (population standard deviation).

With use of a control chart, normal operation limits are set at $\pm 3\sigma$ when $K = 3$, that is, 3-standard deviation from the mean on the basis that 99.7% of the observations occur within $\pm 3\sigma$ of the mean. When a process is stable and in control, it displays only the common cause variation or random variation that is inherent to the process. If the process is unstable, the process displays special cause variation or nonrandom variation from external factors. In this case, abnormal operating points will be outside of 6σ or Six Sigma; ($X_U - X_L = 6\sigma$). Because control limits are calculated from process data, they are independent of customer expectations or specification limits.

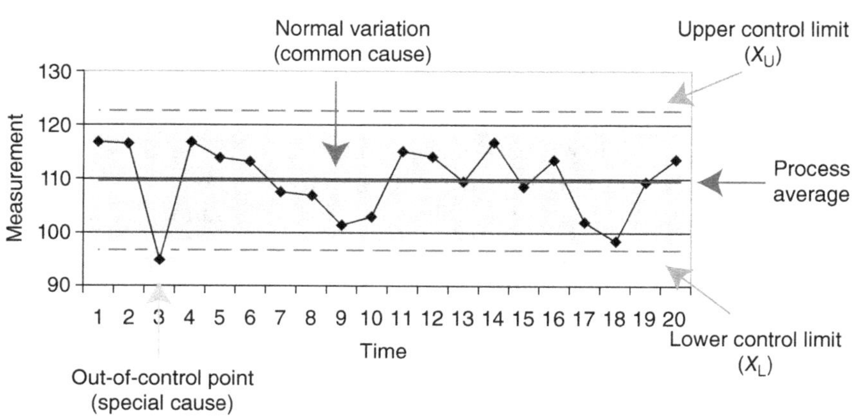

FIGURE 3.2. A typical control chart.

Example 3.4 Statistic Correlation
Many practical phenomena have multiple correlative variables. A person's happiness (y) can be attributed to health (x_1), financial situations (x_2), upbringing (x_3), wealth (x_4), and so on. These variables can be included in a multivariable linear regression model as

$$y = \beta_0 + \beta_1 x_1 + \beta_2 x_2 + \cdots + \beta_m x_m = \beta_0 + \sum_{i=1}^{m} \beta_i x_i \tag{3.11}$$

where x_i's are the independent variable and y the dependent variable; β_i's are the coefficients of x_i's and β_0 is the intercept.

β_0 and β_i's are determined based on the least squares method. The higher values of coefficient β_i translate to high dependence of y on x_i. The goodness of fit for multivariable regression is measured by the regression coefficient, R^2. If a model fits all the data very well (all data points lie exactly on the regression line), R^2 will be 100%. If 100% is unachievable, you should aim to achieve the highest R^2 value as possible by selecting appropriately independent variables x_i's and a large quality sample data set when fitting a model.

Once we have a regression model, we can predict the outcome based on the variables involved. In addition, a regression model can help to identify those variables as significant (or insignificant) when their coefficients are much larger (or smaller) than other coefficients. As a result, the model can guide actions to focus on those significant variables. However, we must be cautious with respect to causality and the range of prediction. By causality, we must avoid use of unrelated variables in a regression model. By a valid range of prediction, we should avoid extrapolation beyond the valid range, which could give completely misleading prediction. For example, a linear growth model will predict that a person will become a giant when he or she grows up to become an adult based on the growth data from infant to teenager.

Many statistic applications are available for guiding operation, and detailed discussions of statistic methods are given in Chapters 4 and 5.

3.7 OPTIMIZATION MODELS

Optimization models are mathematical models, which are different from statistic models and fundamental models. Optimization models define objective functions to optimize independent variables (decision variables) to provide better solutions than the initial values of these variables. In this manner, we want to explore the organized changes that we can act on and improve.

With physical models, such as statistic and fundamental models, the intent is to simulate reality (product yields, properties, energy use, operating conditions, etc.).

Thus, variables can be changed but not optimized because there is no optimization mechanism embedded. Therefore, optimization models are different from physical models in their use of independent or decision variables. However, in most cases, physical models provide the basis for optimization models, and the level of detail in the physical models depends on the application needs.

In general, the goal of an optimization problem is to determine optimal values for n decision variables $x_1, x_2, \ldots, x_n$ within a feasible region in such a way as to optimize (minimize or maximize) a given objective function $f(x_1, x_2, \ldots, x_n)$ of the decision variables. In maximization, the general optimization model is expressed as

$$\begin{aligned} &\text{Maximize } f(x_1,\ldots,x_n) \\ &\text{Subject to } h_i(x_1,\ldots,x_n)=0 \quad i=1,\ldots,m \\ &\qquad g_j(x_1,\ldots,x_n)\le 0 \qquad j=1,\ldots,p \end{aligned} \tag{3.12}$$

As shown in Equation 3.12, an optimization model consists of three essential components:

1. An objective function: Maximize for a profit function; minimize for a cost function.
2. Equality functions: These functions consist of physical models including statistic and fundamental models.
3. Inequality constraints: Providing processing limits and equipment limits.

Thus, the key for building an optimization model is to select the proper decision variables, which are closely related to the problem definition. Optimizing the objective function via changing decision variables gives the optimal solution, which is judged based on the criteria of optimality. The optimized solution gives the combined changes (synergy) to the decision variables simultaneously instead of changing one variable at a time sequentially. The former produces a step change instead of the incremental change by the latter.

Modern manufacturing faces many choices in selection of feedstocks from many sources and generation of products to meet market needs. It is a complex economic optimization problem as it involves varying market conditions as well as consideration of various processes that are interrelated. These processes make different intermediate products, which are blended to make final products that meet market specifications. Optimization is applied across different levels of operation decision making to achieve safe, economic, and reliable operation. Mathematical optimization has been applied to a wide variety of applications, not only in oil refining and petrochemicals but also in airlines, chemicals, utility sectors, and so on.

Example 3.5 Optimize Distillation Reflux Rate
When fuel cost is low, relatively more reboiling duty could be used to generate more reflux and maximize separation for a given distillation column. However, the tower operation involves certain limitations such as tower tray flooding, condensing duty, and reboiler duty. On the other hand, when fuel cost is high, a lower reflux rate may be used to reduce energy cost. Then one may ask: how to determine the optimal reflux rate when fuel cost and product price vary?

There is an economic optimum based on the trade-off between product value and energy cost as shown in Figure 3.3. In reality, operators tend to overflux a column because this strategy will make the tower more stable albeit at the expense of excessive energy. Optimization can help operators be aware of the optimum operation of reflux and thus minimize energy use while achieving the desired product specifications.

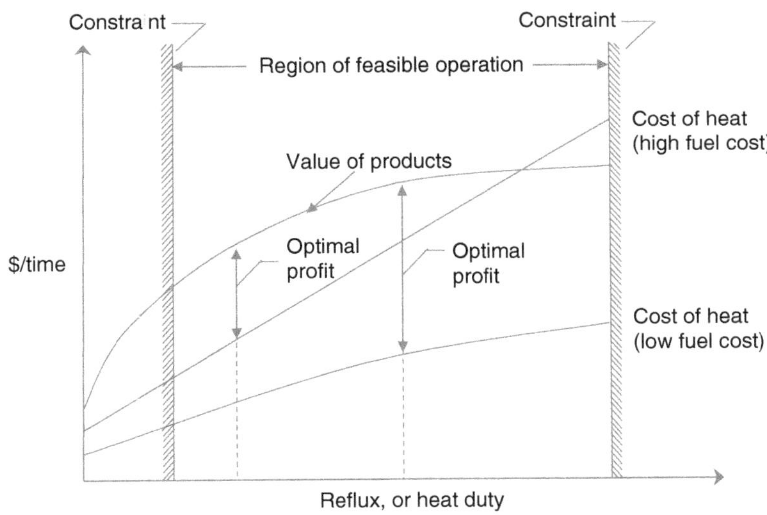

FIGURE 3.3. Optimal column reflux for different fuel costs.

Example 3.6 Multi-plant Product Distribution
Industrial companies must deal with product distribution. Generally, *n* number of products ($P_1, P_2, \ldots, P_n$) are produced from *m* number of plants ($Y_1, Y_2, \ldots, Y_m$), which need to be delivered to *p* number of customers ($Z_1, Z_2, \ldots, Z_p$). The optimization problem should consider both production costs from each plant and transportation costs between plants and customers. The individual plants may vary with respect to design capacity, and also some plants may be more efficient than others. This is a very interesting product distribution optimization problem.

Detailed discussions of linear and nonlinear optimization models are given in Chapters 7 and 8, respectively.

3.8 ARTIFICIAL INTELLIGENCE (AI) AND MACHINE LEARNING (ML) MODELS

With the presence of sustained variabilities and abnormal situations in process operations, it is unrealistic to require that first principle models and optimization models to include all possible scenarios of uncertainties. In these situations, first principle and optimization models represent the known part of the operation, while data analytics discussed below deals with uncertain part of the operation. Therefore, model and data interplay for more accurate prediction and intelligent decision making.

When dealing with problems with high degree of uncertainty and complex hidden relationships, use of physical models alone may fail. That is when AI and ML models can contribute to describe the uncertainty and uncover the hidden relationships.

AI is the ability of computer software to function like a human brain, which is capable of reasoning, abstraction, pattern recognition, learning, natural language processing, and perception. Installed with AI software, robots can think and act like a human being. AI is a reference to the software that manifests intelligence, whereas robots infer a physical equipment equipped with sensing, actuation, and control devices and capable of communications and decision making using AI software. AI can be classified into two broad types: narrow AI (or weak AI) and general AI (or strong AI).

3.8.1 Narrow AI

Narrow AI is what we see around us today. There are a vast number of emerging applications for narrow AI from interpreting video feeds from drones carrying out visual inspections of infrastructure such as oil pipelines, organizing personal and business calendars, responding to simple customer-service queries, coordinating with other intelligent systems to carry out tasks like booking a hotel at a suitable time and location, helping radiologists to spot potential tumors in X-rays, flagging inappropriate content online, and detecting wear and tear in elevators from data gathered by IoT devices—the list goes on and on. Unlike humans, these systems can only learn or be taught how to do specific or narrow tasks, which they do extremely well often surpassing human capabilities, hence the term narrow AI.

3.8.2 General AI

General AI has human-level intelligence with stronger intelligence compared with narrow AI. General AI can achieve high-level AI that can understand and reason its environment as a human would. In other words, it is the type of adaptable intellect found in humans, a flexible form of intelligence capable of learning how to carry out vastly different tasks, anything from haircutting to building spreadsheets, or to reason about a wide variety of topics based on its accumulated experience.

General AI has always been elusive. We have been saying for decades that it's just around the corner. But the more we delve into it, the more we realize that it is hard to achieve—and the more we come to appreciate the miracle that is behind the human brain. It is really hard to define what a human-level AI would be. You just need to look at how you perceive things, juggle between multiple unrelated thoughts and memories when making a decision. It is very hard to teach a computer to reach the level of human intelligence. We believe the smarter AI gets, the better humans will become at saving the environment, curing diseases, exploring the universe, and understanding themselves. AI researchers dream to develop general AI systems, which is the holy grail of AI. Due to the complexity of teaching a software to think, behave, and act like a human being, general AI could be decades away from commercialization.

3.8.3 Machine Learning (ML)

ML is a key component of AI since ML gives a computer the ability to learn from data—analyze and find patterns in data and then make inference or prediction based on the data. In other words, ML provides ways of learning from data to achieve AI for decision making.

DL (deep learning) is a subfield of ML that uses algorithms called artificial neural networks (ANNs), which are capable of self-learning. ANNs are brain-inspired networks of interconnected layers of algorithms, called neurons, that feed data into each other and can be trained to carry out specific tasks by modifying the weighting factors attributed to input data as it passes between the layers. ANNs can be trained to learn or identify hidden patterns and relationships rather than being explicitly instructed to solve a problem.

ANNs were invented many years ago but remained in obscurity until just recently because they require extensive training. The amount of training historically exceeded the computing power and data size available. With the explosion in parallel computing power and big data readily available, though, DL via ANNs has emerged as the biggest breakthrough for AI research. From self-driving cars to robots, DL is fueling what many believe will be the biggest technology revolution the world has ever seen.

Although DL is a subset of ML, its capability goes way beyond normal ML. An ML model can become better progressively, but the model still needs guidance and calibration. If the ML model generates an inaccurate prediction, the programmer must modify the software code to fix the problem explicitly. In contrast, the DL model can make the correction by itself. An autonomous car driving system is a good example of DL.

3.8.4 ML Applications in the Process Industries

All ML methods can be divided into supervised learning and unsupervised learning. The supervised learning uses input data X to predict or interpret target or label data Y, while unsupervised learning analyzes X without supervision from Y.

Supervised ML is widely used for industrial analytics. Working with labeled historical data, one can develop classification model for future event predictions and regression model to predict key process and product properties. Advanced pattern recognition model can also be developed for anomaly detection for machinery reliability and process safety improvement. Time series model can be developed to improve demand forecasting, optimize staffing levels, manage inventory, and more.

One can build clustering model using unlabeled data by using unsupervised ML to uncover patterns and trends in the data that were not previously distinguishable. This creates group of similar events (or clusters) and explicitly express what feature is decisive in these results.

In most cases, AI and ML split data into training data, validation data, and holdout data to perform cross validation and to ensure models are both highly accurate and generalize well to future data. Training set is used to generate and optimize solutions, validation data is used to measure model performance, and holdout data is used to test how well those models generalize to new data. With cross validation, model was validated in several iterations with several different portions of the data. When sampling size is small, cross validation is more important and necessary.

To deal with collinearity in the data, for example, PCA is unsupervised learning, and PLS is supervised learning. For categorical data such as modes of operations, grades of products, and classes of faults, the corresponding analytics are classification for supervised learning and clustering for unsupervised learning. Deep neural networks can be used for supervised learning or unsupervised learning, depending on the availability of target or label data. RL (reinforcement learning), with the goal to learn optimal actions based on data, also performs supervised learning (Badgwell et al., 2018; Hoskins and Himmelblau, 1992; Sutton et al., 1992).

There are straightforward applications of these methods to process monitoring, fault diagnosis, risk management, and inferential sensors. For example, any supervised learning methods such as DL can be used as inferential sensors. Unsupervised learning methods, especially those that are nonlinear extensions of PCA, can be used for process monitoring and fault diagnosis. It is critical to demonstrate why the chosen method shows advantages for the specific application.

3.8.4.1 Deep Learning Applications There are two essential aspects of data analytics: prediction and interpretation. To achieve accurate prediction, often complex models such as deep neural networks are more capable (Qin and Chiang, 2019). On the other hand, in order to achieve interpretability, simpler models are preferred. Interpretability is critical for industrial adoption, since operators and decision makers must establish a level of trust in the analytics algorithms. DL has been applied to derive inferential sensors from data (Shang et al., 2014). Although the main goal for inferential sensors is prediction accuracy, it is necessary to provide a trade-off between model complexity and maintainability. The increased complexity of deep neural networks together with the requirement of massive amounts of data needs to be justified against the incremental accuracy gain compared with other simpler models (Qin and Chiang, 2019). Besides inferential sensors, deep neural networks have also been applied for image classification for a combustion furnace (Zhu et al., 2019b). This is directly applying image analysis methods to process operations.

3.8.4.2 Deep Reinforcement Applications Besides supervised and unsupervised learning, RL involves learning by trial and error, solely from rewards or punishments. RL is a special DL approach. Neural networks based on RL can start from a blank slate, and under the right conditions, they achieve superhuman performance based solely on knowledge about how to value the quality of outcomes (rewards) (Figure 3.4). The RL approach combines intuition to make an initial prediction based on general rules and adjustments based on the gap between the initial prediction and the goal or target. Over the process of self-learning, the RL prediction will be very close to the target. RL is regarded as a game changer in AI.

RL involves an agent, a set of states {S}, a set of actions {A} per state, and the environment (Figure 3.4). By performing an action $a \in A$, the agent transitions from state to state. Executing an action in a specific state provides the agent with a reward (a numerical score).

The goal of the agent is to maximize its total (future) reward. It does this by adding the maximum reward attainable from future states to the reward for achieving its current state, effectively influencing the current action by the potential future reward. This potential reward is a weighted sum of the expected values of the rewards of all future steps starting from the current state.

The most known learning method in RL is Q-function that estimates the impact of a certain action at a given state. Q-learning is a model-free RL algorithm. The central concept of Q-learning is that of a function $Q(s, a)$ that calculates the maximum future reward for performing action a in state s. It is called Q-function because it represents the quality of a certain action in a given state. Given a Q-function, the strategy for picking the best action (e.g., to open or close a valve) is simply to pick the action with the highest Q-value.

The Q-value of state s and action a can be stated in terms of the Q-value of the next state s:

$$Q(s,a) = r + \gamma \max Q(s',a') \tag{3.13}$$

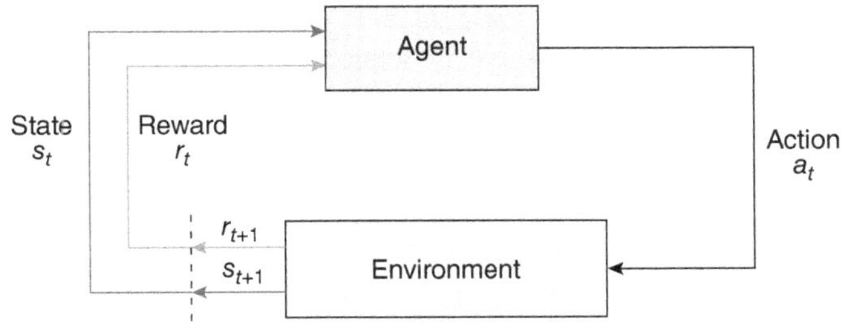

FIGURE 3.4. In reinforcement learning, the agent's goal is to maximize the total amount of reward it receives over time. The agent accepts the state and reward as input and outputs an action. Meanwhile, the environment sees the action, changes the state, and gives a reward.

where $Q(s, a)$ is the expected future reward of performing action a at state s; r is the reward, if any, of taking action a at state s; and max $Q(s', a')$ is the maximum future reward. In other words, if taking action a at state s leads to state s' and there are multiple actions a' that can be taken in s', max $Q(s', a')$ is the maximum among those. γ is called the discount factor. If set to a value near 0, the immediate reward is given preference, and, if set near 1, future rewards are valued. Therefore, Q-function calculates the maximum future rewards for performing different actions at state s, and the action with the highest Q-value will be selected.

One area where RL can be applied is process control (Spielberg et al., 2017), where features describing the current state of a plant can represent states, controller inputs to the plant are actions, and the objective is to make the output reach the setpoint while meeting certain constraints. For example, in the manufacture of paper, the steam flow rate can be the control action, the current moisture content can represent states, and rewards can be associated with how close the current moisture is to the desired moisture content of a sheet of paper. Given such a set of rules, RL can come up with control policies without the need for handcrafted feature descriptors, controller tuning, deriving control laws, and developing mathematical models.

The problems of interest in RL have also been studied in the theory of optimal control, which is concerned mostly with the existence and characterization of optimal solutions, and algorithms for their exact computation, and less with learning or approximation, particularly in the absence of a mathematical model of the environment. Another problem of interest is simulation-based optimization (Gosavi, 2015). Using RL, various operating performance of equipment can be assumed, and the data can be used for RL training to generate different clusters of healthy, failure, or near failure.

Example 3.7 Predicting Valve Failure (Pandya et al, 2018)
In the Shell company, studies were carried out at the Shell Pernis, Netherlands, as well as Shell Martinez, CA, manufacturing sites with the objective of predicting valve failure. Experiments for the use case were performed on multiple control valves. The aim of these studies was to verify whether machine learning (ML) methods are capable of distinguishing between normal and abnormal valve behavior. The difference between the predicted and measured output (i.e., error rate) should be as low as possible for the normal valves and as high as possible for the abnormal valves.

Multiple solutions were developed. The solution developed at the Martinez site is the ML-based predictive deterioration model that relies on integration of first principles and statistical multivariable regression to augment and validate traditional mass balances. The model was tested well using a traditional mass balance, with meters upstream and downstream of a target flow element. The model verified that the mass balance approach provided acceptable meter accuracy and allowed engineers to track predicted flow meter performance against actual metering.

The solution developed at the Pernis site involves ANNs and statistic models to model the normal behavior of the valves. Mismatches with predictions of the modeled systems were then used to predict failure. The ANNs could predict failure up to a month in advance.

As a result, ML algorithms for ANNs have been developed, which can avoid unexpected failures and unnecessary maintenance and could save millions of dollars per year in optimized maintenance and deferment avoidance. Going forward, these models provide confidence and direction toward predictive maintenance in the Shell company.

Detailed discussions of AI and ML are given in Chapter 10.

3.9 PUT ALL TOGETHER: DIGITAL TWIN AS A DATA SCIENCE PLATFORM

Industrial operational problems manifest in two sides. On one side, they are governed by the known part based on the nature of chemistry, transport, and flow mechanics, which can be described by first principles. This transparency is the reason why a fundamental model is termed a white model. On the unknown or dark side, time variance, randomness, and uncertainty are featured, and use of data analytics and ML models is more efficient. These kinds of models are called black models as they cannot reveal fundamental relationships behind the model. The duality of industrial problems makes them different from other data analytics in business, marketing, gaming, etc.

The following discussions based on Qin and Chiang (2019), shed lights on good opportunities for applying model thinking in process industries.

3.9.1 Integrate First Principles with ML Models

With the presence of operation variability and abnormal situations, the strength of data analytics and ML models is in describing this dark side of the situational data, but there is a fundamental issue at stake here. The glaring failure in purely data science models, such as deep neural networks, is their lack of "understanding" of the underlying knowledge. For instance, a self-driving car can navigate impressively through traffic, but does it "know" and "understand" the concepts of mass, momentum, acceleration, force, and Newton's laws, as we do? It does not; at the best, the current AI and ML systems have not gained deeper "understanding" than most humans do (Venkatasubramanian, 2019).

On the other hand, first principle models can capture the white side or known part of industrial problems. The integration of both kinds of modeling (white and black models) leads to hybrid models. One possible hybrid modeling method is to use first principle models to describe the fundamental relationships between process variables while using neural networks or PCA to model the residuals of the first principle models (Psichogios and Ungar, 1992; Wachs and Lewin, 1998).

Furthermore, the well-known Kalman filter is a systematic framework to use first principle models for the deterministic subsystem while using data analytics to model the residuals (Odelson et al., 2006). There are other hybrid modeling approaches available, which are discussed by Glassey and von Stosch (2018). These methods can be applied for operation monitoring, optimization, and control.

3.9.2 Troubleshooting and Predictive Maintenance

Process monitoring requires good data. This requirement, however, is not always easy to fulfill because of the following two situations:

(a) Data contain normal records as the majority, but there are many outliers or faulty values mixed in with them.
(b) The very task is to figure out if the current operation is normal or not by troubleshooting the data to uncover undesirable patterns or variabilities.

When data are contaminated with outliers or bad data, data analytics are used to provide a descriptive model for the normal situation. The robust low-rank matrix approximation via ML modeling is applicable (Liu et al., 2013). In theory and practice, it is shown that a data matrix can be expressed as a low-rank matrix plus a sparse matrix of outlying values. Robust PCA and matrix completion models are proposed and applied effectively to video data and text data analysis.

3.9.3 Supervised Monitoring and Diagnosis

Most work in process monitoring are focused on PCA and unsupervised learning methods. Although PLS has been used for process monitoring, the quality-relevant diagnosis has only been recently developed (Li et al., 2010; Qin and Zheng, 2013; Zhu et al., 2017). Unlike diagnosis based on unsupervised learning models where one can analyze abnormal changes in process variables, quality-relevant diagnosis refers to the use of supervised learning models for the diagnosis of a quality anomaly due to relevant abnormal process variations. Since this approach aims to monitor and diagnose quality anomalies before they are measured and confirmed, this approach is referred to as supervised monitoring and diagnosis (Zhu and Qin, 2019). Supervised process monitoring has received increased attention since early work on total PLS and current PLS (Ge and Chen, 2016; Huang and Yan, 2017; Li et al., 2010; Liu et al., 2014; Qin and Zhao, 2018). Supervised monitoring and diagnosis deserves much more attention since it is predictive and can reduce quality-irrelevant alarms significantly.

3.9.4 Deep Learning for Model Sustainability

Spielberg et al. (2017) reported that deep RL is applied to process control problems. The authors demonstrated that if reward hypothesis functions are formulated properly, they can be used for industrial process control. The controller setup follows the RL setup, whereby an agent (controller) interacts with an environment (process) through control actions and receives a reward in discrete time steps. Deep neural networks serve as function approximators and are used to learn the control policies. Once trained, the learned network acquires a policy that maps system output to control actions. Though the policies are not explicitly specified, the deep neural networks were able to learn policies that are different from the traditional controllers. This approach was evaluated on single-input single-output (SISO) systems and multi-input multi-output (MIMO) systems and tested it successfully under various scenarios.

3.9.5 Digital Twins: Integrate Data, Analytics, and Modeling into One System

To capture both the known and unknown parts, digital twins are adopted as a data science platform for industrial applications. The digital twin system is a software representing the virtue asset that mimics the physical asset. In addition to virtue representation, digital twin contains intelligence via simulation and built-in predictive analytics. Process simulation is used to mimic the design, conditions, and limits of actual processes and equipment based on real-time operation and lab data—known part of the process behavior based on first principle models. Once the known part of the process and equipment is described by simulation, fault detection and diagnostics (FDD) can be conducted by applying data analytics and ML techniques to predict what is going to happen and provide recommendations for what actions to take and what outcomes may result—this becomes the basis for human operators to make informed and intelligent decisions.

Therefore, the cornerstone for our digital twin is a set of prediction models with high accuracy. Good models can help us to categorize, explain, and most importantly predict. Indeed, experience and information can be good teachers, but many times in operation and asset management, we simply cannot afford to learn on the job. One cannot go through several incidents to learn how to protect critical equipment. Suboptimal service providers are unable to offer high-fidelity prediction models and thus cannot provide coordinated, integrated, and timely maintenance.

This is why data analytics and models become so valuable in digital twin methodology: accurately predicts what will happen before it occurs and interprets the status of the installed base, thus enabling operators to quickly initiate maintenance measures when irregularities occur.

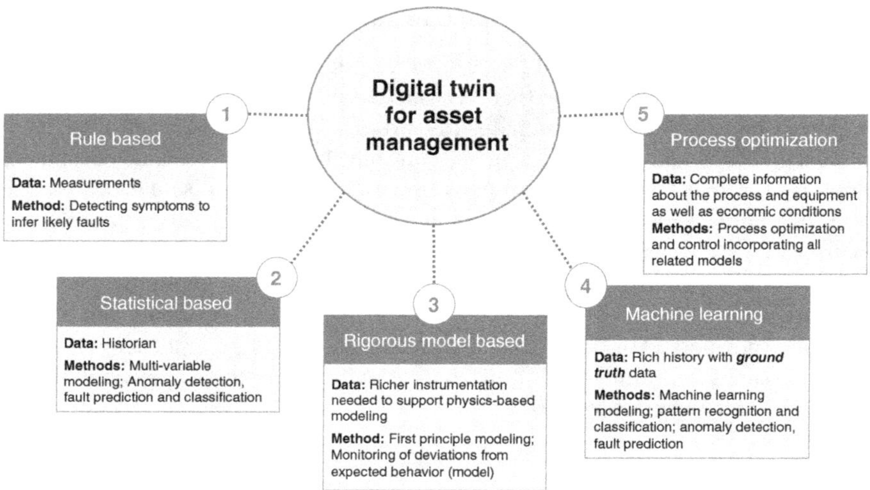

FIGURE 3.5. Model thinking in the APM digital twin.

Figure 3.5 provides an overview of modeling techniques used in digital twins for achieving operation excellence:

- **Rigorous models**: It is converted from process design simulation that consists of mass and energy balances, proprietary rigorous reaction models, product fractionation and impurity separation (Zhu et al., 2019a), rotating equipment, fired heaters, heat exchangers, operating limits, and many more. In this manner, process design is integrated with feed conditions and production requirement. Process changes and technology improvement often provide business opportunities that enable more effective maintenance schedule and deployment of resources in the context of production needs.
- **Fault detection and diagnostics** (what is wrong): FDD is the first natural opportunity for applying data analytics. Due to relatively small number of sensors and limited availability of historical data for reliability, it has been recognized that integrated model and rule-based approach is the most easily deployable. Generalized rule sets were specified based on industry expertise and domain knowledge and can be easily incorporated and used working together with equipment models.
- **Descriptive models** (what may go wrong): Descriptive models identify a problem, and then predictive models can determine possible damage levels and maintenance options. The descriptive models can be those models for defining the operating windows for fired heaters, distillation columns, compressors, pumps, heat exchangers, etc.
- **Predictive maintenance** (what is likely to go wrong and what will be the consequences): The FDD approach is usually reactive—that is, signs of faults are

detected (what's wrong). Then the next step is moving toward predictive maintenance. The predictive analytic models can be developed based on rigorous models and statistic models such as PCA and PLS. In situations when there is enough ground truth data about various equipment faults, it should be possible to apply ML methods to train models for fault prediction. Predictive maintenance can leverage FDD-based algorithms to continuously monitor individual control and equipment performance. Predictive modeling is then applied to a selected condition indicator that deteriorates in time.

- **Prescriptive maintenance** (what is the right solution): This is an emerging analytic technology that started to be embraced by oil and gas companies. Prescriptive analytics help to determine what is the best maintenance option and when it should be performed based on actual conditions rather than a time-based maintenance schedule. The prescriptive analytics moves beyond descriptive and predictive analysis. Once a problem is identified via descriptive analytics, then possible damage, failure, or maintenance options can be predicted by predictive analytics, followed by prescriptive analytics to select the right solution based on the company's capital, operational, and other requirements. The objective for prescriptive maintenance is to enable effective parts inventory and delivery and expert allocation with the ability to "fix it on the first visit." High fidelity of failure forecasting models is critical for maintenance in a timely fashion, extend the useful life of critical machinery, keep maintenance costs down, develop optimal service centers, remotely fix troubled equipment, and maximize the cost efficiency.
- **Process optimization** (what are the optimal conditions?): It is about determining optimal conditions via adjusting setpoints and schedules. This approach can provide cost savings in the operation of equipment and reduce performance drifts. The optimized setpoints and schedules can be communicated directly to the underlying controllers, which allows an automated closing of the loop from analytics to control.

REFERENCES

Badgwell TA, Lee JH, Liu KH (2018), Reinforcement learning overview of recent progress and implications for process control. In: Eden MR, Ierapetritou MG, Towler GP (Eds.), 13th International Symposium on Process Systems Engineering (PSE 2018), Vol. 44 of Computer Aided Chemical Engineering. Elsevier, pp. 71–85.

Bardin B (2017), Industry 4.0 challenge and opportunity for the chemical industry. In: 2017 AIChE Spring Meeting, San Antonio, TX.

Erickson KT, Hedrick JL (1999), Plantwide Process Control, John & Wiley Sons, Inc.

Ge Z, Chen X (2016), Supervised linear dynamic system model for quality related fault detection in dynamic processes, J. Process Control **44**, 224–235.

Glassey J, von Stosch M (2018), Hybrid Modeling in Process Industries, CRC Press, Taylor & Francis Group.

Gosavi A (2015), Simulation-based Optimization – Parameter Optimization Techniques and Reinforcement Learning, 2nd ed., Springer.

Hoskins J, Himmelblau D (1992), Process control via artificial neural networks and reinforcement learning, Comput. Chem. Eng. **16** (4), 241–251.

Huang J, Yan X (2017), Quality relevant and independent two block monitoring based on mutual information and kpca, IEEE Trans. Ind. Electron. **64** (8), 6518–6527.

Li G, Qin S, Zhou D (2010), Output relevant fault reconstruction and fault subspace extraction in total projection to latent structures models, Ind. Eng. Chem. Res. **49**, 9175–9183.

Liu G, Lin Z, Yan S, Sun J, Yu Y, Ma Y (2013), Robust recovery of subspace structures by low-rank representation, IEEE Trans. Pattern Anal. Mach. Intell. **35**, 171–184.

Liu Q, Qin S, Chai T (2014), Quality-relevant monitoring and diagnosis with dynamic concurrent projection to latent structures, IFAC Proc. **47** (3), 2740–2745.

Odelson BJ, Rajamani MR, Rawlings JB (2006), A new autocovariance least-squares method for estimating noise covariance, Automatica **42** (2), 303–308.

Pandya D, Suursalu S, Lam B, Kwaspen P (2018), Digital twins for predicting early onset of failure flow valves. In: 2018 AIChE Spring Meeting, Paper 37a, April.

Psichogios DC, Ungar LH (1992), A hybrid neural network-first principles approach to process modeling, AIChE J. **38** (10), 1499–1511.

Qin S (2014), Process data analytics in the era of big data, AIChE J. **60** (9), 3092–3100.

Qin S, Badgwell TA (2003), A survey of industrial model predictive control technology, Control Eng. Pract. **11** (7), 733–764.

Qin SJ, Chiang LH (2019), Advances and opportunities in machine learning for process data analytics, Comput. Chem. Eng. **126**, 465–473.

Qin Y, Zhao C (2018), A quality-relevant monitoring method for closed-loop industrial processes with dual consideration of static and dynamic analysis. In: 2018 10th International Conference on Modelling, Identification and Control (ICMIC). IEEE, pp. 1–6.

Qin S, Zheng Y (2013), Quality-relevant and process-relevant fault monitoring with concurrent projection to latent structures, AIChE J. **59** (2), 496–504.

Shang C, Yang F, Huang D, Lyu W (2014), Data-driven soft sensor development based on DL technique, J. Process Control **24** (3), 223–233.

Sheward WA (1986), Statistical Method from the Viewpoint of Quality Control, Dover Publications, Inc.

Spielberg SK, Gopaluni RB, Loewen PD (2017), Deep reinforcement learning approaches for process control. In: 6th International Symposium on Advanced Control of Industrial Processes, May 28–31, Taiwan.

Sutton RS, Barto AG, Williams RJ (1992), Reinforcement learning is direct adaptive optimal control, IEEE Control Syst. Mag. **12** (2), 19–22.

Venkatasubramanian V (2019), The promise of artificial intelligence in chemical engineering. Is it here, finally?, AIChE J. **65** (2), 466–478.

Wachs A, Lewin D (1998), Process monitoring using model-based PCA, IFAC Proc. **31** (11), 87–92.

White D (2016), Big Data – What Is It?, 32–35, March, CEP.

Zhu Q, Qin S (2019), Supervised diagnosis of quality and process faults with statistical learning models, I&EC Research Revised.

Zhu Q, Liu Q, Qin S (2017), Concurrent quality and process monitoring with canonical correlation analysis, J. Process Control **60**, 95–103.

Zhu FXX, Johnson JA, Ablin DW, Ernst GA (2019a), Efficient Petrochemical Processes – Technology, Design and Operation, (ISBN-13: 978-1119487869), Wiley, Hoboken, NJ.

Zhu W, Ma Y, Benton M, Romagnoli J, Zhan Y (2019b), DL for pyrolysis reactor monitoring: from thermal imaging toward smart monitoring system, AIChE J. **65** (2), 582–591.

PART 2

MODEL THINKING FOR SMART OPERATIONS

Part 2 is about model thinking, which consists of seven chapters. In this part, individual modeling techniques including first principles, statistics, optimization, artificial intelligence, and machine learning are discussed in details. Intelligent models can empower digital solutions to convert data to new insights and actions. Mastery of models improves your ability to reason, explain, design, predict, communicate, and explore.

This part also explains model application—how to apply them where and why. Some models provide tractable results that allow one to act directly. For example, when a compressor is operating near a surge threshold, it is in a dangerous operation zone and the compressor will collapse quickly if no urgent action is taken. Some models may provide intractable prediction that requires one to interpretable results with experience. For example, a machine learning model indicates the possibility of failure for a key equipment in the next few weeks. This alerts the subject matter expert to analyze it in more granule details and thus can predict how long it can continue to run without failure until the next turnaround. Some models are economics based. For example, a planning optimization model suggests the best feed mixture to use—applying the results in a timely manner could lead to many million-dollar profit for the company.

Each individual model has their own strength and weakness, and integrating different models allow us to avoid narrowness inherent in each individual model and avoid their blind spots. That is the essence of model thinking—for a given phenomenon, define the problem, select a model system, and run the model to search for insights that lead to an effective solution.

Digitalization and Analytics for Smart Plant Performance: Theory and Applications,
First Edition. Frank (Xin X.) Zhu.

4

STATISTICS BASICS

4.1 INTRODUCTION

Real-world information is expressed through data, and the use of data as evidence in support for decision making is nothing new. Statistics was originally understood as the sole method for analysis and interpretation of data. In the age of digitalization, people take advantage of much increased computing power and Internet in order to achieve strong connectivity among people, asset, and workflows. Thus, the amount of data exists in unprecedented dimensionality and granularity, resulting in so-called big data.

Conventionally, statistics is the technique for analyzing data. Today with data science, the data analytics have extended to data mining and machine learning with the aim to derive insights and actionable decisions for value creation. Let's do a little refresher of statistics starting from normal distribution as follows.

4.2 NORMAL DISTRIBUTION

Fitting data to a distribution is a statistic modeling exercise as it can be used to obtain core knowledge about the pattern of events. A normal distribution not only indicates the average value of the events, but also it tells the standard deviation (SD): the distance of the data to the average. The normal distribution is a very

Digitalization and Analytics for Smart Plant Performance: Theory and Applications,
First Edition. Frank (Xin X.) Zhu.

common distribution to represent measurements of real data as it usually provides a good approximation for data with random error. The normal distribution assumes that obtaining values at the mean are most probable, while values away from the mean are less probable. Moreover, the probabilities of obtaining values away from the mean are unbiased, meaning the data are equally centered around the mean. The normal distribution can be described by a probability density function as

$$f(x) = \frac{1}{\sigma\sqrt{2\pi}} e^{-\frac{1}{2}\left(\frac{x-\mu}{\sigma}\right)^2} \tag{4.1}$$

where μ is the mean or average indicating the location of the peak and σ is the SD indicating the shape of the curve.

All normal distribution curves look similar as they are characterized by its mean (μ) and SD (σ). They all look symmetric about mean with the probability around mean being high but low for those far away from the mean. They look identical in distribution with approximately 68% of all outcomes within $\pm 1\sigma$, 95% of all outcomes within $\pm 2\sigma$, and more than 99% of all outcomes within $\pm 3\sigma$. In other words, normal distributions allow for any size outcome or event to occur. The outcomes near the mean have the highest probability with events far away being rare.

As a recap, mean indicates the peak location in the center of the distribution curve. A large σ implies a wide variation featuring a short and fat distribution curve, while a small σ indicates a narrow variation corresponding to a tall and thin distribution curve. The outcomes or events around the mean in a small σ normal distribution (tall and thin) have a higher probability than that of a large σ distribution.

4.2.1 Central Limit Theorem

The central limit theorem states that the sampling distribution of an average tends to be normal as the sample size gets larger—regardless of the population distribution shape. In other words, as you take more samples, especially with a sample size greater than 30, the distribution of the sample means will look similar to a normal distribution. In practice, this means that if you add up the means from all of your samples and take the average, then this is the actual population mean. Similarly, if you find the average of all of the SD in your sample, you'll find the actual SD for your population. It's a very useful phenomenon that can help accurately predict characteristics of a big data set. For example, suppose that a sample containing a large number of observations is obtained, each observation being randomly generated in a way that does not depend on the values of the other observations, and that the arithmetic mean of the observed values is computed. If this procedure is performed many times, the central limit theorem says that the distribution of the average will be closely approximated by a normal distribution.

The normal distribution is by far the most important distributions in statistics, and its applications are endless. The population of IQ scores is modeled with normal distributions, so does the test scores in schools. People's health, height, size, and wealth also follow normal distributions, and the list goes on and on. In industrial applications, random measurement errors and process conditions around setpoints also present as normal distributions. To verify and prove if a certain data set follows normal distribution, *Z*-test or *p*-test may be applied, and details can be seen in Section 4.5. In the following, we will explain how to estimate data average, SD, variance, and covariance.

4.2.2 Estimate Mean and Standard Deviation

4.2.2.1 Mean and Median Mean (μ) is the same as average $(\overline{x})$ but has a subtle caveat: mean is for large population of data, while average for sample data. It is calculated the same as the arithmetic average expressed as

$$\mu = \overline{x} = \frac{\sum_{i=1}^{N} x_i}{N} \tag{4.2}$$

Median is a little different in meaning, which describes the central location of the distribution. To calculate the median, you sort the data in either ascending or descending order; then the one in the middle is the median.

Unfortunately, the mean doesn't tell us a lot about the data except for the average. For example, the following two data sets have the same mean of 10, but are obviously quite different:

$$[600, 640, 650, 660] \text{ and } [645, 647, 648, 650]$$

What is different about these two sets? It is the spread of the data that is different. The SD of a data set is a measure of how spread out the data is.

4.2.2.2 Standard Deviation and Variance The English definition of the standard deviation (σ) is "The average distance from the mean of the data set to the data points." Variance (var) is another measure of the spread of data in a data set. Both var and σ can be expressed as

$$\sigma = \sqrt{\frac{1}{N-1}\sum_{i=1}^{N}\left(x_i - \overline{x}\right)^2} \tag{4.3}$$

$$\text{var} = \sigma^2 \tag{4.4}$$

Figure 4.1 shows the mean and SD for a data set with normal distribution. Variance is similar to SD in meaning because variance is simply the SD squared. Mean and SD are meaningful if they are calculated based on large samples. Small samples ($N<30$) could give misleading estimation of mean and SD.

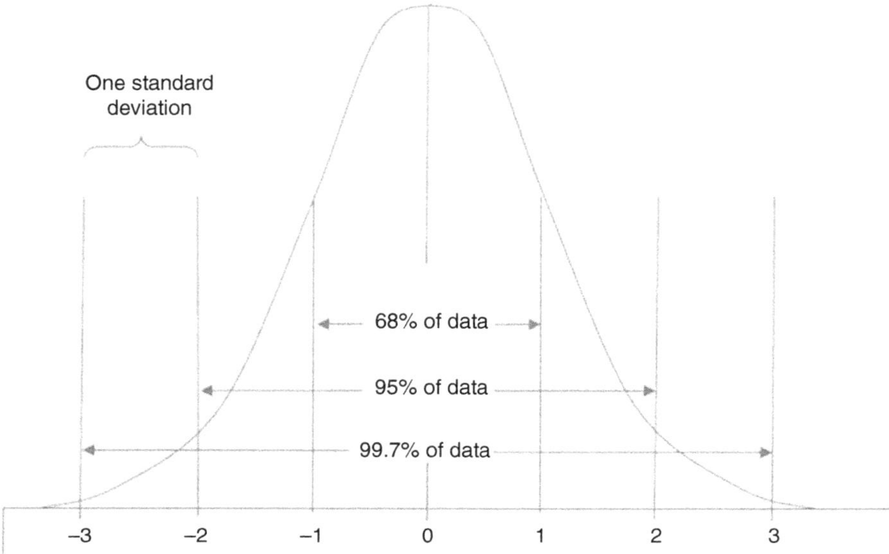

FIGURE 4.1. Normal distribution curve with a bell shape.

4.2.3 Covariance: Determine Variable Dependence

SD and variance only operate on single data set independently of the other data sets. However, sometimes it is useful to determine the relationship in the mean between multiple data sets. Covariance is such a measure and can be measured for two data sets. The covariance for a single data set is just its variance, which is discussed above. So, if you had a three-dimensional data set (x, y, z), then you could measure the covariance between the x and y dimensions, the x and z dimensions, and the y and z dimensions, respectively. Measuring the covariance between x and x or y and y or z and z would give you the variance of the x, y, and z dimensions, respectively.

The formula for covariance is similar to the formula for variance. The expression 4.4 for variance can be rewritten as

$$\text{var}(x)=\sqrt{\frac{1}{N-1}\sum_{i=1}^{N}(x_i-\bar{x})(x_i-\bar{x})} \tag{4.5}$$

Covariance (cov) between x and y is expressed as

$$\text{cov}(x,y)=\sqrt{\frac{1}{N-1}\sum_{i=1}^{N}(x_i-\bar{x})(y_i-\bar{y})} \tag{4.6}$$

Let us consider one example to illustrate the calculation of covariance and explain its meaning a bit more clearly. This example includes two data sets with the number of hours versus exam scores. We want to use covariance to identify the relationship between them.

Step 1: Calculate mean or average:

	Hours (*H*)	Mark (*M*)
Data	9	39
	15	56
	25	93
	14	61
	10	50
	18	75
	0	32
	16	85
	5	42
	19	70
	16	66
	20	80
Totals	167	749
Averages	13.92	62.42

Step 2: Calculate covariance:

H	M	$(H_i - \bar{H})$	$(M_i - \bar{M})$	$(H_i - \bar{H})(M_i - \bar{M})$
9	39	−4.92	−23.42	115.23
15	56	1.08	−6.42	−6.93
25	93	11.08	30.58	338.83
14	61	.08	−1.42	−.11
10	50	−3.92	−12.42	48.69
18	75	4.08	12.58	51.33
0	32	−13.92	−30.42	423.45
16	85	2.08	22.58	46.97
5	42	−8.92	−20.42	182.15
19	70	5.08	7.58	38.51
16	66	2.08	3.58	7.45
20	80	6.08	17.58	106.89
Total				1,149.89
Average				104.54

What does the calculated covariance above tell us? The exact value is not as important as the sign (i.e., positive or negative). If the value is positive, then the number of hours of study positively influences the final scores. On the other hand, if the value is negative, one parameter has a negative impact on the other. If the covariance is zero, it indicates that the two dimensions are independent of each other.

4.2.4 Control Limits

Control chart (Figure 4.2) is a statistical process control tool, which is proposed by Shewhart (1939), and it is used to detect abnormal events. The essential idea is to determine if a manufacturing or business process is in a state of stable control. Control charts are simple, robust tools for understanding process variability. Control charts take advantage of the normal distribution curve. As such, data should be normally distributed (or transformed) when using control charts, or the chart may signal an unexpectedly high rate of false alarms.

With use of control chart, normal operation limits are set to $\pm 3\sigma$ (3 SD from the mean) on the basis that 99.7% of the observations occur within ± three SD of the mean. When a process is stable and in control, it displays only the common cause variation or random variation that is inherent to the process. If the process is unstable, the process displays special cause variation, nonrandom variation from external factors. In this case, operating points will be outside of Six Sigma (6σ) (= UCL − LCL).

Mathematically, the calculation of control limits can be expressed as

$$UCL = \mu + 3\sigma \tag{4.7}$$

$$LCL = \mu - 3\sigma \tag{4.8}$$

Because control limits are calculated from process data, they are independent of customer expectations or specification limits.

4.2.4.1 Stable and Unstable Operation Controlled variation is characterized by a stable and consistent pattern of variation over time (Figure 4.3) and is associated with common causes. A process operating with controlled variation has an outcome that is predictable within the bounds of the control limits. If some operating points are outside of the control limits, the process is under uncontrolled variations that

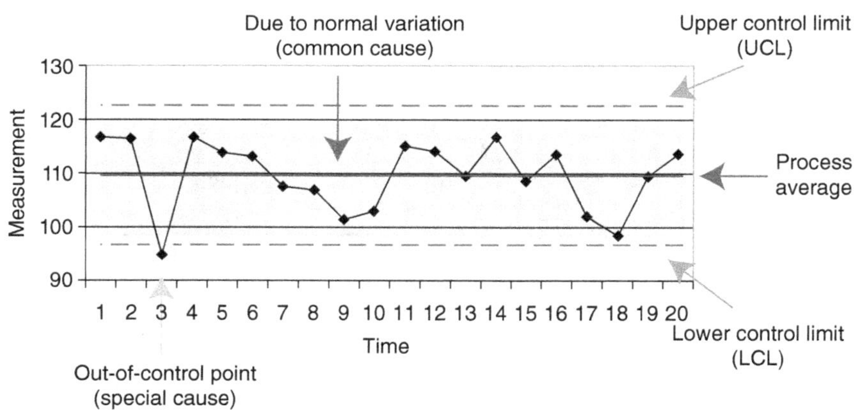

FIGURE 4.2. A typical control chart.

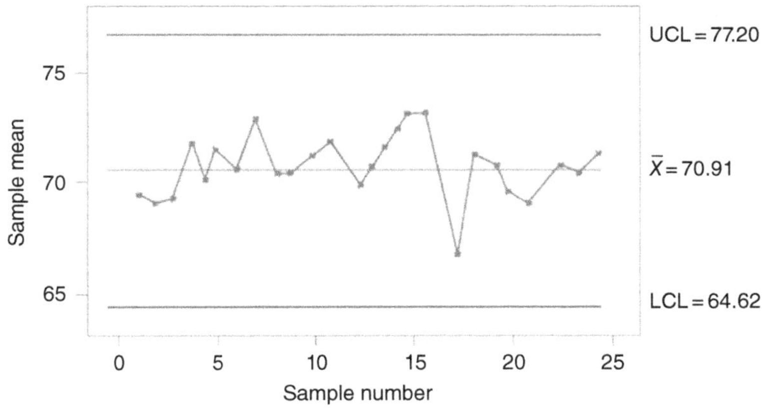

FIGURE 4.3. An example of stable operation.

usually are associated with special causes. The outcomes of this process are unpredictable.

4.2.4.2 Multivariate Control Chart The control chart discussed above is for a single variable. Multivariate analysis is needed in a statistical process control when there is more than one process variable. Multivariate analysis is usually helpful when multiple parameters are not independent of each other. For example, both temperature and pressure can impact the product yields in a hydrocracking process. Applying univariate control charts separately is possible but inefficient and can lead to erroneous conclusions. Multivariate methods that jointly monitor these variables are an effective approach.

Here is a simple hypothetical two-variable example to illustrate how an erroneous conclusion may be made. Based on single variable analysis, the control charts for X_1 (Figure 4.4) and X_2 (Figure 4.5) look stable; however, there is more to the story. When the multivariate model, namely, Hotelling's T^2 statistic based on the

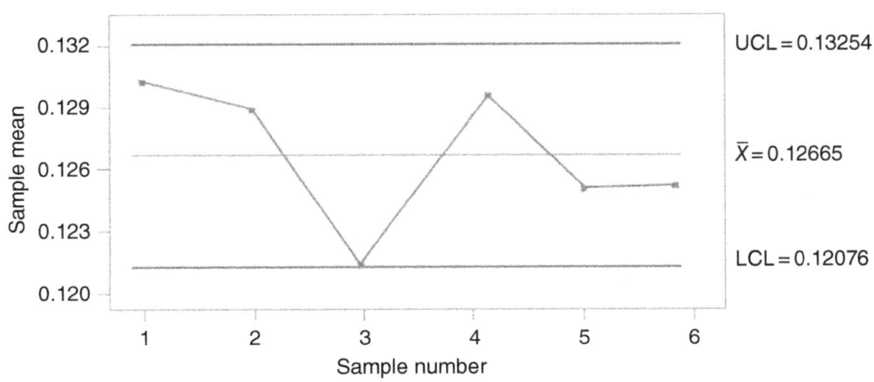

FIGURE 4.4. Control chart for variable x_1.

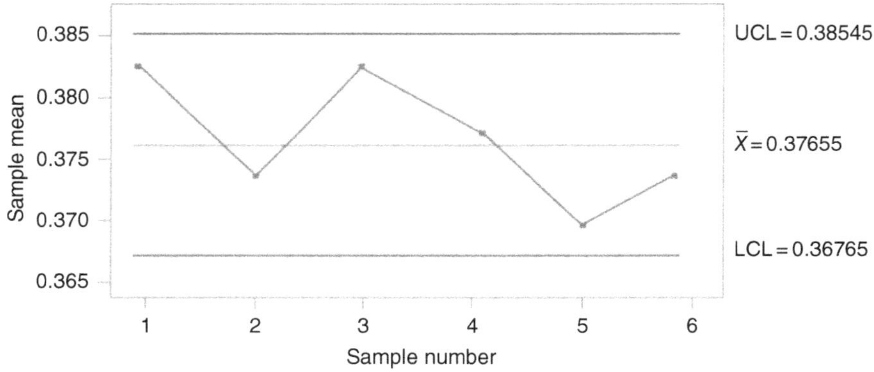

FIGURE 4.5. Control chart for variable x_2.

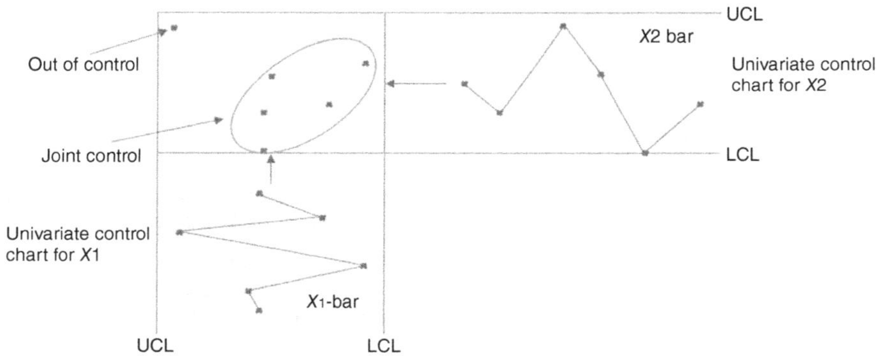

FIGURE 4.6. Control plot for two variables x_1 and x_2: stable operation area (joint control) and out of control.

covariance matrix, is applied, a control ellipse for two variables is generated as shown in Figure 4.6 that provides greater sensitivity to out-of-control points as highlighted in circle. The detailed explanation for T^2 statistic can be seen in Hotelling (1931).

4.3 CONDITIONAL PROBABILITY

Conditional probability is to find the probability of one event happening assuming another event already happened. In mathematical notation, we use $P(A, B)$ to represt the probability of the events A and B occurring although they are independent of each other. On the other hand, we use $P(B \mid A)$ to represent the probability of B given A already happened with probability $P(A)$. Thus, $P(B \mid A)$ can be expressed as

$$P(B|A) = \frac{P(A,B)}{P(A)} \tag{4.9}$$

The probability of B given A is equal to the probability of A and B occurring together over the probability of A occurring alone, which gives the probability of B being dependent on A occurrence. Let us use a simple example to explain. Assume a student is going to takes two tests A and B. Test A is easier and his pass rate is 80%, that is, $P(A) = 0.8$. His pass rate for both A and B, irrespective of each other, is 60%, that is, $P(A, B) = 0.6$. We want to predict his pass rate for test B $P(B|A)$ given A as $P(A)$. Thus,

$$P(B|A) = \frac{P(A,B)}{P(A)} = \frac{0.6}{0.8} = 75\% \tag{4.10}$$

Under the probability of 60% passing both tests and 80% passing test A, the probability of passing test B given that the student passed A works out to be 75%.

4.4 BAYES' PROBABILITY

On the basis that you understand the conditional probability, we are ready to discuss **Bayes' theorem**, which is based on conditional probability. The meaning of Bayes' theorem at the high level is that the probability of event A given B is equal to the probability A times the probability of B given A over the probability of B. It can be expressed as

$$P(A|B) = \frac{P(A,B)}{P(B)} = \frac{P(A) \cdot P(B|A)}{P(B)} \tag{4.11}$$

The key insight of Bayes' theorem is that the probability of A, which depends on B, depends much on the base probability of both A and B. This fact is commonly ignored in classical statistics.

In probability theory and statistics, **Bayes' theorem** describes the probability of an event, based on prior knowledge of conditions that might be related to the event. For example, if cancer is related to age, then, using Bayes' theorem, a person's age can be used to more accurately assess the probability that they have cancer, compared with the assessment of the probability of cancer made without knowledge of the person's age. It must be pointed out that the probability of B given A is not the same thing as the probability of A given B. That is, the probability of being a drug user given that you tested positive is very different from the probability of tested positive given that you are a drug user. These are two different things, and Bayes theorem allows us to quantify the difference clearly. In some cases, even if $P(B|A)$ is high, it does not necessarily mean $P(A|B)$ is high and vice versa.

Let us use an example (Kane, 2017) to drive home, and this example is about a drug test applying Bayes' theorem. Even a highly accurate drug test can produce more false positives than true positives. So in our example here, we're going to come up with a drug test that can accurately identify users of a drug 99% of the time and accurately has a negative result for 99% of nonusers, but only 0.3% of the overall population actually uses the drug in question. So we have a very small probability of actually being a user of a drug. What seems like a very high accuracy of 99% isn't actually high enough, right?

We can work out the math as follows:

- Event A = is a user of the drug
- Event B = tested positively for the drug

So let event A mean that you're a user of some drug and event B the event that you tested positively for the drug using this drug test.

We need to work out the probability of testing positively overall. We can work that out by taking the sum of probability of testing positive if you are a user and the probability of testing positive if you're not a user. So, $P(B)$ works out to 1.3% $(0.99 \times .003 + .01 \times 0.997)$ in this example. So we have a probability of B, the probability of testing positively for the drug overall without knowing anything else about you.

Let's do the math and calculate the probability of being a user of the drug given that you tested positively:

$$P(A|B) = \frac{P(A)P(B|A)}{P(B)} = \frac{.003 \times 0.99}{.013} = 22.8\% \tag{4.12}$$

So the probability of a positive test result given that you're actually a drug user works out as the probability of being a user of the drug overall ($P(A)$), which is 0.3% (you know that 0.3% of the population is a drug user) multiplied by $P(B \mid A)$ that is the probability of testing positively given that you're a user divided by the probability of testing positively overall, which is 1.3%. Again, this test has what sounds like a very high accuracy of 99%. We have 0.3% of the population that uses a drug multiplied by the accuracy of 99% divided by the probability of testing positively overall, which is 1.3%. So the probability of being an actual user of this drug given that you tested positive for it is only 22.8%. So even though this drug test is accurate 99% of the time, it's still providing a false result in most of the cases where you are testing positive.

Even though $P(B \mid A)$ is high (99%), it doesn't mean $P(B \mid A)$ is high. People overlook this all the time, so if there's one lesson to be learned from Bayes' theorem, it is to always take these sorts of things with a grain of salt. Apply Bayes' theorem to these actual problems and you'll often find that what sounds like a high accuracy rate could actually be yielding very misleading results if you're dealing with a low overall incidence of a given problem. We see the same thing in cancer screening and other sorts of medical screening as well. That's a very real problem; there's a lot of people getting very, very real and very unnecessary surgery as a result of not understanding Bayes' theorem. If you're going into the medical profession with big data, please, please, please remember this theorem.

So that's Bayes' theorem. Always remember that the probability of something given something else is not the same thing as the other way around and it actually depends a lot on the base probabilities of both of those two things that you're measuring. It's a very important thing to keep in mind, and always look at your results with that in mind. Bayes' theorem gives you the tools to quantify that effect. I hope it proves useful.

4.5 STATISTIC TESTS

A statistic test is one of the most important concepts in statistics as it is applied for any hypothesis or claim that we want to verify for acceptance or rejection based on sample data in hand. It is like this: you have an idea or proposal and you want to prove it by using the data available—statistic test just does that.

In life, it is often that one wants to challenge a common wisdom or practice. Statistically, the common wisdom is named the null hypothesis, and the new idea is called the alternative hypothesis. Statistic test is about calculating the statistical parameters based on sample data and compare them to the statistical parameters derived from the statistical distribution. This comparison can lead to the decision for rejecting or accepting the common wisdom based on certain statistic criteria.

Null Hypothesis

H_0—the hypothesis currently accepted. For example, Newton's law of gravity was widely accepted in the past.

Alternative Hypothesis

H_a—alternative hypothesis different from H_0, which is to be tested. For example, Einstein claims that gravity is not a force and it is a distortion in the shape of space-time. From experiments, it was demonstrated that Einstein's hypothesis was correct and that Newton's law of gravity was a simplified version of gravity phenomenon.

Statistic Test

Use sample data to calculate the sample statistic parameters and apply distribution theory to calculate theoretic statistic parameters. Then compare them to decide between H_0 and H_a based on a confidence or significance level.

Statistical Significance

- Level of confidence (c): It is about how confident are you in making a decision. Typical confidence is 95% or a very high confidence as 99%.
- Level of significance (α): $\alpha = 1 - c$; typical $\alpha = 5\%$ ($c = 95\%$) or 1% ($c = 99\%$).

Possible Outcomes from the Test

- Reject the null hypothesis.
- Fail to reject the null hypothesis.

4.5.1 Statistic Tests for a Given Confidence Level

Let me use a simple example to walk through. Assume a factory makes chocolate bars with a mass of 5 g on average. After the machine is repaired, some workers point out that the machine no longer makes 5 g chocolate bars. The management wants to find out if it is true or not.

Step 1. Establish the hypotheses and assume level of significance. Statistically, we can set up the above debate statistically as

$$H_0 : \mu_0 = 5\,\text{g}; \quad H_a : \mu_0 \neq 5\,\text{g}. \quad \alpha = .05 \tag{4.13}$$

As stated in equation 4.13, the null hypothesis is that the bar weight should be 5g on average with 95%confidence.

The alternative hypothesis is that weights could be larger or less than 5 g and therefore a two-tailed test is used. If the null hypothesis is true, the statistic test value should fall within 95% of central area around the average. Note that the statistic test values follow normal distribution. To reject the null hypothesis, the statistic test value must fall into the two tails of the normal distribution, which is a rare event, because each tail only counts for 2.5% (e.g. $\alpha/2$) of the total area. The sample data includes 50 chocolate bars selected randomly, and the mean and SD are $\bar{x} = 5.7\,\text{g}$ and $s = 2.3$.

Step 2. Select the appropriate test statistic and decision rule. Because the sample size is large (N>30), the Z-test statistic is appropriate. Otherwise, a t-test would be selected.

Z-statistic is defined as

$$Z = \frac{\bar{x} - \mu_0}{s / \sqrt{N}} \tag{4.14}$$

μ_0 is the SD in null hypothesis. Z follows a normal distribution with the mean of the Z statistic equal to zero. α is the area (or probability) under the normal distribution curve at the far end of the right or left side. These areas are called rejection zone, which are located far away from the mean based on Z statistic. As a note, if the sample size is small (say, less than 30), t-test statistic may be selected (details are given below for why).

Since the alternative hypothesis is ($u \neq 5\,\text{g}$), this requires a two-tailed test as it is equivalent to $u \geq 5\,\text{g}$ and $u < 5\,\text{g}$. To make a prediction for a confidence level at 95%, α = 5% and $\alpha/2$ = 2.5% for both tails. There are two possibilities for the Z value, which can be in either side of the distribution curve (Figure 4.7.). The null hypothesis (u = 5 g) should be rejected, meaning the alternative hypothesis ($u \neq 5\,\text{g}$) is accepted, if the sample $Z \leq -Z_{\alpha/2}$ or $Z \geq Z_{\alpha/2}$.

Step 3. Calculate the test statistic based on Z-distribution. In this example, we are performing a two-tailed test. The critical Z value (or $Z_{\alpha/2}$ value) is equal to 1.96. The calculation is explained in Figure 4.7.

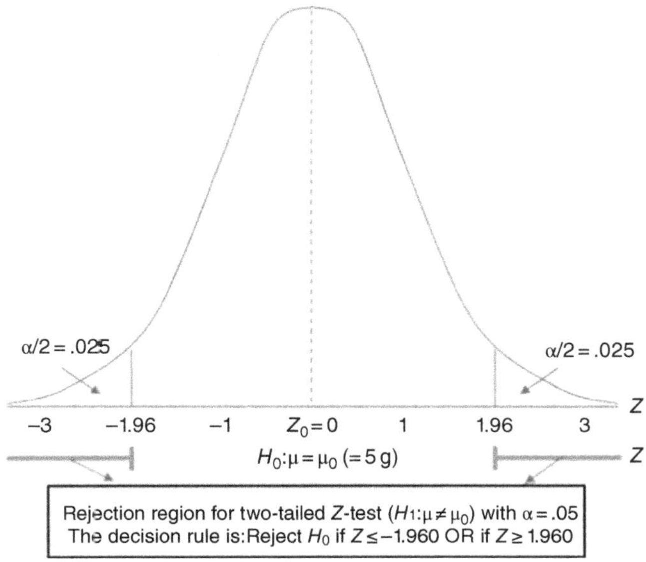

FIGURE 4.7. Two-tailed test.

Step 4. Compute the test statistic based on sample data. We now substitute the sample data into the test statistic equation 4.10 identified in Step 2:

$$Z = \frac{\bar{x} - \mu_0}{s / \sqrt{N}} = \frac{5.7 - 5}{2.3 / \sqrt{50}} = 2.15 \tag{4.15}$$

Step 5. Conclusion. We reject H_0 because 2.15 > 1.96. We have statistically significant evidence at $\alpha = .05$ to show that the mean weight in chocolate bar is not 5 g.

Remarks for the statistic test: If the Z values fall into the rejection areas, the null hypothesis is rejected or the claim in the alternative hypothesis is accepted. Statistically, the chance for the Z value falling into the rejection zone is very small, that is, with α probability (say, 5%) for both right/left-tail tests and $\alpha/2$ (e.g., 2.5%) probability for each tail test. When such a small probability event occurs, the null hypothesis must not true and thus has to be rejected.

In case of the alternative hypothesis as $u > 5$ g (the weight of the chocolate bar heavier than 5 g), this is a right-tail test. This alternative hypothesis is true (or null hypothesis is rejected) if the Z value based on the sample data falls into the rejection zone in the right-hand side (Figure 4.8). In contrast, for the left-tail test with the alternative hypothesis as $u < 5$ g, this alternative hypothesis is true (or null hypothesis is rejected) if the Z value based on the sample data falls into the rejection zone in the left-hand side (Figure 4.9).

4.5.2 *P*-Test: Determine a Required Confidence Level

P-test is a very common type of statistic test, which is applied when we don't know the confidence level needed to test the hypothesis. This situation is very common in

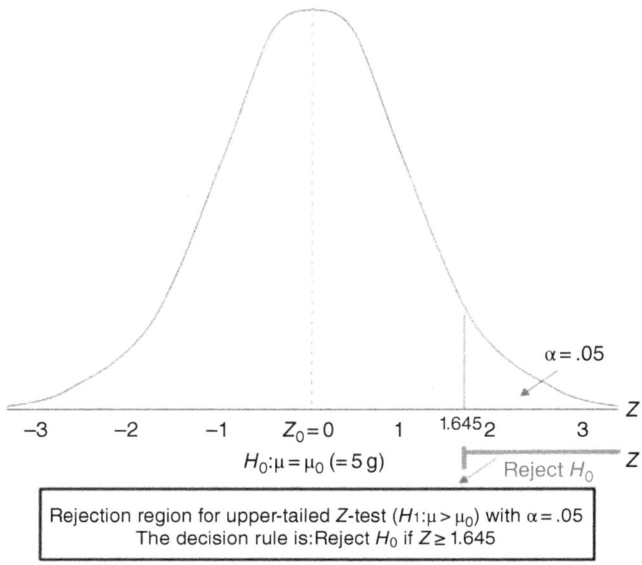

FIGURE 4.8. Right-tail test.

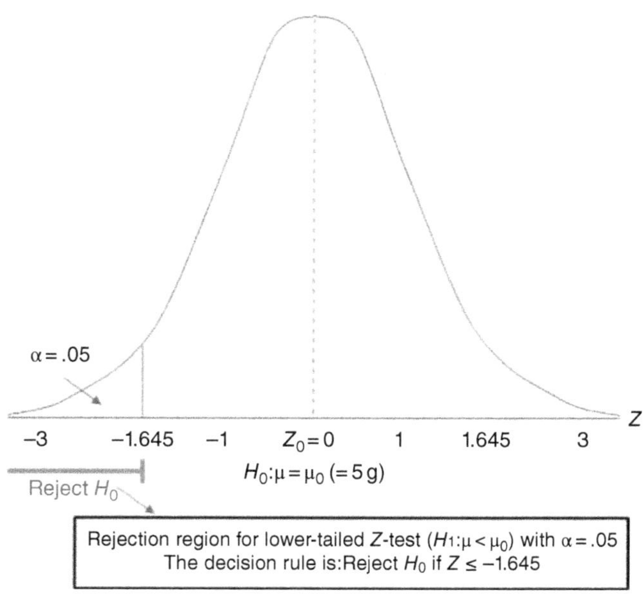

FIGURE 4.9. Left-tail test.

real life. You may dispute for the normal practice commonly accepted and may have your own proposal (the alternative hypothesis). To convince other people, you need to provide evidence that support your idea. Statistically, you need to tell under what circumstances (or at what confidence level) the common practice (the null hypothesis) is rejected. The p-value is designed for this purpose.

In the example above, we rejected the null hypothesis at 95% confidence level. If we define the problem statement slightly differently for the purpose of simplicity for illustration as follows, the complaint from customers is that the chocolate bars are lighter than 5 g which is the product specification claimed by the company. To verify this feedback, 50 bars are selected randomly, and the mean and SD are calculated as $\overline{x} = 4.6\,\text{g}$ and $s = 1.52$. We wish to determine if the customers' complaint is valid.

Step 1. Establish the hypotheses. Statistically, we can set up the above test problem as

$$H_0 : \mu_0 = 5\,\text{g}; \quad H_\text{a} : \mu_0 < 5\,\text{g}. \tag{4.16}$$

Please notice that the confidence or significance level is not specified, which is to be determined as the p-value.

Step 2. Select the appropriate test statistic and decision rule. Z-test is selected as defined in Equation 4.10.

Step 3. Compute the test statistic based on sample data. This is the same as that previously calculated based on the sample data:

$$Z = \frac{\overline{x} - \mu_0}{s / \sqrt{N}} = \frac{4.6 - 5}{1.52 / \sqrt{50}} = -1.86 \tag{4.17}$$

Step 4. Compute the p-value test statistic based on sample data:

$$p(Z < -1.86) \approx .03 \tag{4.18}$$

Thus, the p-value for this example is .03. The above calculation can be explained graphically by Figure 4.10, which explains the meaning of p-value. The p-value represents the probability for accepting the alternative hypothesis H_a or the significance level where we can reject H_0. In our example, $p = .03$ and $\alpha = p$. Thus, the confidence level c is 97% $(= 1 - \alpha)$ for accepting H_1: $\mu < 5\,\text{g}$ (or rejecting H_0: $\mu = 5\,\text{g}$).

Step 5. Conclusion. With the evidence in hand, you can go to the board room and state that "we selected 50 chocolate bars randomly to test if we actually make 5 grams on a consistent basis. As the result of these calculations, we can say with 97% confidence that the chocolate bars are lighter than our specification of 5 grams." This statement involves a very convincing argument with a fairly high confidence. The board would appreciate the evidence and ask the plant management to take actions.

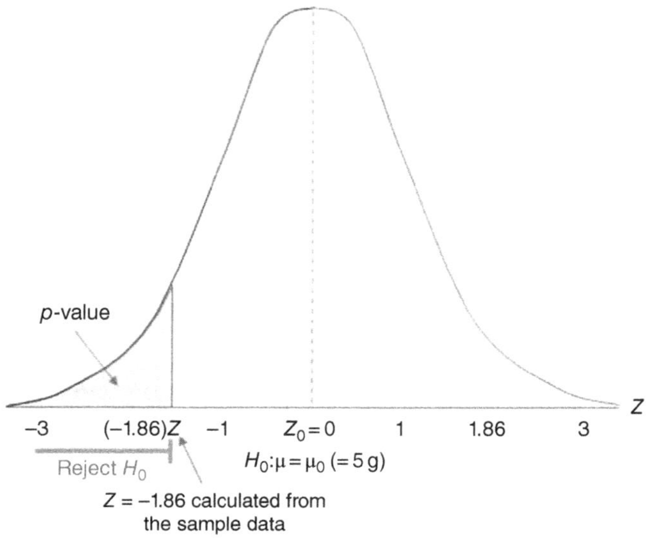

FIGURE 4.10. *p*-Value definition for left-tail test.

Remarks: Thus, a *p*-value is used in hypothesis test to support or reject the null hypothesis. The *p*-value is the evidence against a null hypothesis. The smaller the *p*-value, the more compelling evidence you have to be able to reject the null hypothesis. In other words, the smaller the *p*-value, the more important ("significant") your results are; thus you have higher confidence to reject the null hypothesis.

Similarly, a right-tail *p*-test can be explained as shown in Figure 4.11. If you're running a two-tailed test, *p*-value is equal to the sum of two far-end areas from the left and the right-hand sides.

Decision criteria in *p*-test: When you conduct the p-test, the test will give you a *p*-value. Compare that value to your chosen α level. For example, let's say you chose an alpha (α) level of 5% (.05). If the results from the test give you:

- A small *p*: if $p \le \alpha \rightarrow$ Reject H_0. This is strong evidence that the null hypothesis is invalid.
- A large *p*: if $p > \alpha \rightarrow$ Do not reject H_0. This means the alternate hypothesis is weak, so you do not reject the null.

The above criteria and explanations (including Figure 4.12) for the left-tail test can be applied to the case of a right-tail test (Figure 4.11).

The *p*-value decision criteria can be generalized as:

- If $p > 0.10 \rightarrow$ not significant;
- If $p \le 0.10 \rightarrow$ marginally significant;
- If $p \le .05 \rightarrow$ significant;
- If $p \le .01 \rightarrow$ highly significant.

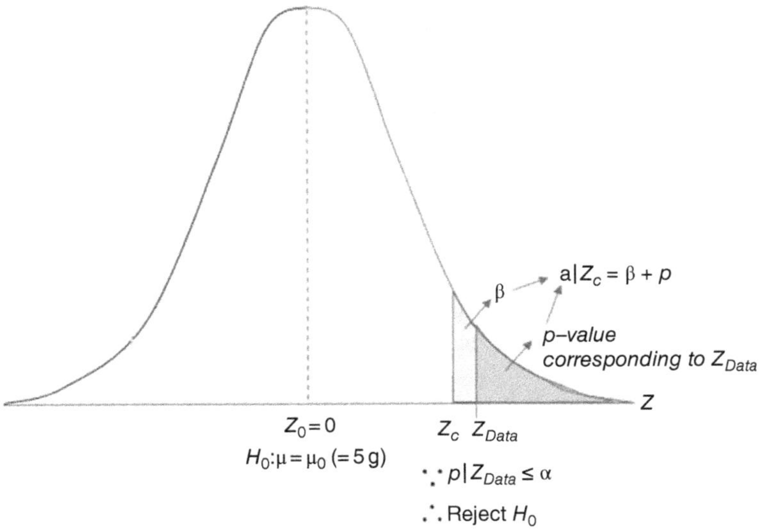

FIGURE 4.11. *p*-Value definition for right-tail test.

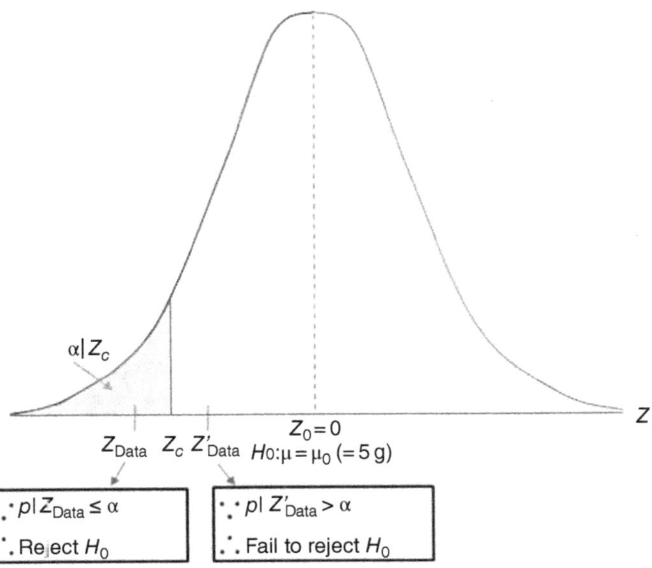

FIGURE 4.12. *p*-Test decision criteria for left-tail test.

4.5.3 Selection of Statistic Test Methods

Beside *Z*-test as discussed above, *t*-test and *F*-test are two common statistic test methods. They are discussed as below although more detailed discussions can be seen in any statistics books.

4.5.3.1 ***t-Statistic*** It was mentioned above. *Z*-test and *t*-test are both applied when the SD is known to determine if the mean of the data set differs from the population mean. *Z*-test is used when the sample size is large, that is, $N > 30$, and *t*-test is appropriate when you do not know the populations' SD and when you have a limited sample data ($N < 30$).

Similar to the *Z* statistic expression, the *t*-statistic is defined as

$$t = \frac{\overline{x} - \mu_0}{s / \sqrt{N}} \tag{4.19}$$

where $\overline{x}$ is the sample average, μ_0 is the population mean, s is the sample SD, and N is the sample size ($N < 30$).

t-Test follows the same five-step procedure and adopts the same decision criteria as discussed above. The difference is that *t*-test is based on the absolute value of the *t*-statistics ($|t|$). If $|t|$ is greater than the critical t value, then the null hypothesis $\left(H_0 : \overline{x} = \mu_0\right)$ is invalid and should be rejected. In other words, the sample $\overline{x}$ is significantly different from population mean μ_0. Otherwise, the sample $\overline{x}$ is not significantly different from population mean μ_0.

4.5.3.2 ***F-Test*** An *F*-test is used to test if the variances of two data sets are equal.

F-test statistic is defined as

$$F = \frac{s_1^2}{s_2^2} \tag{4.20}$$

where s_1^2 and s_2^2 are the sample variances. The greater this ratio deviates from 1, the stronger the evidence for unequal population variances.

F-Test follows the same five-step procedure and adopts the same decision criteria as discussed as above. This test can be a two-tailed test or a one-tailed test. The two-tailed test examines the alternative hypothesis that the variances are not equal. The one-tailed version only tests in one direction, that is, the variance from the first data set is either greater than or less than (but not both) the variance from the second data set.

4.5.4 Caution for Statistic Tests: Type I and Type II Errors

In all tests of hypothesis, there are two types of errors that can be committed. The first is called type I error and refers to the situation where we incorrectly reject H_0 when in fact it is true. This is also called a false positive result (as we incorrectly conclude that the alternative hypothesis is true when in fact it is not). Type II error refers to the situation when we fail to reject the null hypothesis when it is actually false.

TABLE 4.1 Conclusions in A Hypothesis Test

	Do Not Reject H_0	Reject H_0
H_0 is true	Correct decision	Type I error (false positive)
H_0 is false	Type II error (false negative)	Correct decision

In step 1, we purposely select a small level of significance (α) in order to control the probability of committing a type I error. For example, if we select $\alpha = .05$, and our test tells us to reject H_0, then there is a 5% probability that we commit a type I error ($\alpha = p(\text{Type I error})$). Most investigators are very comfortable with this and are confident when rejecting H_0 and accepting the alternative hypothesis at the same time (as it is the more likely scenario when we reject H_0).

When we run a test of hypothesis and decide not to reject H_0 (e.g., because the test statistic is less than the critical value in a right tailed test), then either we make a correct decision of accepting the null hypothesis because it is true or we commit a type II error by failing to reject the null hypothesis when it is actually false.

When we do not reject H_0, it may be very likely that we are committing a type II error (i.e., failing to reject H_0 when in fact it is false). Therefore, when tests are run and the null hypothesis is not rejected, we often make a weak concluding statement allowing for the possibility that we might be committing a type II error. That is the reason why we do not conclude that H_0 is true in the case of not rejecting H_0. In this case, we say "we fail to reject the null hypothesis" instead of saying "accept the null hypothesis."

Table 4.1 shows the different conclusions from a statistic test.

4.5.5 Analysis of Variance (ANOVA)

ANOVA is a statistical hypothesis testing heavily used in the analysis of sample data. A statistically significant result, when a probability (p-value) is less than a specified threshold (significance level), justifies the rejection of the null hypothesis, but only if the a priori probability of the null hypothesis is not high. In the typical application of ANOVA, the null hypothesis is that all groups of sample data come from the same population. For example, when studying the effect of different treatments on similar samples of patients, the null hypothesis would be that all treatments have the same effect. Rejecting the null hypothesis implies different effects between treatment groups (Wikipedia, 2020).

By constructing AOVA assessment, hypothesis testing intends to limit the rate of Type I errors (false positives) to a significance level (e.g. 0.05). Assessment also wishes to limit the rate of Type II errors (false negatives). The rate of Type II errors depends largely on three factors, namely sample size (the rate is larger for smaller samples), significance level (when the standard of proof is high, the chances of overlooking a discovery are also high), and effect size (a smaller effect size is more prone to Type II error).

REFERENCES

Hotelling H (1931), The generalization of student's ratio, Ann. Math Stat. **2** (3), 360–378.

Kane F (2017), Hands-on Data Science and Python Machine Learning, Packt Publishing, Mumbai.

Shewhart WA (1939), Statistical Method from the Viewpoint of Quality Control, ISBN 978-0-486-65232-0, Dover Publications Inc., New York.

Wikipedia (2020), Analysis of variance, https://en.wikipedia.org/wiki/Analysis_of_variance.

5

ADVANCED STATISTIC MODELING

5.1 INTRODUCTION

A statistical model is a mathematical model that is generated from historical data sets. Statistical modeling is a major part of data analytics together with other data analytics methods such as data mining and machine learning.

The objective of building statistical models is to identify key variables and develop the relationship between the decision variable and independent variables. We take the perspective of modelers rather than statistician. As a modeler, we are more interested in what insights that the statistical models can reveal and how important these insights are in terms of value.

5.2 DISTRIBUTION MODELS

When you're dealing with discrete data, the first thing you'd like to know is: what kind of distribution does this data set present? If you know it follows a certain distribution, you can then determine important parameters associated with the distribution. We are concerned about two questions: what can we learn from the distributions and why do distributions matter?

To address the first question, a distribution can teach us the average, variation (difference within a type), and diversity (difference across types). For example, a

Digitalization and Analytics for Smart Plant Performance: Theory and Applications,
First Edition. Frank (Xin X.) Zhu.

normal distribution curve takes the bell shape with symmetrical around their means and does not include extreme large or small events. The central limit theorem explains the prevalence of normal distribution in real life. On the other hand, a Solow growth model shows diminishing return, while exponential decay model provides half-time decayed activity. A power distribution model gives insight for the 80-20 rule.

Let us start with normal distribution.

5.2.1 Normal Distribution Model

To identify the distribution behind the data set, the first thing is to generate a histogram for the real data to visualize the distribution pattern. To do so, one needs to define the lower and upper limits for the data sets and then divide the overall internal into a certain number of smaller intervals (e.g., $N = 30$). We can then calculate the count $\left(N_{\text{Interval}_i}\right)$, which is the number of the occurrences of values within each interval and the frequency for each interval as

$$f_{\text{Interval}_i} = \frac{N_{\text{Interval}_i}}{N_{\text{total}}} \tag{5.1}$$

With the count for each interval calculated, we can plot the histogram with counts versus the operating parameter of interest. As an example, Figure 5.1 shows the histogram for the steam rates going through a letdown valve based on 26 intervals. This histogram based on count can be converted to a frequency plot

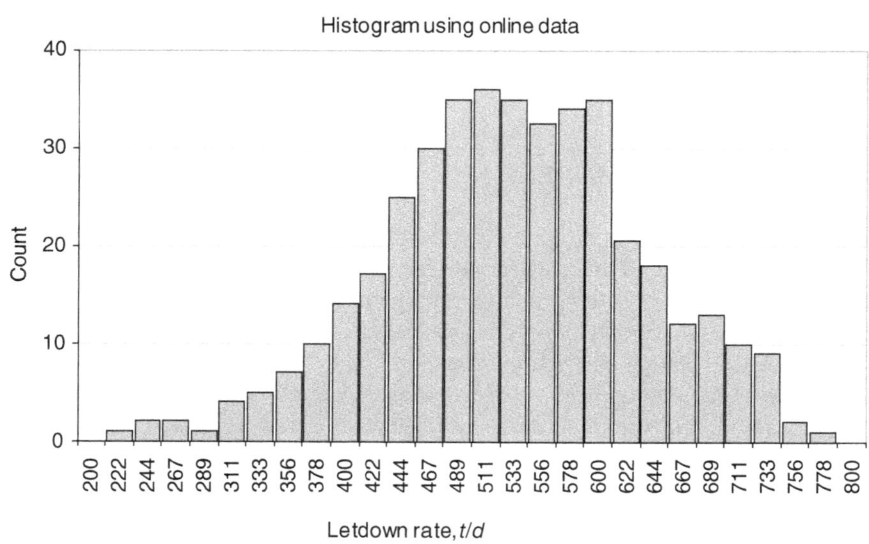

FIGURE 5.1. Online data distribution.

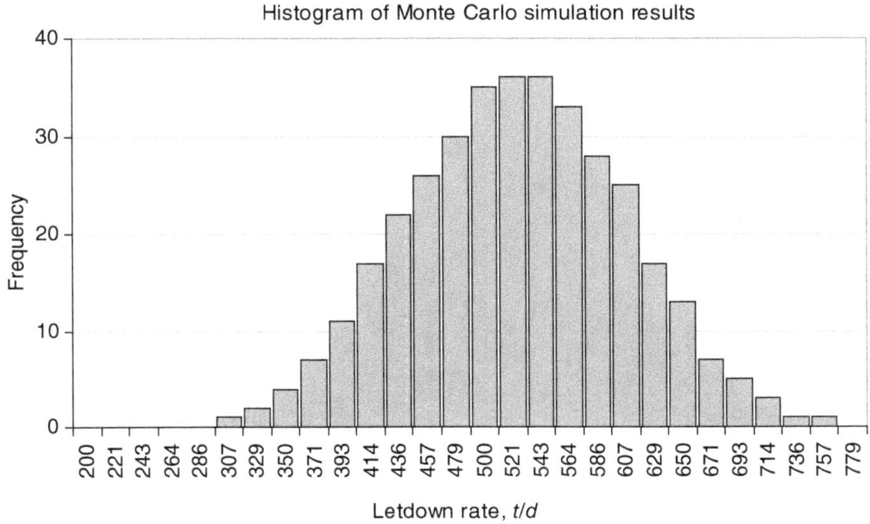

FIGURE 5.2. Normal distribution based on the same average, SD, and intervals.

using Equation 5.1. At the same time, the histogram for a normal distribution based on the same average, SD, and intervals is generated in Figure 5.2. You can compare the sample data histogram in Figure 5.1 with the normal histogram in Figure 5.2.

If the abnormality is not caused by bad data, the most important question then becomes: is it justified to use normal distribution to model the actual distribution? In general, it is rare that normal distribution fits the online data perfectly. As long as the variation pattern mimics the bell curve of normal distribution, it is reasonable to use the normal distribution to model the variation of real data.

Another method to visually inspect a distribution and to ensure that it follows a normal distribution is to generate a quantile–quantile (Q-Q) plot as in Figure 5.3. This is achieved by plotting the data versus the normal quantiles, which is an automated functionality that most statistical software packages have. In the figure, the solid red line indicates a perfectly normal distribution, and the black dots are the data set. Here, it is apparent that the data follow a normal distribution closely.

In additions, other statistical methods (such as *F*-test, Student's *t*-test, and χ^2-test) may be used to determine if a normal distribution appropriately models the data. Usually, *F*-test is applied first to determine if a data set follows the normal distribution. Student's *t*-test is applied to determine the means assuming a common variance between a small set of sample data points and a large data pool. On the other hand, χ^2-test can be applied to determine the variance for a small set of sample data points by assuming a common means between a small set of sample data points and a large data pool. For readers who are interested in this topic, please read Schmuller (2009).

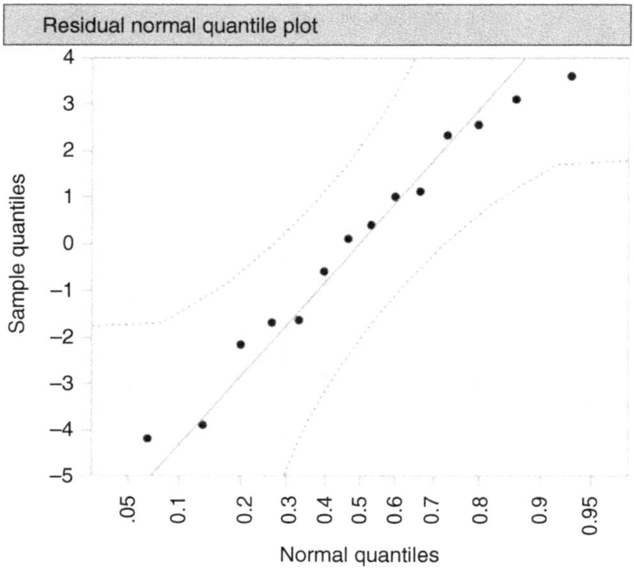

FIGURE 5.3. Q-Q plot of an example data set that is roughly normal.

With a normal distribution, you can determine the following:

- The mean (average) (Equation 4.2);
- The standard deviation (Equation 4.3);
- Evaluate if an assumption of the average value is correct or not (*Z*-test and *p*-test in Chapter 4).

5.2.2 Normal Distribution For Six Sigma

Six Sigma is the most well-known method for minimizing variability, which is based on the assumption that process variability follows a normal distribution. Limits are imposed at ±3σ from the mean, which is how this method receives its name: *Six Sigma* (6σ) (Figure 5.4). Six Sigma states that the process is under good control if each key operating variable falls within 6σ. Within the 6σ (or ±3σ) limits, the process displays only random variation that is inherent to the process. If the process is unstable, the process displays special cause variation, nonrandom variation from external factors. In this case, operating points will be outside of the 6σ limits.

Mathematically, the calculation of control limits can be expressed as

$$\text{UCL} = \mu + 3\sigma \tag{5.2}$$

$$\text{LCL} = \mu - 3\sigma \tag{5.3}$$

where UCL = upper control limit and LCL = lower control limit.

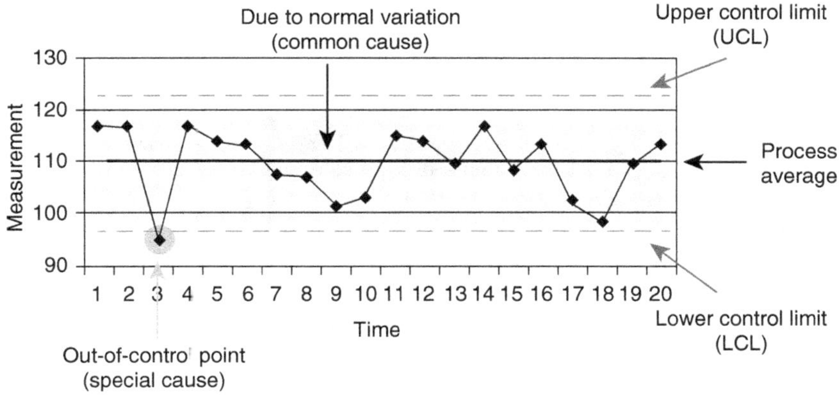

FIGURE 5.4. Definition of Six Sigma.

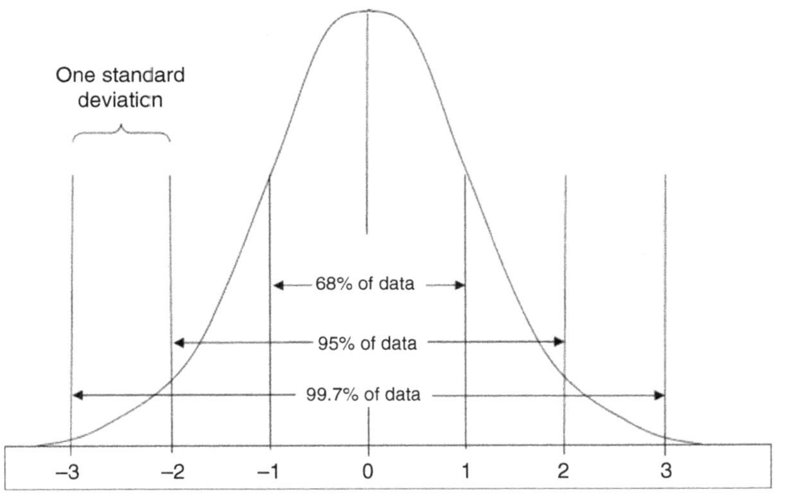

FIGURE 5.5. Normal distribution curve.

Because control limits are calculated from process data, they are independent of customer expectations or specification limits.

Let us walk through an example to explain the rationale of using Six Sigma. Imagine a plant producing bolts for door handles that must fit into knobs made by another manufacturer. The specification calls for the bolts to be 14 mm in diameter with precision ±1 mm allowable. In other words, bolts between 13 and 15 mm in diameter will function properly. Assume the diameters of the bolts follow a normal distribution with a mean of 14 mm and standard deviation (σ) of 0.5 mm. Any bolts that have a variation that is greater than 2σ (or ±1 mm) from the mean will fail the specification. According to the normal distribution as shown in Figure 5.5, events larger than 2σ occur 5% of the time—this failure rate is too high for the plant.

To reduce the failure rate, the variability of production or standard deviation must be reduced. In the bolt example, the standard deviation of the bolt's diameter must be reduced to one-sixth of mm (or $\sigma = 1/6\,\text{mm}$). Thus, any bolts with production error less than 6σ would still meet the precision requirement $\leq\pm1$ mm. According to the normal distribution, the probability for a 6σ event is about 2/billion, which is an impossible event—thus the 6σ methodology can guarantee very high quality of product.

Six-Sigma concept was introduced by Bill Smith while working as an engineer at Motorola in 1980. Jack Welch made it central to his business strategy at General Electric in 1995. A Six-Sigma process is one in which 99.99966% of all opportunities to produce some feature of a part are statistically expected to be free of defects.

5.2.2.1 Minimizing Variability in Operation Operation variability is a major cause of operation inefficiency. In general, there are two kinds of variability. The first one is the large variability, while the second shortcoming is too conservative in reaching the specification limit. If the operation or control strategy improves, the variability could be minimized to achieve more consistent operation (Figure 5.6b) but still far away from the limit. The limit is usually set by a physical limit such as product purity specification, maximum temperature or pressure, maximum valve opening, maximum vapor loading in a separation column, maximum space velocity in a reactor, and so on. The operation can be improved further (Figure 5.6c) by moving the average closer to the limit by adopting a better control strategy.

5.2.3 Exponential Decay Distribution

Another distribution you see fairly often is the exponential decay probability distribution, where the frequency of an event falls off in an exponential manner. When you talk about an exponential fall off, you expect to see a curve, where it's very likely for something to happen, but then, as time moves on, the probability of

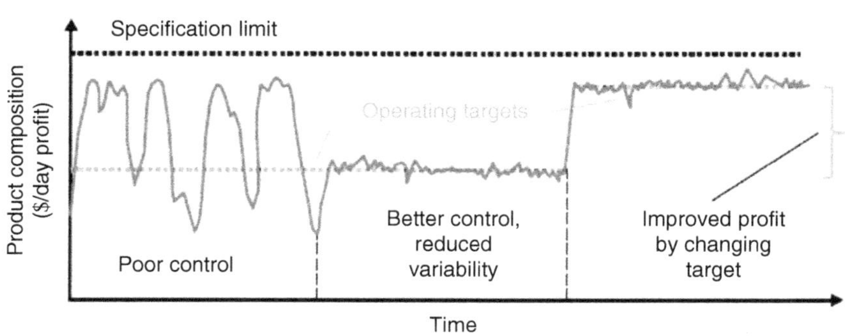

FIGURE 5.6. Operation performance patterns: (a) current operation, (b) reduced variability, and (c) increased profit.

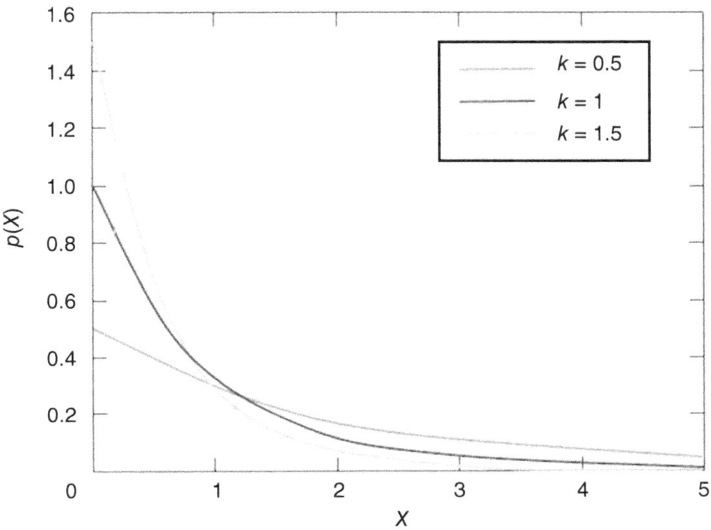

FIGURE 5.7. Exponential decay distribution.

the event drops off very quickly (Figure 5.7). There are many things in nature that behave in this manner.

The probability density function for exponential decay is expressed as

$$y = p(x) = A_0 e^{-kx} \quad x \geq 0; \quad k > 0. \tag{5.4}$$

where y is the probability in relation to x. A_0 is initial value of y at $x = 0$; k is the rate of decay.

5.2.3.1 Half-time Prediction In the model equation (5.4), we can derive the half-life time when the initial quantity A_0 is reduced to half of its initial quantity. The term is most commonly used in relation to atoms undergoing radioactive decay but can be used to describe other types of decay such as catalyst deactivation etc.

The half-life time $\left(x_{1/2A_0}\right)$ can be expressed as

$$x_{1/2A_0} = \frac{\ln(1/2)}{-k} \tag{5.5}$$

The decay rate ($k_{1/2}$) at the half-time is

$$k_{1/2} = \frac{\ln(1/2)}{-x_{1/2A_0}} \tag{5.6}$$

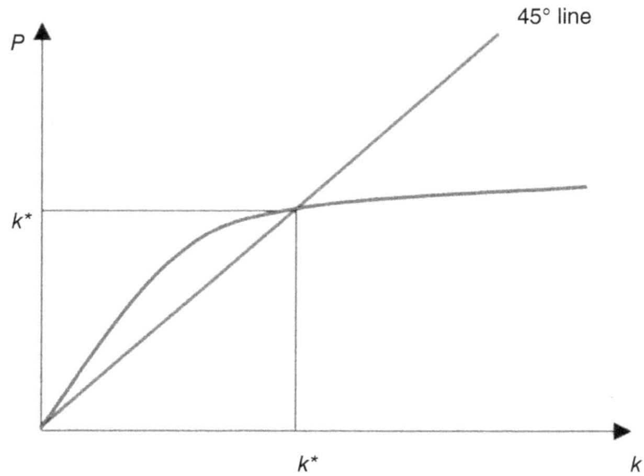

FIGURE 5.8. A typical growth model with diminish return. P = production rate; k = growth rate; k^* = steady-state value.

5.2.3.2 ***Diminishing Return*** A typical growth model is shown as Figure 5.8, which is the opposite of the decay model. The concave nature of the growth model exhibits the law of diminishing return: in all productive processes, adding more of one factor of production, while holding all others constant, will at some point yield lower incremental per-unit returns. The diminishing return is present in many improvement projects.

5.2.4 Power Law Distribution

A power law is a functional relationship between two quantities, that is, x and y, where a relative change in x results in a proportional relative change in y varying as a power of x. Mathematically, it is expressed as

$$y = C \cdot x^{-k} \quad x \geq x_{\min}; \quad k > 1 \tag{5.7}$$

where constant C is expressed as

$$C = (k-1)\, x_{\min}^{k-1} \tag{5.8}$$

where $x_{\min}$ is the minimum value for which the power law holds. $x_{\min}$ is needed; otherwise the distribution has infinite area as x approaches 0. The constant C is a scaling factor to ensure that the total probability (the total colored area in Figure 5.9) is 1, as required by a probability distribution. The value of k determines the likelihood and size of large events.

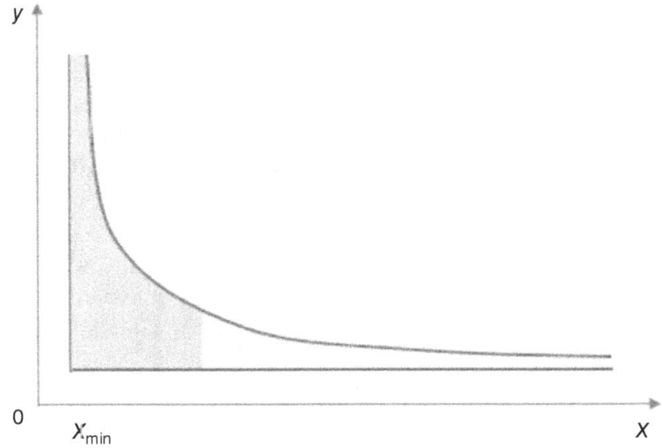

FIGURE 5.9. Power law distribution.

5.2.4.1 80–20 Rule The power law distribution as shown in Figure 5.9 explains the ranking of popularity. The dark gray represents the few events that dominates the total area (80%), while the light gray represents the majority that only contributes to 20%. This is also called the 80–20 rule. For example, given that car exhaust is distributed according to a power law among cars (very few cars contribute to most pollution), it would be sufficient to eliminate those very few cars from the road to reduce total exhaust substantially.

5.2.4.2 Small Probability for Large Events When $k = 2$, an event of size 100 has a probability proportional to $1/100^2$, or 1 in 10,000. When $k = 3$, an event of size 100 has a probability proportional to $1/100^3$, or 1 in million.

However, much of the recent interest in power laws comes from the study of probability distributions: the distributions of a wide variety of quantities seem to follow the power law form, at least in their upper tail (large events). The behavior of these large events connects these quantities to the study of theory of large deviations (also called extreme value theory), which considers the frequency of extremely rare events like stock market crashes and large natural disasters. It is primarily in the study of statistical distributions that the name "power law" is used.

5.2.4.3 Mean and Variance Most identified power laws in nature have exponents k falls in the range $2 < k < 3$ because the mean is well defined although standard deviation is not. When $k \leq 2$, the power law lacks a well-defined mean.

The equivalence of power laws with a particular exponent can have a deeper origin in the dynamic processes that generate the power law relation. In physics, for example, phase transitions in thermodynamic systems are associated with the emergence of power law distributions of certain quantities, whose exponents are referred to as the critical exponents of the system. Diverse systems with the same critical exponents—that is, which display identical scaling behavior as they approach criticality—can be shown, via renormalization group theory, to share the same fundamental dynamics. For instance, the behavior of water and CO_2 at their boiling points falls in the same universality class because they have identical critical exponents. In fact, almost all material phase transitions are described by a small set of universality classes. Similar observations have been made, though not as comprehensively, for various self-organized critical systems, where the critical point of the system is an attractor. Formally, this sharing of dynamics is referred to as universality, and systems with precisely the same critical exponents are said to belong to the same universality class.

5.2.5 The Difference Between the Exponent Law and Power Law

The essential difference is that an exponential function has its variable in its exponent, but a power function has its variable in its base. Mathematically, the difference can be expressed as

$$\text{Exponential law}: \ y \propto (\text{constant})^x \tag{5.9}$$

$$\text{Power law}: \ y \propto x^{(\text{constant})} \tag{5.10}$$

Do you notice the location of constant and variable x? That is the difference! In general, the exponential decay model goes to zero much faster than the power distribution model.

5.3 CORRELATION MODELS

The term "correlation" refers to a mutual relationship between independent and dependent variables. For example, product yield is a function of reaction temperature, pressure, feed composition, catalyst selectivity, and so on. Often, correlation is the first step to understanding these relationships and subsequently building more comprehensive models.

5.3.1 Linear Correlation Models

Linear regression models have played prominent roles in scientific research and industrial applications because they are easy to develop and straightforward to interpret. With the increased availability of data, they have become even more widely used. Once the data is fit to a linear regression model, it can make predictions that steer engineers toward action. Linear models are intuition based as the models tell us the direction (up or down in a linear fashion) of changes and the magnitude of changes (coefficients of each variable). The reason why linear regressions are considered in statistics is because the validity of a linear regression model can be provided by statistic methods.

5.3.1.1 Single-variable Linear Correlations A linear correlation model can be expressed as

$$y = \beta_0 + \beta x \tag{5.11}$$

where x is the independent variable and y the dependent variable; β is the slope of the line, and β_0 is the intercept. By applying the least squares regression method to N sample data points, β_0 and β can determined as

$$\beta = \frac{\sum_i^N x_i \cdot (y_i - \bar{y})}{\sum_i^N x_i \cdot (x_i - \bar{x})} \tag{5.12}$$

$$\beta_0 = \frac{\sum_i^N y_i - \beta \sum_i^N x_i}{N} \tag{5.13}$$

The difference between predicted y by the regression model and observed y from actual data points is termed as a prediction error (ε). Least squares regression (Equations 5.12 and 5.13) minimizes the overall squared distance (ε^2) between the model and data points (Figure 5.10).

A common method to measure the fit of a model is the regression coefficient, R^2. If a model fits all the data very well (all data points lie exactly on the regression line), R^2 will be 100%. If 100% is unachievable, you should aim to achieve the highest R^2 value as possible when fitting a model.

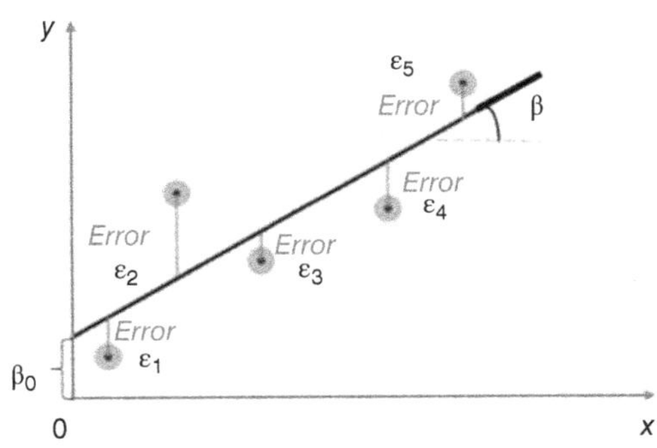

FIGURE 5.10. Least squares regression.

A linear regression model teaches three things:

- Sign of slope: Positive or negative between y and x. This determines the **direction** of the relationship between y and x.
- **Magnitude** of coefficient β: The effect of a change in x toward y.
- Regression coefficient R^2: The **goodness of fit** at a certain significance level, the higher the R^2 value, the better the fit to the data by using a linear model.

In a single-variable regression model, the closer fit to the line or the higher R^2, the more confidence we can place in the sign and magnitude of β. The larger the β, the more impact that the independent variable x has toward the dependent variable y.

In some cases, the error is large between the regression model and actual data. The error ε (Figure 5.10) can be incorporated into the regression model as

$$y_i = \beta_0 + \beta x_i + \varepsilon_i \quad i = 1,\ldots,N \tag{5.14}$$

There are several reasons for large errors. First, there is no correlation at all between y and x. Second, the relationship may not be linear. In this case, you may consider explicitly allowing nonlinear correlations. Lastly, other related variables are not included in the regression mode, which can be addressed using multivariable correlation as discussed below.

5.3.1.2 Multivariable Linear Correlation Models We have discussed single-variable linear regression above. However, most practical phenomena have multiple correlative variables. A person's happiness can be attributed to health, material situations, children upbringing, wealth, and so on. These variables can be included in a multivariable linear regression model as

$$y = \beta_0 + \beta_1 x_1 + \beta_2 x_2 + \cdots + \beta_m x_m = \beta_0 + \sum_j^m \beta_j x_j \tag{5.15}$$

Same as above, β_0 and β_i are determined based on the least square method. The goodness of fit for multivariable regression is also measured by R^2.

Once we have a regression model, we can predict the outcome based on variables involved. In addition, a regression model can help to identify those variables as significant (insignificant) when their coefficients are much larger (smaller) than other coefficients. As a result, the model can guide actions to focus on those significant variables. However, we must be cautious as the model equation (5.15) cannot infer causality automatically.

Correlation versus causality Regression only reveals correlation among variables but not causality. Even if the regression coefficient R^2 is close to 1 with the outcome prediction is supported by the data, we do not prove causality. For example, two unrelated variables (e.g., cancer patients and the amount of ice cream taken) may be correlated using data mining with high R^2. The effective way to avoid putting unrelated variables together in one correlation is to determine the causal variables based on practical knowledge first and then use statistic tools to determine the parameters such as β_0 and $\beta_i (i = 1, \ldots, m)$. Furthermore, creating training sets and testing sets can effectively help rule out spurious correlations. A correlation found on the training set that also holds on the testing set is more likely to be true.

Big coefficient thinking Linear regressions reveal the magnitude of independent variables toward the response or dependent variable. When considering taking actions, it is better to choose those variables with big coefficients as changes to these variables will have large effects. These variables are termed as significant variables. But caution must be given for the valid range of prediction based on linear correlations. Extrapolation beyond the valid range could give completely misleading prediction. For example, a linear growth model will predict a person will become a giant when he or she grows up to be an adult based on the growth data from infant to teenage.

Mathematical expressions of β_0 and β_i The following discussions are intended for readers who are interested in determination of β_0 and β_i for multivariable correlations.

Considering N data points on a dependent variable y, with m independent variables x_j, $(j = 1, 2, \ldots, m)$, the multiple linear regression model is written in algebra form as

$$y_i = \beta_0 + \sum_j^m \beta_j x_{ij} \quad i = 1, 2, \ldots, N \tag{5.16}$$

or in a matrix form as

$$Y = X\beta \tag{5.17}$$

where

$$X = \begin{bmatrix} 1 & x_{11} & x_{12} & \cdots & x_{1m} \\ 1 & x_{21} & x_{22} & \cdots & x_{2m} \\ \vdots & \vdots & & & \vdots \\ 1 & x_{N1} & x_{N2} & \cdots & x_{Nm} \end{bmatrix} \quad Y = \begin{bmatrix} y_1 \\ y_2 \\ \vdots \\ y_N \end{bmatrix} \quad \beta = \begin{bmatrix} \hat{\beta}_0 \\ \hat{\beta}_1 \\ \vdots \\ \hat{\beta}_m \end{bmatrix} \tag{5.18}$$

where β matrix contains the estimates of the coefficients β_j for variables $x_j (j = 1, \ldots, m)$. The mathematical expressions for β_j and β_0 in the form of matrices are given below based on the least squares method, that is, minimizing the sum of the squares residual ε (or the distance from the sample points):

$$\beta = (X'X)^{-1} X'Y \tag{5.19}$$

where X' is the transpose of matrix X and $(\cdot)^{-1}$ is the matrix inverse. The details for multivariable linear regression can be seen in any textbooks for statistics.

5.3.1.3 *Pearson Correlation Coefficient*

Pearson is the most widely used correlation coefficient, namely R, which measures the linear association between dependent variables y's and independent variables x's.

The correlation coefficient (R) formula, developed by Pearson, uses raw data and the means of variables, $\bar{x}$ and $\bar{y}$:

$$R(x,y) = \frac{\sum_{i=1}^{N}(x_i - \bar{x})(y_i - \bar{y})}{\sqrt{\sum_{i=1}^{N}(x_i - \bar{x})^2 \sum_{i=1}^{N}(y_i - \bar{y})^2}} \tag{5.20}$$

In this formulation, raw observations are centered by subtracting their means and rescaled by a measure of standard deviations.

Notice that the numerator is identical to the definition of covariance in Equation 4.6. Dividing the covariance between two variables by the product of standard deviations ensures that correlation coefficient R will always fall in the range $[-1, 1]$. This makes interpreting the correlation coefficient much easier.

There is a clear connection between Pearson correlation coefficient and the slope of the regression line. The regression line is optimal, as it minimizes the distance of all points to the line. Because of this property, Equation 5.20 can be related to the slope (β) of the regression line and the ratio of two standard deviation as

$$R(x,y) = \beta \frac{S_x}{S_y} \tag{5.21}$$

In other words, correlation reflects the association and amount of variability between x and y. There are three possible cases for Pearson correlation coefficient. The closer R is to 1, the more an increase in one type of variable associates with an increase in the other (Case 1). On the other hand, the closer R is to -1, the increase in one type of variable would result in decrease in the other (Case 2). These two cases show why linear correlation method is so widely used in predictive modeling because a high R^2 value indicates a good correlation between x and y. In Case 3, if x and y are independent, then R is close to 0, but not vice versa! In other words, Pearson correlation can be small even if there is a strong relationship between x and y. We will see shortly why this can be in Figure 5.11.

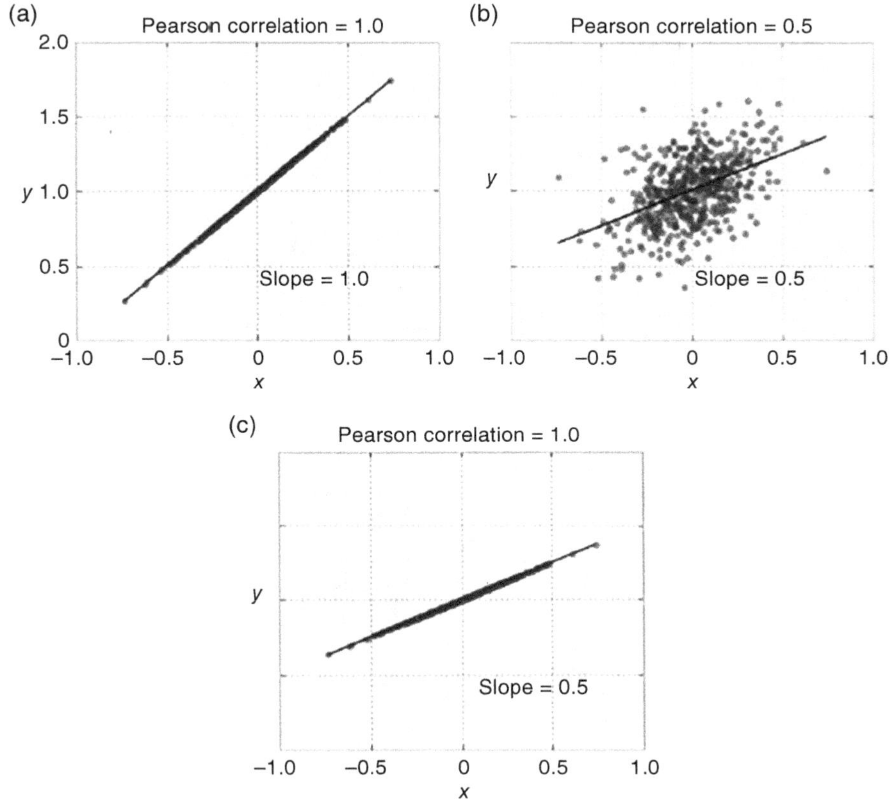

FIGURE 5.11. Pearson correlation coefficient: (a) large slope and high coefficient, (b) scattered data and low coefficient, and (c) small slope but low variability resulting in high coefficient.

It must be pointed out that correlation coefficient R is not equal to the slope—rather, it is standardized by a measure of data variability. It is possible to have a very small magnitude of slope but large correlations between variables. In Figure 5.11c, the line describing this relationship is relatively flat, but correlation coefficient R is 1 since variability S_y is very small.

However, the correlation coefficient is by no means an overall indicator of strong relationships within the data. For example, in Figure 5.12a, although Pearson correlation coefficient is relatively high due to the partial linearity of the relationship, it cannot reveal the nonlinear relationship. On the other hand, Figure 5.12b case shows a small correlation coefficient, but it has a strong relationship between x and y, although it is nonlinear. Therefore, it is always a good idea to use visualization techniques and multiple statistical data summaries to get a better picture of how x and y relate to each other.

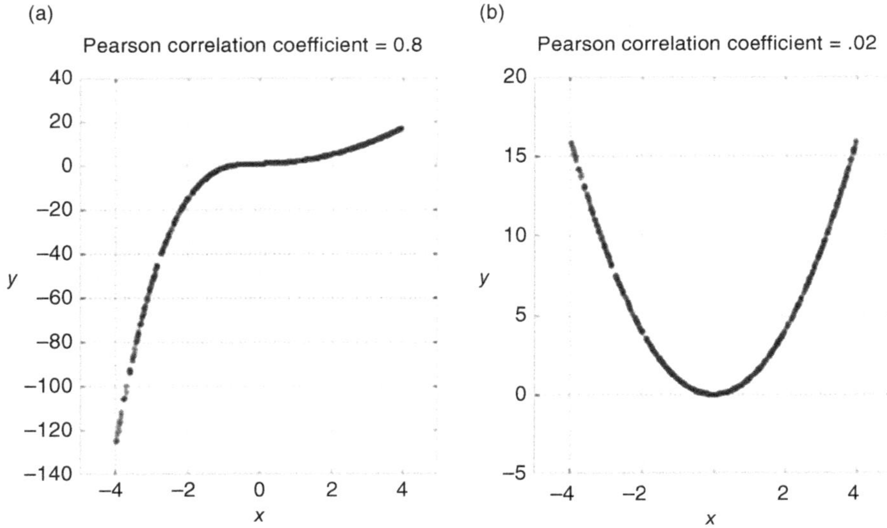

FIGURE 5.12. Pearson correlation coefficient cannot capture nonlinear relationship: (a) large R but with a nonlinear portion and (b) small R but the data has a strong relationship.

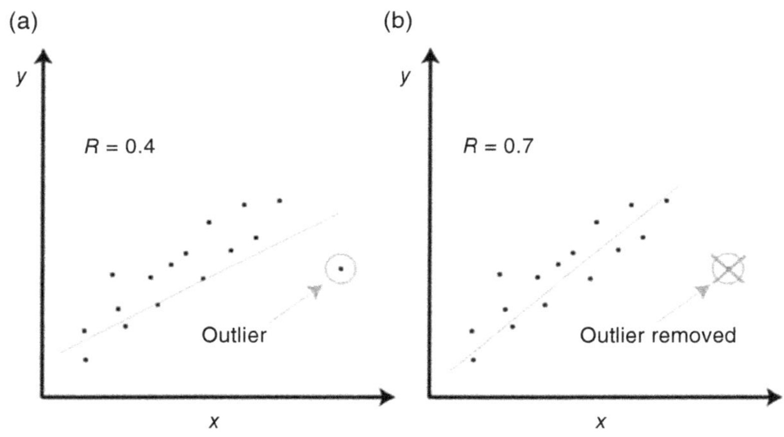

FIGURE 5.13. Effect of outliers on correlation coefficient.

Outlier There should be no significant outliers when generating correlations. Outliers are simply single data points within your data that do not follow the usual pattern. The scatterplots highlight the potential impact of outliers (Figure 5.13).

5.3.2 Nonlinear Correlations

It must be pointed out that linear models become insufficient once nonlinearity is involved. For example, a linear growth model will predict that a person will become a giant based on the growth data from infant to teenage. However, a nonlinear growth model will indicate the diminishing effect in a person's height increase after a certain age. To account for nonlinearity, we can conduct nonlinear transformation to make linear models end to end, for example, the piecewise linearized models can approximate a nonlinear model, but it loses the nonlinear curvature information in terms of second-order derivative for each nonlinear variable or Hessian matrix for all nonlinear variables. When nonlinearity is strong and complex, it is better to build nonlinear models as discussed below.

For modeling many nonlinear patterns, polynomial regression could be a good starting point. A general polynomial regression is a Taylor expansion of $f(x)$ at nth order around the initial point x_o as

$$f(x) = f(x_o) + \frac{f'(x_o)}{1!}(x - x_o) + \frac{f''(x_o)}{2!}(x - x_o)^2 + \frac{f'''(x_o)}{3!}(x - x_o)^3 + \cdots + \frac{f^{(n)}(x_o)}{n!}(x - x_o)^n \qquad (5.22)$$

Remember that the linear regression is a special case of the general polynomial regression equation 5.22, which is the first-order Taylor expansion with higher-order derivatives assumed to be zero. If there is some nonlinearity associated with data set, we could consider using a second-order polynomial to approximate. If it does not fit well, we can then use increasingly higher-order polynomial until it fits the data well. However, the higher order you choose, the more complex the fitting curve becomes. Therefore, the order of a polynomial function depends on the nature of the original data. If the order of a polynomial is higher than the nature of the data pattern, it is called overfitting.

There are several ways to avoid overfitting. Always start by visualizing the data set of interest to determine the nature of the curve by the data and compare with the fitness offered by a certain order of polynomial function. Second, a high R^2 (e.g., >0.8) indicates a good fit of the order you choose. Statistic tools are available for this purpose beyond visualization.

5.4 ADVANCED MODELING TECHNIQUES

It makes sense now to talk about some more advanced topics such as autocorrelation, generalized linear mixed models (GLMM), multilevel models, and principal component analysis (PCA), data mining and so on. These modeling techniques provide powerful statistic models that can deal with more complex problems and give significant insights. Although we are not going to dive into details, the intension here is to explain the concepts so that you can think about them when you search for solutions.

5.4.1 Autocorrelation

With correlation models, there are errors between predicted values and sample data. In the case of multivariable correlation, Equation 5.16 can be extended as

$$y_i = \beta_0 + \sum_j^m \beta_j x_{ij} + \varepsilon_i \quad i = 1, 2, \ldots, N \tag{5.23}$$

where ε_j is the error for the predicted outcome y_i against the data point i. When the regression coefficient R^2 is not high enough, autocorrelation can be applied to model the trend of the error terms, which will give additional insights that, in some cases, can be very useful in predicting future values more accurately.

Autocorrelation is normally seen in the time series data. Plotting the error terms over a time period in the chronological order can give an indication of a positive or negative autocorrelation. Various methods such as Durbin–Watson statistic can be used to detect the autocorrelation in the data sets. A simpler technique can be employed when looking for the autocorrelation pattern: looking at a run chart of the data set can give an indication of the autocorrelation within the data set.

Once it is determined that error terms ε_t is autocorrelation, an ARMA (p, q) [auto-regression (AR) with p-order and moving average (MA) with q-order] model can be generated for ε_t as

$$\varepsilon_t = \rho_1 \varepsilon_{t-1} + \rho_2 \varepsilon_{t-2} + \cdots + \rho_p \varepsilon_{t-p} + u_t + \phi_1 u_{t-1} + \cdots + \phi_q u_{t-q} \tag{5.24}$$

where

u is mean;

ρ is coefficients of the p-order AR model;

ϕ is coefficients of the q-order MA model.

The autocorrelation coefficient can be calculated as

$$R_k = \frac{1}{N-1} \sum_{t=1}^{N-k} \frac{(\varepsilon_t - \bar{\varepsilon})(\varepsilon_{t+k} - \bar{\varepsilon})}{s^2} \tag{5.25}$$

where

R_k is the autocorrelation coefficient

N is the number of ε data points

k is the time difference (lag) between the ε data points

$\bar{\varepsilon}$ is the mean of the errors

s is the sample standard deviation

As a rule of thumb, the autocorrelation coefficients R_k within $\pm 2/\sqrt{N}$, where N is the number of data points, are not usually significant. Since $2/\sqrt{N}$ corresponds roughly to a two-sigma interval, the rule of thumb uses two standard deviations as cutoff points in deciding if a true autocorrelation coefficient is none zero (Pruett and Schneider, 1992). The amount of time in between the data points will also influence the autocorrelation coefficient. Typically, the higher the lag times for a given set of data, the lower the autocorrelation. Thus, autocorrelation can be used a measure to set the time intervals of the data points, depending on whether autocorrelation is beneficial or needs to be in prediction of subsequent values.

5.4.2 Multi-effects or Mixed Models

A mixed model (or more precisely multi-effects model) is a statistical model containing both fixed effects and random effects. These models are useful in a wide variety of disciplines in the chemical, physical, biological, and social sciences. They are particularly useful in settings where repeated measurements are made on the same statistical units or where measurements are made on clusters of related statistical units. Because of their advantage in dealing with missing data, mixed effects models are often preferred over more traditional approaches such as repeated measures analysis of variance (ANOVA) in many disciplines where multiple correlated measurements are made on each unit of interest.

5.4.3 Generalized Linear Mixed Model (GLMM)

As discussed previously, multivariable correlation can be expressed as

$$y_i = \beta_0 + \sum_{j}^{m} \beta_j x_{ij} + \varepsilon_i \tag{5.26}$$

The correlation model above is also called generalized linear model (GLM). GLM considers only the fixed or deterministic effects (x_j), but not random effects, which can be addressed by GLMM. GLMM includes both fixed and random effects (hence mixed models).

The major shortcoming with a GLM model is overdispersion—the variances higher than what would be predicted from the model due to missing covariates (similar to bilinear terms). A GLMM solution to overdispersion is to add a new group of variables z_{ik}'s for random effects for every observation in the data set. The general expression for a GLMM is given as

$$y_i = \beta_0 + \sum_{j}^{m} \beta_j x_{ij} + \sum_{k}^{n} \lambda_k z_{ik} + \varepsilon_i \quad i = 1, \ldots, N \tag{5.27}$$

where y_i is the outcome or dependent variable for data point i, x_{ij}'s are the predictive or independent variables for fixed effects, and z_{ik}'s are variables for random effects that are complement to x_{ij}'s (fixed effects). ε_i's are the prediction error terms.

5.4.4 Multilevel Models

The rationale of multilevel models is that operation activities occur at different levels in the hierarchy. For example, business planning for decisions of purchasing crudes for different refinery plants based on the plant capacity happens at company headquarters. Then product planning is conducted at the plant level to determine crude blending, operating targets for individual processes, and final products for a given crude mixture. Consequently, to meet the planning targets, the production scheduling concerns about crude delivery time and storage schedule as well as crude mixing schedule for achieving the crude mixture determined in the planning. In the operation level, advanced control and online optimization translate the goals set by planning and scheduling to real-time process targets or setpoints, which are executed by regulatory control. Finally, product blending determines the most economical mixing to make products on specs, while logistic optimization sets the timeline for transporting the products to wholesale and customers.

The need for multilevel models is that there is a hierarchy of the events occurring in sequence and the effects that influence each other. The challenge to modelers is: "well, how do I model all these interdependencies? How do I model these events and effects with interactions?"

The challenge here is to identify the factors in each level that have significant impact on the levels above and below. You can think of any number of factors that may affect profit and processing feasibility. If planning decides to buy crude with too much metal content, which could be critical for a hydrocracking unit, that crude may not be processed effectively to meet product specification. There are other factors that may affect the process feasibility in other process units. So the question is: should all these factors be introduced as variables in the planning model? I use planning as an example, and this question is valid for other levels of modeling.

Another challenge is to find the right level of model granularity. In other words, what granularity needs to be covered in each level of modeling? Most modelers would seek the correct level of granularity. The question is: what is the correct and distinct granularity needed for each level of modeling? Finer granularity is not necessarily better. Models may suffer with too much or too coarse level of details.

Then you are just looking at the effects at different levels, you may think: should I address some effects together at one level and other effects at another level. Basically, one could come up with different grouping of effects in each level that leads to different hierarchies for making a prediction. With different grouping, those different layers might have different factors that interact with each other. Thus, the question is: how many levels need to be designed for modeling and why?

These questions can be addressed by multilevel modeling. At this stage, you only need to understand the concept, which itself is pretty simple as it reflects the common sense. Multilevel modeling tries to take account of all those different hierarchies and factors and how they interplay with each other.

5.4.5 Principal Component Analysis

For most real-world problems, there are so many variables that affect a response that it becomes very difficult, sometimes impossible, to understand the relationship between each variable. For cases like this, PCA can help to find patterns in the data by reducing the number of dimensions. With PCA, linear combinations of the original variables are used to generate a new set of dimensions, or components, to describe the system. The data are then projected onto the new dimensions. The dimensions that have the most amount of variance are considered the principal components and will describe the system with the least amount of error. For example, if you want to predict the next "hot tech stock," there is a seemingly endless number of variables you could use such as revenue, operating expenses, taxes, size of market, market share, number of products offered, stock price movement, number of followers of CEO's Twitter account, and many more.[1] After performing PCA, you may find that many of these variables are correlated and can be "grouped" together into more general dimensions such as profitability, market acceptance, and social influence. Looking at the problem this way may show that a company is an outlier among the other companies in terms of these new dimensions.

PCA is commonly used as one step in a series of analyses. For example, you can use principal components before you perform a regression analysis in order to avoid multicollinearity or to reduce the number of predictors relative to the number of observations.

5.4.6 Partial Least Squares Analysis

Partial least squares (PLS) regression is a technique that reduces the predictors to a smaller set of uncorrelated components and performs least squares regression on these components, instead of based on the original data. PLS regression is especially useful when your predictors are highly collinear or when you have more predictors than observations and ordinary least squares regression either produces coefficients with high standard errors or fails completely. PLS does not assume that the predictors are fixed, unlike multiple regression. This means that the predictors can be measured with error, making PLS more robust to measurement uncertainty.

In PLS, components are selected based on how much variance they explain in the predictors and between the predictors and the response(s). If the predictors are highly correlated, or if a smaller number of components perfectly model the response, then the number of components in the PLS model may be much less than the number of predictors. Thus the least squares regression can be applied on these uncorrelated components (Minitab.com, 2019).

Unlike least squares regression, PLS can fit multiple response variables in a single model. PLS regression fits multiple response variables in a single model.

1 This is not intended to be investment advice.

Because PLS regression models the response variables in a multivariate way, the results can differ significantly from those calculated for the response variables individually. You should model multiple responses separately only if the responses are uncorrelated.

5.4.7 Combine PCA and PLS for Fault Detection

With the large amount of data available in a manufacturing plant, one can apply PCA to analyze the process data to visualize variation and strong patterns. To reveal the relationship between quality data and process operating variables, quality data from laboratory and quality prediction with inferential sensors can be analyzed jointly with process data via PLS.

For condition-based maintenance, PCA can be used to identify the most important contributor to characterize the conditions of equipment, while PLS is used to predict the system state and detect abnormal behavior, which helps to determine maintenance tasks. For more complex fault detection, advanced machine learning techniques such as Q-learning technique (Chapter 10) can be applied.

5.5 DATA MINING

Data mining is a technique of discovering patterns in large data sets involving methods at the intersection of machine learning, statistics, and database systems (Han et al, 2011).

The actual data mining task is the semi-automatic or automatic analysis of large quantities of data to extract previously unknown, interesting patterns such as groups of data records (cluster analysis), unusual records (anomaly detection), and dependencies (association rule mining, sequential pattern mining). This usually involves using database techniques such as spatial indices. These patterns can then be seen as a kind of summary of the input data, and may be used in further analysis or, for example, in machine learning and predictive analytics. For example, the data mining step might identify multiple groups in the data, which can then be used to obtain more accurate prediction results by a decision support system. Neither the data collection, data preparation, nor result interpretation and reporting is part of the data mining step, but do belong to the overall KDD process as additional steps.

The difference between data analysis and data mining is that data analysis is used to test models and hypotheses on the dataset, e.g., analyzing the effectiveness of a marketing campaign, regardless of the amount of data; in contrast, data mining uses machine learning and statistical models to uncover hidden patterns in a large volume of data.

5.6 SUMMARY

In this chapter, we discussed statistical distribution models including normal distribution, exponential distribution, and power law distribution. With identifying a certain distribution for a sample data set, we can determine the pattern or behavior and thus can predict the future events with probability.

We also learned regression models, which intend to fit a line or curve to a data set and then use the model to predict new events. There are different forms of regression models: single linear models, multivariable linear models, and nonlinear model. Multivariable regression is more complicated, which is applied when multiple factors play roles together to affect the outcome or dependent variable. We then realized that linearity may be a strong and unrealistic assumption although it offers a good place to start to test our intuitions and to see the relationships between independent (predictive) variables and dependent (outcome) variables. We can then construct more elaborate nonlinear models so that we can see the effect of the large versus small events (the 80-20 role in power law) and the effect of a variable damping as it increases (diminishing return in exponential law).

Finally, we briefly discussed advanced topics such as autocorrelation, mixed-effects modeling, GLMM, multilevel modeling, PCA, PLS and data mining. If we can understand the problems to be addressed and modeling methods, we can figure out the best match between a problem and the right modeling approach to solve the problem.

REFERENCES

Gellaboina MK, Lattanzio LA (2019), Catalyst transfer pipes plugging prediction. US Patent: US2019/C01586 A1.

Han JW, Kamber M, Pei J (2011), Data Mining: Concepts and Techniques (3rd ed.), Morgan Kaufmann. ISBN 978-012-3814791.

Minitab 18 (2019), What is partial least squares regression? Minitab.com.

Pruett JM, Schneider H (1992), Essentials of SPC in the Process Industries, ISA (International Society of America), USA.

Schmuller J (2009), Statistical Analysis with Excel® For Dummies®, 2nd ed., John Wiley & Sons, USA.

6

RIGOROUS PROCESS MODELING

6.1 INTRODUCTION

Without doubt, obtaining rigorous process models is an exciting effort as it strips away "confusion factors" of the problem and gets to the core of the system. In essence, process modeling is about establishing mathematical equations based on fundamental theory to describe the cause-and-effect relationships between variables and thus can predict what is going to happen when certain things change. In general, all models can be categorized into three classes, namely, theoretical, empirical, and semiempirical models, based on how the equations are derived.

Theoretical models describe the laws of nature that govern chemistry, physics, transport phenomena, reaction kinetics, thermodynamics, flow dynamics, and so on. These types of models are appealing because they can be extrapolated to a wide variety of situations with few assumptions. Theoretical models are also called "first principle models," "fundamental models," or "mechanistic models," which mean the same thing virtually.

A first principle model is a basic, fundamental, self-evident proposition. In sciences, theoretical work is said to be from first principles if it starts directly at the level of established science and key parameters are determined based on fundamental principles, which are verified by experiment if necessary. A first principle model is not constrained like an empirical model in terms of fitted data ranges, so it can provide better extrapolation than the interpolation provided by empirical models.

Empirical models are developed with key parameters determined based on statistical data fitting. This type of models can only be reliably applied for identical

Digitalization and Analytics for Smart Plant Performance: Theory and Applications, First Edition. Frank (Xin X.) Zhu.

conditions as those considered during formation of the relationship or where verification was conducted.

Semiempirical models are developed with theoretical expressions and some of the key parameters determined based on fundamental principles but others from statistical data fitting. These types of models are better than empirical models as they can be applied in similar conditions to those of the experiments used to create them, but are more limited than first principle models.

This chapter provides detailed explanation for these three modeling approaches. Subtle differences will become self-evident through discussions below.

6.2 REACTION KINETIC MODELING

Reaction is the core of chemical conversion processes. It requires deep understanding of reaction fundamentals to develop first principle models so that the hidden relationships of yields against feed composition and properties, catalyst effects, and process conditions can be revealed. However, it could be extremely difficult to derive a purely first principle kinetic model in some cases due to the level of molecular details involved. Thus, semiempirical modeling is the most common approach adopted in the industry. In addition, empirical modeling approach is widely used as well due to its simplicity for intended applications.

6.2.1 Reaction Kinetic Modeling Basics

The general form of a reaction rate model can be expressed in the form of three terms as

$$r = \frac{dC_i}{dt} = k(T) \times f(c_1, \dots, c_i, \dots, c_{nc}) \times \eta \tag{6.1}$$

Let us investigate these terms, namely, reaction constant k, concentration profile, and catalyst effectiveness η, in more details as below.

6.2.1.1 Reaction Constant k is the reaction constant although it is not really a constant because it includes all the kinetic parameters that affect reaction rate, except for concentration, which is explicitly considered. Of all the parameters influencing reaction rates, temperature is normally the most important one and is accounted for by the Arrhenius equation as

$$k = k_0 e^{-E_a/RT} \tag{6.2}$$

The rate constant k is a function of activation energy (E_a) and temperature (T) where R is the universal gas constant. k_0 is the pre-exponential factor, a constant for each reaction. As denoted in Equation 6.2, the higher the temperature, the larger the k value and thus the higher rate. This is because with a high temperature, the kinetic energy of the reactants increases, allowing far more collisions between the molecules. This, therefore, allows for products to be formed faster. On the other hand, use of catalyst in a reaction reduces E_a and thus increases k for speeding up a reaction as it enables a reaction starting at lower temperature.

6.2.1.2 *Arrhenius Activation Energy (E_a)*

The activation energy term E_a from the Equation 6.2, which is also called Arrhenius equation, defines the threshold barrier that a reaction must overcome via raising the reactants' temperature. After climbing over the threshold, a reaction starts and proceeds to make products while going down the hill. For an exothermic reaction, the products have the lower energy state than reactants (Figure 6.1a), while the products have the higher energy state than reactants for an endothermic reaction (Figure 6.1b). However, most chemical reactions that take place in reactor vessels are like hydrocarbon combustion: the activation energy is too high for the reactions to proceed significantly at ambient temperature.

6.2.1.3 *Order of Reaction*

Another important aspect of the reaction rate law is concentration profile in terms of reaction order. The order of a reaction is important in that it enables us to classify specific chemical reactions easily and efficiently. Reaction order can be calculated from the rate law by adding the exponential values of the concentrations of the reactants.

Here is an example of how you can look at this. For a reaction, the rate equation is expressed as

$$r = k(T)[\mathrm{A}]^n[\mathrm{B}]^m \tag{6.3}$$

where [A] and [B] are the concentrations of each reactant. For gas-phase reactions, the rate is often alternatively expressed by partial pressures.

If the reaction orders with respect to [A] and [B] are $n = 2$ and $m = 1$, respectively, that basically means the concentration of [A] is decreasing by a factor of 2, while the concentration of [B] is decreasing by a factor of 1 during reaction. The overall reaction order is equal to 3 ($= n + m$).

It is important to note that the reaction order can be determined from experimental data as part of a rate law equation. If the correlated reaction order is the same as the stoichiometric coefficients in the chemical equation, the reaction is called an elementary reaction. Otherwise, the reaction is a nonelementary reaction.

6.2.1.4 *Reaction Catalyst*

To increase the rate of reaction, the threshold barrier, that is, activation energy (E_a), must be lowered so that more collisions among reactants can occur. An alternative route with lower threshold is to add catalyst to reaction because the catalyst lowers the activation energy (Figure 6.2). It does not change the energies of the original reactants or products. Rather, only the activation energy for the reaction taking place on the catalytic route is lowered.

During reaction, the catalyst is not consumed although the activity may be reduced due to deactivation over time. Relatively small quantities of catalysts can catalyze relatively large masses of reactants and thus can be used repeatedly over

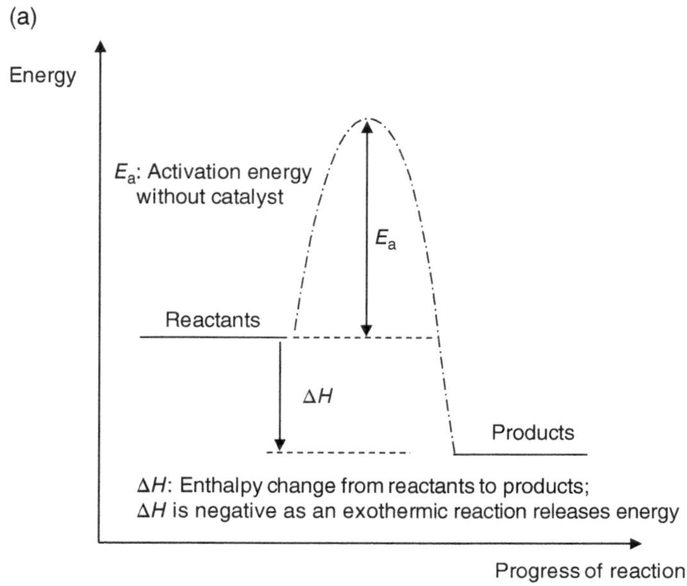

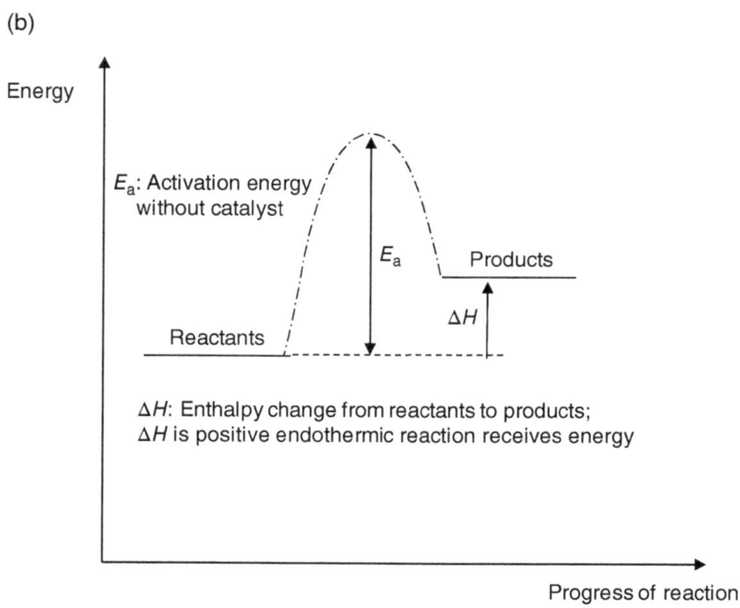

FIGURE 6.1. Activation energy without catalyst: (a) for exothermic reactions and (b) for endothermic reactions.

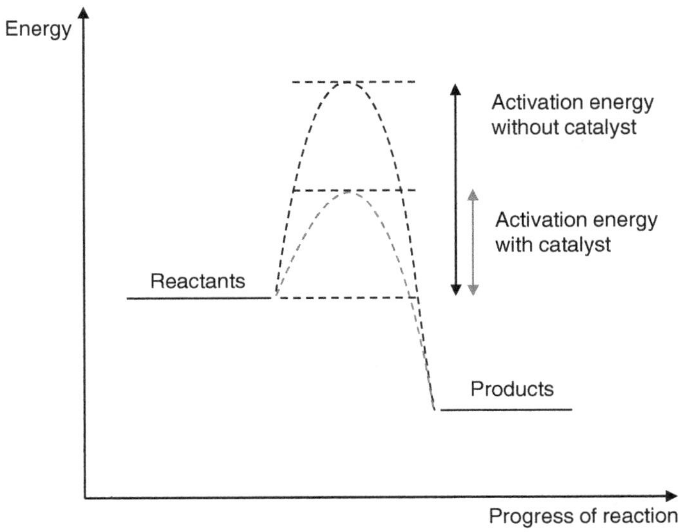

FIGURE 6.2. Activation energy with catalyst in exothermic reaction.

their lifetimes. Catalysts are developed to suit the needs of specific reactions. Catalyst performance is described by catalyst effectiveness η.

6.2.2 Reaction Kinetic Modeling

In practice, reaction kinetics is determined via experiment based on the reaction mechanism.

6.2.2.1 Elementary Reaction Rate Law It is an elementary reaction if the order of reaction for each species in the rate expression is the same as the stoichiometric coefficients. Consider a single reaction with stoichiometric equation as

$$O_2 + 2NO \rightarrow 2NO_2 \tag{6.4}$$

The rate expression derived from experimental data is

$$-r_{NO} = k_{NO} C_{NO}^2 C_{O_2} \tag{6.5}$$

Since the order of reaction for each species matches with the stoichiometric coefficient of each species, it is an elementary reaction.

6.2.2.2 Reversible Reaction The net reaction rate is equal to the difference between the rate of formation in the forward reaction and the rate of formation in the reverse reaction:

$$r_{net} = r_{forward} - r_{reverse} \tag{6.6}$$

At equilibrium, the rate of forward reaction is equal to that of reverse reaction, that is, $r_{forward} = r_{reverse}$ or $r_{net} = 0$, which is thermodynamically described by the equilibrium constant K for the reaction.

*6.2.2.3 **Nonelementary Reaction Rate Law*** When there is no direct correspondence between the order of reaction for each species and stoichiometry, then the reaction is nonelementary. For example, consider the reaction as

$$A + B \rightarrow C + D \tag{6.7}$$

The rate expression derived from experimental data is

$$-r_A = kC_A^2 C_B \tag{6.8}$$

Thus, it is a nonelementary reaction because the reaction is second order for species A and first order in species B, which does not have direct correspondence with the reaction stoichiometry. Nonelementary reaction often involves the presence of intermediate reaction steps and intermediate products. To derive reaction rate law for a nonelementary reaction, the reaction mechanism containing intermediate reaction steps and products needs to be devised, which will be discussed below.

*6.2.2.4 **Steady-State Approximation*** When a reaction involves one or more intermediates, the concentration of one of the intermediates remains constant at a certain stage of the reaction. Thus, the system has reached a steady state for which the method for deriving the rate law is called steady-state approximation. The method assumes that one intermediate in the reaction mechanism is consumed as quickly as it is generated. Its concentration remains the same throughout duration of the reaction.

*6.2.2.5 **Reaction Mechanism*** A reaction mechanism describes all intermediate steps from reactants to products. An intermediate is a species that is neither a reactant nor a product. Devising a reaction mechanism requires a broad understanding of properties of reactants and products, and this is usually done by experienced chemists. The steady-state approximation is a technique for deriving a rate law from the proposed mechanism.

Let us consider the following example:

$$2N_2O_5 \leftrightarrow 4NO_2 + O_2 \tag{6.9}$$

The mechanism for this reaction is devised based on elementary reaction steps as below:

- Step 1: Let k_f and k_r be forward and reverse rate constants for Equation 6.10:

$$N_2O_5 \leftrightarrow NO_2 + NO_3 \tag{6.10}$$

- Step 2:

$$NO_3 + NO_2 \rightarrow NO + NO_2 + O_2 \quad (\text{rate constant } k_2) \tag{6.11}$$

- Step 3:

$$NO_3 + NO \rightarrow 2NO_2 \quad (\text{rate constant } k_3) \tag{6.12}$$

NO and NO_3 are intermediate products and we try to replace them with final products.

The production rate of NO in step 2 and consumption rate of NO in step 3 are

$$r_{NO} = k_2[NO_3][NO_2] \tag{6.13}$$

$$-r_{NO} = k_3[NO_3][NO] \tag{6.14}$$

where the brackets indicate the concentration of a component.

Based on the steady-state assumption, the rate of production of an intermediate is equal to the rate of its consumption. Thus, we have

$$k_2[NO_3][NO_2] = k_3[NO_3][NO] \tag{6.15}$$

Solving Equation 6.15 for the concentration of NO ([NO]) gives

$$[NO] = \frac{k_2[NO_2]}{k_3} \tag{6.16}$$

Now let us turn to the other intermediate, that is, NO_3. The rate law for both production and consumption of NO_3 from steps 1–3 can be derived as

$$r_{NO_3} = k_f[N_2O_5] \tag{6.17}$$

$$-r_{NO_3} = k_2[NO_3][NO_2] + k_3[NO_3][NO] + k_r[NO_3][NO_2] \tag{6.18}$$

Applying the steady-state assumption gives

$$k_f[N_2O_5] = k_2[NO_3][NO_2] + k_3[NO_3][NO] + k_r[NO_3][NO_2] \tag{6.19}$$

Solving Equation 6.19 for the concentration of NO_3 ($[NO_3]$) gives

$$[NO_3] = \frac{k_f[N_2O_5]}{k_2[NO_2] + k_3[NO] + k_r[NO_2]} \tag{6.20}$$

Take a pause to review the three reaction steps: (a) Step 1 is an equilibrium reaction and thus cannot give a rate expression or $r_{NO} = 0$ under equilibrium. (b) Step 2 gives the production of both intermediate and final products. The former leads to further production of NO_2 in step 3. From step 2, the production rate for O_2 is

$$\frac{d[O_2]}{dt} = k_2[NO_3][NO_2] \tag{6.21}$$

Integrating [NO] in Equation 6.16 and [NO_3] in Equation 6.20 with Equation 6.21 yields

$$\frac{d[O_2]}{dt} = \frac{k_2 k_f}{k_r + 2k_2}[N_2O_5] = k_0[N_2O_5] \tag{6.22}$$

where

$$k_0 = \frac{k_2 k_f}{k_r + 2k_2} \tag{6.23}$$

The reaction rate for product O_2 in Equation 6.22 agrees with the experimental results, which means the reaction mechanism assumed as above is correct. Otherwise, a different reaction mechanism needs to assumed and verified against experiment data until they both matches.

6.2.3 Rate Kinetic Modeling for Solid Catalyst

In many reactions, a catalyst is used to speed up a reaction or slow a reaction (negative catalyst). Most catalysts are solid with porous structure. Reactants must interact with catalyst for reaction to occur, and thus three major steps need to occur successfully (Levenspiel, 1999):

Step 1 (adsorption): A molecule is absorbed onto the surface and is attached to an active site ready for reaction.

Step 2 (reaction): The molecule reacts with another molecule on the active site.

Step 3 (desorption): Products are desorbed from the surface and then free the site.

All species of molecules in above steps are assumed to be in equilibrium.

Rate expression for catalytic reaction is of the form

$$\text{Reaction rate} = \frac{(\text{kinetic term})(\text{concentration terms})}{(\text{resistance terms})} \tag{6.24}$$

Consider the following reaction:

$$A + B \leftrightarrow R + S, \quad \text{with reaction equilibrium constant } K \tag{6.25}$$

The rate equation when adsorption of A controls in step 1 is

$$-r_A = \frac{k(P_A - P_R P_S / K P_B)}{(1 + K_A P_R P_S / K P_B + K_B P_B + K_R P_R + K_S P_S)^2} \tag{6.26}$$

When reaction between adjacent site-attached molecules of A and B controls (step 2), the rate equation is

$$-r_A = \frac{k\left(P_A P_B - P_R P_S / K\right)}{\left(1 + K_A P_A + K_B P_B + K_R P_R + K_S P_S\right)^2} \tag{6.27}$$

where P_A, P_B, P_R, and P_S are partial pressures of gaseous species A, B, R, and S. K_A, K_B, K_R, and K_S are the adsorption equilibrium constants, while K is the reaction equilibrium constant.

But when desorption of R controls (step 3), the rate equation becomes

$$-r_A = \frac{k\left(P_A P_B / P_S - P_R / K\right)}{\left(1 + K_A P_A + K_B P_B + K \cdot K_R P_A P_B / P_S + K_S P_S\right)} \tag{6.28}$$

The above shows that each controlling reaction has its own resistance and thus reaction rate equation based on surface kinetics involving reaction equilibrium constant K and several adsorption equilibrium constants.

6.2.3.1 Catalyst Effectiveness Factor The above reaction kinetic modeling does not take into account the effects of internal diffusion resistance on reaction rate. To measure how much the reaction rate is lowered because of diffusion resistance, the effectiveness factor, η, is defined as

$$\eta = \frac{\text{actual reaction rate}}{\text{rate without diffusion resistance}} = \frac{r_{A,\text{with diffusion}}}{r_{A,\text{without diffusion resistance}}} \tag{6.29}$$

or

$$r_{A,\text{with diffusion}} = \eta \cdot r_{A,\text{without diffusion resistance}} \tag{6.30}$$

Effectiveness factor η is determined based on a parameter called the Thiele modulus, M_T. M_T is a very important parameter for describing the effects of internal pore diffusion on reaction rate in the porous catalyst pellets with no mass transfer limitations.

6.2.3.2 Determine Catalyst Effectiveness Factor Expressions for M_T are given below, and details of the derivation can be seen in Levenspiel (1999):

For first-order reversible reactions:

$$M_T = L\sqrt{\frac{k}{D_e X_A}} \tag{6.31}$$

For nth-order reversible reactions:

$$M_T = L\sqrt{\frac{(n+1)kC_{As}^{n-1}}{2D_e}} \tag{6.32}$$

where D_e is effective diffusion coefficient of fluid within catalyst pores, L is the characteristic size of catalyst particle, X_A is the conversion of component A, and C_{As} is the concentration of species A on the surface of the catalyst particle. Figure 6.3 provides the definition of C_{As} and L for the pore length with a cylinder shape.

As indicated in Equations 6.31 and 6.32, the two parameters for M_T are diffusion coefficient (D_e) and catalyst size (L). Let us pay attention to effective diffusivity D_e first. The effective diffusivity accounts for the following facts: (a) Not all of the area is available (i.e., the area occupied by solids) for the molecules to diffuse. (b) The internal paths are tortuous. (c) The pores are of varying cross-sectional areas. These facts can be described graphically by Figure 6.4. Thus, D_e is defined below to address these facts:

$$D_e = \frac{D'_{AB}\phi_p\sigma_c}{\tau} \tag{6.33}$$

where

$$\tau = \text{tortuosity} = \frac{\text{actual distance a molecule travels between two points}}{\text{shortest distance between two points}} \tag{6.34}$$

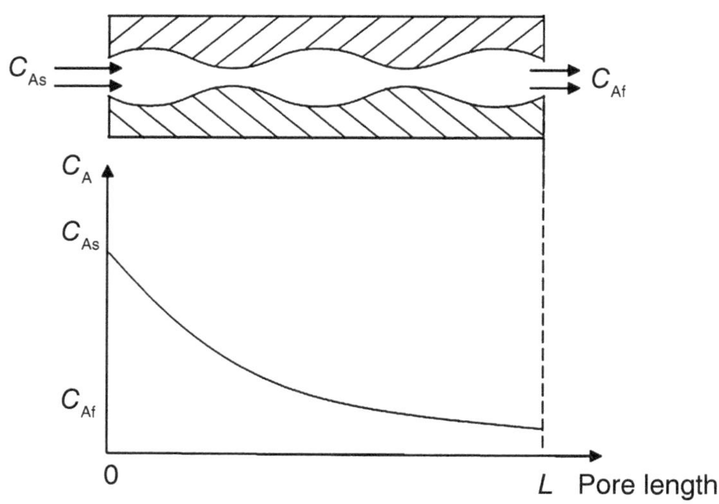

FIGURE 6.3. Concentration profile within a cylinder pore.

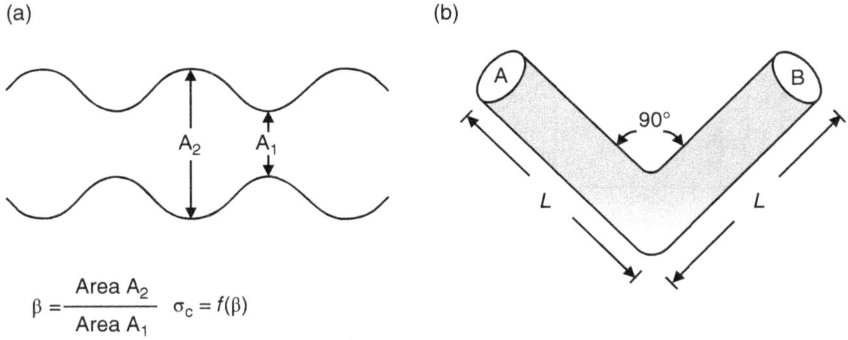

FIGURE 6.4. Catalyst geometry: (a) pore constriction and (b) pore tortuosity.

$$\phi_p = \text{pellet porosity} = \frac{\text{volume of void space}}{\text{total volume (voids and solids)}} \tag{6.35}$$

$$\sigma_c = \text{constriction factor} \tag{6.36}$$

The constriction factor σ_c accounts for the variation in the cross-sectional area that is normal to the direction of diffusion. It is a function of the ratio (β) of maximum and minimum pore areas (Figure 6.4a).

The second parameter is characteristic size L. To obtain the effective distance penetrated by reactants to reach all internal surfaces, the characteristic size of catalysts, L, is defined as

$$L = \frac{\text{volume of particle}}{\text{exterior surface available for reactant penetration}},$$
$$\text{e.g., } L = \delta/2 \text{ for flat plates; } L = R/2 \text{ for cylinders; } L = R/3 \text{ for spheres} \tag{6.37}$$

δ is the thickness of a plat; R is the radius of cylinder or sphere. To find how the pore resistance affects reaction rate, D_e and L are calculated first from Equations 6.33 and 6.37, and then M_T can be determined based on Equations 6.31 or 6.32. The $\eta - M_T$ charts in Figure 6.5 (Levenspiel, 1999) are used to determine η and finally calculate the actual reaction rate based on Equation 6.30.

The $\eta - M_T$ charts show that for $M_T < 0.4$, η is close to 1, and the concentration of reactant does not decrease appreciably within the pore; thus, pore diffusion resistance has negligible effect on reaction rate. A small M_T implies a short pore and rapid diffusion rate. For large M_T, $\eta \ll 1$, reactant concentration drops rapidly to zero within the pore; hence diffusion affects the reaction rate significantly. This is a regime of strong pore resistance.

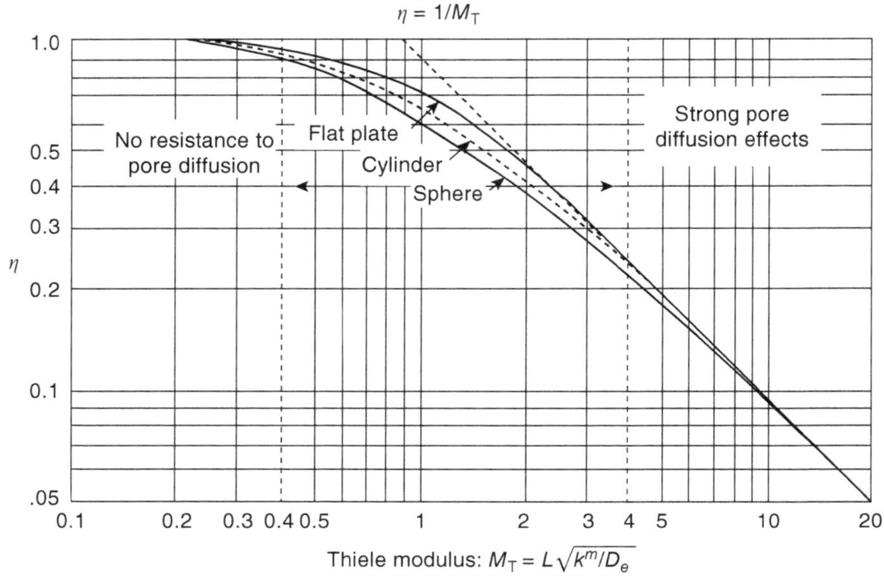

FIGURE 6.5. Effectiveness factor η versus M_T for porous particle with various shapes. *Source*: From Levenspiel (1999). © 1999 John Wiley & Sons. Reprinted with permissions of John Wiley & Sons.

6.2.4 Review of Kinetic Modeling Methods

For petroleum processes, hydrocarbon feeds contain thousands of basic components. Modern analytic technologies can provide measurement of composition of refining streams to support molecular modeling. For example, gas chromatography (GC) can be used to provide sufficient composition information for naphtha range materials. For heavier streams, such as distillate, vacuum gas oil (VGO), and vacuum residue, GC cannot provide enough molecular details due to their compositional complexity. Advanced analytical tools are available for measuring compositions for heavier streams, which include comprehensive two-dimensional gas chromatography (GC×GC) and mass spectrometry (MS).

The composition details measured via analytical tools can be represented in an appropriate structure suitable for kinetic modeling. Basically, there are three kinetic modeling approaches, namely, lumping approach (empirical based), pathway approach (semiempirical based), and mechanistic approach (first principle based).

6.2.4.1 Lumping Kinetic Modeling (Empirical Based) Traditionally, hydrocarbon streams are characterized based on lumped compounds where individual molecules in a hydrocarbon feedstock are lumped into a small number of pseudocomponents measured based on narrow boiling range. The lumping method has been applied to major reaction processes by Smith (1959), Weekman and Nace (1970),

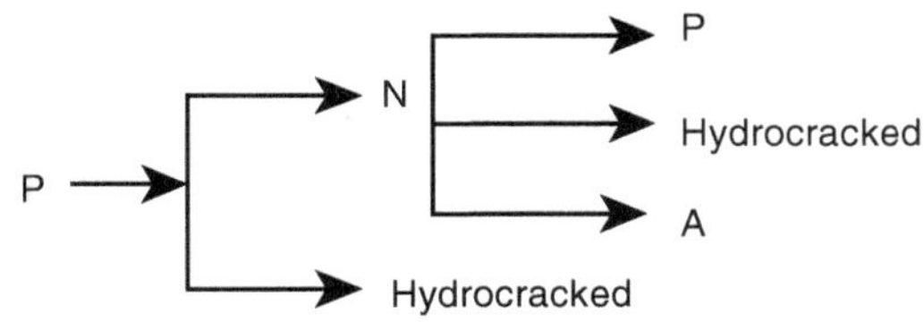

FIGURE 6.6. PNA lumps for naphtha reforming reaction.

Stangeland (1974), Taskar and Riggs (1997), and so on. These lumping methods are implemented in commercial software including HYSYS, UniSim, Aspen Plus, and PRO/II.

Let us use a naphtha reforming reaction as an example where three lumps are used to describe the reaction chemistry, namely, paraffins (P), naphthenes (N), and aromatics (A). The reaction network is shown in Figure 6.6. Many different molecules are placed into each group or lump. In the lumping approach, the reaction kinetic model assumes that all species in a lump behave identically.

For naphtha reforming reactions based on the three lumps (P, N, A), four dominant reactions are identified, and the empirical rate equations (Rase, 1977) are given below:

(1) Dehydrogenation (N → A) where naphthene is converted to aromatics:

$$r_1 = \alpha_1 k_{p_1}\left(P_{\mathrm{N}} - P_{\mathrm{A}} P_{\mathrm{H}}^3 / K_{p_1}\right) \tag{6.38}$$

where

$$k_{p_1} = 9.87 e^{(23.21 - 34750/1.8T)}$$

(2) Ring closure where naphthene is transformed into paraffin:

$$r_2 = \alpha_2 k_{p_2}\left(P_{\mathrm{A}} P_{\mathrm{H}} - P_{\mathrm{p}} / K_{p_2}\right) \tag{6.39}$$

where

$$k_{p_2} = 9.87 e^{(35.98 - 59600/1.8T)}$$

(3) Hydrocracking of paraffin:

$$r_3 = \alpha_3 k_{p_3} P_{\mathrm{p}} / P_{\mathrm{T}} \tag{6.40}$$

where

$$k_{p_3} = e^{(42.97 - 62300/1.8T)}$$

(4) Hydrocracking of naphthene:

$$r_4 = \alpha_4 k_{p_4} P_E / P_T \tag{6.41}$$

where

$$k_{p_4} = e^{(42.97 - 62300/1.8T)}$$

where k_{p_1}, k_{p_2}, k_{p_3}, and k_{p_4} are the reaction rate constants corresponding to r_1, r_2, r_3, and r_4, respectively, with units of kmol/(h·kg·cat). K_{p_1} is the equilibrium constant of the naphthene-to-aromatics and aromatics-to-naphthene reactions, MPa^3; K_{p_2} is the equilibrium constant of the naphthene-to-paraffin and paraffin-to-naphthene reactions, MPa^{-1}; P is pressure of system or component partial pressure, MPa. $\alpha_1 - \alpha_4$ are model tuning parameters.

It needs to be mentioned that the kinetic parameters above in the lumping approach are determined based on experimental data, which explains the reason why the kinetic models in Equations 6.38–6.41 are empirical. Although the lumping-based approach is relatively straightforward, the major shortcoming for lumping models is that it does not address chemical reactions in details and does not capture catalyst details and individual molecule conversions; thus opportunities for optimizing selectivity for these molecules get lost. Furthermore, the lumping-based models have limited prediction power as they are developed based on experiments for a limited feedstock. When a feed is outside of experimental data, the kinetic parameters must be reestimated based on new experiments.

6.2.4.2 Pathway Kinetic Modeling (Semiempirical Based) Quann and Jaffe (1996) introduced the structure-oriented lumping (SOL), which is a pathway approach. A large number of key molecules are modeled by assembling 25 structural groups in various ways so that the SOL synthetic feed could satisfy the observable characteristics, both chemical and physical. In other words, the SOL approach uses structural or functional groups to characterize the composition of complex feedstock and generates kinetic models with details at the reaction pathway level. In this manner, the SOL method can track molecules in the feed based on the composition and structure in terms of number of aromatic rings, number of nitrogen and sulfur substituents, etc.

Another pathway approach (Hu et al., 2002) is based on molecular-type homologous series (MTHS) matrix. An MTHS is used to represent molecular compositions of a petroleum mixture to describe different homologous series corresponding to functional groups such as normal paraffins (nP), isoparaffins (iP), olefins (O), naphthenes (N), and aromatics (A). Different sulfur and nitrogen compounds are also shown in this matrix. A homologous series has the same base structure but varying carbon number, for example, through the addition of alkyl chains of increasing length. With this MTHS representation, the important assumption is that the molecules belonging to a homologous series with the same carbon number have the same reactivity and similar physical and chemical properties, and thus they are lumped into a single component as one element of the matrix. Thus an MTHS defines a stream in functional molecular types and carbon number, which can be used as the basis for developing reaction pathways and reaction kinetic models.

As an example of MTHS, Figure 6.7 represents a naphtha stream based on five different types of homologous series including normal paraffins (NP), isoparaffins (IP), olefins (O), naphthenes (N), and aromatics (A) together with different carbon numbers. The molecules that belong to a homologous series with the same carbon number have similar physical properties, and thus, they are lumped into a single component as an element in the matrix. The reactions involved in the reforming process include hydrogenation and dehydrogenation, isomerization, paraffin cracking, naphthene and aromatic side-chain cracking, ring opening, and ring closure. Based on the MTHS pathway approach, the reforming reaction pathways can be represented as shown in Figure 6.8, which involves 21 components (or composition groups) and 51 reaction steps. This representation can capture more detailed information versus the traditional method based on lumping and pseudo-components.

The homologous series (Figure 6.8) also form the basis for developing reaction kinetic modeling. The reaction kinetics is modeled by Ancheyta-Juarez and Aguilar-Rodriguez (1994) with the effective reaction rate constant k_j for the reaction j ($j = 1, \ldots, 51$) in the network correlated with temperature (T) and pressure (P) as

$$k_j = k_j^0 \exp\left[\frac{E_j}{R}\left(\frac{1}{T_o} - \frac{1}{T}\right)\right]\left(\frac{P}{P_o}\right)^{\alpha_j} \quad j = 1, \ldots, 51 \tag{6.42}$$

where k_j^0 is the effective rate constant, E_j is the activation energy, α_j is the pressure effect parameter, and T_o and P_o are the base reaction temperature and pressure, respectively.

	NP	IP	O	N	A
C_4	1.16	0.15	0.36	0	0
C_5	2.72	4.79	5.33	0.36	0
C_6	2.03	6.74	6.16	1.46	5.18
C_7	1.09	5.73	3.88	1.70	8.83
C_8	0.45	8.95	2.73	1.28	9.97
C_9	0.34	2.83	1.56	0.84	5.34
C_{10}	0.18	1.40	1.66	0.11	2.17
C_{11}	0.23	1.00	0.99	.04	.06

FIGURE 6.7. An MTHS matrix for gasoline stream.

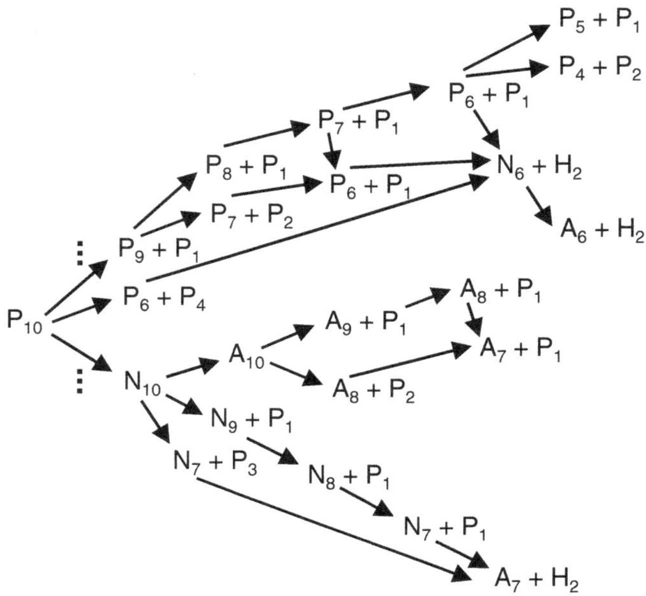

FIGURE 6.8. Reaction network of catalytic reforming.

For the reaction network of Figure 6.8 and the kinetic parameters described above, the rate equations are in the form of differential equations, which can be expressed as below:

$$r_i = \frac{dc_i}{dt} = f_i\left(k_1,\ldots,k_j,\ldots,k_{nr},c_1,\ldots,c_i,\ldots,c_{nc}\right)\times\eta \tag{6.43}$$

where r_i is the reaction rate for component i in the reforming product; c_i is the mole fraction of component i ($i = 1, \ldots, n_c$) and n_c (=21) is the number of reforming components modeled; f_i is the reaction rate equation for component i; k_j is the kinetic constant for reaction j ($j = 1, \ldots, n_r$) defined in Equation 6.42, and n_r (=51) is the number of reforming reactions; and t is the time. η is the effectiveness of catalyst pore diffusion describing the effects of diffusion on the reaction rate.

Although the pathway approach can capture more details of molecular information than the lumping approach, it is still not completely independent of feedstock because fundamentally the method is still based on function lumps.

6.2.4.3 Micro-Kinetic Modeling (MKM) (First Principle Based) Recent advances in kinetic modeling have already made a paradigm shift toward micro-kinetics approaches, of which the single-event approach by Froment (2014) is the most well known. This approach can explore fundamental chemistry of the conversion of each individual hydrocarbon molecule, which is achieved by modeling elementary reactions and using molecules and intermediates explicitly. With this micro-kinetics approach, the kinetic models are independent of feed type and compositions as few assumptions are required and more accurate predictions can be obtained.

Due to the nature of the micro-kinetics approach, the modeling for the naphtha reforming reactions involves a reaction network of as many as $O(10^{15})$ species and reactions and thus requires estimation of many kinetic parameters, which is beyond manual capability. To reduce the number of kinetic parameters to be estimated, the single-event concept is applied to formulate rates using only a reasonable number of independent kinetic parameters based on structure–activity relationships. Experimental data may be used to fit the remaining parameters. This single-event approach has been successfully applied to a variety of processes. For example, the reaction network for naphtha reforming contains 116 components and 546 reactions (Wei et al., 2008).

6.2.4.4 Comparison of Different Kinetic Modeling Approaches The prediction capability for the three kinetic modeling methods and their comparison are shown in Figure 6.9. The comparison is based on the complexity of the methods and model fidelity. The traditional lumping method is the simplest since fewer lumps are used to obtain kinetic parameters by fitting experimental data. However, the models derived have the lowest prediction capability as they are strongly dependent on feedstocks. When feed type changes, the kinetic parameters must be reestimated based on new experiments.

The pathway method requires more detailed feed data to develop molecular representations and estimate numerous kinetic parameters. The models can be

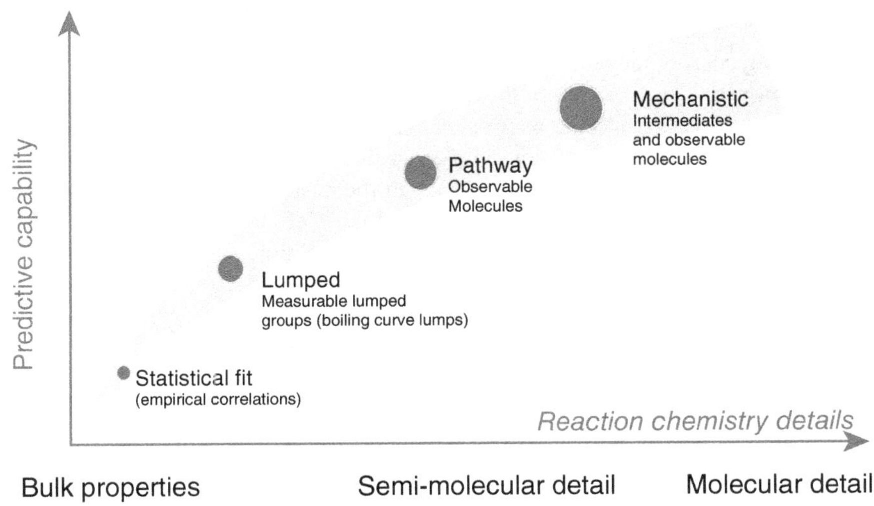

FIGURE 6.9. Comparison of prediction power of different kinetic modeling methods.

extrapolated for a range of feedstocks and process conditions. However, the pathway models are still not completely independent of feedstock because fundamentally the method is still based on function lumps.

In contrast, a mechanistic method such as single-event approach is feed independent because it is based on fundamental chemistry of transformation of each individual hydrocarbon. As a result, the kinetic parameters are invariant with feedstock, and prediction capability is much more accurate for a wide range of feed compositions and operating conditions.

However, the challenge is that the MKM approach involves too many reaction kinetic parameters and is challenging from an experimental point of view, and the quality of the model cannot be guaranteed because too many parameters need to be fitted against experiment data. Any errors in experiments and data fitting can cause the MKM kinetic model to be incorrect. The question is: do we need such complexity for developing kinetic models that can capture the key issues for the reactions?

Which approach is selected depends on the nature of the problems that we are dealing with and the objective that we want to accomplish. For determining process conditions for yield optimization, the pathway approach may be good enough, while the mechanistic approach is applied for feed selection and catalyst selection purposes. For reaction pathway selection for new catalyst and process development, a MKM model may be desired.

6.3 REACTOR TYPES AND MODELING

Reactor design is the most important factor in determining the overall production rate, environmental performance, and economics to achieve the reaction rate and yield predicted by kinetic modeling. Reactor design is first about reactor selection to match the reactor type with the reaction needs and catalyst and then internal design in order to achieve optimal reaction rates and product yields while minimizing capital and operating costs of the reaction system as well as downstream separation system.

Reaction rate is the measure of the speed of product formulation or the amount of reactant consumed per unit time of reaction. There are mainly four factors that affect reaction rate, which are concentration, temperature, catalyst, and the nature of reaction.

We shall see that reactor design involves all the basic principles of engineering including chemical kinetics, mass transfer, heat transfer, and fluid flow, and complications arise when, as so often is the case, interaction occurs between these transfer processes and the reaction itself. In designing a reactor, it is essential to weigh up all the various factors involved and, by an exercise of judgment, to place them in their proper order of importance. If one wishes the design is to be successful, the major decisions taken at the outset must be correct.

6.3.1 Reactor Types

Reactor design is basically concerned with selection of the right type of reactor and the size as well as what method of operation should be employed for a given

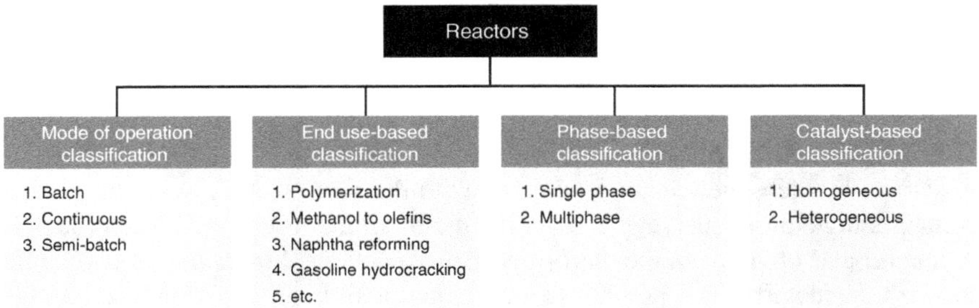

FIGURE 6.10. Reactor-type classifications.

conversation. Different types of reactors offer distinctive strengths and weaknesses. It is essential to find the right type of reactor to match the chemistry, catalyst, and production needs.

Reactor types can be categorized based on mode of operation, phases, catalyst, and the end-use applications as shown in Figure 6.10.

6.3.1.1 Batch Reactors A batch of reactant is fed into the batch reactor that is operated at the same conditions until the target conversion is reached. They are typically tanks in which stirring of reactants is achieved using impellers, gas bubbles, or a pump-around loop. Temperature is regulated via internal cooling surfaces (coils or tubes), jackets, reflux condensers, or pump-around loops through heat exchangers. Batch reactors are suited for small production rates, reactions that require long times to achieve desired selectivity, and for flexibility in campaigning different products. Typical applications include fermentation of beverage products and water treatment.

6.3.1.2 Continuous Flow Reactors In these reactors, reactants are added and products are removed continuously. They are used for large production rates. There are two basic types of continuous flow reactors, namely, continuous stirred tank reactors (CSTR) and tubular flow reactors (or plug flow reactors [PFR]). In the tubular flow reactor, the aim is to pass the reactants along a tube so that there is as little intermingling between the reactants entering the tube. The tubular reactor is the natural choice for gas-phase reactions although is sometimes used for liquid-phase reactions. Usually catalyst is used, and temperature is raised to obtain high reaction rates, in which case a relatively small tubular reactor is sufficient to produce a high volumetric flow rate of gas. Typical applications of tubular reactors are in gasoline production, oil cracking, synthesis of ammonia, and oxidation of sulfur dioxide to sulfur trioxide.

A CSTR is also called a backmix reactor and is by its nature well suited to liquid-phase reactions. In a CSTR, an agitator is used deliberately to disperse the reactants thoroughly into the reaction mixture immediately after entering the tank in order to obtain a uniform mixture with the same composition everywhere within the reactor. The product stream is drawn off continuously with the same composition as the contents of the tank. A series of several CSTRs could approach a tubular reactor in terms of reaction rate versus concentration.

6.3.1.3 Semi-Batch Reactors These reactors are essentially batch reactors. At the start of a batch, the reactor is charged with the first reactant, with the second reactant being continuously added over the period of the reaction. This approach is the natural and obvious way to carry out many reactions. For example, if a liquid must be treated with a gas, perhaps in a chlorination or hydrogenation reaction, the gas is normally far too voluminous to be charged all at once to the reactor, so it is instead fed continuously at the rate at which it is used up in the reaction. A semi-batch reactor may be chosen for the following scenarios: (a) reacting a gas with a liquid, (b) controlling a highly exothermic reaction, and (c) improving product yield in suitable circumstances.

6.3.1.4 Reactor Phases Single-phase reactors contain either gas (plasma) or liquid only. Two-phase reactors may be gas–liquid, liquid–liquid, gas–solid, and liquid–solid phases. The common type of multiphase reactor is a gas–liquid–solid reactor; however, liquid–liquid–solid reactors are also used in the industry.

6.3.1.5 Multifunction A reactor may perform multiple functions that provide both reaction and mass transfer (e.g., reactive distillation, reactive crystallization, reactive membrane, etc.). The purpose is to reduce capital and operating costs via exploitation of cross-functional synergy. The coupling could lead to additional operating constraints on some of the functions. Multifunction reactors are a hot topic for process intensification.

6.3.1.6 Homogeneous and Heterogeneous In homogeneous reactors, only one phase (a gas or a liquid) is present, and the catalyst is in the same phase as the reactants. If more than one reactant is involved, mixing the reactants is the way of initiating the reaction.

In heterogeneous reactors, multiple phases are present. Examples are gas–liquid, gas–solid, liquid–solid, and liquid–liquid reactions. Usually a solid catalyst is present. For example, gas–solid catalytic reactors particularly form an important class of heterogeneous reaction systems. Generally, heterogeneous reactors have a greater variety of configurations and contacting patterns than homogeneous ones. Practical heterogeneous reactors include fixed bed reactors, trickle bed reactors, moving bed reactors, fluidized bed reactors, and slurry reactors.

6.3.1.7 Fixed Bed Reactors A fixed bed reactor is a cylindrical tube filled with catalyst pellets with reactants flowing through the bed and being converted into products. The catalyst may have multiple configurations including one large bed, several horizontal beds, several parallel packed tubes, and multiple beds in their own shells. The various configurations may be adapted depending on the need to maintain temperature control within the system. The flow of a fixed bed reactor is typically downward.

6.3.1.8 Trickle Bed Reactors A trickle bed reactor is a fixed bed where liquid flows without filling the spaces between particles. Like with fixed bed reactors, the liquid typically flows downward. At the same time, gas is flowing upward. This reactor is often utilized to handle feeds with extremely high boiling points.

6.3.1.9 Moving Bed Reactors A moving bed reactor has a fluid phase that passes up through a packed bed. Solid is fed into the top of the reactor and moves down to the bottom, where it is removed. Moving bed reactors require special control valves to maintain close control of the solids.

6.3.1.10 Fluidized Bed Reactors A fluidized bed reactor suspends small particles of catalyst by the upward motion of the fluid to be reacted. The fluid is typically a gas with a flow rate high enough to mix the particles without carrying them out of the reactor. The particles are much smaller than those for the above reactors and are typically on the scale of 10–300 μm. One key advantage of using a fluidized bed reactor is the ability to achieve a highly uniform temperature in the reactor. In addition, for cases with rapid catalyst deactivation, catalyst can be treated and regenerated easily in a fluidized bed, which offers overwhelming advantages over a fixed bed design.

6.3.1.11 Slurry Reactors A slurry reactor contains the catalyst in a powdered or granular form. This reactor is typically used when one reactant is a gas and the other a liquid while the catalyst is a solid. The reactant gas is passed through the liquid and dissolved. It then diffuses onto the catalyst surface. Slurry reactors can use very fine particles, and this can lead to problems of separation of catalyst from the liquid. Trickle bed reactors have the advantage of avoiding this problem, although unfortunately the large particles in trickle bed result in a much lower reaction rate. Overall, the trickle bed is simpler but has a slow reaction rate, the slurry reactor usually has a high reaction rate but is more complex, and the fluidized bed is somewhat in between.

The major features of the above reactors and the guiding principles for reactor selection can be seen in Zhu et al. (2019) and Levenspiel (1999).

6.3.2 Reactor Design Modeling

After a reactor type is selected, the engineer will determine the process conditions including temperature, pressure, feed rate, and composition at the reactor inlet, which form the basis for determining the reactor conversion, selectivity, and yield versus reactor volume and residence time. This is probably the most critical step in reactor design.

Step 1: Determine reaction conversion and yield.
Economic optimization determines the conversion (X_f) and product yield.

Step 2: Determine reactor size and volume.

After the conversion (X_f) is decided, for a given mole content (N_{A_0}) and feed rate (F_{A_0}) for component A with reaction rate as r_A, the reaction time t for a batch reactor can be determined by

$$N_{A_0}\frac{dX}{dt} = -r_A \cdot V \tag{6.44}$$

or in the integral form to give the reaction time as

$$t = N_{A_0}\int_0^{X_f}\frac{dX}{-V \cdot r_A}$$

For a CSTR, the reactor volume can be determined via

$$V = \frac{F_{A_0} \cdot X_f}{-r_A} \tag{6.45}$$

For a PFR, the reaction volume can be calculated via

$$F_{A_0}\frac{dX}{dV} = -r_A \tag{6.46}$$

or in the integral form

$$\frac{V}{F_{A_0}} = \int_0^{X_f}\frac{dX_A}{-r_A}$$

For a packed catalytic reactor, the reaction volume is obtained as

$$F_{A_0}\frac{dX}{dW} = -r_A' \tag{6.47}$$

or in the integral form

$$\frac{W}{F_{A_0}} = \int_0^{X_f}\frac{dX_A}{-r_A'}$$

where catalyst weight W and reactor volume (V) are related to catalyst bulk density ρ as

$$W = \rho \cdot V \tag{6.48}$$

For illustration, the above relationships can be explained in Figure 6.11. The details for reactor modeling can be seen in Levenspiel (1999) and Farrauto and Bartholomew (1997).

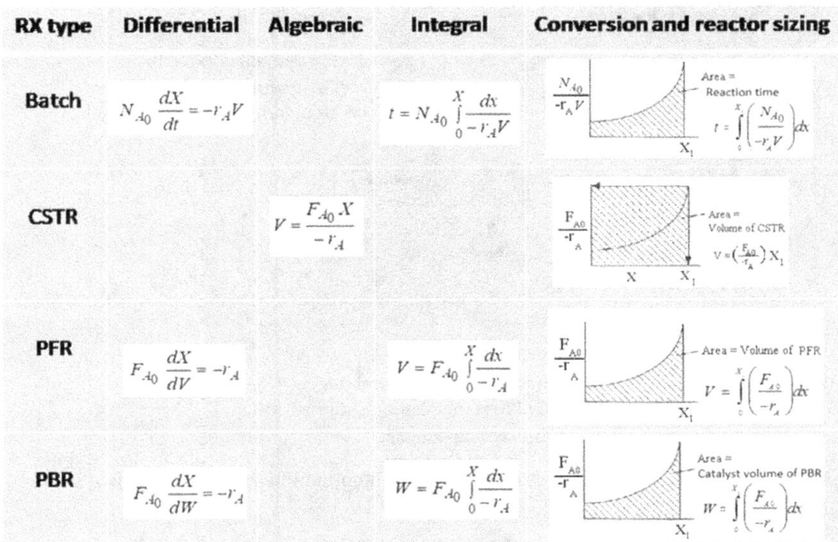

FIGURE 6.11. Design equations and plots for a single reactor. Reaction conversion (X), volume (V), reaction time (t), with concentration of A (N_{A_0}), and feed rate with composition A (F_{A_0}).

6.4 INTEGRATED KINETICS AND REACTOR MODELING

After choosing the kinetic model and reactor design equation, we can develop an integrated model for kinetics and reactor design. In general, the integrated model includes the kinetic model, catalyst performance, mass and energy balances, and heat and mass transfer models (Figure 6.12).

Let us consider the naphtha reforming reaction as an example for illustration. The kinetic models shown in Equations 6.38–6.41 indicate that four dominant reactions take place in naphtha reforming:

(1) r_1 modeling the dehydrogenation where naphthene is converted to aromatics (very fast);

(2) r_2 modeling the ring closure where naphthene is transformed into paraffin (slow);

(3) r_3 modeling the hydrocracking of paraffin (mild);

(4) r_4 modeling the hydrocracking of naphthene (fast).

A radial flow fixed bed reactor is selected, and thus we are ready to derive the integrated reaction model (Liang et al., 2005). The structure of a catalytic reforming radial reactor is shown in Figure 6.13. The following assumptions are

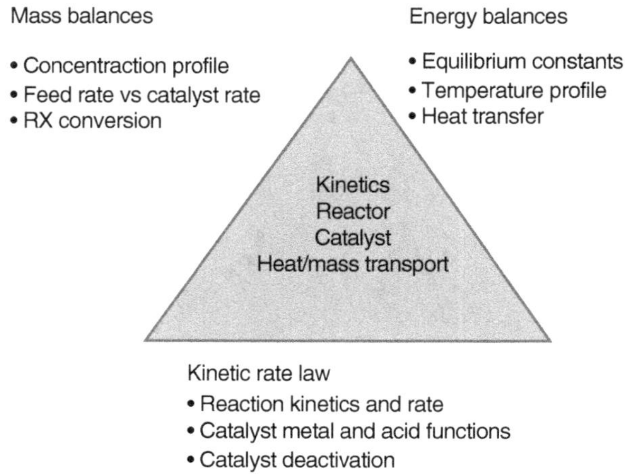

FIGURE 6.12. A triangle representation of an integrated model for reaction system.

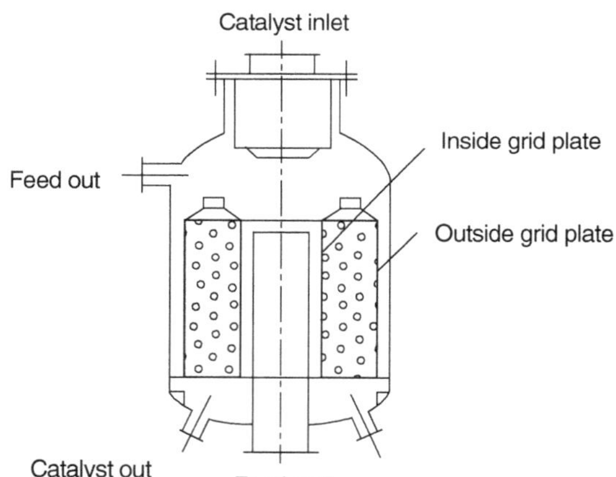

FIGURE 6.13. Plane section view of a radial flow reactor.

proposed based on the features of catalytic reforming reactions, reactor, and operating conditions:

- The temperature along the axial direction (R) of the radial flow reactor is homogeneous, that is, $dT_A = 0$.
- The concentration (C) along the axial direction of the radial flow reactor is homogeneous, that is, $dC_A = 0$.
- Reactions are in homogeneous phase. Based on the above assumption, the temperature and concentration gradients along the axial direction can be neglected, and only radial direction variables are considered. In fact, the fixed bed reactor is considered as a homogeneous reactor.

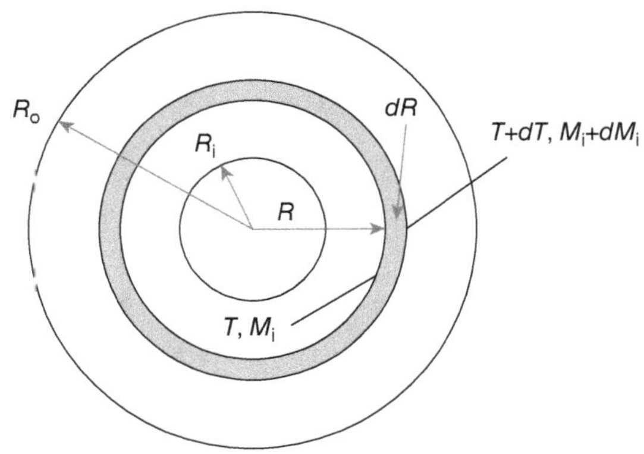

FIGURE 6.14. Cross-sectional view of a radial flow reactor.

The plane section view for a radial flow reactor is shown in Figure 6.14. The outer radius of the catalyst bed is R_o, and the inner radius of the catalyst bed is R_i. For an element dR at $(R, R+dR)$, the inlet material stream is M_i (with concentration C_i), and the temperature is T. The outlet material stream is $M_i + dM_i$, and the temperature is $T + dT$. Combining mass and enthalpy balances for the element with kinetic and thermodynamic equations, the concentration and temperature distributions along the reactor radius can be derived as follows:

(1) Aromatics generation from reaction 1 (dehydrogenation):

$$\frac{dM_{\mathrm{A}}}{dR} = \frac{2\pi RL\rho_b \cdot r_1}{(P_{\mathrm{H}}\theta)^2} \tag{6.49}$$

(2) Naphthene disappearance from reactions 1, 2, and 4:

$$\frac{dM_{\mathrm{N}}}{dR} = \frac{2\pi RL\rho_b \cdot (-r_1 - r_2 - r_4)}{\phi} \tag{6.50}$$

(3) Paraffin generation from reaction 2 and disappearance from reaction 3:

$$\frac{dM_{\mathrm{P}}}{dR} = \frac{2\pi RL\rho_b \cdot (r_2 - r_3)}{\phi} \tag{6.51}$$

(4) Hydrogen use and generation from reactions:

$$\frac{dM_{\mathrm{H}}}{dR} = \frac{2\pi RL\rho_b \cdot \left(3r_1 - r_2 - \frac{n-3}{3} r_3 - \frac{n}{3} r_4\right)}{\phi} \tag{6.52}$$

Temperature gradient in the radial direction:

$$\frac{dT}{dR}=\frac{2\pi RL\rho_b}{\sum_i F_i Cp_i}\cdot\left[3r_1\left(-\Delta H_1\right)+r_2\left(-\Delta H_2\right)+\frac{n-3}{3}r_3\left(-\Delta H_3\right)+\frac{n}{3}r_4\left(-\Delta H_4\right)\right] \tag{6.53}$$

(5) Catalyst metal function:

$$\theta = 1 + K_{\mathrm{MCH1}}P_{\mathrm{MCH}} + K_{\mathrm{MCH2}}P_{\mathrm{MCH}}/P_{\mathrm{H}}^2 \tag{6.54}$$

(6) Catalyst acid function:

$$\phi = P_{\mathrm{H}} + K_{\mathrm{C_{6-}}}P_{\mathrm{C_{6-}}} + K_{\mathrm{P_7}}P_{\mathrm{P_7}} + K_{\mathrm{N_7}}P_{\mathrm{N_7}} + K_{\mathrm{T}}P_{\mathrm{T}} \tag{6.55}$$

Equations 6.49–6.52 describe changing rates of aromatics, naphthenes, paraffins, and hydrogen along the bed layer radius. Temperature distribution for the adiabatic fixed bed radial flow reactor is represented by Equation 6.53. F is feedstock, kmol/h; M is flow rate, kmol/h; L is height of catalyst bed in reactor, m; ρ_b is bulk density of catalyst, kg/m^3; n is the number of carbon atoms. ϑ represents the catalyst metal function, while ϕ the acid function.

The model treats the process as a bifunctional catalyst system containing a metal function (ϑ) and an acid function (ϕ) based on the surface reaction chemistry and kinetics. The metal function performs via platinum sites and promotes only the dehydrogenation (desired), which converts naphthene to aromatics. The acid function is produced via chloride on the alumina base, which contributes to hydrocracking reactions in making light end products such as LPG and fuel gas (undesired). The rate parameters of dehydrogenation on the metal function are very large compared with those of cracking reactions on the acid function, a situation preferred for high selectivity of aromatics in catalytic reforming.

What can we learn from this integrated model? If it is for process development, the integrated model can evaluate different types of reactors by changing the reactor equations with the purpose to evaluate the product yield, reactor volume, and reactor system costs as well as heat transfer methods required. If it is for operation optimization purposes, the integrated model can be used to evaluate the effects of different operating conditions on the product yield. For the former application, process simulation software should be used as it embeds all the first principle models for reaction kinetics, reactor types, and related equipment models including reactors, separators, heat exchangers and rotating machines, treating systems, and so on. For the latter, either simulation software or real-time modeling software with model simplifications should be used for fast speed, albeit at the expense of accuracy. Recent experience with real-time optimization (RTO) indicates that molecular-based kinetic models can reveal new insights for directing molecules to their desired destiny that can unlock hidden value (Zhu and Hu, 2000; Hu et al., 2002).

A complete process model must include not only the reaction system but also the separation system that facilitates separation of desired products from byproducts.

6.5 CATALYST DEACTIVATION ROOT CAUSES AND MODELING

Catalyst deactivation is that the catalyst will lose activity and selectivity over time. The plant has to regenerate or replace the catalysts in a reactor when the deactivation reaches a certain level. It is important to understand the deactivation mechanism.

There are three principal causes of deactivation:

(1) The gradual buildup of reaction byproducts, such as coking tar, that obstruct the active pores and eventually lead to the extinction of the catalyst.

(2) Catalyst poisoning. Poisoning is provoked by substances contained in the feed being processed, even when they exist in small, and sometimes undetectable, concentrations.

During their passage through the reactor, these substances are adsorbed on active sites, and the cumulative effect of this preferential adsorption eventually leads to a significant reduction of the activity of the catalytic mass.

(3) The slow structural transformation of the catalytic species or the agglomeration and growth of micro-crystallites leading to a gradual decrease in the number of active sites per unit surface or—which amounts to the same thing—to a decrease of the active surface area. This is possibly the most harmful of the three causes of deactivation.

To overcome catalyst poisoning, the substance responsible for poisoning must be identified and the feed treated to eliminate the unwanted compounds, or a catalyst better resistant to poisoning can be used, or a portion of the catalytic bed located ahead of the actual bed can be sacrificed.

Structural transformation must be attacked at the root, for it is during the development of catalysts that the means to prevent transformation of the crystalline phases or recrystallization are tested. When designing reactors, engineers must obtain the maximum benefit from commercially available catalysts, whatever their advantages and disadvantages. Together with the equations describing the chemical kinetics of the process, at least one deactivation equation must be taken into consideration, regardless of the ultimate causes of deactivation.

Deactivation of the catalyst can be described via the use of kinetic equations, and the simplest and most general version is proposed by Levenspiel (1999). It is based on empirical laws without concern for any of the underlying mechanisms. It turns out that deactivation processes, which most frequently cannot be analyzed because of their complexity, can generally be represented by relatively simple kinetic equations.

The initial rate of a process can be expressed by

$$r_A = kC_A^n \tag{6.56}$$

where the rate at time t, all other conditions being equal, can be written as

$$r_A = \alpha \cdot kC_A^n \tag{6.57}$$

where α is defined as a relative activity.

Different rate laws may be used, for example,

$$r_{\mathrm{A}} = \frac{kC_{\mathrm{A}}^{n}}{1 + k_{\mathrm{d}}t} \quad \text{(hyperbolic form)} \tag{6.58}$$

or

$$r_{\mathrm{A}} = kC_{\mathrm{A}}^{n} e^{-k_{\mathrm{d}}t} \quad \text{(exponential form)} \tag{6.59}$$

or

$$r_{\mathrm{A}} = kC_{\mathrm{A}}^{n} \left(1 + k_{\mathrm{d}}t\right)^{p} \quad \text{(power law form; } p < 0\text{)} \tag{6.60}$$

The variation of α with time (*dα/dt*) is a function of concentration, temperature, and the activity α at time *t*, according to a general expression:

$$-\frac{d\alpha}{dt} = k_{\mathrm{d}} \sum_{i} C_i^m \alpha^d \tag{6.61}$$

where k_{d} is the deactivation rate constant and $\bullet\ C_i^m$ cumulative concentration of one or more compounds.

The most frequently found values for *d*, *m*, and the number of compounds included in $\bullet\ C_i^m$ are:

- *d* comprised between 1 and 3, and generally 1.
- *m* comprised between 0 and 2, and generally 1.
- The number of compounds included in $\bullet\ C_i^m$ is between 0 and 2 and is generally 1.

6.6 DISTILLATION MODELING

Modeling of distillation is discussed here because it is the most common separation technology used in chemical processes. If readers are interested to know other separation technologies, please refer to Seader et al. (2011) and Zhu et al. (2019).

Distillation converts multicomponent streams into desirable products based on the difference in boiling points between key components. Figure 6.15 shows the configuration of a typical distillation column.

There are two modeling approaches for distillation columns, namely, rate based and equilibrium-stage based. The rate-based approach is based on rigorous heat and mass transfer rates between vapor and liquid phases and equilibrium between the vapor and liquid phases occurring at the vapor–liquid interface. This approach can be highly accurate and account for different operation conditions including physical

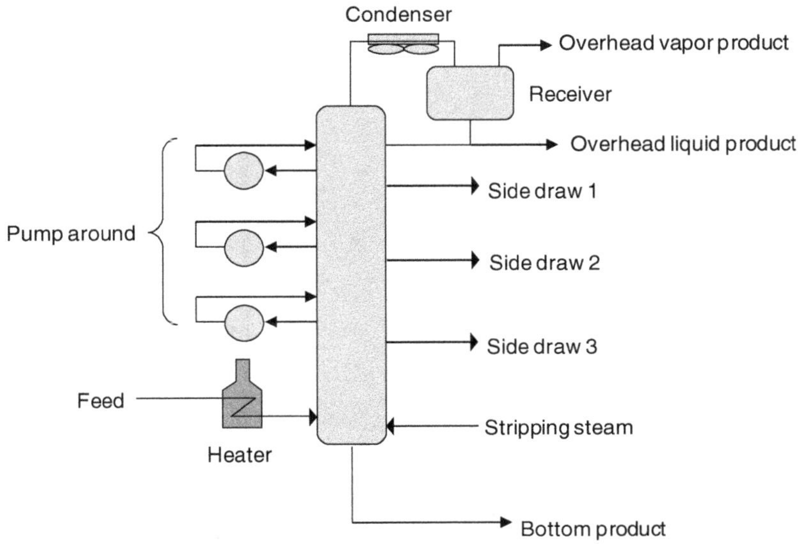

FIGURE 6.15. A distillation column.

layout of the column. However, this approach requires many parameters to make a reasonable model.

The more commonly applied approach is the equilibrium-stage approach, which is based on MESH modeling where M = material balances, E = equilibrium relationships between the vapor and liquid leaving the stage, S = summation of mole fractions equal to unity, and H = heat or enthalpy balances. MESH equations are provided below, and Figure 6.16 shows an equilibrium stage.

Based on Figure 6.16, mass balance for stage j can be expressed:

$$L_{j-1}x_{i,j-1}+V_{j+1}y_{i,j+1}+F_jz_{i,j}-\left(L_j+U_j\right)x_{i,j}-\left(V_j+W_j\right)y_{i,j}=0 \quad (6.62)$$

Equilibrium between vapor and liquid compositions leaving the stage is related by a K-value:

$$y_{i,j}=K_{i,j}x_{i,j} \quad (6.63)$$

Equilibrium between vapor and liquid compositions using Murphree efficiency η:

$$E_{i,j}=\left(y_{i,j}-y_{i,j+1}\right)-\eta_j\left(K_{i,j}x_{i,j}-y_{i,j+1}\right)=0 \quad (6.64)$$

Summation of mole fractions equal to unity:

$$\sum_{i=1}^{N}x_i-\sum_{i=1}^{N}y_i=0 \quad (6.65)$$

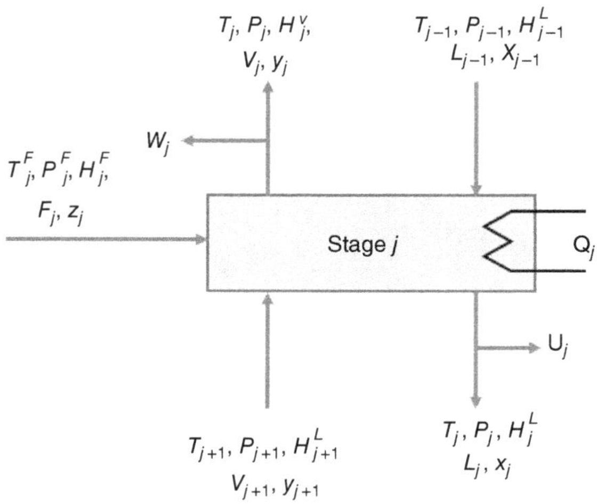

FIGURE 6.16. Equilibrium-stage scheme.

Heat or enthalpy balances:

$$L_{j-1}H_{L,j-1} + V_{j+1}H_{V,j+1} + F_j H_{F,j} - \left(L_j + U_j\right)H_{L,j} - \left(V_j + W_j\right)H_{V,j} - Q_j = 0 \tag{6.66}$$

K value, liquid enthalpy H_L, vapor enthalpy H_V, and feed enthalpy H_F are expressed as a function of temperature, pressure, and liquid and vapor mole fractions leaving stage j.

Equation 6.64 involves the stage efficiency factor η, so-called Murphree tray efficiency, to deal with nonideal stages, which is defined as

$$\eta = \frac{y_j - y_{j+1}}{y_j^* - y_{j+1}} \quad \text{or} \quad \eta = \frac{x_j - x_{j+1}}{x_j^* - x_{j+1}} \tag{6.67}$$

where x_j and x_{j+1} represent mole fraction of a given component in the liquid leaving actual stages j and j+1, respectively. y_j and y_{j+1} refer to the vapor mole fraction of a given component in the vapor leaving actual stages j and j+1, respectively. y_j^* and x_j^* refer to the vapor and liquid mole fractions of a given component leaving theoretical stages j, respectively.

6.7 PROCESS SYSTEM MODELING AND SIMULATION

After choosing a kinetic model, reactor design model, and separation model, we are ready to generate a truly integrated model for the process unit of interest. Let us consider the naphtha reforming unit as an example for illustration. Our modeling objective is to evaluate the cause-and-effect relationships between process conditions and product yields and thus enable prediction of the effects of changing process

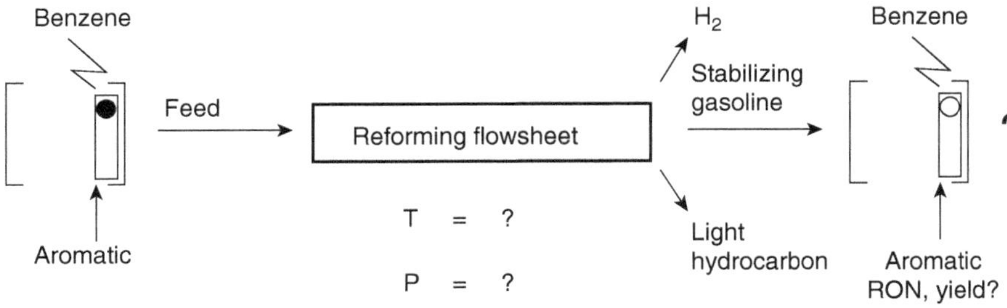

FIGURE 6.17. Process modeling for the naphtha reforming process.

conditions (Figure 6.17). The modeling of the overall process starts from providing feed composition data to system modeling, reaction model tuning, and finally model results verification. Let us discuss them one by one below.

6.7.1 Feed Composition Data

Accurate feed composition data must be provided to the kinetic model. Without an accurate and up-to-date feed composition, the kinetic model fails to make good predictions of product yields.

There are several methods commonly used for providing feed composition data. One method is to estimate composition lumps based on limited composition information, bulk properties such as specific gravity and viscosity, and distillation curves (Taskar and Riggs, 1997). With this method, the base composition may be adjusted based on measured bulk properties and converted into desired composition lumps for the kinetic model to use. This step is quite simple as the kinetic model chosen includes the total PNA content. A key assumption in this method is that each class of molecules, that is, paraffins, naphthenes, and aromatics, is statistically distributed around a certain mean value. This assumption is good for the naphtha feed in the reformer for this example because a significant portion (80+wt%) lies in the C6–C9 range.

Alternatively, if detailed feed composition information with a rich database is available, the method will select the appropriate set of composition to use based on the measured bulk properties versus the ones in the database (Hu et al., 2002; Klein, 2006).

6.7.2 Process System

Figure 6.18 gives an overview of the key components for a four-heater reformer. This overview applies to both semi-regeneration fixed bed and continuous catalytic reformers (CCR). The task is to select either lumping pathway or MKM-based approaches for the kinetic models as well as the reactor type to use (e.g., PFR or radial flow) for the naphtha reforming. Then we need to model interstage heaters, process separators and compressors, and other supporting equipment.

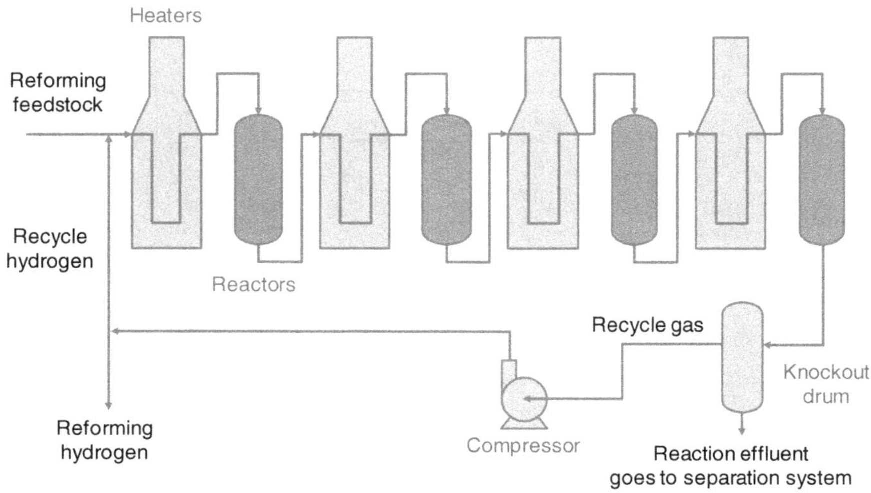

FIGURE 6.18. Block flow diagram of a naphtha reforming unit.

When implementing the process scheme in simulation software, one needs to select the appropriate thermodynamic package and then select the composition management method. If the lumping approach is applied, the feed is broken into many lumps for each chemical group, and the composition management in the simulation software relies on a base composition profile that is adjusted based on the changes in measured properties. Typically, these measured properties are the distillation curves and PNA contents.

With simulation software, all the models are embedded. By selecting the desired models, the reaction models take the feed lumps and then convert to product composition in the reaction effluent, which are separated into desired products in separators modeled based on *K*-values (equation 6.63) as discussed above.

6.7.3 Model Calibration

Incoming operating data may provide a misleading picture of current operation as they may include process transience or measurement errors. If wrong data is used as the basis to tune the reactor and separation models, the model prediction could be misleading, which results in the axiom "garbage in, garbage out."

To make sure the process models are tuned properly to mimic operation, the overall calibration (Raftery et al., 2019) is automatically conducted in two sequential steps. The first step is selecting steady-state data, followed by a combination of gross error detection, reaction model tuning, and parameter estimation in the second step.

The first step is data selection, where we perform outlier detection, steady-state detection, and necessary balance and yield calculations and apply our data quality metric to the data. The second step is performing the parameter estimation, data reconciliation, and gross error detection simultaneously. This second step is generic and can be applied to separation or reactor models.

In particular, kinetic model tuning is the determination of a set of best parameters (α's in Equations 6.38–6.41) that minimizes the difference between the kinetic model and actual process performance. These parameters are determined as a solution to the sum of least squares error minimization that uses well-controlled pilot plant data or historical commercial operating data as a basis for comparison. However, tuning strategies relying on raw input data are subject to both random and systematic errors that can distort the resulting parameters, that is, poor-quality input data leads to poor tuning parameters.

As a rule of thumb for required accuracy, the product yields predicted from the properly turned model should be within 1% of actual plant yields. If not, likely the material balance and hydrogen balance are not closed sufficiently. The reactor outlet temperature predicted should be less than 3 °C from measured values.

Raftery et al. (2019) reported a robust model tuning strategy that has the following advantages over the currently available methods:

(a) Integrates reaction chemistry and detailed process topology insights to more accurately represent overall process performance;
(b) Capitalizes on the synergy between parameter estimation and data cleansing techniques such as steady-state detection, data reconciliation, and gross error detection to address data quality issues;
(c) Utilizes a newly developed software component for parallel computing to synergize process simulation and optimization capabilities and the large database of available data.

6.7.4 Simulation Methods

Once process models are developed, these models can be applied via simulation to represent a process design configuration. UniSim, HYSYS, and Aspen Plus are simulation software that are widely used for simulating chemical processes. There are two kinds of process simulation methods, namely, modular-based simulator (sequential mode) and equation-oriented (EO) simulator (simultaneous mode).

With the EO approach, all the process related to equations representing the process are assembled to form the equation system, which is solved simultaneously. If it is for single-point solution, the equation system is solved via matrix operation to determine the response (e.g., conversion) for the given value(s) of parameter(s) (e.g., reaction temperature). If it is for optimization, the best response in the range of parameters is selected based on the objective function provided. The run time is very fast (in seconds or minutes at most).

With the simulation (or modular) approach, equations are organized based on modules, with each module representing a certain subsystem (e.g., reaction system) and the overall process is described by a set of modules that are arranged in a certain sequence. The interconnections for the modules are set up in such a manner that data (streams, flow rates, conditions, compositions, properties, etc.) can be transferred from one module to another. The master routine calls the modules in the proper order, transmits information from a library of calculational subroutines, and picks information on physical properties from a thermodynamic package. The module calculations are executed in sequence and may be repeated based on recycles. For example, the module representing a reaction system is solved first, and then the reaction effluent consisting of composition and properties becomes the input for the module representing downstream separators, which is solved afterward with some recycles back to reactors, which are solved again. The final solution is obtained after several iterations when recycles are converged.

This sequential simulation approach is popular for use in process design due to the natural design procedure and the accuracy requirement, which is accomplished by use of accurate data concerning thermodynamics, flow-dynamics, characterization, and physical/chemical properties. Extensive calculations and longer convergence times than the EO mode are required.

In contrast, EO-based simulator is more common for use in online modeling and optimization with model simplification because it allows for faster solution and more flexibility in using more sophisticated optimization solution strategies.

With a simulation software, all the required fundamental models are embedded. Thus, it is much more efficient and effective to construct a simulation case for a process than to develop mathematical equations from scratch. That is the reason why the simulation approach is more popular with engineers versus open-equation software such as GAMS, MATLAB, etc., which are good tools as an aid for research and development.

6.8 SEPARATION TECHNOLOGY OVERVIEW

For completeness, the major separation technologies are reviewed briefly here, while the details for modeling these technologies can be seen in Seader et al. (2011).

6.8.1 Distillation

Although distillation was discussed, it is briefly mentioned here for the completeness of this section. Distillation is the separation of a mixture into its components on the basis of differences in relative volatility. Distillation requires a source of liquid to the top stage and a source of vapor to the bottom stage. The top liquid may be obtained by condensing some of the vapor rising from the top stage.

There are some variations to distillation, which include extractive distillation, absorbers, and strippers. Extractive distillation units add extraction material to improve the relative volatility of the key components. Absorbers may add an additional absorption liquid to remove desired relative heavy components from a vapor feed stream. Strippers may add a (non-feed) stripping vapor to remove desired light components from a liquid feed stream. In each of these cases, some of the added material ends up in each of the product streams. Subsequently additional separation columns are required to recover and reuse this material.

6.8.2 Liquid–Liquid Extraction

Liquid–liquid extraction involves a descending heavy liquid stream is repeatedly contacting with a rising light liquid stream. This is very similar to distillation but involves the use of two liquid phases. A heavy liquid phase is fed onto the top stage, and a light liquid phase is fed into the bottom stage. Both flooding calculations and efficiency calculations are much more difficult for extraction as compared with distillation.

6.8.3 Crystallization

Crystallization is a chemical solid–liquid separation technique in which mass transfer of a solute from the liquid solution to a pure solid crystalline phase occurs. Crystallization occurs in two major steps. The first is nucleation, the appearance of a crystalline phase from either a supercooled liquid or a supersaturated solvent. The second step is known as crystal growth, which is the increase in the size of particles leading to a crystal state. An important feature of this step is that loose particles form layers at the crystal's surface lodge themselves into open inconsistencies such as pores, cracks, etc. Attributes of the resulting crystal depend largely on factors such as temperature, air pressure, and, in the case of liquid crystals, time of fluid evaporation.

6.8.4 Adsorption

This method uses particles to preferentially adsorb some components from a feed stream. A bed of solid particles is used for this purpose. Non-regenerated adsorbers can be used to adsorb very small amounts of material out of a feed stream. The adsorbent material is replaced rather than regenerated.

Regenerated adsorbers are much more common than non-regenerated adsorbers. The regeneration is sometimes done by lowering the pressure or sweeping a regenerant material over the adsorbent. The regenerant is frequently heated to assist with regeneration. It is common that multiple bed configurations will be used so one or more beds can be adsorbing while one of the beds is being regenerated. In cases where large amounts of high purity products are needed, simulated moving bed (SMB) design can be used.

6.8.5 Membrane

Gas separation membranes work according to the principle of selective permeation through the membrane surface. The permeation rate of each gas depends on its solubility in the membrane material and on the diffusion rate of the gas.

Gases with high solubility and small molecules pass through the membrane very quickly. Less soluble gases with larger molecules take more time to permeate the membrane. In addition, different membrane materials separate differently. The driving force needed to separate gases is achieved by means of a partial pressure gradient.

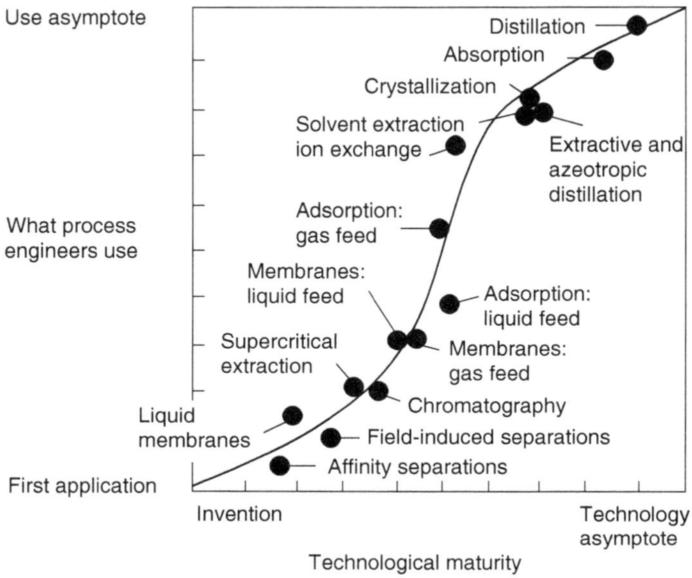

FIGURE 6.19. Technological maturity of separation methods.

The driving force for a gas to permeate through a membrane is the partial pressure difference—in other words, the partial gradient between the inside of the hollow fiber (retentate side) and the outside of the hollow fiber (permeate side). The greater the difference, the more gas permeates through the membrane. For example, if carbon dioxide and methane are being separated, as is the case with biogas upgrading, carbon dioxide permeates through the membrane very quickly, while the methane tends to be held back.

Figure 6.19 shows the comparison based on the technology maturity in relation to commercial applications. Of all factors influencing the decision to choose one process in preference to another, operation reliability is the most important factor. This is because the constructed plant must work properly to produce an acceptable product that can be sold at a profit. The economic consequences of a design failure are too dramatic to accept any potentially malfunctioning separation process design. It is therefore not surprising that those separation operations that are well understood can be readily designed from first principles and can be demonstrated prior to installation are favored in an industrial environment.

REFERENCES

Ancheyta-Juarez J, Aguilar-Rodriguez E (1994) New model accurately predicts reformate composition. Oil Gas J. **92** (5), 93–95.

Farrauto RJ, Bartholomew CH (1997), Fundamentals of Industrial Catalytic Processes, Chapman & Hall.

Froment GF (2014), On fundamental kinetic equations for chemical reactions and processes, Curr. Opin. Chem. Eng. **5**, 1–6.

Hu SY, Towler G, Zhu XX (2002), Combine molecular modeling with optimization to stretch refinery operation, Ind. Eng. Chem. Res. **41**, 825–841.

Klein MT (2006), Molecular Modeling in Heavy Hydrocarbon Conversions, CRC Press, Boca Raton, FL.

Levenspiel O (1999), Chemical Reaction Engineering, 3rd ed., John Wiley & Sons, Inc., New York.

Liang KM, Guo HY, Pan SW (2005), A study on naphtha catalytic reforming reactor simulation and analysis, J. Zhejiang Univ. Sci. B **6** (6), 590–596.

Quann RJ, Jaffe SB (1996), Building useful models of complex reaction system in petroleum refining, Chem. Eng. Sci. **51** (10), 1615.

Raftery J, Keels J, Bowman D, Ulas-Acikgoz S, Zhu FX (2019), Developing data cleansing methodologies for honeywell connected plant solutions, Honeywell Mag. Exponent **11** (2).

Rase HF (1977), Chemical Reactor Design for Process Plants, Vol. **2**. John Wiley & Sons, Inc.

Seader JD, Henley EJ, Roper DK (2011), Separation Process Principles, 3rd ed., John Wiley & Sons, Inc.

Smith RB (1959), Kinetic analysis of naphtha reforming with platinum catalyst, Chem. Eng. Prog. **55**, 76–80.

Stangeland BE (1974), Kinetic model for prediction of hydrocracker yields, Ind. Eng. Chem. Process Des. Dev. **13**, 72.

Taskar UM, Riggs JB (1997), Modeling and optimization of a semiregenerative catalytic naphtha reformer, AICHE J. **43**, 740–753.

Weekman VW, Nace DM (1970), Kinetics and catalytic cracking selectivity in fixed, moving, and fluid-bed reactors, AIChE J. **16**, 397.

Wei W, Bennett CA, Tanaka R, Hou G, Klein Jr MT, Klein MT (2008), Computer aided kinetic modeling with KMT and KME, Fuel Process. Technol. **89**, 350–363.

Zhu XX, Hu SY (2000), Molecular modeling for overall refinery optimization. AICHE Annual Meeting, Los Angeles, CA.

Zhu XX, Johnson JA, Ablin DW, Ernst GA (2019), Efficient Petrochemical Processes, John Wiley & Sons, Inc.

7

LINEAR OPTIMIZATION MODELING

7.1 INTRODUCTION

Optimization models are mathematical models, which are different from statistic models (see Chapter 5) and physical models (Chapter 6). Optimization models define objective functions to optimize independent variables (decision variables) to provide better solutions than the initial values of these variables. In this manner, we want to explore the organized changes that we can act on and improve. With physical models, the intent is to simulate reality (product yields, properties, energy use, capital costs, etc.) based on a set of data. In other words, variables can be changed but cannot be optimized because there is no optimization mechanism embedded. Therefore, optimization models are different from physical models in their use of the objective function connected with independent or decision variables. However, in most cases, physical models provide the basis for optimization models, and the level of detail in the physical models depends on the application needs. These aspects will become evident later in this chapter.

Thus, the key for optimization model building is to select the proper decision variables, which are closely related to the problem definition, while the level of details required is determined based on the objective of the task at hand. The general structure of an optimization model consists of two parts, namely, an objective function (the intended purpose, e.g., profit) and constraints (the limits for the variables, e.g., reactor temperature). Optimizing the objective function via changing decision variables gives the optimal solution, which is judged based on the criteria of optimality. The optimized solution gives the combined changes (synergy) to the decision

Digitalization and Analytics for Smart Plant Performance: Theory and Applications, First Edition. Frank (Xin X.) Zhu.

variables simultaneously instead of changing one variable at a time sequentially. The former produces a step change instead of the incremental change by the latter.

Modern manufacturing faces many choices in selection of feedstocks from many sources and generation of products to meet market needs. It is a complex economic optimization problem as it involves varying market conditions as well as consideration of various processes that are interrelated. These processes make different intermediate products, which are blended to make final products that meet market specifications. Optimization is applied across different levels of operation decision making to achieve safe, economic, and reliable operation. Mathematical optimization has been applied to a wide variety of applications, not only in oil refining and petrochemicals but also in airlines, chemicals, utility sectors, and so on.

Optimization models are built for multiple purposes, which include:

- Operation optimization: Finding the optimal solution for decision variables;
- Performance prediction: Evaluating what-if scenarios and sensitivity;
- Control: Determining setpoints for regulatory controllers;
- Insights: Providing better understanding of relationships in the systems;
- Justification: Identifying evidence for supporting important decisions.

Operation optimization models usually consist of three types:

- Nonlinear programming (NLP) problems:
  - Either objective functions or constraints are nonlinear or both are nonlinear.
  - Limited use for online and closed loop due to computer execution time and model maintenance issues.
  - NLP can be converted to SLP problems and solved sequentially.
- Quadratic programming (QP) problems:
  - Quadratic objective function plus constraints.
  - Computationally efficient and rigorous algorithms are available.
- Linear programming (LP) problems:
  - Both objective function and constraints are linear.
  - Computationally efficient and rigorous algorithms are available.

Let us start with LP optimization first as it is the simplest form but also most widely applied in industry. We then will discuss more complex topics such as nonlinear and integer programming in Chapter 8.

7.2 LINEAR OPTIMIZATION FOR PLANNING

Most companies have a planning group with specialists who have both economic knowledge and mathematical optimization skills to conduct economic analysis.

They usually rely on LP methods due to their simplicity in model building and easily explained results.

LP involves a linear objective function subject to linear constraints for all the variables involved. The variables affecting the objective function are typically the raw feeds available in the market, product slates, key operating conditions, utility usage, and so on.

Let us use a simple example to walk through the steps of building a linear model.

Example 7.1

A 110,000 bbl/day refinery wishes to determine the optimal mixture of two crudes in order to make products in the most economic manner. The feed costs and product sale prices are shown in Figure 7.1, while product yields and processing costs together with production limits are given in Table 7.1.

Solution

Step 1: Identify decision variables, which are:

- x_1 = bbl/day for crude 1
- x_2 = bbl/day for crude 2

Step 2: Identify the constraints for the variables.

In this problem the constraints are the production rates for gasoline, kerosene, diesel, and fuel oil. The production rates can be calculated based on the product yields given in Table 7.1 as follows:

$$\text{Gasoline} = 0.5x_1 + 0.3x_2, \quad \text{bbl / day} \tag{7.1}$$

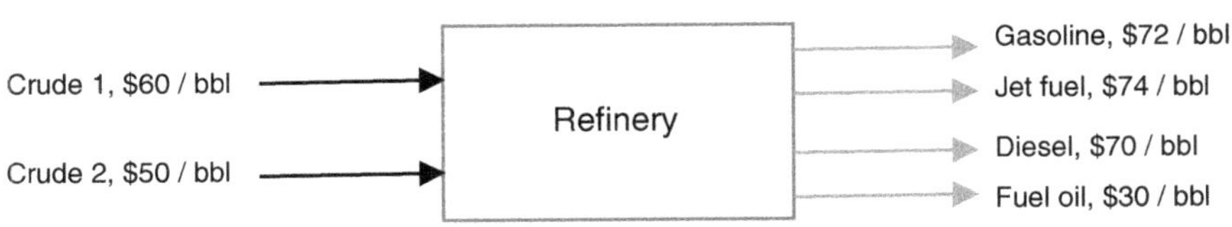

FIGURE 7.1. Refinery feeds and products schematic.

TABLE 7.1. Product Yields and Limits as Well as Operating Costs

	Volume Percent Yield, %		
	Crude 1	Crude 2	Market Demand (bbl/day)
Gasoline	50	30	40,000
Jet fuel	10	5	7,500
Diesel	30	40	39,000
Fuel oil	10	25	21,000
Process cost ($/bbl)	4.0	6.0	

$$\text{Jet fuel} = 0.1x_1 + .05x_2, \quad \text{bbl / day} \tag{7.2}$$

$$\text{Diesel} = 0.3x_1 + 0.4x_2, \quad \text{bbl / day} \tag{7.3}$$

$$\text{Fuel oil} = 0.1x_1 + 0.25x_2, \quad \text{bbl / day} \tag{7.4}$$

Thus, the maximum allowable production based on Table 7.1 can be expressed as

$$0.5x_1 + 0.3x_2 \le 40,000 \quad (\text{gasoline}) \tag{7.5}$$

$$0.1x_1 + .05x_2 \le 7,500 \quad (\text{jet fuel}) \tag{7.6}$$

$$0.3x_1 + 0.4x_2 \le 39,000 \ (\text{diesel}) \tag{7.7}$$

$$0.1x_1 + 0.25x_2 \le 21,000 \quad (\text{fuel oil}) \tag{7.8}$$

Step 3: Define the objective function. The objective is to maximize the total profit, which can be expressed as

$$\text{Maximize profit} = \text{revenue} - \text{feed costs} - \text{operating cost} \tag{7.9}$$

To express the objective function mathematically, let us address revenue calculation first as

$$\text{Gasoline} = 72 \cdot (0.5x_1 + 0.3x_2), \quad \$/\text{day} \tag{7.10}$$

$$\text{Jet fuel} = 74 \cdot (0.1x_1 + .05x_2), \quad \$/\text{day} \tag{7.11}$$

$$\text{Diesel} = 70 \cdot (0.3x_1 + 0.4x_2), \quad \$/\text{day} \tag{7.12}$$

$$\text{Fuel oil} = 30 \cdot (0.1x_1 + 0.25x_2), \quad \$/\text{day} \tag{7.13}$$

Combining Equations 7.10–7.13 gives

$$\text{Revenue} = 67.4x_1 + 60.8x_2 \tag{7.14}$$

And raw materials and processing costs can be derived as

$$\text{Feed costs} = 60x_1 + 50x_2 \tag{7.15}$$

$$\text{Operating costs} = 4x_1 + 6x_2 \tag{7.16}$$

Based on Equation 7.9, the objective function as profit can be expressed as

$$\text{Profit} = (67.4x_1 + 60.8x_2) - (60x_1 + 50x_2) - (4x_1 + 6x_2) \tag{7.17}$$

Simplifying Equation 7.17 yields

$$\text{Profit} = 3.4x_1 + 4.8x_2 \tag{7.18}$$

The goal of the LP model for the refinery optimization problem is to find optimal values for two decision variables, x_1 and x_2, which can maximize Z *or* profit. The overall LP model can be expressed as below:

$$\begin{aligned} \text{Maximize} \quad & Z = 3.4x_1 + 4.8x_2 \\ \text{Subject to:} \quad & 0.5x_1 + 0.3x_2 \le 40,000 \\ & 0.1x_1 + 0.05x_2 \le 7,500 \\ & 0.3x_1 + 0.4x_2 \le 39,000 \\ & 0.1x_1 + 0.25x_2 \le 21,000 \\ & x_1 + x_2 \le 110,000 \\ & x_1 \ge 0, \quad x_2 \ge 0 \end{aligned} \tag{7.19}$$

An LP solver can run the LP model in Equation 7.19. The optimal solutions for decision variables, x_1 (crude 1 feed rate) and x_2 (crude 2 feed rate), can be generated as

$$x_1^{opt} = 38,571; \; x_2^{opt} = 68,571 \quad \text{bbl / day} \tag{7.20}$$

$$Z^{opt} \; (\text{profit}) = 3.4x_1^{opt} + 4.8x_2^{opt} = 460,286 \quad \$ / \text{day} \tag{7.21}$$

Under this optimal solution, the refinery makes 39,857 bbl/day gasoline, 7,285 bbl/day jet fuel, 39,000 bbl/day diesel, and 21,000 bbl/day fuel oil, which almost achieves the overall market demands.

To compare with nonoptimal operation, assume the refinery uses crude 2 alone. In this case, the refinery throughput will be 84,000 bbl/day, while the profit is 403,200 \$/day. The throughput is restricted by fuel oil production, which reaches the market demand at this crude 2 rate. On the other hand, if the refinery uses crude 1 alone, the throughput will be only 75,000 bbl/day, while the profit is 255,000 \$/day. The throughput is limited by jet fuel production, which reaches the market demand at this crude 1 rate. In both cases, the profit is much lower than the optimal solution above with the case of using crude 1 alone being the worst.

7.3 HOW TO DEAL WITH NONLINEAR TERMS?

The major assumption in linear optimization is that all the equations, including the objective function and relationships between variables, must be linear. This assumption is very confining and may not be true in some cases. For example, in refinery optimization, the product properties such as octane number and vapor pressure are nonlinear functions of the mole fractions of each component in the blend. Another example is that reaction temperature in hydrocracking has a nonlinear contribution to the product yield.

Therefore, we must convert nonlinear equations describing the process system into linear equations if we wish to use LP techniques. If variables are expressed in the form of square roots, squares, exponentials, bilinear, etc., these terms are nonlinear.

Linearization is quite straightforward. All we need to do is to expand a nonlinear function via Taylor series around the current operating point and neglect all terms with higher order than the first partial derivatives.

Assume that $f(x)$ is a nonlinear function and x_0 is the current operating point. Now we expand this function around feasible point x_0:

$$f(x) = f(x_0) + \left.\frac{df}{dx}\right|_{x=x_0} (x - x_0) + \left.\frac{d^2 f}{dx^2}\right|_{x=x_0} \frac{(x - x_0)^2}{2} + \cdots \tag{7.22}$$

Neglecting all but the first two terms, the linearized approximation of the nonlinear equation $f(x)$ within the range of validity is obtained as

$$f(x) \approx f(x_0) + \left.\frac{df}{dx}\right|_{x=x_0} (x - x_0) \tag{7.23}$$

In this way, we are approximating the real nonlinear function by a linear function. The procedure is illustrated in Figure 7.2.

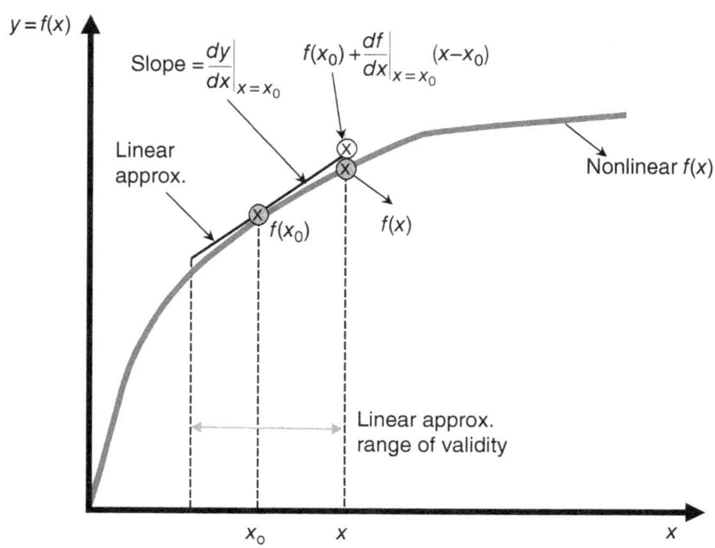

FIGURE 7.2. Linearization about the current operating point.

Example 7.2

Linearize the nonlinear function for the temperature dependence of specific reaction rate k:

$$k(T) = k_0 e^{-E/RT} \tag{7.24}$$

Expanding $k(T)$ around T_i via Taylor series as

$$k(T) = k(T_i) + \left.\frac{dk}{dT}\right|_{T=T_i} (T - T_i) + \left.\frac{d^2k}{dT^2}\right|_{T=T_i} \frac{(T - T_i)^2}{2} + \cdots \tag{7.25}$$

By neglecting the high-order terms, we obtain linearized function around T_i as

$$k(T) \approx k(T_i) + \left.\frac{dk}{dT}\right|_{T=T_i} (T - T_i).$$

$$k(T) \approx k(T_i) + k_0 \left(\frac{E}{RT_i^2} e^{-E/T_i^2} \right) (T - T_i) \tag{7.26}$$

$$k(T) \approx k(T_i) + \left[\frac{E}{RT_i^2} k(T_i) \right] (T - T_i)$$

Example 7.3

A nonlinear optimization model is expressed below and convert it to a linear model:

$$\begin{aligned} &\text{Maximize} && f(x) && \\ &\text{Subject to} && h_i(x) = 0 \quad i = 1,\ldots,m && (\text{equality constraints}) \\ & && g_j(x) \le 0 \quad j = 1,\ldots,n && (\text{inequality constraints}) \end{aligned} \tag{7.27}$$

where x is a vector of independent variables.

Assume x_k is a vector of variables at a certain feasible point. Linearization of this nonlinear function around x_k via Taylor series gives

$$\begin{aligned} &\text{Maximize} && f(x) \approx f(x_k) + (x - x_k)\nabla f(x_k) \\ &\text{Subject to} && h_i(x) \approx h_i(x_k) + (x - x_k)\nabla h_i(x_k) = 0 \quad i = 1,\ldots,m \\ & && g_j(x) \approx g_j(x_k) + (x - x_k)\nabla g_j(x_k) \le 0 \quad j = 1,\ldots,n \end{aligned} \tag{7.28}$$

The specific linearization method, so-called delta yield method, applied to LP planning for the refining and petrochemical industries is discussed below.

7.4 DELTA VECTOR AS LINEAR APPROXIMATION OF NONLINEAR YIELD MODELS

The requirement for using LP methods is that relationships between variables must be linear. This assumption is very confining. In the 1950s, an effective way to tackle nonlinearity with an LP model was found by taking advantage of the fact that in real operation, many process units feature a small operating window or a narrow range of variations for most operating variables. This reality allows us to linearize highly nonlinear processes around the small operating window. The method for the linearization of nonlinear behavior in this manner is so-called delta vector and is widely applied for oil refining operation planning using LP models. Hence, we explain this delta vector method through a refining FCC example.

The procedure starts with the base vector, which represents the product yields for the base case operation. An example of the base vector is provided in Table 7.2.

Delta yield vectors are used to cover the neighboring operating points around the base operating point. When feed properties change, the yields change as well. Delta yield vectors are developed to account for these changes within a small range, and they are generated from a one-unit incremental change (or small delta change) to the property parameter of interest. Using the change in specific gravity (SG) as an example, a linear yield correlation is given as

$$\text{yield}_i^{\text{N}} = \text{yield}_i^{\text{B}} + \Delta\text{yield}_i^{\text{SG}} \cdot \Delta\text{SG} \tag{7.29}$$

where

$\text{yield}_i^{\text{N}}$ = the new yield for product i for the feed that is different from the base feed

TABLE 7.2. Base Yield Structure

	Yield Vector
Feedstock	
Butylene	−1.0000
Isobutane	−1.2000
Product	
n-Butane	0.1271
Pentane	.0680
Alkylate	1.5110
Alky bottoms	0.1190
Tar	.0096
Utilities	
Steam, 1 lb	7.28
Power, kWh	2.45
Cooling water, m·gal	2.48
Fuel, MMBtu	0.69

TABLE 7.3. Base and Delta Vectors for a Max Gasoline FCC Unit

Product	Base, %	$\Delta \text{yield}^{\text{SG}}\|_i$
Feed	1.00	—
Dry gas	−.04	−.01
LPG	−.18	.02
Gasoline	−.40	.01
Diesel	−.30	−.01
Coke and losses	−.08	−.02

$\text{yield}_i^{\text{B}}$ = the base yield for product i for the base feed

$\Delta \text{yield}_i^{\text{SG}}$ = the delta yield for product i per unit change in SG

ΔSG = the difference of SG between the new feed and the base feed

An example of a vector yield structure (Gary et al., 2007) is shown in Table 7.3. The base yield vector from the base feed is shown in the second column in Table 7.3. The third column shows the delta yield vector of SG for different products. With one unit change of SG, dry gas will reduce by .01% (or $\Delta \text{yield}^{\text{SG}}|_{\text{DG}} = -.01\%$), LPG will increase by .02% ($\Delta \text{yield}^{\text{SG}}|_{\text{LPG}} = .02\%$), gasoline goes up by .01% ($\Delta \text{yield}^{\text{SG}}|_{\text{Gas}} = .01\%$), etc. For every feed property (SG, UOP K, concarbon, sulfur content, etc.), a delta yield vector can be generated like the third column in Table 7.3.

As yield is represented in percentage, the total yield should be equal to 100%. Thus, the yields of a process unit can be represented as

$$\sum_{i=1}^{N} \text{yield}_i = 1.0 \ (\text{normalized feed}) \tag{7.30}$$

Equation 7.30 can be demonstrated from the second column in Table 7.3. Note the feed yield features a positive sign, while the product base yields feature negative signs. In LP for planning, the base yields are referred to as base vectors.

Mathematically, the general yield correlations to address the effects of changes in feed properties can be expressed as

$$\sum_{i=1}^{N} \text{yield}_i^{\text{B}} + \Delta \text{SG} \cdot \sum_{i=1}^{N} \Delta \text{yield}_i^{\text{SG}} + \Delta \text{Sul} \cdot \sum_{i=1}^{N} \Delta \text{yield}_i^{\text{Sul}} + \cdots = 1.0 \tag{7.31}$$

$$\text{yield}_i^{N} = \text{yield}_i^{\text{B}} + \Delta \text{yield}_i^{\text{SG}} \Delta \text{SG} + \Delta \text{yield}_i^{\text{Sul}} \Delta \text{Sul} + \cdots \tag{7.32}$$

where

$\Delta \text{yield}_i^{\text{Sul}}$ = the delta yield for product i per unit change in sulfur amount

ΔSul = the difference of sulfur between the new feed and the base feed

As can be observed, yield Equations 7.31 and 7.32 include a base yield vector and multiple delta yield vectors. The delta yields are used to adjust the base yield based on changes in feed properties and process conditions. Thus the method is called delta–base yield modeling.

In a similar manner, we can apply delta yield vectors to address the changes in process conditions such as riser temperature, pressure difference, and cat/oil ratio in FCC. Therefore, delta yields can model the changes in FCC feed properties and process conditions.

The question is: how to determine delta yields? The answer is that we can determine delta yields via FCC simulation in which the first principle FCC kinetic model is embedded. The kinetic model can accurately predict FCC yields for varying feed and process conditions, but it is a large and complex nonlinear model. We can generate delta yields based on certain feed and process conditions by making changes to properties such as SG or sulfur content as well as process conditions in the simulation.

For example, a change in Conradson carbon can be obtained via changing the amount of vacuum residue (VR) in the VGO. Simulation can predict the yield effects (delta yields) due to the change in VR blending with VGO. Delta yields for SG can be generated similarly. A delta yield can be thought of as a slope that modifies the base yield of each product due to the change in this parameter from the base feed conditions. In simulation software such as UniSim and HYSYS, the delta yields can be generated automatically and sent to an LP planning model directly. By incorporating the base–delta yield model (Equations 7.31 and 7.32) into the LP planning model, optimization can guide the feed selection and key operating parameters based on the overall objective function.

7.4.1 Incorporating Delta–Base Vectors into LP model

Basically, what has been done above is generation of the approximate first-order derivatives. Using the SG change as an example, rearranging Equation 7.29 gives

$$\text{yield}_i^N - \text{yield}_i^\text{B} = \Delta\text{yield}_i^\text{SG} \cdot \Delta\text{SG}$$

or

$$\Delta y = \Delta\text{yield}_i^\text{SG} \cdot \Delta x \tag{7.33}$$

where $\Delta y = \text{yield}_i^N - \text{yield}_i^\text{B}$ and $\Delta x = \Delta\text{SG}$.

Equation 7.33 can be rearranged to become

$$\frac{\Delta y}{\Delta x} = \Delta\text{yield}_i^\text{SG} \tag{7.34}$$

where $\Delta y/\Delta x$ represents the ratio of delta yield and delta change in a certain parameter, which is the approximation of the first-order derivatives (or Jacobian) for each individual variable (Figure 7.3).

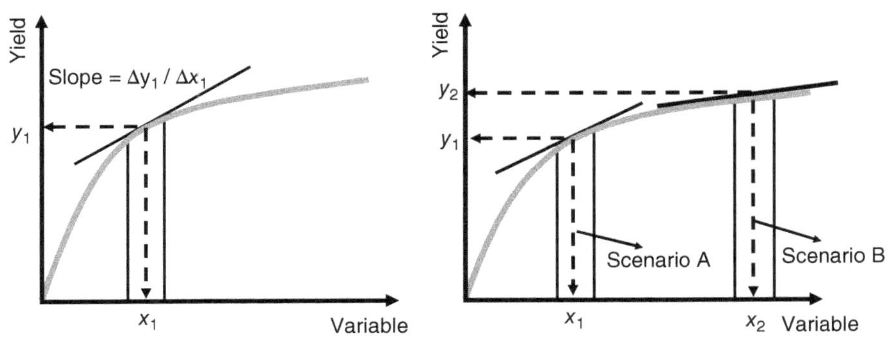

FIGURE 7.3. Delta yield—linearization of product yield around the vicinity of the base case. (a) Linearization based on delta yield. (b) Linearization for two operating modes.

Now we are ready to incorporate the delta yield concept into the LP planning model. To do so, let us consider a **y**-vector that represents the updated yields prediction based on the changes in feed and operating parameters (Δx) and a $\hat{y}$-vector that represents the base yields. Matrix $\Delta y/\Delta x$ is the Jacobian matrix representing the ratios of delta yields (Δy) and the selected parameters (Δx). Equation 7.35 defines the connection between $\hat{y}$, $\Delta y/\Delta x$, and Δx as

$$\begin{bmatrix} y_1 \\ y_2 \\ \vdots \\ y_m \end{bmatrix} = \begin{bmatrix} \hat{y}_1 \\ \hat{y}_2 \\ \vdots \\ \hat{y}_m \end{bmatrix} + \begin{bmatrix} \Delta y_1 / \Delta x_1 & \cdots & \Delta y_1 / \Delta x_m \\ \Delta y_2 / \Delta x_1 & \cdots & \Delta y_2 / \Delta x_m \\ \vdots & & \\ \Delta y_m / \Delta x_1 & \cdots & \Delta y_m / \Delta x_m \end{bmatrix} \begin{bmatrix} \Delta x_1 \\ \Delta x_2 \\ \vdots \\ \Delta x_m \end{bmatrix} \tag{7.35}$$

or writing Equation 7.35 in matrix form as

$$Y = \hat{Y} + J \cdot \Delta X \tag{7.36}$$

Thus, a nonlinear process model is converted to the linear mode as Equation 7.36. The independent variables are (Δx_1, Δx_2, . . ., Δx_m), which are incremental changes to feed properties (e.g., SG, sulfur content) and key process conditions (e.g., reaction temperature). Similarly, other process units in addition to the FCC unit can be modeled linearly in the form of Equation 7.36. Optimization of the LP model can then guide the selection of crude feeds for the overall refinery and process conditions for various process units and ultimately maximize the overall profit for the refinery.

A change of 3–5% ($=\Delta x/x_{base}$) is a reasonable range to avoid overextrapolation from the base condition using delta change (a linear approximation). Estimation of an accurate Jacobian ($\Delta y/\Delta x$) is essential to generate correct yield predictions for changes in process conditions. These measures are used to make sure that the unit representation calculated by base + delta (or shift) vectors is close enough to the representation calculated from the simulator directly for the new conditions. If the optimal solution cannot be reached in a single step of the delta vector-based linear approximation, the next step of linear approximation (new delta yield vector) based on the most recent optimization run will be triggered, and this procedure is repeated until the optimal solution is found. This updating procedure is so-called successive linear programming (SLP), which is widely applied in planning optimization. Detailed discussions for SLP will be given below.

It must be pointed out that using a rigorous process model (see Chapter 6) is the right way to obtain the delta yield matrix or Jacobian in Equation 7.36. This ensures that product yields can be predicted accurately when considering changes in key parameters.

7.4.2 Build LP Planning Model Based on Delta Vectors

The LP planning model consists of a linear objective function and many linear constraints. The process units are modeled based on the delta–base vector structure (Equation 7.36). A typical LP planning model contains around 1,000–5,000 equations, which describe process yields, utility requirements, and blending operations. Because the primary purpose of the LP planning model is to determine the overall profit through optimization of purchased feeds, product slates, and key operation conditions, the LP requires good relationships between product yields and feed and process conditions in order to make the economic decisions pertaining to feed selection as well as the conditions to process the selected feed. That is why the concept of delta–base yields is applied as shown in Equation 7.29. The connection between different processes can be made via a site mass and energy balance, while capacity and operation limits for process units can be specified in the LP model.

In mathematical terms, the objective function in the LP model is expressed as

$$\text{Maximize profit} = \text{product sales} - \text{material costs} - \text{operating costs} \quad (7.37)$$

subject to

- Process operation (feed rate, kinetics, reaction temperature, pressure, etc.);
- Site mass balance for crude selection and material distribution among processes;
- Site energy balance for utility distribution among processes (power, steam, hydrogen, etc.);
- Limitations (capacity, demand, product specifications, equipments, regulations, etc.).

By using linear delta–base yields as discussed above, nonlinear processes are approximated as linear models. LP has achieved success in applications and become widely used in the process industries since the 1950s. In fact, LP is still the state-of-the-art technology for refinery planning purposes (e.g., optimizing selections of raw materials and product slates, product blending). One thing that must be pointed out is the popularity of LP in the industries lies with the maturity of LP applications over the years and the complete and transparent value structure in terms of marginal prices that LP generates, which provides meaningful information to facilitate explanation of solutions and sensitivity analysis.

Seemingly, linear equations would be inadequate to model complex refinery processes. The delta vector method may be inapplicable for certain nonlinear terms such as nonlinear feed blending, process characterization, and so on. Modern LP software include techniques to linearize such nonlinear terms.

The objective function for refinery LP models is often defined as in Equation 7.37 as the product revenue less feedstock costs and utility costs. In some cases, capital charges and other factors are also incorporated into the objective function. LP planning model is currently used for refinery-wide applications that include:

- Crude oil evaluation;
- Resources allocation;
- Monthly and daily operation planning;
- Integration with APC applications;
- Product blending;
- Shutdown planning;
- Multi-refinery supply and distribution;
- Investment studies;
- Technology selection and capacity expansion.

7.5 SUCCESSIVE LINEAR PROGRAMING (SLP) APPROACH

In the SLP technique, the solution of the original NLP problem is obtained by solving a series of LP problems. Each LP problem is generated by linearizing the nonlinear objective and constraints at the most recent step of iteration. Figure 7.4 shows the progression from initial point to the optimal point for the nonlinear yield profile via piecewise linear approximations.

Assume that x_k is a feasible solution to the nonlinear problem in Equation 7.27. Linearization of each nonlinear function via Taylor's theorem around x_k gives Equation 7.28.

In SLP, original objective functions $f(x)$ as well as equality constraints $h(x)$ and non-equality constraints $g(x)$ are linearized at an initial point (see Equation 7.28). The resulting LP is solved to give an initial optimized solution (x_k, f_k). Then $f(x)$, $h(x)$, and $g(x)$ are linearized at the x_k, and the new x_{k+1} is generated as the reference point for the next linearization until the difference between two consecutive objective function values is within the specified tolerance, upon which the optimal solution is reached.

Recursion techniques in LP are commonly used (Baker and Lasdon, 1985), which require sequential execution of successive linear optimizations where a linear model is solved several times based on piecewise approximation of nonlinear relationships. The solution x_0 from the current LP is used as a starting point for the next LP (Figure 7.4); then the new x_{k+1} is used as the reference point for the next linearization. However, if the new x_{k+1} is infeasible (often occurs to equality constraints),

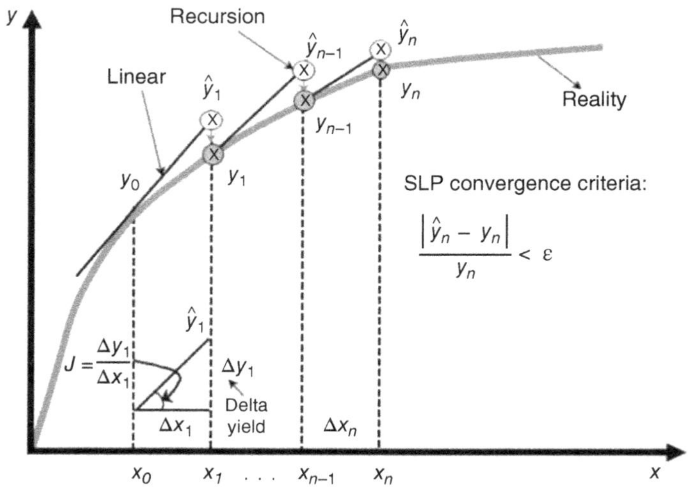

FIGURE 7.4. SLP recursion procedure.

the new x_{k+1} is not accepted, and the SLP is repeated for a narrower range of upper and lower bounds on x_{k+1} to search for a feasible solution. Several heuristics are employed to make x_{k+1} feasible. This technique can alleviate many nonlinear correlations especially from yield estimation, product blending, and property estimation.

SLP methods have the advantage of being able to handle large numbers of variables and constraints; they will achieve rapid convergence when the optimum lies at the vertex of the constraints. Typical examples are planning problems. The disadvantage is that SLP cannot handle significant nonlinear constraints, so exceedingly small search steps may be required, resulting in very slow convergence. For complex nonlinear problems, SLP not only increases computing time but also reduces the transparency of LP's value structure. Thus, successive quadratic programming (SQP) methods should alternatively be applied, which is discussed in Chapter 8.

It must be pointed out that LP models are built based on simplified correlations (e.g., delta–base yields in Equation 7.35), which are just a rough approximation for many processes that are highly nonlinear. Thus, LP models are restricted to long-term and medium-term planning and are not suitable for day-to-day and real-time operation optimization because they cannot accurately describe the nonlinear aspects of these applications.

REFERENCES

Baker TE, Lasdon LS (1985), Successive linear programming at Exxon, Manage. Sci. **31** (3), 264–274.

Gary JH, Handwerk GE, Kaiser MJ (2007), Petroleum Refining—Technology and Economics, 5th ed., Taylor & Francis.

8

NONLINEAR OPTIMIZATION MODELING

8.1 INTRODUCTION

Many phenomena behave in a nonlinear fashion. With nonlinearity, we see diminishing and increasing returns of investment based on nonlinear growth functions, and we see catalyst decay over time based on nonlinear decay functions. With nonlinear models, we have better understanding of the rate of changes—where is a flat region, where is a steep slope, and what is the shortest path leading to the deepest valley and to the highest mountain top. Nonlinear models improve our capacity to reason, predict, explain, and act in the complex world.

In general, the goal of a nonlinear optimization problem is to determine optimal values for n decision variables $x_1, x_2, \ldots, x_n$ within a feasible region in such a way as to optimize (minimize or maximize) a given objective function $f(x_1, x_2, \ldots, x_n)$ of the decision variables. The problem is called a *nonlinear programming* (NLP) *problem* if the objective function is nonlinear and/or constraints are nonlinear. In maximization, the general nonlinear program is stated as

$$\begin{aligned} &\text{Maximize} \quad f\left(x_1,\ldots,x_n\right) \\ &\text{Subject to} \quad h_i\left(x_1,\ldots,x_n\right)=0 \quad i=1,\ldots,m \\ &\qquad g_j\left(x_1,\ldots,x_n\right)\leq 0 \qquad\quad j=1,\ldots,p \end{aligned} \tag{8.1}$$

Digitalization and Analytics for Smart Plant Performance: Theory and Applications, First Edition. Frank (Xin X.) Zhu.

8.2 SUCCESSIVE QUADRATIC PROGRAMMING (SQP) APPROACH

Sequential quadratic programming (SQP) is used for solving constrained nonlinear optimization problems. Analogous to SLP with piecewise linearization, we approach NLP problems by simply replacing each nonlinear function by its local quadratic approximation (convex or concave; a second order of Taylor series) at x_k and solving the resulting series of approximate quadratic subproblems. With local quadratic approximation, the Hessians of the functions (objective function and constraints) are constant, which makes the search for optimum fast and effective.

Mathematically, quadratic programming concerns the maximization of a quadratic objective function as

$$f(x) \approx f(x_k) + \nabla f(x_k)(x - x_k) + \frac{1}{2}(x - x_k)\nabla^2 f(x_k)(x - x_k); \qquad (8.2)$$

then the subproblem becomes one of maximizing a quadratic objective function subject to linear or quadratic constraints. An optimal solution at step k becomes the initial point for the next step $k + 1$ until the true optimal solution is found, when convergence criteria based on Karush–Kuhn–Tucker (KKT) conditions are reached. It must be pointed out that while SQP cannot guarantee a global optimum, finding an arbitrary local minimum is relatively straightforward.

8.3 LOCAL VERSUS GLOBAL OPTIMUM

Geometrically, nonlinear problems behave much differently from linear ones, even for problems with linear constraints. One such distinction is with local and global optima as shown in Figure 8.1, in which Figure 8.1a shows a curve of a multimodal function of one variable with two minima while Figure 8.1b illustrates the contours of a multimodal objective function of two variables with two minima.

Example 8.1 Local Optimum

Suppose that we are to minimize $f(x)$ in this example, with $0<x<10$. The point $x = 7$ is optimal. However, within the indicated bracketed interval in Figure 8.1a, the point $x = 0$ is the best feasible point; that is, it is an optimal point in the local vicinity of $x = 0$. This example illustrates that an optimal solution in a local sense may not be optimal for the overall problem. Two types of solution must be distinguished. A global optimum is a solution to the overall optimization problem. Its objective value is as good as any other point in the feasible region. A local optimum, on the other hand, is optimal only with respect to feasible solutions close to that point. Points far removed from a local optimum play no role in its definition and may actually be preferred to the local optimum.

The concept of a local optimum is extremely important. As we shall see, most general-purpose nonlinear programming procedures are nearsighted and can do no better than determine a local optimum.

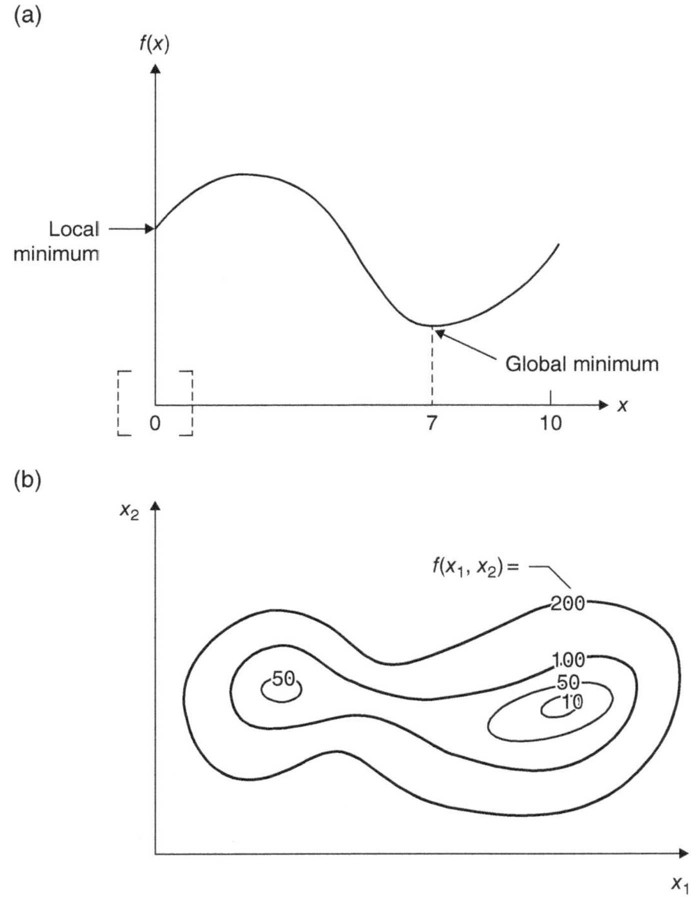

FIGURE 8.1. Local and global minima. (a) A single variable function. (b) Contours of a multimodal function.

8.3.1 Convexity and Nonconvexity

Under certain circumstances, local maxima and minima are known to be global. Whenever a function "curves upward" as in Figure 8.2a, a local minimum is also global. These functions are called *convex*. Whenever a function "curves downward" as in Figure 8.2b, a local maximum will be a global maximum. These functions are called *concave*. For this reason, we usually wish to minimize convex functions and maximize concave functions as they provide unique optimal solutions. On the other hand, if a function presents to be both convex and concave (Figure 8.2c), this kind of functions are called non-convex and non-concave or multimodal.

Because of their pivotal role in model formulation and their convenient mathematical properties, certain functional forms predominate in mathematical

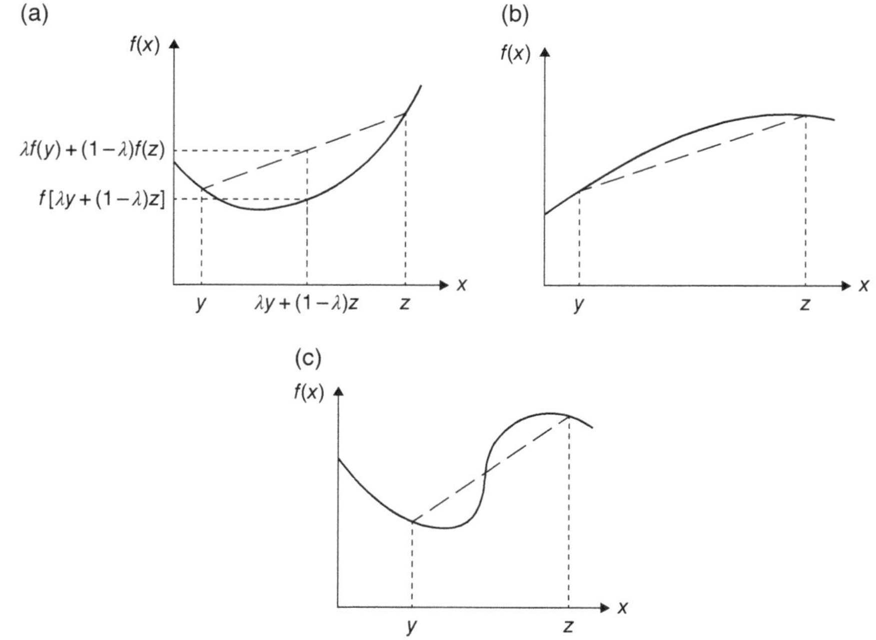

FIGURE 8.2. Convex and concave functions. (a) Convex function. (b) Concave function. (c) Nonconvex and nonconcave.

programming. Linear functions are by far the most important. Next in importance are functions, which are convex or concave. Determining the convexity or concavity will help to establish whether an optimization problem is unimodal, i.e. local optimal solution is also the global optimum. Otherwise, it is a multimodal optimization problem with multiple optima. Computation for finding the unimodal solution can be accelerated by using the appropriate algorithm.

8.3.2 Convex and Concave Functions

An essential assumption in a linear programming model for profit maximization is constant returns to scale for each activity. This assumption implies that if the level of one activity doubles, then that activity's profit contribution also doubles; if the first activity level changes from x_1 to $2x_1$, then profit increases proportionally from say \$20 to 40. In many instances, it is realistic to assume constant returns to scale over the range of the data. At other times, though, due to economies of scale, profit might increase disproportionately, for example, jump to \$45; or, due to saturation effects, profit may increase incrementally to only \$35. In the former case, marginal returns are increasing with the activity level, and we say that the profit function is *convex* (Figure 8.2a).

In the second case, marginal returns are decreasing beyond a certain activity level, and we say that the profit function is *concave* (Figure 8.2b), which is the opposite of the convex function. Of course, marginal returns may increase over parts of the data range and decrease elsewhere, giving functions that are neither convex nor concave (Figure 8.2c).

An alternative way to view convex and concave functions is to observe the linear interpolation. In the convex case, the linear interpolation overestimates its values. That is, for any points y and z, the line segment joining $f(y)$ and $f(z)$ lies above the function (see Figure 8.2a). In the concave case, the linear interpolation underestimates the function (Figure 8.2b).

In mathematics, convexity is used as a measure of the curvature or the degree of the curve of a convex or concave function.

8.3.3 Convex Space

A notion intimately related to convex and concave functions is that of a convex space. The feature of a convex space is that, whenever y and z are contained in the set, every point on the line segment joining these points is also in the set (see Figure 8.3a). A set of points C is called a convex space if the line between any two points lies in C. Mathematically, for all y and z in C, λ in the interval $0 \le \lambda \le 1$, it has

$$\lambda y + (1-\lambda) \times z \subseteq C \tag{8.3}$$

A nonconvex space does not satisfy Equation 8.3; there exist some points along the line connecting two internal points (e.g., y–z in Figure 8.3b) that appear outside of the space.

8.3.4 Convex and Nonconvex Optimization Problems

What is the implication of convexity on optimization? A convex optimization problem is a problem where all of the constraints are convex functions as shown in Figure 8.3a, and the objective function is a convex function if minimizing

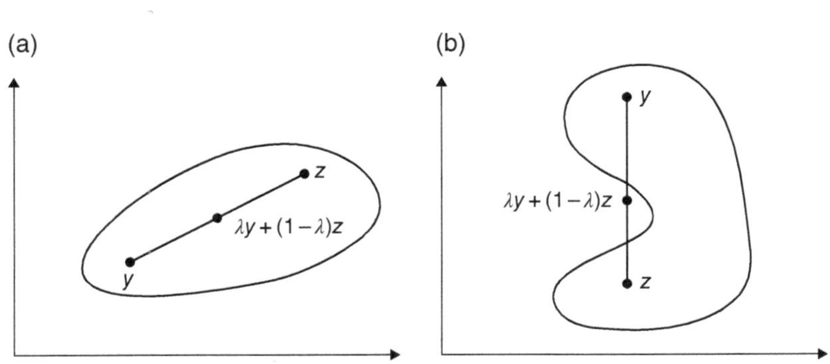

FIGURE 8.3. Convex space. (a) Convex space. (b) Nonconvex space.

(Figure 8.2a) or a concave function if maximizing (Figure 8.2b). With a convex optimization problem, there can be only one optimal solution, which is globally optimal. Convex problems can be solved efficiently up to very large sizes. A nonconvex optimization problem is any problem where the objective function is nonconvex and nonconcave (Figure 8.2c) or any of the constraints are non-convex, as pictured in Figure 8.3b. Such a problem may have multiple locally optimal solutions. No mathematical search algorithms exist that can guarantee a global optimal solution for nonconvex problems.

8.4 OPTIMALITY CONDITIONS

In mathematical optimization, the KKT conditions (Gill et al., 1981) provide the necessary and sufficient conditions for an optimal solution in NLP. The KKT approach allows inequality constraints. In general, many optimization algorithms solve the system of equations to identify potential optimal solutions first based on the KKT necessary conditions and then apply KKT sufficient conditions to verify if each is a true local optimum. For a nonconvex optimization problem, KKT conditions cannot guarantee if an optimal solution is global or not.

8.4.1 KKT Necessary Conditions

8.4.1.1 KKT necessary conditions: The first-order derivative equals zero at stationary point

If x* is an optimal solution, thus with a non-negative μ_0 in front of $\nabla f(x^*)$ and $(\mu, \lambda) \neq 0$, the KKT stationary conditions turn into

$$\mu_0 \nabla f\left(x^*\right) + \sum_{i=1}^{m} \mu_i \nabla g_i\left(x^*\right) + \sum_{j=1}^{\ell} \lambda_j \nabla h_j\left(x^*\right) = 0, \tag{8.4}$$

$$\mu_j g_i\left(x^*\right) = 0, \ i = 1, \ldots, m, \tag{8.5}$$

This optimality condition holds without constraint qualifications and is the first-order optimality necessary conditions.

8.4.2 KKT Sufficient Conditions

8.4.2.1 KKT sufficient conditions: Second-order derivative or Hessian is positive (negative) at a minimum (maximum) stationary point

The solution x^*, λ^*, μ^* found is a constrained local minimum if for the Lagrangian,

$$L(x,\lambda,\mu) = f(x) + \sum_{i=1}^{m} \mu_i g_i(x) + \sum_{j=1}^{\ell} \lambda_j h_j(x) \tag{8.6}$$

then

$$s^T \nabla_{xx}^2 L(x^*, \lambda^*, \mu^*) s \geq 0 \tag{8.7}$$

where ∇_{xx}^2 is the second derivative or Hessian and s ($\neq 0$) is a vector satisfying the following:

$$\left[\nabla_x g_i(x^*), \nabla_x h_j(x^*)\right]^T s = 0 \tag{8.8}$$

where only those active inequality constraints $g_i(x)$ corresponding to strict complementarity (i.e., where $\mu_i > 0$) are applied. The solution is a strict constrained local minimum in the case that the inequality is also strict.

8.4.3 The Significance of Second Derivative (Hessian) in KKT Conditions

In the KKT sufficient conditions above, the second derivative is used to test whether a stationary point for a function [i.e., a point where the first derivative $f'(x^*) = 0$] is a true optimal solution via the following criteria:

- If $f''(x^*) < 0$, then $f(x)$ has a local maximum at x^*.
- If $f''(x^*) > 0$, then $f(x)$ has a local minimum at x^*.
- If $f''(x^*) = 0$, then x^* is an inflection point at which non-maximum or non-minimum occurs.

Inflection point is defined based on the second derivative being equal to zero. Let us prove this concept qualitatively. Consider the inflection point in Figure 8.4. The function switches from concave at the left side of x^* where $\left(f''\left(x_-^*\right) < 0\right)$ to convex at the right side of x^* where $\left(f''\left(x_+^*\right) > 0\right)$. Thus, assuming the second derivative is continuous, the second derivative must take a value of zero at the inflection point $\left(f''\left(x^*\right) = 0\right)$.

8.5 NONLINEAR PROCESS OPTIMIZATION MODEL

Let us consider a process optimization problem based on a reaction system. Most reaction models are highly nonlinear. Here, a naphtha reforming reaction is used, for example.

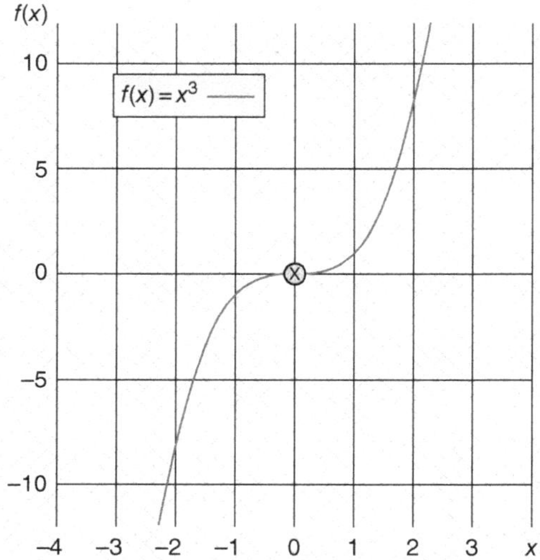

FIGURE 8.4. The highlighted point is an inflection point ($f''(x^*) = 0$) at which concave appears on the left side $\left(f''\left(x_-^*\right) < 0\right)$ and convex on the right side $\left(f''\left(x_+^*\right) > 0\right)$.

8.5.1 Reaction Kinetic Modeling

Basically, a kinetic model is a set of rate expressions. To construct a kinetic model, first reaction families must be generated, and then kinetic parameters must be determined for each reaction type by fitting experimental data. The objective of kinetic modeling is to predict product yield when feed composition and process conditions change.

Example 8.2 For a naphtha reforming reaction, four dominant idealized reactions (Smith, 1959) were identified, which are expressed as follows:

$$\text{Reaction 1 } (r_1): \text{Cycloalkane} \rightleftarrows \text{Aromatics } (A) + 3H_2 \text{ } (H) \tag{8.9}$$

$$\text{Reaction 2 } (r_2): \text{Cycloalkane } (E) + H_2 \rightleftarrows \text{Alkane } (P) \tag{8.10}$$

$$\text{Reaction 3 } (r_3): \text{Hydrocracking of alkane } (P) \tag{8.11}$$

$$\text{Reaction 4 } (r_4): \text{Hydrocracking of cycloalkane } (A) \tag{8.12}$$

The corresponding reaction rate equations (Rase, 1977) are developed below.

For reaction 1 (Equation 8.9) where cycloalkane is transformed into aromatics, the rate equation is

$$r_1 = k_{p_1}\left(P_E - P_A P_H^3 / K_{p_1}\right) \tag{8.13}$$

where

$$k_{p_1} = 9.87e^{(23.21 - 34,750/1.8T)}$$

For reaction 2 (Equation 8.10) where cycloalkane is transformed into alkane, the rate equation is

$$r_2 = k_{p_2}\left(P_E P_H - P_p / K_{p_2}\right) \tag{8.14}$$

where

$$k_{p_2} = 9.87e^{(35.98 - 59,600/1.8T)}$$

For reaction 3 (Equation 8.11) for hydrocracking of alkane, the rate equation is

$$r_3 = k_{p_3} P_p / P_T \tag{8.15}$$

where

$$k_{p_3} = e^{(42.97 - 62,300/1.8T)}$$

For reaction 4 (Equation 8.12) for hydrocracking of cycloalkane, the rate equation is

$$r_4 = k_{p_4} P_E / P_T \tag{8.16}$$

where

$$k_{p_4} = e^{(42.97 - 62,300/1.8T)}$$

where r_1, r_2, r_3, and r_4 are the reaction rates for the reactions 1–4 as above; k_{p_1}, k_{p_2}, k_{p_3}, and k_{p_4} are the reaction rate constants corresponding to r_1, r_2, r_3, and r_4, respectively; K_{p_1} and K_{p_2} are the equilibrium constants of cycloalkane transformed into aromatics and alkane (and vice versa), respectively; P_T is system pressure; and P_A, P_E, P_p, and P_H are component fractional pressures.

8.5.2 Process System Nonlinear Optimization

After the kinetic model (Equations 8.13–8.16) is verified by experiments and is proven to be accurate enough for yield prediction, the reaction system and the product separation system are modeled to form an overall optimization model. We must make reasonable predictions of reactant concentration under system pressure, reaction temperature changes in three reactors, K-values for the product separators, and heating duties for heaters.

An optimization model (Zhu and Hu, 2000) for the overall process can be generated as shown in the equation system (8.17) with the objective to optimize hydrocarbon fractions in the product and thus maximize profit via optimization of process conditions:

$$
\begin{aligned}
&\underset{T,P}{\text{Maximize}} \sum_{i=1}^{n_c} p_i \cdot (c_i \cdot m_i \cdot F) \\
&\text{s.t.} \quad \frac{dc_{i,j}}{dt} = r_j \text{ (Equations 8.13–8.16)} && j = 1,2,3,4 \\
&\qquad k_j = k_j^0 \exp\left[\frac{E_j}{R}\left(\frac{1}{766} - \frac{1}{T_j}\right)\right] && j = 1,2,3,4 \\
&\qquad c_{i,j}(0) = c_{i,j-1}^t && j = 1,2,3,4 \qquad (8.17) \\
&\qquad \text{Component mass balances} \\
&\qquad \text{Overall mass balance} \\
&\qquad \text{Utility balances (hydrogen, steam, power)} \\
&\qquad \text{Operating limits (rate, size, severity, etc.)}
\end{aligned}
$$

$$\text{RON} = \sum_{i=1}^{n_c} V_i \cdot \text{RON}_i$$

$$\text{BEN} = \sum_{i}^{n_c} v_i \cdot \text{BEN}_i$$

$$T^{\text{L}} \le T \le T^{\text{U}}$$

$$P^{\text{L}} \le P \le P^{\text{U}}$$

$$\text{RON}^{\text{L}} \le \text{RON} \le \text{RON}^{\text{U}}$$

$$\text{BEN}^{\text{L}} \le \text{BEN} \le \text{BEN}^{\text{U}}$$

where F is the total mass rate of the product, p_i is the price of component i, c_i is the product mole fraction for component i, m_i is the ratio of molar mass of component i and average molar mass of the product, and n^c is the number of product fractions, and k_j is the kinetic constant for reaction j as defined in Equations 8.13–8.16. The initial concentration of reactor j is the terminal concentration from the prior reactor $j-1$. RON_i and BEN_i are the research octane number and benzene content of component i, while V_i and v_i are the volume fractions of component i contributing to RON_i and BEN_i, respectively. Superscript L indicates the lower bound, and superscript U the upper bound applied to temperature (T), pressure (P), benzene content (BEN), and octane (RON).

Temperature (T) and pressure (P) are the independent variables, while mole fractions (c_i) are dependent variables. Temperature and pressure influence the reaction kinetics and the compositions, while flow rates and composition determine the profit of the process defined in the objective function. The profit could be increased by 3% from the base case as indicated in Figure 8.5. Benzene content affects gasoline RON, but a maximum limit must be observed due to environmental concerns.

As can be observed, the optimization model involves highly nonlinear terms including exponential variable T and power law variable P as well as bilinear terms of property blending for RON and BEN even though the objective function is linear. More effective nonlinear solvers (e.g., SQP) other than SLP should be applied to solve such a problem.

8.5.3 Mixed Integer Optimization

A mixed integer programming problem is a mathematical optimization in which some of the variables are restricted to be integers and others continuous variables. For example, some problems in operation, design, and scheduling involve integer variables (i.e. 0, 1, 2, 3, . . .). An important special case is a decision variable that must be either 0 or 1. Such variables are called binary integer variables and can be used to model yes/no decisions.

In many settings an optimization problem is featured as mixed integer linear programming (MILP), in which the objective function and the constraints (other than the integer constraints) are linear. In other settings, it refers to mixed integer nonlinear programming (MINLP), in which the objective function and/or the constraints (other than the integer constraints) are nonlinear. The use of integer variables greatly expands the scope of optimization problems. At the same time, introducing integer variables makes an optimization problem nonconvex and therefore far more difficult to solve.

8.6 STOCHASTIC PROGRAMMING

There are two main reasons why stochastic optimization is applied in the process industries. The first reason is that stochastic methods are intended to tackle problems featuring uncertainty and random variables. It is difficult to formulate and to solve such problems with deterministic optimization based on LP, NLP or MINLP models.

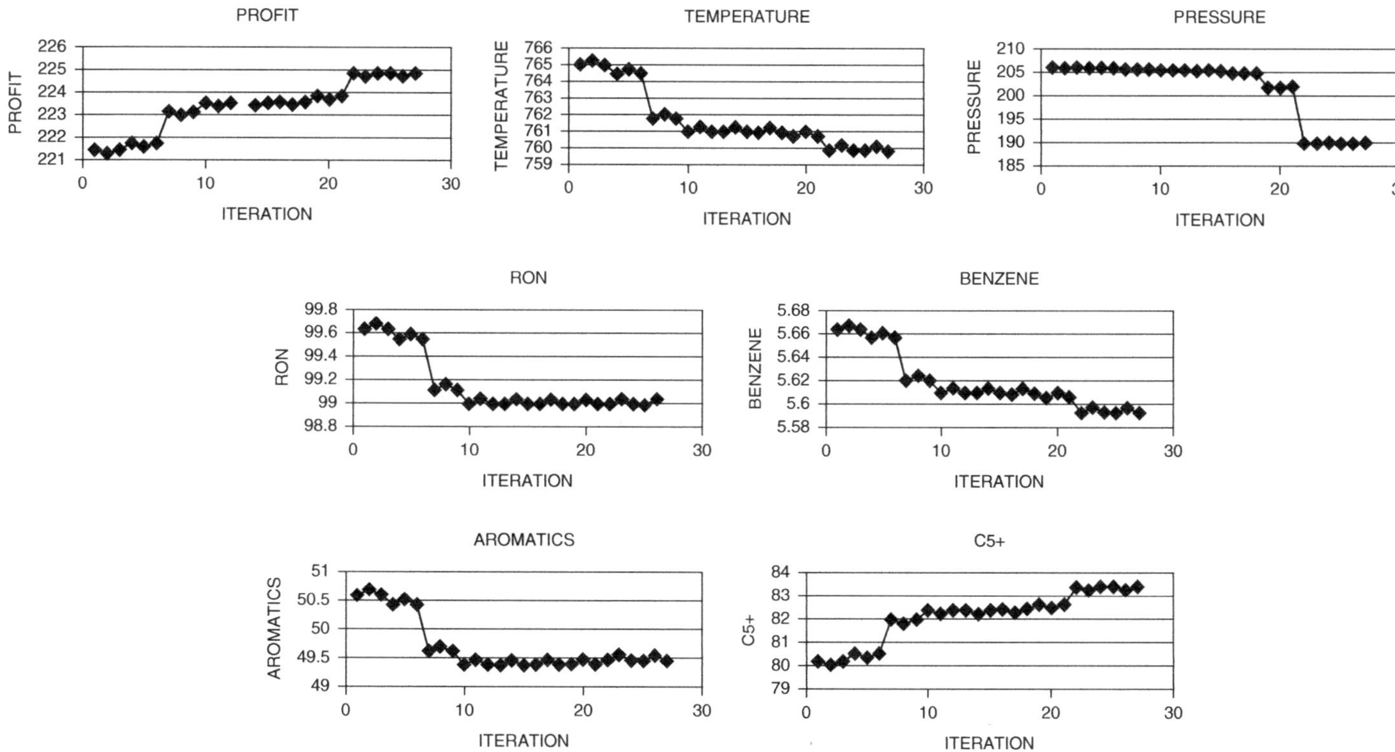

FIGURE 8.5. Optimization of T and P for the reforming process.

Example 8.3 MINLP optimization problem (Zhu, 2018)

The NLP model in equation set (8.17) is extended to address catalyst selection in the reaction sections by introducing integer variable (z). Thus, an MINLP optimization model for the overall process can be generated as shown in the equation system (8.18) with the objective to maximize profit via optimization of catalyst selection, product yields, and process conditions:

$$\underset{z,T,P}{\text{Maximize}} \quad \sum_{l=1}^{n_x}\sum_{i=1}^{n_c} p_i \cdot (c_{i,l} \cdot m_{i,l} \cdot F_l) \cdot z_l$$

$$\text{s.t.} \quad \frac{dc_{i,j,l}}{dt} = r_{j,l} \ (\text{Equations } 8.13-8.16) \qquad j = 1,2,3,4$$

$$k_{j,l} = k_{j,l}^0 \exp\left[\frac{E_{j,l}}{R}\left(\frac{1}{766} - \frac{1}{T_j}\right)\right] \qquad j = 1,2,3,4$$

$$c_{i,j+1,l}(0) = c_{i,j,l}^t \qquad j = 1,2,3,4 \quad (8.18)$$

$$1 \le \sum_l z_l \le n_x$$

Component mass balances

Overall mass balance

Utility balances (hydrogen, steam, power)

Operating limits (rate, size, severity, etc.)

$$\text{RON} = \sum_i^{n_c}\sum_{l=1}^{n_x} z_l \cdot V_i \cdot \text{RON}_{i,l}$$

$$\text{BEN} = \sum_i^{n_c}\sum_{l=1}^{n_x} z_l \cdot v_i \cdot \text{BEN}_{i,l}$$

$$T^{\text{L}} \le T \le T^{\text{U}}$$

$$P^{\text{L}} \le P \le P^{\text{U}}$$

$$\text{RON}^{\text{L}} \le \text{RON} \le \text{RON}^{\text{U}}$$

$$\text{BEN}^{\text{L}} \le \text{BEN} \le \text{BEN}^{\text{U}}$$

$$i = 1,\ldots,n_c, \ j = 1,\ldots,n_r, \ l = 1,\ldots,n_x$$

The type of catalyst (z_l), temperature (T), and pressure (P) are the independent variables, while mole fractions (c_i) are dependent variables. Catalyst type, temperature, and pressure influence the reaction kinetics and the molecular compositions, while molecular flow rates and composition determine the profit of the process defined in the objective function. While temperature and pressure are treated as continuous variables, catalyst selection is modeled as an integral variable.

With the sum of z_l less than the number of catalysts (n_x) considered, three scenarios for optimization are established: (a) only one catalyst is selected from the given number of catalysts (n_x), thus setting $\Sigma z_l = 1$; (b) two catalysts are selected at the same time for mixed catalysis to explore synergetic interactions, thus setting $1 \le \Sigma z_l \le 2$; and (c) more than two catalysts are selected, then $n_x > 2$; this scenario is not common. If the optimization model selects the option of two catalysts in coexistence, it could lead to a novel catalyst bed design. One such an example is to stack these two catalysts in layers of packing, which features linear interaction of two catalysts. When two different catalysts are in mixture that are stacking in layers of packing, both the linear and the nonlinear interactions between catalysts must be considered with mathematical expressions.

Two catalysts were considered for selection in this case, which have different yield performance in terms of C_{5+}, aromatics, and H_2. Type B catalyst has relatively lower C_{5+} yield but higher aromatics yield because of higher benzene content. On the other hand, type A catalyst has relatively high C_{5+} yield, which is important for reformate production.

For simplicity, it was assumed that the reaction system consists of four plug flow reactors. Each reactor is modeled via division into five continuous stirred tank reactor (CSTR) zones, and thus four reactors form 20 CSTR zones. The outlet of a previous reaction zone becomes the feed for the next zone. The optimization model is solved sequentially from zone 1 to zone 20. Through catalytic reforming, the naphtha feed is converted into three products, namely, hydrogen, light hydrocarbon products, and reformate for gasoline (main product). The optimization model expressed in equation set (8.18) will determine the optimal catalyst, temperature, and pressure that will maximize the profit under constraints of process conditions (T, P), gasoline quality in terms of octane (RON), etc. If necessary, other specifications, such as MON, RVP, etc., can also be added to the model as constraints.

Overall, the optimization determines the operation that can maximize stabilized gasoline rate while maintaining a sensible RON value. For comparison purposes, two different objective functions are investigated under the same constraint RON$\geq$99 and within the temperature range 750–790 K and the pressure range 190–200 psia. Optimization results are given in Table 8.1.

For the objective function of maximizing profit, the benzene content limit is provided in model (8.18) to reflect the market specification for gasoline. Type A catalyst with the lower benzene yield but higher C_{5+} yield was selected by the optimization method in order to meet the market requirement for the benzene content in gasoline. From the optimization results, it is observed that raising the temperature can make the RON higher for the gasoline via increasing benzene and aromatics content. However, the yield of the stabilizing gasoline decreases significantly due to more hydrocracking of valuable components (converting to light hydrocarbon products), which generally has a negative effect on the profit. On the other hand, reducing pressure promotes the production of reformate for the same RON.

Now assume the catalytic reforming process will be utilized to produce an intermediate product as the feed for aromatics production. To reflect this operation, a different objective function, for example, maximizing RON, is used. In this case, the benzene limit is removed from the model as the operation mode is to produce intermediate feed for aromatics production. A different catalyst (type B) with higher aromatics yield was selected by the optimization method. At the same time, the reaction severity (e.g., temperature and pressure) is increased in order to increase the benzene and aromatics contents as much as possible. In this case, profit will be calculated based on the ultimate products (i.e., benzene, toluene, and paraxylene) instead of the stabilized gasoline.

Example 8.4 Sometimes we can treat integer variables as if they are continuous by rounding the optimal solution to the nearest integer variables. This may lead to a suboptimal solution, yet one that is acceptable from a practical point of view. In some cases, rounding up may lead to infeasible solution. Let us look at Figure 8.6 (Wikipedia, 2019) that depicts the following problem:

$$\begin{aligned} &\text{maximize } y \\ &-x + y \leq 1 \\ &3x + 2y \leq 12 \\ &2x + 3y \leq 12 \\ &x, y \geq 0 \\ &x, y \in \mathbb{Z} \end{aligned} \tag{8.19}$$

The feasible integer points are shown in red, and the dashed lines indicate their convex hull, which is the smallest polyhedron that contains all of these points. The solid lines together with the coordinate axes define the polyhedron of the integer relaxation, which is given by the inequalities without the integrality constraint. The goal of the optimization is to move the dotted points as far upward as possible while still touching the polyhedron. The optimal solutions of the integer problem are the points ($x = 1$, $y = 2$) and ($x = 2$, $y = 2$) that both have an objective value of 2. The unique optimum of the relaxation is ($x = 1.8$, $y = 2.8$) with objective value of 2.8. Note that if the solution of the relaxation is rounded to the nearest integers ($x = 2$, $y = 3$), it leads to an infeasible solution.

TABLE 8.1. Optimization Results for the Catalytic Reforming Process

	Catalyst	Profit	T	P	Octane	G_a	G_b	G_c
	(Type)	($/hr)	(K)	(psia)	RON	(wt%)	(wt%)	(wt%)
Initial value		72,662	765	206	99.6	50.61	5.66	80.22
Obj: maximize RON	Catalyst B	79,726	790	204.2	103.5	57.75	5.68	62.73
Obj: maximize profit	Catalyst A	80,953	760	190.1	99.0	49.47	5.59	83.41

G_a (wt%) is the weight percentage of total aromatics in the stabilizing gasoline.
G_b (wt%) is the weight percentage of total benzene in the stabilizing gasoline.
G_c (wt%) is the weight percentage of stabilizing gasoline in the corresponding product.

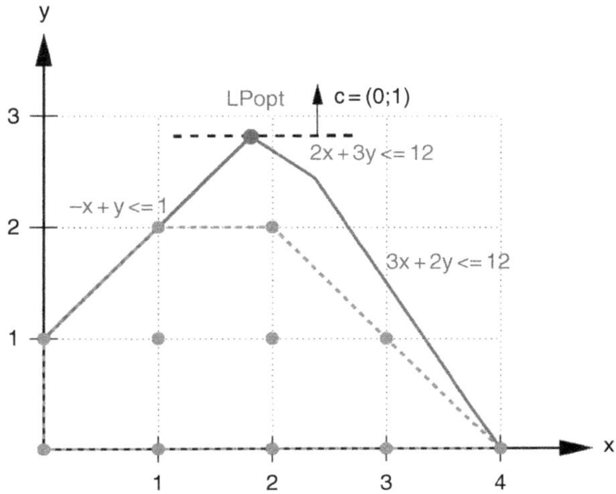

FIGURE 8.6. An integer optimization profile (Fanosta, 2019). *Source*: From Fanosta, IP polytope with LP relaxation, 2019. Licensed under CC-BY-SA-4.0.

The second reason is that stochastic optimization is not limited by a local optimum, as it identifies multiple optima via random search and does not rely on initial points. In contrast, SQP relies on initial points, and the search stops whenever an optimal solution is found.

Stochastic optimization methods are optimization methods that generate and use random variables. For stochastic problems, the random variables appear in the formulation of the optimization problem itself, which involves random objective functions or random constraints. Stochastic optimization methods also include initial points and random iterates.

Partly random input data arise in such areas as real-time estimation and control, simulation-based optimizations where Monte Carlo simulations are run as estimates of an actual system, and problems where there is experimental (random) error in the measurements of the criterion. In such cases, knowledge that the function values are contaminated by random "noise" leads naturally to algorithms that use statistical inference tools to estimate the "true" values of the function and/or make statistically optimal decisions about the next steps. The methods include stochastic approximation (SA) by Robbins and Monro (1951) and simultaneous perturbation SA by Spall (1992).

On the other hand, even when the data set consists of precise measurements, some methods introduce randomness into the search process to accelerate progress. Such randomness can also make the method less sensitive to modeling errors. Further, the injected randomness may enable the method to escape a local optimum and eventually approach a global optimum. Indeed, this randomization principle is known to be a simple and effective way to develop algorithms with high certainty of good performance across many data sets and for many sorts of problems. The simulated annealing and generic algorithm methods are the most popular stochastic optimization methods used.

8.7 SIMULATION-BASED OPTIMIZATION

Process simulation-based optimization is suitable for process optimization which involve complex calculations including thermodynamics, hydraulics, kinetics, separation, heat transfer, and rotating equipment. These calculations are embedded within simulation software. Thus, one can build a flow scheme for the process readily in simulation software and then use the embed optimizers to optimize the process.

8.7.1 Simulation-Based Optimization using the Embed Optimizer

Once the process is simulated, the engineer can define an objective function and operating limits, then select an optimization solver (e.g., SQP) to run the optimization within the simulation and generate an optimized feasible solution.

UniSim, HYSYS, and Aspen Plus are software widely used for simulating chemical processes. There are two kinds of process simulation methods, namely, modular-based simulator (sequential mode) and equation-oriented (EO) simulator (simultaneous mode). With the EO simulator, all the process-related equations are assembled and solved simultaneously. The run time is very fast (in seconds or minutes at most). With the modular-based simulator, calculations involve first-principle models for unit operations, thermodynamics, and kinetics and they are self-contained as modules. In addition, high quality data are used which include detailed molecular information, chemical, thermodynamical and transport data. The module calculations are executed in sequence and may be repeated due to recycles. This approach is also known as a sequential simulator and is popular in use for process design for high accuracy. In contrast, EO-based simulators are more commonly used for online optimization for high calculation speed.

8.7.2 Simulation-Based Optimization Integrating with Microsoft Excel

Different optimization solvers for solving LP, NLP, MINLP, and stochastic problems are available in Excel. Some of these tools, particularly the latest versions, may not be available in process simulation software, so it is necessary to integrate simulations with Excel for complex nonlinear problems to find the global optimal solution. Engineers are familiar with Excel and can easily use worksheets to create what-if scenarios for simulation and optimization, data analysis, results plots, and so on.

Interfacing a process simulator with Excel requires use of VBA coding (Microsoft Visual Basic for Applications), which can facilitate (a) transfer of decision variables from Excel to the simulator, (b) activation of simulation runs from Excel, and (c) transfer of selected data from the simulator to Excel, and d) plotting of optimization results in charts in Excel.

A good use case is conducting what-if analysis from end users. The user can provide the decision variables, change limits on variables and constraints, and select a more relevant and effective optimization solver in Excel. Below is an example involving optimization of heat exchanger cleaning cycles to combat fouling in a preheat train in the crude distillation unit in an oil refinery plant (Zhu, 2014).

Fouling is accumulation of undesirable materials as deposits on heat exchanger surfaces. Fouling deposits have different causes and come in many forms. Regardless of what material these deposits are composed of, similar consequences arise, which are reduction in thermal performance and increase in pressure drop. There are complex factors causing heat exchanger fouling such as physical and chemical properties of process streams, operating conditions, heat exchanger design, and operation. Since complex factors affect the choice of methods to reduce and prevent fouling, more effective solutions can be derived through identification of root causes.

Although fouling involves very complex mechanisms, the objective from the fouling mitigation point of view is to determine a few fouling control variables to control a fouling process. These variables are fluid velocity, fluid and heat transfer surface temperatures, physical and chemical properties of the fluid, and geometry of the fluid flow passage. Other important variables are concentration of foulant or precursor, impurities, heat transfer surface roughness, surface chemistry, fluid pressure, fluid chemistry (pH level, oxygen concentration, etc.), and more (Shah and Sekulić, 2003).

An important part of mitigating a fouling problem in operation is to determine the time interval for cleaning of heat exchangers. The cleaning interval is referred to as a cleaning cycle. Once a cleaning cycle is decided, exchangers are subsequently cleaned by using the most appropriate cleaning method(s).

In general, there are three types of cleaning cycle strategies: fixed time based, reliability based, and cost based. With the fixed time-based strategy, exchangers are cleaned based on a regular maintenance schedule during process shutdowns. With the reliability-based cleaning strategy, exchanger performance for critical services is restored according to a certain maintenance schedule. The cost-based cleaning strategy considers the costs associated with cleaning versus the financial penalty and additional fuel/power consumption due to fouling. In situations where the cost of operation and maintenance of critical services is an important factor, maintenance decisions can be optimized by developing cost as a function of reliability and then searching for a minimum cost-based solution. This cost-optimized maintenance solution will also result in an optimal level of heat exchanger reliability of critical services.

A simulation-based optimization method was developed for this purpose (Zhu, 2014). The crude unit with multiple distillation columns and preheat train consisting of 20+ heat exchangers is simulated. A clean cycle optimization model is built in Excel and is run based on the relevant data extracted from the simulator. The model can be adjusted to reflect different situations (a) where only the most fouled exchangers are selected to clean, (b) where time variant margin is estimated instead of average margin, and (c) where other significant costs are introduced. A view of data transfer and function design is shown in Figure 8.7. Data is collected

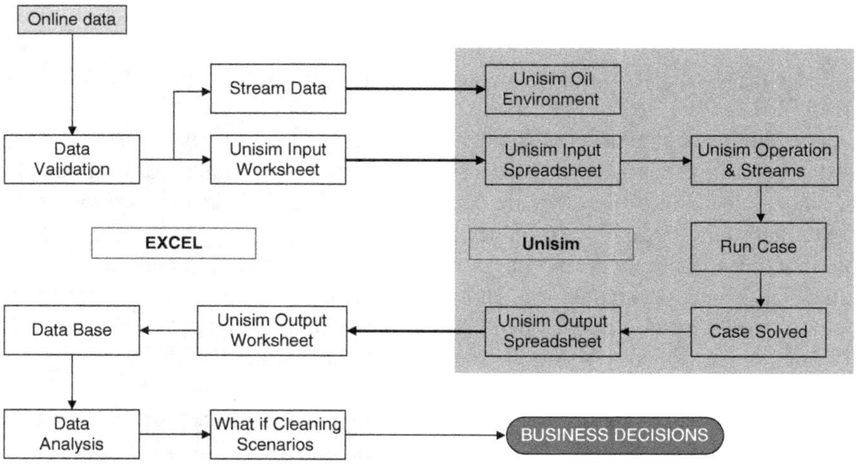

FIGURE 8.7. Heat exchanger clean cycle optimization using Excel integrating with simulation. *Source*: From Zhu XX (2014), reprinted with permissions of John Wiley & Sons.

in Excel and pushed to the simulation, while the simulation sends converged data back to Excel where optimization is conducted. Results are displayed in Excel for root cause analysis, trending, and reporting.

8.8 A CASE STUDY FOR PROCESS OPTIMIZATION

Example 8.5 This example (Loe and Pults, 2001) provides a case of how operational improvements for a single deisopentanizer fractionation tower are identified, implemented, and sustained. To do so, the current operation is simulated and assessed. Then improvement opportunities are identified, and the limiting factors are determined. Optimal solutions are obtained by optimizing tower ISBL conditions (Inside System Battery Limits) and OSBL conditions (Outside System Battery Limits). The improvements on this single tower have generated over $500,000 as compared with historical operation over the first 6 months of implementation.

8.8.1 Define the Project Basis

The deisopentanizer (DIP) tower, shown in Figure 8.8, processes light straight-run naphtha from two sources. The debutanizer bottoms consist of primarily isopentane

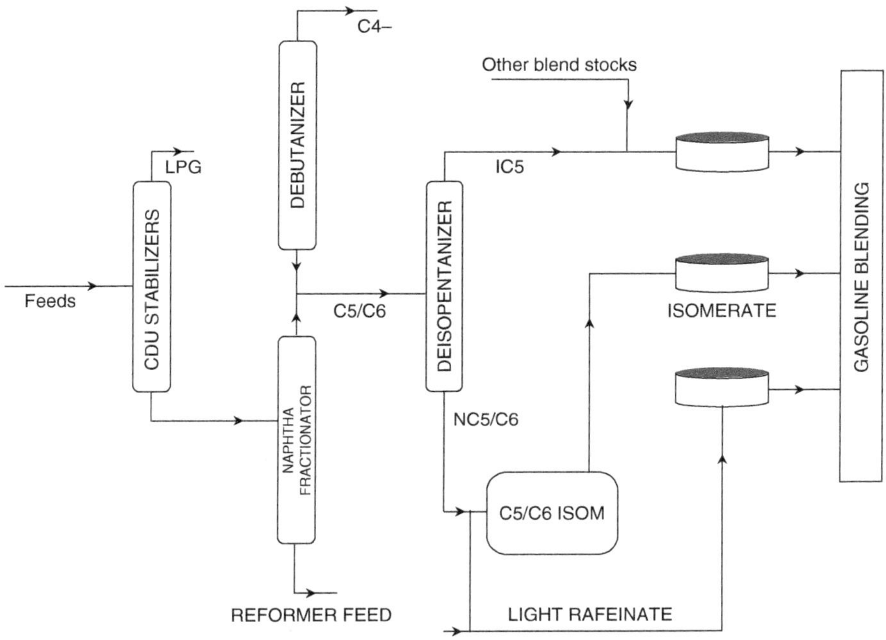

FIGURE 8.8. Deisopentanizer flow scheme. *Source*: From Loe B, Pults J (2001), reprinted with permissions of American Institute of Chemical Engineers.

and normal pentane (iC_5 and nC_5), with small fractions of butane and C_{6+} components. The overhead from the naphtha fractionator tower contains mostly C_5 and C_6 paraffin compounds, with some benzene and C_6 naphthenes and a small amount of butane. The combined feed to the DIP is typically in the range of 9,000–15,000 bbl/day.

The DIP overhead product is normally rich in isopentane and is routed to gasoline blending, along with other high-octane, low Reid vapor pressure (RVP) gasoline components. The DIP bottoms, which is rich in nC_5 and C_6 paraffins, is routed to the light naphtha isomerization unit, along with light raffinate from the aromatics extraction unit.

The DIP tower under consideration has 50 trays in comparison with 70+ trays used in a typical DIP. This tower consequently often has a difficult time making a good split between isopentane and normal pentane components.

8.8.2 Current Operation Assessment

Historically, the tower had been operated with a target of 10% nC_5 in the overhead and 20–30% iC_5 in the bottoms product. The tower was reported to be limited by reboiler and condenser duty. One of the two steam reboilers had been out of service for some time, and 20–30% of the condenser fin fan motors were not operating and in need of repair. The DIP equipment had not been a maintenance priority, in part because no economic penalty had been calculated for having a reboiler or condenser out of service.

The DIP process control was accomplished with a DCS system equipped with an advanced process control (APC). The controller was set to target 10% nC_5 in the overhead and 20–30% iC_5 in the product and would increase reboiler steam and reflux rate until reaching the maximum limits. The tower pressure was also controlled within a specified range by the APC, and this could indirectly limit the reboiler duty as well, if the tower pressure increased beyond its set maximum. Inferential estimates for product iC_5 and nC_5 qualities were calculated based on tower temperatures and pressures, and a bias for these values was continually updated based on daily laboratory data.

To establish the historic performance, operating and laboratory data were collected from the past year, with periods of known equipment failure or poor unit volume balance eliminated. The data showed an average of 19% nC_5 in the DIP overhead and 27% iC_5 in the bottoms product, with a wide variation in the product quality, as shown in Figure 8.9. An average of 18% of normal butane was observed in the overhead product, indicating poor debutanization in the upstream fractionation towers.

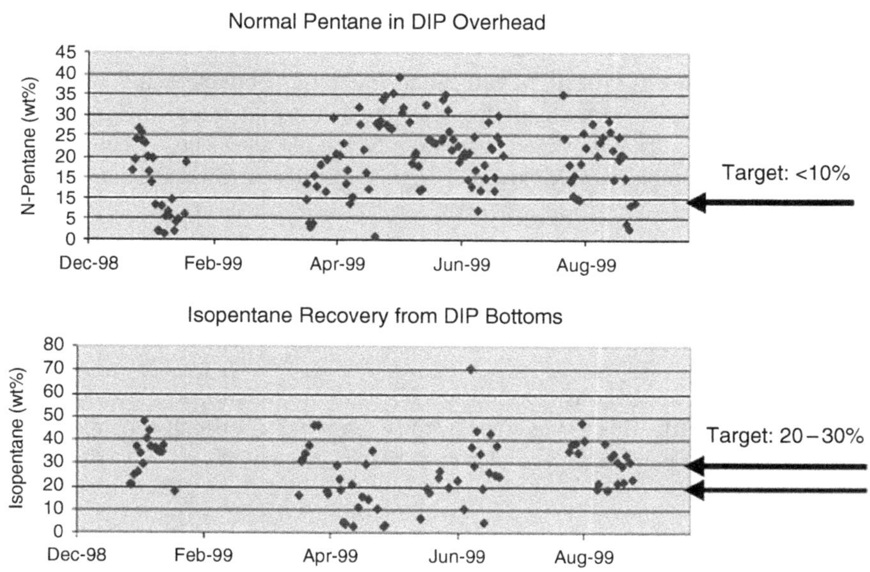

FIGURE 8.9. Variation of DIP performance. *Source*: From Loe B, Pults J (2001), reprinted with permissions of American Institute of Chemical Engineers.

The process design for the tower showed an available reboiler duty of 51,000 lb/hr of steam, but the average for the data collected showed only 27,000 lb/hr. The data furthermore showed an erratic variation in reboiler duty.

8.8.3 Simulation to Establish the Base Case for Optimization

A simulation model for the DIP was developed using the average process and laboratory data. The feed rate and composition to the tower were fixed, as well as the reflux rate, tower pressures, and overhead rate. The model results for reboiler duty, tower temperatures, and compositions compared favorably with the unit operating data, as shown in Table 8.2.

The calibrated model was used to simulate DIP performance at the design reboiler duty to determine if available condenser duty and tower tray capacity would be adequate for this operation. As shown in Table 8.3, the predicted condenser duty for this operation was only slightly above the design value, and tower tray parameters indicated that flooding was unlikely. Also the separation of iC_5 and nC_5 improved dramatically versus the historical operation, as would be expected with the increased tower traffic.

8.8.4 Define the Objective Function

The profitability of the DIP column was determined based on the value of separating iC_5 for direct blending to gasoline and nC_5 to be used as feed for the C_5/C_6 isomerization unit, less the utility and downstream unit opportunity costs incurred to do so. Lighter feed components, such as n-butane, were assumed to always be

TABLE 8.2. Simulation Results Versus DIP Operating Data

Result	Data	Model
Reboiler duty (MMBtu/hr)	23.6	26.7
Top temperature (°F)	143	153
Bottom temperature (°F)	179	183
nC_5 in overhead (wt%)	18%	16%
iC_5 in bottoms (wt%)	31%	30%

TABLE 8.3. Simulation Results At Designed Reboiler Duty

Result	Design	Model
Reboiler duty (MMBtu/hr)	48.8	48.8
Condenser duty (MMBtu/hr)	45.0	48.1
Max jet flood	85%	74%
Max downcomer backup	50%	37%
nC_5 in overhead (wt%)	NA	5%
iC_5 in bottoms (wt%)	NA	18%

fractionated into the DIP overhead, and components heavier than nC_5 were assumed to always be found in the DIP bottoms stream. Thus only the disposition of iC_5 and nC_5 components was considered in the profitability calculation.

Therefore, the objective function for optimizing the DIP tower is defined as below:

$$\text{DIP upgrade value} = \text{overhead value} + \text{bottoms value} - \text{feed value} - \text{reboiler steam cost} - \text{Isom operating cost} - \text{Isom capacity penalty}$$

The DIP feed and overhead values are calculated as the gasoline blending value of the iC_5 and nC_5 in this stream, with corrections for road octane and RVP of these components versus those of conventional regular gasoline. The DIP bottoms stream is normally processed at the isomerization unit, where 75% of the exiting C_5 components are assumed to be iC_5. After this equilibrium conversion, the value of the resulting iC_5/nC_5 stream is calculated at the gasoline blending value as described above for the feed and overhead. The reboiler steam cost is calculated assuming a 70% generation efficiency from refinery fuel gas, and the isomerization unit operating cost (for fuel, power, and catalyst) was taken to be the same value per barrel as used in the refinery planning model.

During some periods, the refinery isomerization unit has more feed available than can be processed. If additional DIP bottoms are produced, less capacity is available to process light raffinate from the aromatics extraction unit. The isomerization unit capacity penalty, or the opportunity cost for processing additional DIP tower bottoms at this unit, was therefore estimated by evaluating the octane upgrade of light raffinate.

8.8.5 Verification and What-If Analysis

Once the economic evaluation criteria were determined, they were implemented in the simulation model. Numerous case studies were conducted via process simulation to determine the optimum operating point for the DIP tower under different scenarios. It quickly became apparent that in nearly all economic and operating situations, maximizing the DIP reboiler duty up to its design limit resulted in the highest profitability.

For subsequent case studies, the simulation was completed with maximum reboiler duty, and tower pressure and nC_5 content of the overhead product were also fixed. These constraints completely specified the tower operating conditions.

The DIP profitability was first examined for scenarios assuming available capacity in the isomerization unit. The DIP feed rate and overhead nC_5 content were varied, and profitability was calculated, as shown in Figure 8.10. This analysis showed that the optimum target was around 5% of nC_5 in the overhead, regardless of the tower feed rate.

Profitability was then examined assuming the isomerization unit was at its maximum charge rate, and additional production of DIP bottoms would result in

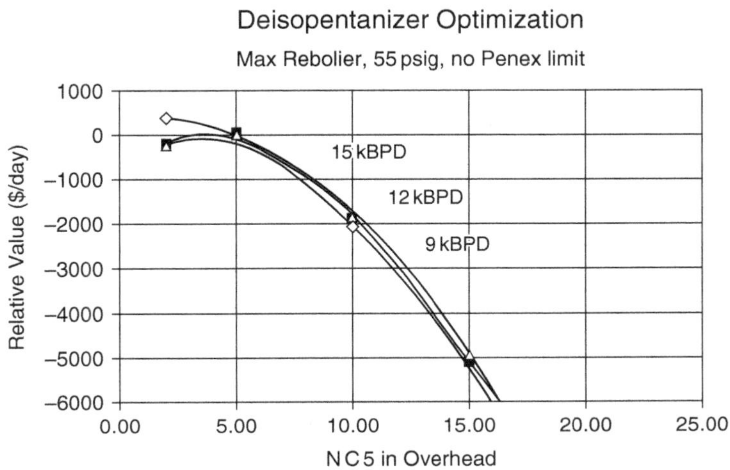

FIGURE 8.10. Optimization without Isom capacity constraint. *Source*: From Loe B, Pults J (2001), reprinted with permissions of American Institute of Chemical Engineers.

bypassing of light raffinate around the isomerization unit, direct to gasoline blending. The cost of losing the light raffinate octane upgrade can vary between \$2 and 5/bbl, so simulation cases were completed for both scenarios as shown in Figure 8.11. In these cases, the optimum nC_5 target in the DIP overhead is dependent on the charge rate to the tower. At low charge rates, the available reboiler duty is sufficient to obtain good separation between iC_5 and nC_5 components so that minimal iC_5 is lost into the DIP bottoms when targeting 5% nC_5 in the overhead. At higher charge rates, more iC_5 is lost to the bottoms stream, and it is more profitable to increase the overhead nC_5 target, reducing the DIP bottoms rate to the isomerization unit and allowing additional raffinate upgrading. Thus the optimal nC_5 in overhead target varies between 5 and 20%, depending on DIP charge rate and the value of light raffinate upgrading.

Based on a comparison of the optimal tower operation as determined above and the historical performance, an incentive of around \$1.5 million/year with improved DIP fractionation was identified.

8.8.6 Optimization Implementation

In order to realize the benefit indicated by the optimization results above, several unit hardware, process control, and operating philosophy changes were needed. Firstly, it was clear that both reboilers would be required, so the spare bundle was leak-tested and returned to service by operations. Several fin fan motors were also quickly repaired to ensure that design condenser duty was available and that overpressurization of the tower would not limit the reboiler duty that could be applied.

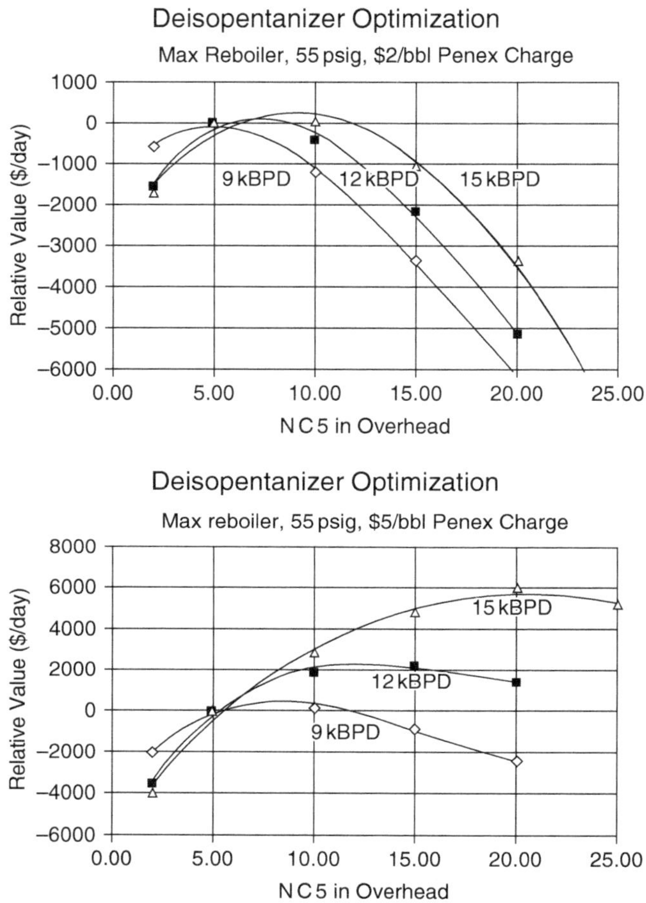

FIGURE 8.11. Optimization with Isom capacity constraint. *Source*: From Loe B, Pults J (2001), reprinted with permissions of American Institute of Chemical Engineers.

Secondly, the optimum target for nC_5 in the DIP overhead was implemented into the APC system. The APC controller on the DIP DCS system was reconfigured to operate at this nC_5 target while maximizing the reboiler duty as governed by the high limit on tower pressure. This APC system allowed the DIP operation to be maintained at an economic optimum, accounting for the isomerization unit capacity and the daily economic scenario.

Thirdly, communication of the new operating philosophy was also critical in improving DIP performance. Operators and unit supervisors were trained on the importance of always maximizing the reboiler duty and setting the nC_5 in overhead target. The iC_5 content of the bottoms stream was still measured by laboratory and inferred analysis, but this was no longer a tower control variable. The reboiler duty and overhead nC_5 were tracked daily, and performance for these key performance indicators (KPIs) was discussed at weekly operations and planning meetings.

8.8.7 Operation Monitoring

The economic benefit of improving the DIP fractionation was tracked on a monthly average basis shortly after implementation of the new optimization strategy. Economic performance versus the baseline operation is shown in Figure 8.12 using actual monthly averaged economics and unit operating and lab data. Monthly benefits of over $100,000 were achieved during the summer months, when octane values were at their highest level. As octane values dipped during spring and fall months, benefits from the improved DIP fractionation dropped off as well.

A significant drop in benefits can be seen in Figure 8.12 for July. This was due to poor operation of the DIP tower, caused by a high butane content in the tower feed from the crude unit stabilizers. The C_4's caused the DIP tower pressure to increase up to its safe operating limit, and the reboiler duty was cut back to avoid overpressuring the tower. This resulted in a reduction in fractionation efficiency and profitability during part of July and represented one of the challenges encountered in sustaining the improved DIP performance.

8.8.8 Sustaining Benefits

Without ongoing attention, optimization improvements from initiatives such as that for the DIP tower tend to fade over time, for a multitude of reasons. Challenges to the new level of performance must be tackled as they arise, whether they result from hardware or control problems, misunderstandings, or operating changes in other parts of the refinery. Tracking the economic benefits of the initiative is a

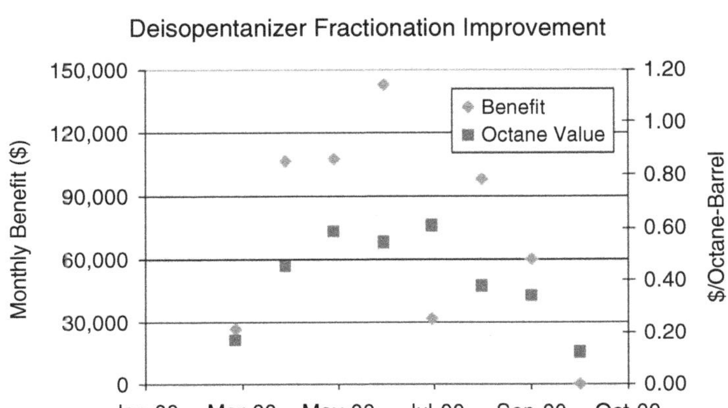

FIGURE 8.12. DIP economic improvements. *Source*: From Loe B, Pults J (2001), reprinted with permissions of American Institute of Chemical Engineers.

critical element of sustaining the change, since knowing the lost profits associated with a loss in performance helps to set work priorities within the refinery. Several problems occurred during the first 6 months of improved DIP operation, which temporarily reduced the profit derived from this tower, but these issues were quickly addressed to sustain the improvement.

During the summer months, a change in routing of a portion of the refinery condensate resulted in a hydraulic constraint on the amount of condensate that could be removed from the DIP reboilers into the condensate header. This caused a reduction in reboiler duty that led to a drop in fractionation efficiency. The economic calculations clearly showed that the benefit derived from the additional reboiler duty was much higher than the value of recovering the condensate. For this reason, condensate was safely discharged into the sewer until normal condensate header operation was restored, and the improved DIP fractionation performance was maintained.

A second challenge occurred when it was noticed that the DIP reflux drum temperature had increased above 140 °F, which was higher than the recommended rundown limit to tankage. Initially reboiler duty was decreased to ensure safe operation. However, after meeting with the tank farm operators about the problem, it was found that the DIP overhead mixed with several other much larger streams before entering a gasoline blending tank. Calculations showed that the effect of the higher DIP rundown temperature on the tank temperature was minimal and DIP reboiler duty was again increased.

Another problem in maintaining reboiler duty was identified when the DIP tower pressure increased due to butanes in the feed, as mentioned above. It was found, however, that the maximum tower pressure was set well below the vessel design pressure, so the safe operating limit of the tower was increased after an appropriate management of change (MOC) review. Simulation modeling showed that operation at the higher pressure did not significantly impact the tower profitability if the reboiler duty could be maintained near the maximum. The pressure limit was increased, and performance of the tower again was restored.

A fourth reduction in DIP profitability was observed when the APC controller began cutting reboiler duty for no apparent reason. Further investigation showed that the inferred nC_5 content of the overhead stream had been deviating from the lab value for a few days because of a computer glitch. This was quickly corrected and DIP operation returned to normal.

Although all of these obstacles reduced profitability for a short period, timely identification and resolution of the constraints averted potentially long periods of underperformance.

8.9 CONCLUDING REMARKS

The profit improvement process should pass through several phases. First, current performance must be assessed, evaluating upstream and downstream constraints and unit equipment limitations. Understanding of how the unit operation can be optimized is discerned by use of an appropriate process model and refinery economics, and a new operating strategy is then developed to improve profitability. Implementation of this strategy requires good communication of its benefits throughout the plant. Upgrading or repair of process equipment may be needed to enable operation under desired conditions. Tracking of key operating parameters as well as economic performance of the unit and distributing these results within the plant is essential to sustaining the improvement as this procedure helps to monitor deterioration of profitability. The profit improvement process requires input from technical, operations, safety, environmental, and economics groups within the plant.

Thus, optimization models are built for multiple purposes, which are listed below.

8.9.1 Operation Optimization: Finding the Optimal Decision Variables

The purpose of the optimization model in this case is to determine optimal values for decision variables such as feed selection, product blending, process conditions, and so on with the objective of maximizing overall profit. We caution against blind use of optimization output; optimization results must be evaluated and confirmed before implementation.

8.9.2 Performance Prediction: Evaluating Step Change What-If Scenarios

"What-if" questions must be answered in performance prediction: what if feed composition changes, what if additional amounts of a specific product are made, what if product quality varies, what if catalyst performance deteriorates, what if a column reboiling exchanger gets severely fouled, etc. We can test these what-if scenarios by providing input, and optimization output can indicate the new economic performance that becomes the basis for further analysis and decision making. The optimization with potential changes for these what-if assumptions can provide the best plan for alternative operation scenarios.

8.9.3 Sensitivity Analysis: Evaluating Incremental What-If Scenarios

Sensitivity analysis from optimization models is also an important part of performance prediction. By varying materials costs, changing equipment scheduling during a production campaign, taking major heat exchangers out for cleaning, and so on, the effect of changes can be assessed. Accordingly, these scenarios should be available for sensitivity analysis over the range of possible changes. One must be aware of the fact the sensitivity may be different if dealing with nonlinear systems. For example, the sensitivity could be diminishing for convex type of behaviors while the effect could increase for concave type of phenomenon.

8.9.4 Control Optimization: Determining Setpoints for APC Controllers

Model-based control may be able to provide better setpoints and thus achieve better performance than simple gain-based control. In recent years, the use of first principles within autocalibrated models has gained much attraction due to their ability to accurately represent molecular information and nonlinear behavior, leading to more informed actions and implementation. This fact is even more compounded by ever-increasing computing power at low cost and capability of data handling and nonlinear solvers for large problems. The situation is double-edged though. On one hand, rigorous models can capture highly nonlinear behavior at a fundamental level and thus can provide new insights and opportunity. On the other hand, it is harder for process engineers to understand the prediction results compared with traditional linear equation-based APC. In addition, it could become very difficult to troubleshoot root causes in the case of wrong predictions. Aggressive model tuning can only shift error from one place to another. Thus, the challenge is how to capture the benefit of rigorous models but avoid the difficulty of applying them.

8.9.5 Insights: Provide Better Understanding of Relationships

Playing with the model can generate the circumstances under which the problems can occur, exacerbate, or go away. Once the source of the problem is established, corrective actions can be taken. The model building process has considerable value. The lessons learned during model building, validation, and calibration provide considerable insight into the real system. In the context of insight, we include the use of models as diagnostic tools for troubleshooting and identification of root causes and thus determine the actions accordingly.

REFERENCES

Fanosta (2019), https://en.wikipedia.org/wiki/Integer_programming. Wikipedia, 2019. Licensed under CC-BY-SA-4.0.

Gill PE, Murray W, Wright MH (1981), Practical Optimization, Academic Press, New York.

Loe B, Pults J (2001), Implementing and sustaining process optimization improvements on a de-isopentanizer tower. In: AIChE Spring Meeting, April, Houston, TX.

Rase HF (1977), Chemical Reactor Design for Process Plants, John Wiley & Sons, Inc., New York.

Robbins H, Monro S (1951), A stochastic approximation method, Annals of Mathematical Statistics **22** (3), 400–407.

Shah RK, Sekulić DP (2003), Fundamentals of Heat Exchanger Design, John Wiley & Sons, Inc., New York.

Smith RB (1959), Kinetic analysis of naphtha reforming with platinum catalyst, Chemical Engineering Progress **55**, 76–80.

Spall JC (1992), Multivariate stochastic approximation using a simultaneous perturbation gradient approximation, IEEE Transactions on Automatic Control **37** (3), 332–341.

Zhu XX (2014), Energy and Process Optimization for the Process Industry, John Wiley & Sons, Inc., New York.

Zhu XX (2018), Molecular analysis for process synthesis, Industrial and Engineering Chemistry Research **57**, 14583–14595.

Zhu XX, Hu SY (2000), Molecular modeling for overall refinery optimization. In: AICHE Annual Meeting, Los Angeles, CA.

9

PROCESS CONTROL AND APC MODELING

9.1 INTRODUCTION

Process modeling plays a central role in process control. The process modeling for control should answer the following questions:

- What are the control variables (CVs)? Proper control of these CVs is the goal of the control system.
- What are the measured variables? The four major measurements are flow, temperature, pressure, and level. The measured variables are related to CVs in the form of control algorithms.
- What are the manipulated variables (MVs)? Control inherently involves MVs that can be adjusted to affect a change in the measured variables.
- How does the system react to changes in the operating point? A sensitive analysis at the expected operating points of the process must be conducted as a part of modeling.

Understanding the process is the key to successful control applications in the process industries as it helps engineers to determine related CVs and MVs and their directional relationships and sensitivity effectively. Furthermore, it also helps to design a control system with the right speed of process response to changes in MVs.

Digitalization and Analytics for Smart Plant Performance: Theory and Applications,
First Edition. Frank (Xin X.) Zhu.

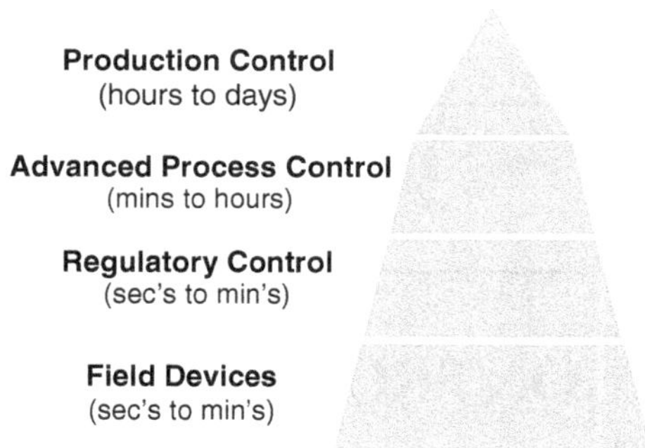

FIGURE 9.1. Process control consists of four tiers: field devices, regulatory control, advanced process control (APC), and production control (real-time integrated optimization and control).

Therefore, developing a sufficient understanding of the principles of automatic control is the core know-how of process engineers.

Process control can be represented as a hierarchy (Figure 9.1). The base of the structure is field devices such as sensors, transmitters, actuators, and valves that measure process variables (PV). Next is regulatory controllers—mostly based on proportional–integral–derivative's (PID) type of controllers, which keep values measured by field devices, such as pressure, temperature, and flow rate, within specified limits. While regulatory controls tackle each variable individually, the advanced process control (APC) layer determines optimal setpoints for each PID controllers to achieve optimal economic performance of an entire process unit. The last layer represents so-called production control, which is responsible for controlling and optimizing production through different process units on the production plant facility. From field devices to production control, each layer depends on the previous layer in the hierarchy, and the control performance of an individual layer directly affects the stability of the process, the quality of the product, and the costs associated with making the product. However, running a single unit at its local optimum is not necessarily the best strategy for achieving overall profitability of the facility. This is where the production control layer steps in to manage the individual units' APCs in concert to accomplish plantwide optimization. Regulatory control and APC are the focus of this chapter, while product control is discussed in Chapter 13 for site-wide optimization and control.

9.2 PROCESS MODELING IN CONTROL

There are several process modeling approaches for control that include dynamic modeling, steady-state modeling, linearization, rigorous modeling, empirical modeling, etc. The method discussed here is dynamic modeling based on first principle with linear approximation.

For the conservation balance equations, a system model is expressed as a set of nonlinear differential equations. In order to design a dynamic control system, there are two options:

1. Solve the nonlinear system directly.
2. Develop linear approximation and model dynamic behavior of the process.

The first option is useful for control design testing before implementing it on the field to the real process, but it can be overly complicated for the purpose for control system design. The second option is most useful for control system design, and it is the common approach adopted in the industry.

Developing a linear model that approximates the dynamic behavior in the vicinity of the operating point involves the following steps (Erickson and Hedrick, 1999):

1. Formulate system differential equations based on first principles.
2. Linearize the differential equations around the operating point.
3. Laplace transformation of the linear equations.
4. Derive a transfer function.

Let us walk through these steps below with examples.

9.2.1 Develop Process Model Based on First Principles

The first step is to develop process modeling based on fundamental heat, mass, and momentum quantities. These balances obey the principle of conservation, which is generally stated as

$$\text{Accumulation} = \text{in} - \text{out} + \text{generation} - \text{consumption} \tag{9.1}$$

For a nonzero accumulation, the above conservation balance results in a differential equation expressed as

$$\frac{\begin{array}{c}\text{Accumulation of } X\\ \text{within the system}\end{array}}{\text{Time period}} = \frac{\begin{array}{c}\text{flow of } X\\ \text{into the system}\end{array}}{\text{time period}} - \frac{\begin{array}{c}\text{flow of } X\\ \text{out of the system}\end{array}}{\text{time period}} + \frac{\begin{array}{c}\text{amount of } X\\ \text{generated in system}\end{array}}{\text{time period}} - \frac{\begin{array}{c}\text{amount of } X\\ \text{consumed in system}\end{array}}{\text{time period}} \tag{9.2}$$

where X is one of the following fundamental quantities: total mass, component mass, energy, and momentum.

Other nonlinear equations are very common such as

$$
\begin{aligned}
&\text{Heat transfer:}\quad Q = hA\Delta T \\
&\text{Reaction rate:}\quad r_{\mathrm{A}} = \frac{dc_{\mathrm{A}}}{dt} = k_0 e^{-E/RT} c_{\mathrm{A}} \\
&\text{Ideal gas law:}\quad \mathrm{PV} = nRT \\
&\text{Fluid flow rate:}\quad F = c_{\mathrm{v}} \sqrt{\Delta P / \rho} \\
&\text{Phase equilibrium:}\quad y_i = k_i x_i
\end{aligned}
\tag{9.3}
$$

The following example explains how to develop a process model based on first principle.

Example 9.1 Consider the continuous stirred tank reactor (CSTR) of Figure 9.2 (Erickson and Hedrick, 1999). It consists of a tank with an inlet and outlet flow. This process is a very simple reactor where chemical *A* in the inlet stream reacts to form chemical *B*. This reaction generates no heat, and the volume of the liquid is constant due to the nature of both chemicals.

Solution

There is no energy transfer in this system, and so the differential equations are developed based on the mass balances:

Overall mass balance:

$$
\begin{aligned}
\frac{d(\rho V)}{dt} = \frac{d(\rho A h)}{dt} &= \rho F_1 - \rho F \\
0 &= \rho F_1 - \rho F \\
F_1 &= F
\end{aligned}
\tag{9.4}
$$

Component mass balance:

$$
\begin{aligned}
\frac{d(n_{\mathrm{A}})}{dt} = \frac{d(c_{\mathrm{A}} V)}{dt} &= c_{\mathrm{A}_1} F_1 - c_{\mathrm{A}} F - r_{\mathrm{A}} V \\
\frac{d(c_{\mathrm{A}} V)}{dt} = c_{\mathrm{A}} \frac{dV}{dt} + V \frac{dc_{\mathrm{A}}}{dt} &= c_{\mathrm{A}_1} F_1 - c_{\mathrm{A}} F - k_0 e^{-E/RT} c_{\mathrm{A}} V
\end{aligned}
\tag{9.5}
$$

where V is the reactor volume. Using Equation 9.4, Equation 9.5 becomes

$$
\frac{dc_{\mathrm{A}}}{dt} = \frac{F}{V}\left(c_{\mathrm{A}_1} - c_{\mathrm{A}}\right) - k_0 e^{-E/RT} c_{\mathrm{A}} \tag{9.6}
$$

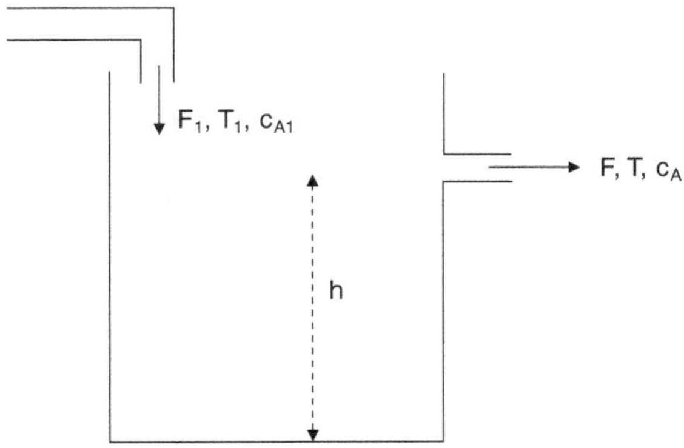

FIGURE 9.2. CSTR for Example 9.2.

Therefore, this system consists of variables c_A and F. c_{A_1} is an external stimulus, and other symbols represent specified constants and depend on the chemicals involved and reactor size. If T is considered as a variable as well, Equation 9.6 is highly nonlinear exponential term as a function of temperature.

However, a system model with nonlinear differential equations is not immediately useful for control system design. One could use a nonlinear control system algorithm, but it is difficult to develop and is usually overly complicated. The common practice is to linearize the nonlinear model first and then solve differential equations in state space form or via Laplace transformation with the purpose of generating a state space equations or transfer function for the system that is very convenient for further control analysis.

9.2.2 Linearization About the Operating Point

Linearization is quite straightforward. All we need to do is to expand a nonlinear function via Taylor series around the current operating point and neglect all terms with higher order than the first partial derivatives.

Assume that $f(x)$ is a nonlinear function and x_0 is the current operating point. Now we expand this function around feasible point x_0:

$$f(x)=f(x_0)+\left.\frac{df}{dx}\right|_{x=x_0}(x-x_0)+\left.\frac{d^2f}{dx^2}\right|_{x=x_0}\frac{(x-x_0)^2}{2!}+\cdots+\left.\frac{d^nf}{dx^n}\right|_{x=x_0}\frac{(x-x_0)^n}{n!} \tag{9.7}$$

Neglecting all but the first two terms, $f(x)$ is linearized around x_0 becomes

$$f(x)\approx f(x_0)+\left.\frac{df}{dx}\right|_{x=x_0}(x-x_0) \tag{9.8}$$

Thus, the linearized approximation of the nonlinear equation $f(x)$ is

$$y = f(x_0) + \left.\frac{df}{dx}\right|_{x=x_0} (x - x_0) \tag{9.9}$$

In this way, we are approximating the real nonlinear function by a linear function within the range of validity. The procedure is illustrated in Figure 9.3.

Since $y_0 = f(x_0)$, Equation 9.9 can be rewritten as

$$y - y_0 = \left.\frac{df}{dx}\right|_{x=x_0} (x - x_0) \tag{9.10}$$

By defining $\bar{y} = y - y_0$ and $\bar{x} = x - x_0$, Equation 9.10 becomes

$$\bar{y} = \left.\frac{df}{dx}\right|_{x=x_0} \bar{x} \tag{9.11}$$

Equation 9.11 is the basis for generating linear approximation in control system design.

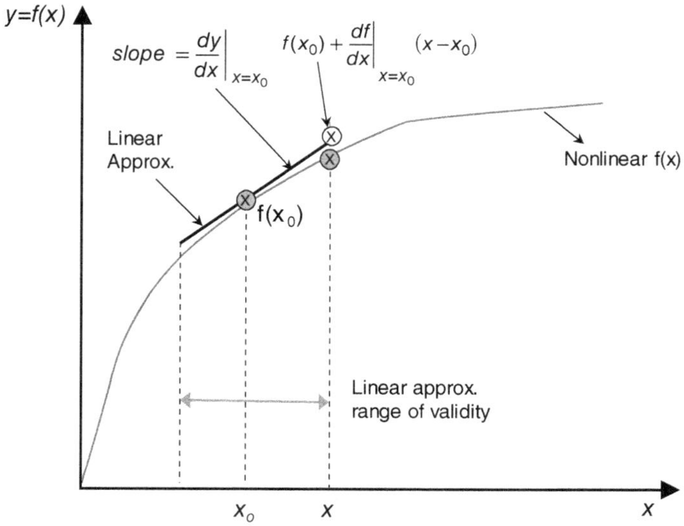

FIGURE 9.3. Linearization about the current operating point.

Example 9.2 Linearize the nonlinear equation for the temperature dependence of specific reaction rate k:

$$k(T) = k_0 e^{-E/RT} \tag{9.12}$$

Expand $k(T)$ around T_0 via Taylor series as

$$k(T) = k(T_0) + \left.\frac{dk}{dT}\right|_{T=T_0}(T - T_0) + \left.\frac{d^2k}{dT^2}\right|_{T=T_0}\frac{(T - T_0)^2}{2} + \ldots \tag{9.13}$$

By neglecting the high-order terms, we obtain linearized function around T_0 as

$$k(T) \approx k(T_0) + \left.\frac{dk}{dT}\right|_{T=T_0}(T - T_0).$$

$$k(T) \approx k(T_0) + k_0\frac{E}{RT_0^2}e^{-E/T_0^2}(T - T_0). \tag{9.14}$$

$$k(T) \approx k(T_0) + \left[\frac{E}{RT_0^2}k(T_0)\right](T - T_0).$$

Therefore, the linear function (9.14) is obtained from the nonlinear equation 9.12 by keeping the constant and the first-order term, while the higher-order terms in this example are neglected.

9.2.2.1 State space model The first-order differential equation (9.14) describes dynamic behaviors of the CSTR system in Figure 9.2. In general, many dynamic systems can be easily characterized by first-order differential equations.

In the control system domain, the general model representation takes the state space form, which is linear time-dependent function of actual state of the system (Friedland, 1986) as

$$x' = \frac{dx(t)}{dt} = f(x, u, t) \tag{9.15}$$

where x is the investigated state variable in a vector form, u is the input variable in a vector form, and t represents time. For a linear system, it can be reformulated as

$$x' = \mathbf{A}(t)x + \mathbf{B}(t)u \tag{9.16}$$

where

$x = [x_1, x_2, x_3, \ldots, x_n]^T$ is the system state vector (CVs)
$u = [u_1, u_2, u_3, \ldots, u_m]^T$ is the input variable vector (MVs)

A is the state or system matrix $n \times n$
B is the input matrix $n \times m$

Equation 9.16 characterizes the state of the system x based on available measurements. In addition if we define output of the investigated system as y, we can describe it as

$$y = \mathbf{C}(t)x + \mathbf{D}(t)u \tag{9.17}$$

where

$y = [y_1, y_2, y_3, \ldots, y_q]^T$ is the system output variable vector (CVs)
C = the output matrix $q \times n$
D = the output matrix $q \times m$; **D** is the "feedthrough (or feedforward) matrix," in cases where the system model does not have a direct feedthrough, **D** = 0

The equation set (9.18) describes system model dynamic in the linear time dependent, and it is called **state space** model of the investigated system:

$$\begin{aligned} x' &= \mathbf{A}(t)x + \mathbf{B}(t)u \\ y &= \mathbf{C}(t)x + \mathbf{D}(t)u \end{aligned} \tag{9.18}$$

State space form that characterizes system of interest is commonly used in control theory due to the fact that the related formulation is a very convenient way of describing system dynamic behaviors, that is:

- Eigenvalues of **A** matrix can help to evaluate system stability.
- Appropriate combination of **A**, **B**, and **C** matrices can evaluate controllability and observability of the system model.

In addition, by applying Laplace transform to the state space equation (9.18), the output function of the system can be obtained as

$$Y(s) = G(s)U(s) \tag{9.19}$$

The transfer function $G(s)$ in the form of matrix $q \times m$ for each input–output combination, with assumption of **D** = 0, can be derived as

$$G(s) = \frac{Y(s)}{U(s)} = \frac{C \times B}{sI - A} \tag{9.20}$$

Therefore, a transfer function describes the dynamics between the input and output, i.e. it "transfers" the input to the output. The details for using Laplace transformation to derive transfer function are provided below.

9.2.2.2 Laplace transformation and transfer function The Laplace transformation is a mathematical way to solve differential equations by converting equations from the time t domain into frequency domain called after transformation s domain. This Laplace method is primarily applied in control system in order to obtain transfer functions in the s domain that allows to investigate the control stability and performance of the system models. In doing so, Laplace transformation changes differential equations into a set of algebraic equations, which are easy to solve and analyze. The solution to the resulting algebraic equations is the transfer function that represents the dynamic behavior of the physical system in the s domain.

The Laplace transformation for a function of time $f(t)$ is

$$F(s) = L[f(t)] = \int_0^\infty f(t) e^{-st} dt \tag{9.21}$$

The inverse Laplace transform is

$$f(t) = \frac{1}{2\pi j} \int_{\sigma - j\infty}^{\sigma + j\infty} F(S) e^{+st} ds \tag{9.22}$$

The commonly used transformation integral pairs are generated in prior and listed in table format for ease of practical applications. For example, Table 9.1 shows some important Laplace transform pairs.

TABLE 9.1. Table of Laplace Transforms

$F(s)$	$f(t)$, $t \geq 0$
1	$\delta(t)$, Dirac delta function
$\frac{1}{s}$	(t), unit step function
$\frac{1}{s^2}$	t
$\frac{2!}{s^3}$	t^2
$\frac{1}{s+a}$	e^{-at}
$\frac{1}{(s+a)^2}$	te^{-at}
$\frac{a}{s(s+a)}$	$1 - e^{-at}$

(Continued)

TABLE 9.1. (Continued)

$\dfrac{a}{s^2(s+a)}$	$\dfrac{1}{a}\left(at-1+e^{-at}\right)$
$\dfrac{a}{s^2+a^2}$	sin (at)
$\dfrac{s}{s^2+a^2}$	cos (at)
$\dfrac{a}{(s+b)^2+a^2}$	e^{-bt} sin (at)
$\dfrac{s+b}{(s+b)^2+a^2}$	e^{-bt} cos (at)
$\dfrac{s+a}{(s+b)^2+a^2}$	$\dfrac{\sqrt{(a-b)^2+a^2}}{a}e^{-bt}\sin(at+\phi),\quad \phi=\tan^{-1}\dfrac{a}{a-b}$

Example 9.3 Use the Laplace transformation to determine the response of the isothermal CSTR of Example 9.1 to a +0.2 step change in the inlet concentration C_{A1} (Erickson and Hedrick, 1999). Also, determine the transfer function of the system where C_{A1} is the input and C_A is the output.

Solution

If we are considering F_1 and T as constant, only concentration C_A as a variable, thus Equation 9.6 does not need to be linearized. So, given F_1 and T, the response to a change in concentration will be the same regardless of the initial concentration. Without loss of generality, the initial concentration will be assumed to be zero because this assumption will allow a transfer function to be obtained for the system. The Laplace transform of the differential equation (9.6) gives

$$sC_A(s)=\frac{F_1}{V}C_{A_1}(s)-\left(\frac{F_1}{V}+k_0e^{-E/RT_1}\right)C_A(s)$$

$$C_A(s)=\frac{F_1/V}{s+F_1/V+k_0e^{-E/RT_1}}C_{A_1}(s) \tag{9.23}$$

Assume the following conditions for the system: $T_1 = 340\,\text{K}$, $k_0 = 1.0\times1.0^{10}$/min, $E/R = 8{,}330.1\,\text{K}$, $V = 10\,\text{m}^3$, $F_1 = 1.0\,\text{m}^3$/min. Now Equation 9.23 becomes

$$C_A(s)=\frac{0.1}{s+0.9894}C_{A_1}(s) \tag{9.24}$$

With the step change, $c_{A_1} = 0.2u(t)$ kmol/m^3 ($u(t)$ is a unit step function), so

$$C_{A_1}(s) = \frac{0.2}{s}$$

Equation 9.24 become

$$C_A(s) = \frac{0.02}{s(s+0.9894)} \tag{9.25}$$

Doing a partial fraction expansion on $C_A(s)$ yields

$$C_A(s) = \frac{0.02}{s} - \frac{0.02}{s+0.9894} \tag{9.26}$$

Taking the inverse Laplace transform gives

$$c_A(t) = 0.02u(t) - 0.02e^{-0.9894}u(t) \tag{9.27}$$

The time response (Equation 9.27) is plotted in Figure 9.4. The final value of the response is 0.02. This final value can also be obtained without plotting the time response by using the final-value theorem on Equation 9.25 to give

$$sC_A(s) = s\left[\frac{0.02}{s(s+0.9894)}\right] = 0.02 \tag{9.28}$$

The transform transfer function of this system is obtained from Equation 9.24 as the ratio of the Laplace transform of the output $C_A(s)$ to the Laplace transform of the input $C_{A1}(s)$:

$$G(s) = \frac{C_A(s)}{C_{A_1}(s)} = \frac{0.1}{s+0.9894} \tag{9.29}$$

The tank model transfer function provides the response of product concentration for any changes in feed concentration C_{A1}, which can be done by manually changing dosage of the concentrated substance (thicken/dilute) or potentially implemented by operators via PID control system.

As you can see starting from the analytical differential equation (9.6) that describes the process in the tank, it is possible to transfer it into the dynamic model represented by Laplace transfer function, $G(s)$, shown in Equation 9.29. For control and automation engineers, the transfer function is the more convenient dynamic solution to the linearized differential equations. In the transfer function form, it is much easier to investigate a specific linearized model of the process and investigate its stability. Also the transfer function form of the process is more suitable for designing an appropriate controller for the process represented by a transfer function.

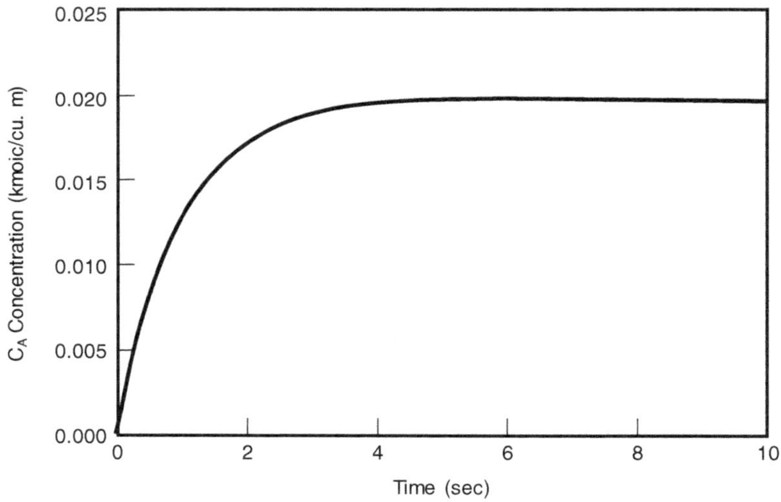

FIGURE 9.4. Response of CSTR to +0.2 change in inlet concentration.

9.2.3 Dynamic Modeling Based on Step Testing

The above discussions explain the procedure how a process model is developed via first principle, then linearized via Taylor Series and converted to a state-space model, and finally solved via Laplace transformation to generate transfer function. However, process modeling by first principle can be highly complicated and involve oversophisticated mathematics for complex problems. Thus, dynamic models are often not constructed based on first principle in practice. Instead, a dynamic model is obtained via step testing, which the most common method used.

When a step change is made to a manipulated or input variable, a response from a control or output variable is observed, and its trend is collected. In doing the step testing, a plant operator makes changes to input variables such as feed rate, column reflux rate, steam rate, etc. from one operating condition to a new condition and then observe the response in output variables such as reaction temperature, column overhead temperature, etc. Four general types of dynamic response models can be observed and used to simulate any processes, which are gain-only, ramp or integrated, first-order, and second-order models (Figure 9.5). These models generated from step testing are based on one-input and one-output.

First-order models have a simple and predictable response, while second- and higher-order models can have multiple peaks and valleys. Ramp or integrating models imitate the level of a tank: if the input and output flows are equal, the tank level remains constant, but if there is a disparity, the tank will overflow or be empty. The response shown in Figure 9.5d can be recognized as an unstable response. As it can be seen that PV as response to step tests is continuously increasing, without

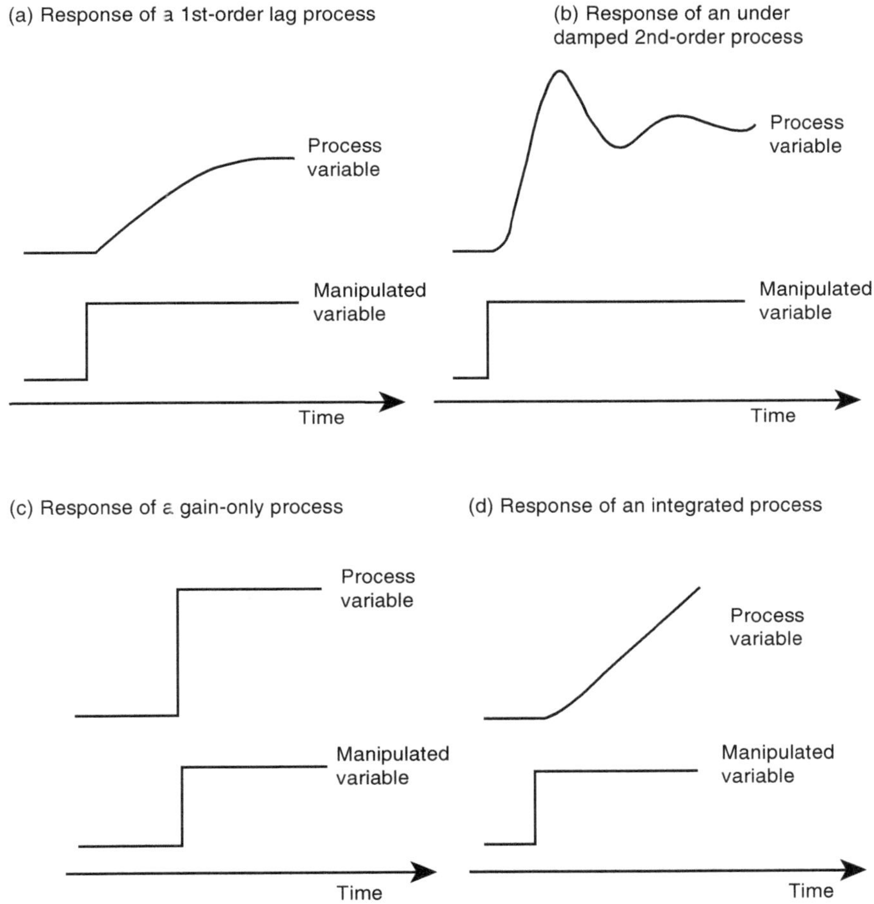

FIGURE 9.5. Four typical dynamic response models. (a) Response of a first-order lag process. (b) Response of an under damped second-order process. (c) Response of a gain only process. (d) Response of an integrated process.

a proper control, action level in the tank or concentration might exceed process safety limits.

There is also possibility to obtain from the step tests oscillated or unstable response; it is common situation that for high-order systems, that is, with wrongly designed PID control system transfer function response could be seen as a merge of responses shown in Figure 9.5b and d, which is increasing unstable oscillation.

Therefore, very often step tests in the real plants are done manually, while the corresponding transfer function can be obtained using dedicated software (i.e., MATLAB). Although step testing was once performed manually, automated stepping software that incorporates decades of experience is now available. The software automates many tasks that were once manual, including stepping the process, collecting data, identifying models, and validating models. The software even makes it possible to conduct these tasks automatically, while a predictive control (MPC) application is controlling the process at the same time, which enables step testing with less impact on process operation. Such automated solution is very convenient for engineers especially when access to the plant is limited, for example, crude platforms on the sea.

9.2.4 Control Parameter Estimation

After the step testing generates a response curve, the next task is to determine model parameters so that the empirical dynamic model is developed and can be used in PID controllers. The parameters for a dynamic model include time constant, dead time, and gain. Let us consider first-order-plus-dead-time (FODT) model as an example for illustration of parameter estimation. Details for other estimation methods can be seen in Erickson and Hedrick (1999).

The transfer function for a FODT model is:

$$G(s) = \frac{Ke^{-s\theta_D}}{\tau_1 s + 1} \tag{9.30}$$

where K = process gain; θ_D = apparent process dead time; and τ_1 = apparent process time constant.

The response function to a step pulse of magnitude A for the FODT transfer function above is

$$y(t) = KA\left(1 - e^{-(t-\theta_D)/\tau_1}\right) \tag{9.31}$$

The response in Equation 9.31 is evaluated for two values of the exponent, 1 (when $t = \theta_D + \tau_1$) and 1/3 (when $t = \theta_D + 1/3 \cdot \tau_1$). When the exponent is 1, the value of the exponent becomes

$$y(t) = KA\left(1 - e^{-1}\right) = 0.632KA \tag{9.32}$$

In other words, the response reaches 63.2% of the step change (A) when $t = \theta_D + \tau_1$. Similarly, when the exponent is 1/3, the value of the response is

$$y(t) = KA\left(1 - e^{-1/3}\right) = 0.283KA \tag{9.33}$$

In this case, the response reaches 28.3% of the step change when $t = \theta_D + 1/3 \cdot \tau_1$.

Using the graphic approach in Figure 9.6 (Ziegler and Nichols, 1942), the model parameters can be identified as follows:

- Step 1: Draw a tangent to the process response curve as its point of maximal slop, the point of inflection is T_2.
- Step 2: The apparent process dead time $\theta_D = T_2 - T_1$. T_1 is the time at which the step change in the MV was imposed; T_2 is the time at which the tangent drawn in step 1 intersects the time axis.

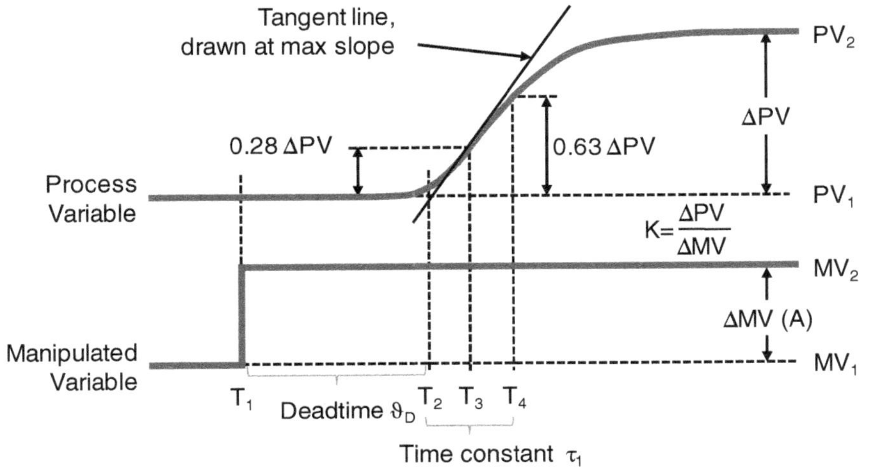

FIGURE 9.6. Parameter estimation for a first-order model from step-testing data.

- Step 3: The apparent process time constant $\tau_1 = T_4 - T_2$. T_4 is the time at which the PV response has reached 63.2% of its change.
- Step 4: The process gain K is the change in the PV divided by the change in the MV as expressed in Equation 9.34:

$$K = \frac{\Delta PV}{\Delta MV} = \frac{\mathrm{PV}_2 - \mathrm{PV}_1}{\mathrm{MV}_2 - \mathrm{MV}_1} \tag{9.34}$$

Applying the method based on the step test response versus step test input, the parameters of the first-order transfer function as Equation 9.30 can be found. In the reality, several step tests with several responses have to be collected, and statistically the best fit has to be calculated to obtain the transfer function of the investigated system.

9.3 REGULATORY CONTROL: MANAGING INDIVIDUAL VARIABLES

So far, we discussed the process model and its transfer function. To control a process via the model, it is common to use a regulatory controller because it can maintain a CV at its setpoint to minimize variability when conditions change. Regulatory control loops can be broadly classified as feedback and feedforward types. Figure 9.7 shows a simplified scheme of the control strategies with feedback and feedforward PID controls for a regulatory controller block:

Feedback control. In feedback control, the controller compares the measured CV to its setpoint (SP) and computes a compensating move based on the deviation between the actual CV and the desired SP values. Think about feedback

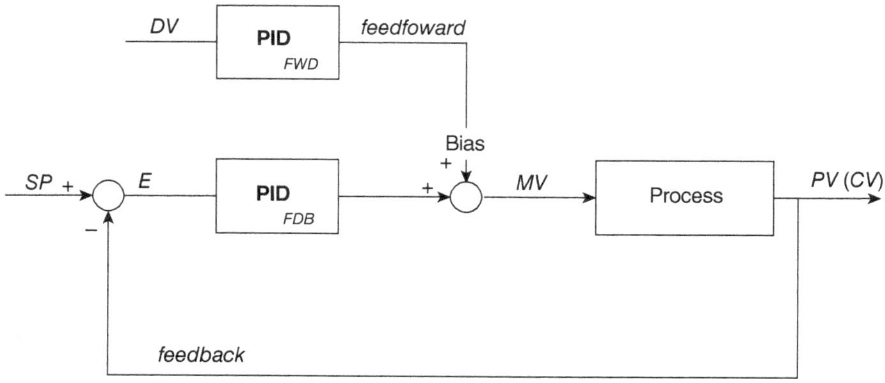

FIGURE 9.7. Simple control scheme.

control—only responding to an error after it happens—like driving while looking in the rearview mirror; only after the car is in the ditch can a correction be made to get back on the road again.

Feedforward control. In addition to monitoring the CV, the controller measures a disturbance variable (DV), such as a change in weather, and tries to predict the effect this change will have on the CV. This requires prior knowledge of the process response patterns. Unlike feedback control, which compensates for errors after they occur, feedforward control aims to combine feedback with another process signal to minimize errors and compensate as they happen.

These two PID controls will be discussed in details in Sections 9.4.5.1 and 9.4.5.2.

In a real system, signals can be received or sent via automation devices such as sensors and actuators. An example of the real system (Smith, 2016a) is used to walk through these concepts, and Figure 9.8 presents a simple temperature control loop, which consists of three essential pieces:

- **A measurement device**—The hot air temperature transmitter. It provides a measure of some PV to be controlled, called the CV. The CV could be a level, temperature, or flow rate, or something more complex, such as a Kappa number (i.e., a measure of the lignin content remaining in wood pulp).
- **A controller**—Typically CPU computer. It targets the desired value of the CV, called the setpoint (SP). Relative to the SP, the controller calculates a compensating control move, or output (OP), and sends that signal to the control actuator. The variables affected by the control move are called MVs.
- **Control actuator**—The control valve. It implements the control move received from the controller (OP) into the process. Control actuators are often valves; other common actuators include louvers and variable-frequency drives on pumps or fans.

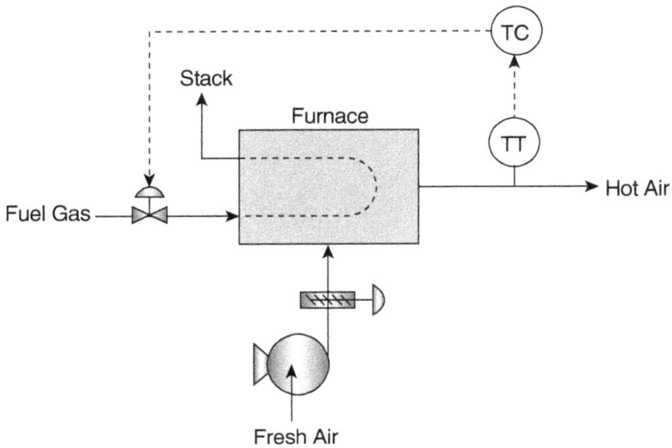

FIGURE 9.8. In this piping and instrumentation (P&I) diagram for a simple temperature control loop. The hot air temperature TT is controlled by controller TC that sends signal command (OP) to actuator that is a valve on the fuel gas supply line. Source: From Smith CL (2016a), reprinted with permissions of American Institute of Chemical Engineers.

Among regulatory controllers, the most common type is the PID loop (Figure 9.9). Figure 9.9 shows a block diagram representation of the PID loop in Figure 9.8. This block diagram contains three function blocks:

- **PV block**—Converts the input from the measurement device to a value in engineering units (i.e., mA into °C for the hot air temperature). The result is usually referred to as a PV or CV.
- **PID block**—Provides the PID control calculations. The result of the control calculations is the output (OP) to the final control element in engineering units of % open.
- **Valve block**—Converts the % open input to whatever signal is required by the final control element (it could be voltage, mA, etc., depending on device converter type).

9.3.1 PID Block

In addition to the PV input, the PID block also requires an additional input—the desired value, or setpoint, for the PV input. Control engineers often use the following terminology for the PV input and the controller output:

- **Controlled variable** (CV)—The PV value is to be maintained and desired to be at or near the setpoint.
- **Manipulated variable** (MV)—The variable whose value is at the discretion of the controller. For the loop in Figure 9.8, the MV is the fuel gas valve opening. Throughout this chapter, the controller output will be designated by M.

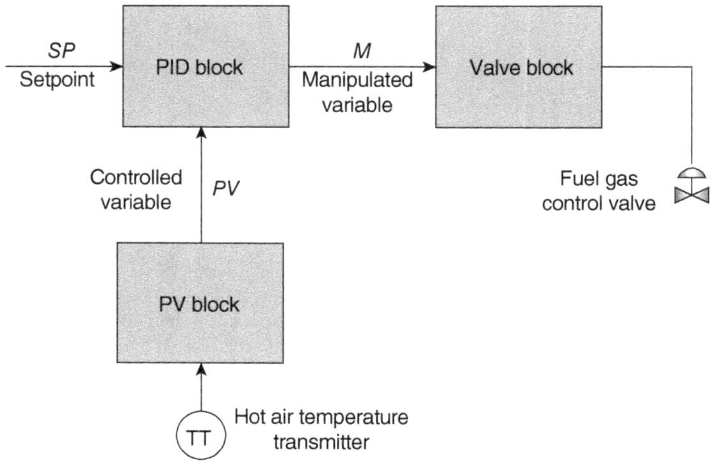

FIGURE 9.9. In this block diagram representation of the PID loop in 2, the setpoint is entered as input by the operator, and the process variable is determined by the hot air temperature transmitter. The PID block provides the PID control calculations. Source: From Smith CL (2016a), reprinted with permissions of American Institute of Chemical Engineers.

Process relationships such as material and energy balances are expressed in terms of flows, not valve or damper openings. For the loop in Figure 9.8, process-minded people sometimes refer to the fuel flow as the MV. Technically, such flows are dependent variables that are largely determined by the valve opening but are affected by upstream pressure, downstream pressure, and fluid properties.

9.3.2 Controller Action

Figure 9.10 illustrates the PID block in more detail. It has three features:

- **Comparator**. The control error (E) is the difference between the setpoint (SP) and the process variable (PV). Figure 9.10 suggests that the control error is computed as SP minus PV. But for some loops, the control error must be computed as PV – SP.
- **PID control attributes**. The differential equation for PID control contains three possible attributes: proportional, integral (also called reset), and derivative. The tuning coefficients are called PID control attributes. These coefficients are set and tuned once but sometimes need to be retuned over the time when valve degrades or process changes.
- **PID control modes.** PID can work in manual mode, auto mode, and program or remote cascade mode.

When drawing a block diagram for a PID loop, control engineers often label the control modes block in Figure 9.10 as the controller. Indeed, this block is an important one.

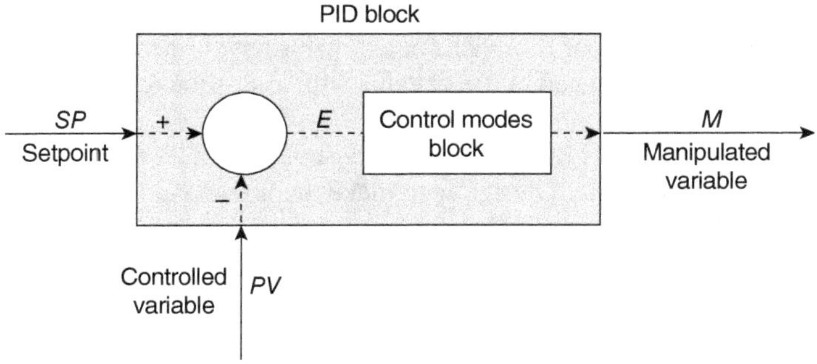

FIGURE 9.10. Within the PID function block, a comparator computes the control error, *E*. The differential equation for PID control is implemented in the control modes block, whose output is OP the manipulated variable, *M*. Source: From Smith CL (2016a), reprinted with permissions of American Institute of Chemical Engineers.

Before examining the individual attributes, an explanation of controller action is essential. Consider the control loop in Figure 9.8. The hot air temperature is controlled by manipulating the fuel gas valve opening. Suppose the hot air temperature is increasing. How should the controller respond? The controller should decrease the fuel gas valve opening. An increase in the controlled variable (hot air temperature) requires a decrease in the MV (fuel gas valve opening or fuel flow).

The above discussions explain the PID controller actions directionally for the system in Figure 9.8. In the following, we will explain how to determine controller tuning parameters P, I, and D that effect on how fast and how big the controller action (MV) should be.

9.4 PID CONTROLLER MODELING

Conventional feedback controllers use several methods to determine the value of the CV as the controller output. These methods, called the modes of control, are defined as follows:

- **Proportional (P):** This mode determines the speed of response for the loop. The higher the control gain, the faster the loop responds. However, high control gain may lead to overshoot and oscillation in the response. For each loop, there is maximal gain or ultimate gain for which the loop is stable.
- **Integral (I):** This mode does not produce a fast response. The contribution of this mode is to ensure that the loop can line out at its setpoint (i.e., there is no droop) and minimize steady-state error (called offset). It accomplishes this by adjusting the value of controller output bias (M_R). Changing the bias too rapidly could result in more oscillation and, in extreme cases, lead to an unstable loop.

- **Derivative (D):** This mode can reduce the overshoot and oscillation. In turn, this mode allows use of a higher control gain, which increases the speed of response. It is mostly used in the systems with long-time response and inertia.

The reason of the emphasis on speed of response is that most processes are slow. The purpose of control system design is to make them respond faster and stable.

A PID controller is a control loop mechanism employing feedback that is widely used in industrial control systems (Figure 9.10). A PID controller continuously calculates an error as the difference between a desired setpoint (SP) and a measured PV and applies a correction based on proportional (P), integral (I), and derivative (D) terms, In other words, PID refers to different types of correction for potential disturbance—hence the name PID.

In every PID cycle, the CV is compared with the SP, and then the proportional, integral, and derivative elements are combined to calculate the move (MV). The combination of proportional, integral, and derivative factors is determined by the type of process, control block used, which depends on the control infrastructure (e.g., type of DCS/PLC controller). For example, a standard PID block adds the contributions of proportional, integral, and derivative to the MV signal, but a PID gap control block additionally allows for tuning within a user-specified range around the SP or PV gap.

9.4.1 Proportional Model

In process applications, almost all PID controllers rely at least to some degree on the proportional mode of control.

A proportional control move (MV) is calculated directly from the current error signal E (the difference between CV and SP) multiplied by the proportional gain as

$$\mathrm{MV}(t) = K_\mathrm{P}E(t) \tag{9.35}$$

where K_P is the constant or proportional gain (a tuning coefficient). It does not consider past error, but only the error in the current cycle.

Take a closer look at Equation 9.35. Consider the hot air temperature loop in Figure 9.8. Suppose that the setpoint is 210 °C and that the controller has acted so that the hot air temperature is also 210 °C. Since PV = SP in this case, the control error (E) is zero. Substituting $E = 0$ into Equation 9.35 suggests that MV = 0, which means that the fuel gas valve is 0% open (fully closed). This makes no sense. The fresh air enters at the ambient temperature, so some fuel is required to heat the air to 210 °C. The fuel valve cannot be fully closed. Thus, a bias is introduced to the proportional mode to address the issue. The equation is best described as proportional-plus-bias:

$$\mathrm{MV}(t) = K_\mathrm{P}E(t) + M_\mathrm{R} \tag{9.36}$$

where M_R is the bias, specifically the controller output bias (the term bias is used in other contexts within process control). The control error E is computed and then

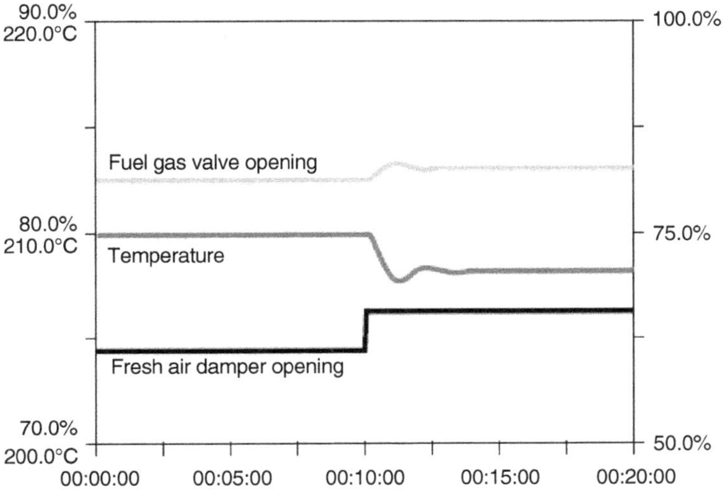

FIGURE 9.11. A proportional-only controller exhibits droop after a 5% increase in the fresh air damper opening. Initially, the controller output (fuel gas valve opening) is 82.5% open. When the fresh air damper opening increases (black line), the temperature drops from its setpoint at 210°C drifted to 207.8 °C (dark gray line), and the fuel gas valve opening increases to 83.2% (light gray line). For the controller to return the temperature to its setpoint, the controller output bias, M_R, must be adjusted. Source: From Smith CL (2016a), reprinted with permissions of American Institute of Chemical Engineers.

multiplied by the controller gain K_P, and the result added to the controller output bias M_R to obtain the MV move.

Figure 9.11 shows the behavior of the hot air temperature controller for a 5% increase in the fresh air damper opening. An increase in the fresh airflow rate causes the hot air temperature to droop (or drift) down (from 210 to 207.8 °C, an error of 2.2 °C); a decrease causes the hot air temperature to ride up.

Following the increase in the fresh airflow rate, the hot air temperature controller opens the fuel gas valve from 82.5 to 83.2%, an increase of 0.7%. This increase is entirely due to the K_PE term in the proportional-plus-bias equation (9.36).

At the higher fresh airflow, a controller output of 83.5% is required to attain a hot air temperature of 210 °C. But should a temperature of 210 °C be attained, the control error would be zero, and the controller output would be 82.5% (the value of the controller output bias, M_R).

A value of 210 °C for the hot air temperature can be obtained by adjusting the controller output bias. Most commercial process controls do not provide features to readily display the value of the controller output bias nor accept changes to its value. This was not always the case. The earliest controllers were proportional only, but provided an adjustment (often labeled manual reset) for the controller output bias, M_R, that permitted the process operator to remove any droop or offset.

9.4.2 Integral Mode

Integral (or reset) control acts on accumulated error over multiple cycles. The longer the CV is out of specifications (above or below the SP), the more integral action is applied.

To eliminate the droop exhibited by the hot air temperature response as shown in Figure 9.11, the controller output bias, M_R, must be increased. This can be achieved by adjusting MV at a rate proportional to the control error, E:

$$\frac{d\text{MV}(t)}{dt} = \frac{1}{T_I} E(t) = K_I E(t) \tag{9.37}$$

where T_I is integral time constant while K_I is the integral gain (a tuning coefficient). For positive values of E, the controller output bias MV increases; for negative values, MV decreases.

Integrating Equation 9.37 gives the integral mode:

$$\text{MV}(t) = \frac{1}{T_I} \int_0^t E(t)\,dt \tag{9.38}$$

In digital controller, T_I repeats per minute or second, and the integration is approximated by a summation as

$$\text{MV}(n) = \frac{T}{T_I} \sum_{i=0}^{n} E(i) = \text{MV}(n-1) + \frac{T}{T_I} E(n) \tag{9.39}$$

where T is the sample period, n is the current sample number, and $E(n)$ is the error at the nth sampling instant.

Addition of the integral mode to the control equation does not assure that a loop will line out, but if it does line out, the PV will be equal to the setpoint. A controller with integral action will not experience the droop depicted in Figure 9.11.

It must be noted that the integral mode is used to force the offset to zero. However, it tends to slow the system down, in contrast to proportional mode.

9.4.3 Derivative Mode

Derivative control is also called rate action, i.e. evaluates the rate of change of the error or PV values from the last cycle to the current cycle. The equation for the rate action is commonly as

$$\text{MV}(t) = T_D \frac{dE(t)}{dt} \tag{9.40}$$

where T_D is the derivative time constant in minute or second and T_D also is a tuning coefficient. Its value suggests how far into the future the PV can be projected using its current rate of change. This depends on the nature of the process. One of the most common applications of the derivative mode is for temperature processes, which are slow to respond.

The derivative mode for the controller is anticipating the process and acting in advance. The larger the rate of change is in the undesired direction, the more derivative control action is applied. Derivative action is generally used sparingly because it can introduce variability.

This is illustrated in Figure 9.12. PV is below the setpoint but moving toward it. The projected value is closer to the setpoint than the current value of PV.

9.4.4 PID Modes Together

In the noninteracting form of PID, all three terms (P, I, D) are calculated independently and summed to obtain the move for MV together with the bias (M_R) as

$$\mathrm{MV}(t) = K_P E(t) + \frac{1}{T_I}\int E(t)\,dt + T_D \frac{dE(t)}{dt} + M_R \tag{9.41}$$

Equation 9.41 can be graphically represented as in Figure 9.13.

- The first term is proportional (P): if the error is large and positive, the control output will be proportionately large and positive, taking into account the gain factor "K." Using proportional control alone will result in an error between the setpoint and the actual process value, because it requires an error to generate the proportional response.

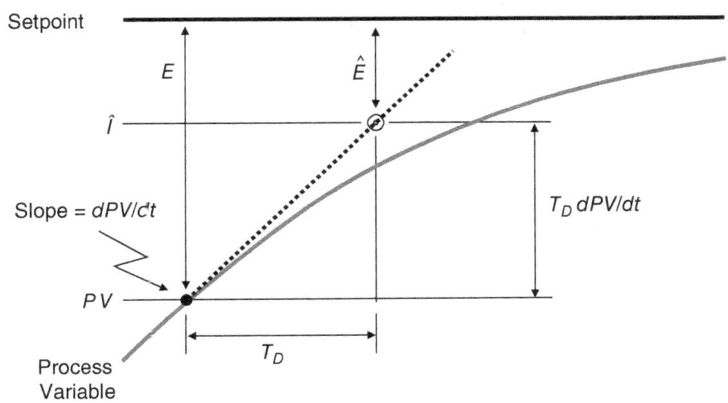

FIGURE 9.12. Based on the rate of change of PV, a projected value, $\hat{I}$, for the process variable can be computed. The control calculations are then based on the projected error, $\hat{E}$, computed from $\hat{I}$.

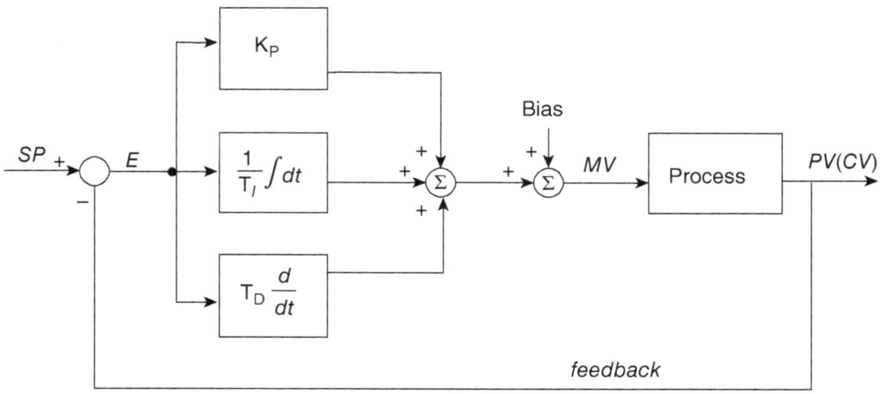

FIGURE 9.13. Noninteracting PID control.

- The second term is integral (I) that depends on the past values of the SP – PV error over time and is responsible for getting rid of accumulated error *E*. When the error is eliminated, the integral term will cease to grow. This will result in the proportional effect diminishing as the error decreases, but this is compensated for by the growing integral effect.
- The third term is derivative (D) that is responsible for speeding up MV reaction. It is a best estimate of the future trend of the SP – PV error, based on its current rate of change. It is sometimes called "anticipatory control," as it is effectively seeking to reduce the effect of the SP – PV error by exerting a control influence generated by the rate of error change. Thus, derivative action on the measurement is preferred in most industrial applications, because the SP changes often do not cause a large and quick enough change in MV (especially for temperature control). Also sometimes derivative part is applied directly to PV value, so instead of acting on *E* error of PID, the D part can start work immediately after a trend of process value changes direction.

The balance of these effects (P, I, and D) is achieved by loop tuning to produce the optimal control function. The tuning constants must be derived for each control application, as they depend on the response characteristics of the complete loop external to the controller. These are dependent on the behavior of the measuring sensor, the final control element (such as a control valve), any control signal delays, and the process itself. Approximate values of constants can usually be initially entered knowing the type of application, but they are normally refined, or tuned, by "bumping" the process in practice by introducing a setpoint change and observing the system response.

Example 9.4 Let us assume we have a simple first-order process model that we would like to control via PID controller with only P parameter (Figure 9.14). Process model transfer function described in Equation 9.29 can be expressed as follows:

$$C_A(s) = \frac{0.1}{s+0.9894} C_{A_1}(s) \tag{9.42}$$

Transfer function of the P controller that will control C_{A1} flow is described by

$$G_p = K_p E(s) \tag{9.43}$$

Open-loop transfer function for the system is expressed as

$$G_{OL} = G_p(s) C_A(s) \tag{9.44}$$

Closed-loop transfer function of the control scheme is expressed as follows:

$$G_{CL} = G_p(s) C_A(s) / \left[1 - G_p(s) C_A(s)\right] \tag{9.45}$$

By evaluating denominator $[1 - G_p(s)C_A(s)]$ of the Equation 9.45, we can find stability of the closed-loop system.

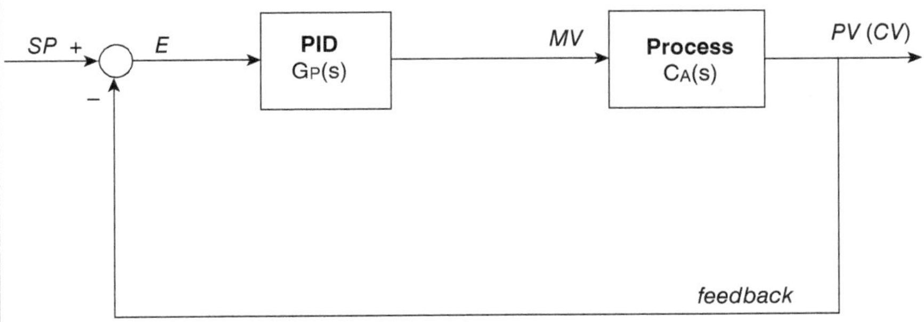

FIGURE 9.14. Control scheme for Example 9.4.

9.4.5 Advanced Regulatory Control

Cascade and feedforward controllers are the most popular designs used in the advanced regulatory controllers.

9.4.5.1 Cascade control Cascade control is the most common type of advanced regulatory control. There are two control loops, the faster one (secondary or slave loop) and the slower one (primary or master loop). The master loop sets the setpoint to the slave loop. The slave loop has the following features:

- Directly affect the PV of the primary loop;
- Minimize disturbance affecting the master loop;
- Have dynamics much faster than the master loop.

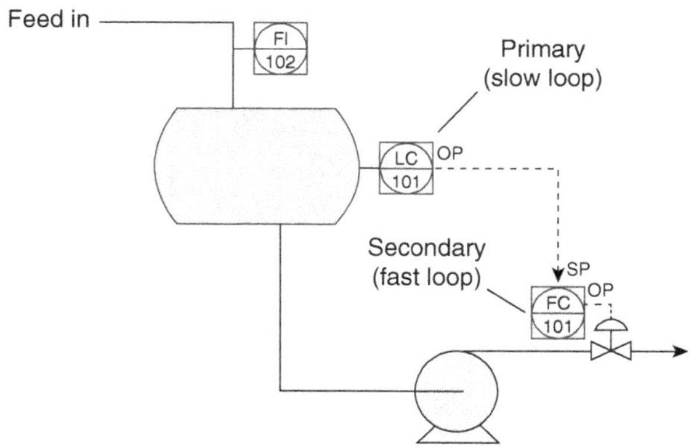

FIGURE 9.15. Level-to-flow cascade control to handle pressure disturbances. Source: From Lehman KA (2018), reprinted with permissions of American Institute of Chemical Engineers.

Level-to-flow is a typical type of cascade control. A single feedback controller controls the level in a vessel by adjusting the outlet valve directly as shown in Figure 9.15 (Lehman, 2018). The slave control loop (FC101), which is faster, is used to deal with pressure disturbance before it affects the master loop (LC101), which is slow.

Also as in advance regulatory control, PID slave controllers can be used to get setpoints not directly measured (like level in the example above in Figure 9.15), but they can be calculated from the external algorithm, that is, higher control level APC or some other calculations that are specific for application.

9.4.5.2 ***Feedforward control*** Feedforward control can significantly reduce process variability and improve stability by responding to disturbances proactively, instead of reacting to their effects. An effective FF variable candidate should:

- be measurable and reliable;
- cause a consistent process response;
- provide enough lead time to be better than feedback alone.

In Figure 9.16, a cascade control loop controls the heater outlet temperature (TC203, the master loop) by adjusting the fuel gas pressure (PC204, the slave loop). In this example, the heater outlet temperature (TC203) uses the PIDFF algorithm additionally to the example in Figure 9.15 to proactively compensate for change in feed rate FR (FI201). The fuel gas pressure PID controller handles disturbances from the fuel gas header, but feed flow changes can also affect the temperature. Thus, the temperature controller (PIDFF) is configured with FF algorithm. It acts like a standard PID controller, with a FF component added to the PID output (OP). The FF component is the incremental change of the FF input (FI201) multiplied by the FF gain (K_{FF}). The FF gain is the change in the PID output (OP) required to offset the effect of the feed rate change ($\Delta OP/\Delta F_R$).

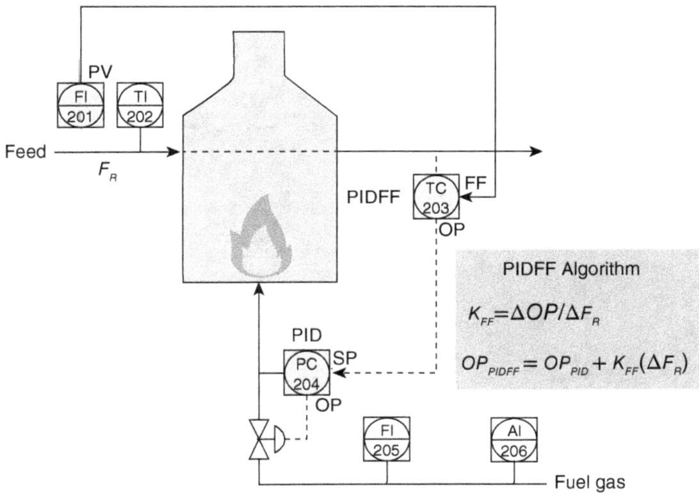

FIGURE 9.16. The PID with FF algorithm (PIDFF): it combines feedback (PID) control (PC204) with feedforward action to proactively compensate for change in feed rate FR (FI201). Source: From Lehman KA (2018), reprinted with permissions of American Institute of Chemical Engineers.

9.4.5.3 PID control performance measures Ultimately, the purpose of the PID controller is to force the PV to follow a setpoint by adjusting MV. The effectiveness of the control system centers on the speed and stability. Does the CV change rapidly or slowly when the given MV varies? How stable the control system behaves? The answer to these questions determines what kind of controller manipulations should be selected. A too aggressive controller response to a change in MV could cause the system response to oscillate and make the system unstable, which is not acceptable. Thus, stability is of vital concern in feedback control systems

Sometimes, the setpoint is not a constant but a trajectory. For most processes, it is impossible to make the PV perfectly follow the setpoint. Hence, quantitative control performance measures are required to judge the quality of the controller, which focus on two external inputs to the system: setpoint and disturbance input changes.

- Setpoint input change: The most common control loop performance is based on changes to the setpoint. In some processes, the setpoint is set as a constant for long period of time. In others, the setpoint is changed at frequent intervals. The common control performance measures are based on a step change in the setpoint, which corresponds to the situation where an

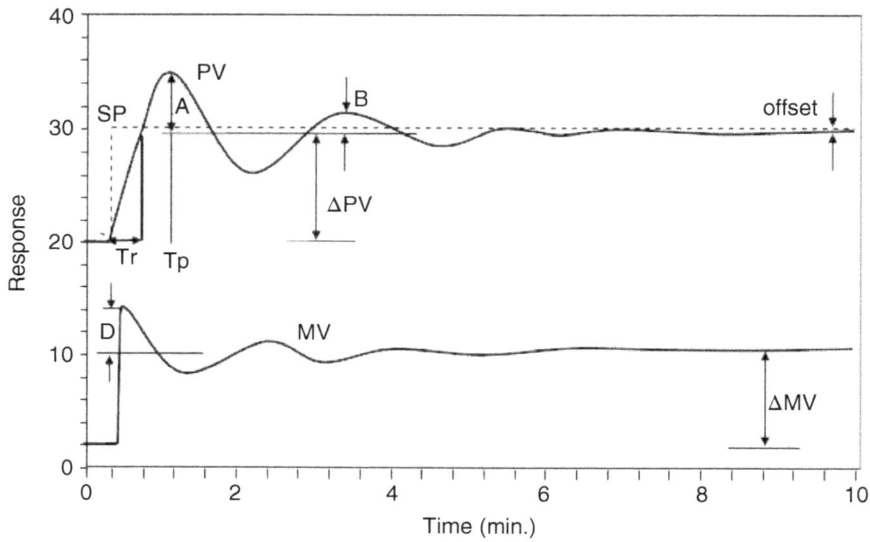

FIGURE 9.17. Typical CV and MV response to a setpoint (SP) change.

operator makes an abrupt change in the setpoint and waits for the PV to reach a steady-state value. A typical MV and PV response is shown in Figure 9.17. The following aspects of these curves are used to evaluate the control system performance:

◦ Offset: The desired offset should be zero.

◦ Rise time: It is the time required for the CV to go from 10 to 90% of the change. A short rise time is should be determined based on the constraints on the magnitude of MVs.

◦ Percent overshoot: It is the amount that CV proceeds beyond its final value. Generally, minimum overshoot is desirable as overshoot could have detrimental effect on downstream processes.

◦ Damping ratio: It is the ratio of the second peak to the first peak in the response (B/A in Figure 9.17).

◦ Settling time: It is the time from the setpoint change to the time that CV response has settled within a small band (usually 2–5%) of the final value. Generally, a short rise time means a short settling time.

- Disturbance input changes: Disturbances are those uncontrolled inputs to the process that affect the controlled variables. These inputs generally cause significant deviation in process value from the setpoint if corrective action is not taken.
- Control parameter tuning: The purpose is to have the control loop performance to match the behavior of the process under control. In other words, control parameter tuning is to get the control characteristics "in tune" with the process characteristics. For P, I, and D modes, each has a tuning coefficient. The

objective of tuning is to obtain a value for each of the tuning coefficients, which would achieve desire performance listed earlier (settling time, offset, overshoot, rise time, etc.). The details of control parameter tuning can be seen in Smith (2016b).

There are other performance measurement methods to evaluate control performance (i.e. sensitivity, Bode approach, Nyquist analysis, sensitivity criteria, etc.). However, in practice those shown in Figure 9.17 are most commonly in use.

9.5 ADVANCED PROCESS CONTROL (APC)

APC is an intelligent layer above the regulatory control layer (Figure 9.1) that treats an entire process unit as a single multivariable system. While PID controllers reduce operation variability thereby improving process stability, APC applications enable operation of the process unit closer to its specification limit—the optimum point at which the process is most profitable (Figure 9.18).

To continuously maximize the profit of a process unit, APC uses dynamic models of the process that show how an MV affects the CV over time. APC coordinates and decouples these effects while keeping the unit within its defined operating window. Many APC controllers include an integrated optimizer that pushes the process unit toward its most profitable state by solving for the maximum value of all economic variables in each iteration.

APC has three significant features, which are discussed in details below:

1. Multivariable optimization—which helps coordinate and decouple the effects of interactions among multiple PVs in parallel finding optimum solution for the multivariable process model.
2. Dynamic modeling—which could capture transition in variations in process parameters, start-up, shutdown, etc.
3. Model predictive—MPC uses dynamic models to predict process behavior in the future, and this information is then used to proactively control the process and correct future errors.

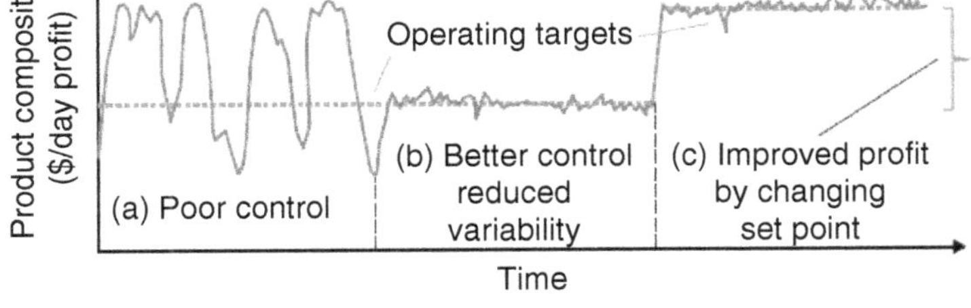

FIGURE 9.18. Operation performance for a process unit. (a) Current operation, (b) reduced variability via PID controllers, and (c) increased profit via APC.

9.5.1 Multivariable Control

Multivariable control is a technique that allows us to deal with more than one CV associated with PID controllers. PID controllers are applied to control one process value via its setpoint, and it is called SISO (single-input single-output) systems. A multivariable system called MIMO (multi-input multi-output) can control at the same time several process values and fulfill control performance objectives at the same time. The reason of such approach is that in reality for a particular piece of equipment or a process unit, two or more PV or CVs must be kept at their target values or setpoints although there are interactions between them (coupling). To be able to achieve this, several MVs are used to control CVs at the optimal values.

To make this a bit more practical, let us consider a simple blending problem, that is, blending of two material streams, where both streams consist of pure components A and B (Figure 9.19). The task is to deliver a final product in the required quantity (flow rate) and concentration of component A with considerations of disturbance in the pressure of streams A and B.

The two CVs are the total flow rate (F_1, CV_1) and the concentration of component A in the final product (A_1, CV_2). There are two MVs, namely, the flow rate of component A (F_A, MV_1) and the flow rate of component B (F_B, MV_2). Now let us see how this system behaves.

When the flow rate of component A (MV_1) changes, then the total flow CV_1 will change, but the concentration CV_2 will also change. The similar interaction happens when MV_2 is varied. In other words, when one MV is changed to correct for a deviation in one control objective, it causes unavoidable and unwanted change in the second one. So, clearly a different solution from the single loop control is needed.

In this case, multivariable control is required as it provides the mechanisms that allow us to change the setpoint of one CV without disturbing the other ones, that is, without driving them away from their setpoints. This exercise is called decoupling. We can therefore also say that MV control is a technique that provides the decoupling

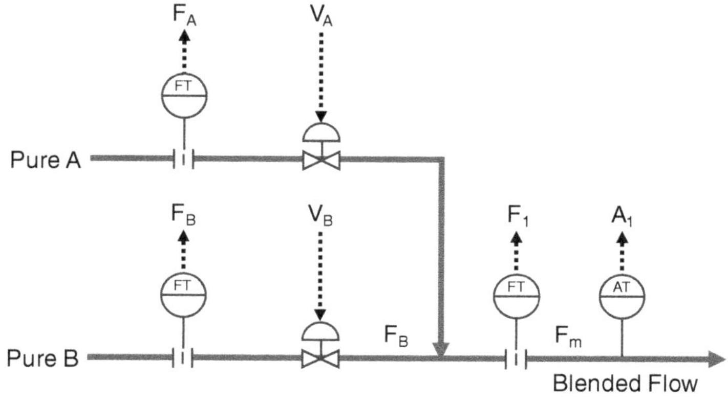

FIGURE 9.19. Blending control problem.

of interacting variables. To achieve this, the control actions cannot be computed individually, as in the case of a PID controller where there is only one single input and one single output (SISO) of the controller. Instead, the actions of all MVs must be determined in a coordinated way, in one global control scheme that considers all the CVs and MVs, a scheme with multiple inputs and multiple outputs (MIMO).

For illustration of multivariable modeling, this blending problem can be modeled by linear programing. We can formulate the relationship between the MVs and CVs as

$$\begin{aligned} \mathrm{CV}_1 &= a_1 + a_{11}\mathrm{MV}_1 + a_{12}\mathrm{MV}_2 \\ \mathrm{CV}_2 &= a_2 + a_{21}\mathrm{MV}_1 + a_{22}\mathrm{MV}_2 \end{aligned} \tag{9.46}$$

where the parameter a_{11} represents the process gain of the effect of MV_1 on CV_1, a_{12} the gain for the effect of MV_2 on CV_1; a_1 and a_2 are bias terms that describe initial concentration for CV_1 and CV_2.

If rearranging Equation 9.46 by moving the CVs and bias terms a_1 and a_2 to the left-hand side while maintaining the gains and MVs on the right-hand side, we obtain

$$\begin{aligned} \mathrm{CV}_1 - a_1 &= a_{11}\mathrm{MV}_1 + a_{12}\mathrm{MV}_2 \\ \mathrm{CV}_2 - a_2 &= a_{21}\mathrm{MV}_1 + a_{22}\mathrm{MV}_2 \end{aligned} \tag{9.47}$$

The LP formulation can be shown in a table format as shown in Figure 9.20 (Eder, 2007). Its core is the matrix that contains all the process gains. For every MV there is one column and for every CV, one row. Additional "artificial" variables can and will be used, and thus we normally have more rows and columns present. The upper (UBV) and lower limits (LBV) for the two MVs are indicated as well. The solution vector (SOLN) will give us the needed answer, i.e. the values of the MVs that satisfy the CVs (setpoints).

On the top of Figure 9.20, we have the objective vector that guides the solution as multiple solutions are available, which is manifested by the fact that the attainable solution space for MV_1 and MV_2 is shown in Figure 9.21. This is the way to drive the solution into the desired direction when extra degrees of freedom exist. The objective function can be expressed as

$$\text{Minimize} \quad J = \sum_{j=1}^{2} \left\{ \sum_{i=1}^{H} \gamma_j^2 \left(\mathrm{CV}_{k+i,j}^{\mathrm{sp}} - \mathrm{CV}_{k+i,j}^{\mathrm{mp}} \right)^2 + \sum_{i=1}^{\mathrm{m}-1} \lambda_j^2 \Delta MV_{k+i,j}^2 \right\}$$

$$\begin{aligned} \text{Subject to} \quad & \mathrm{CV}_j^{\mathrm{L}} \le \mathrm{CV}_{i,j} \le \mathrm{CV}_j^{\mathrm{U}} \quad (j = 1,2) \\ & \Delta MV_j^{\mathrm{L}} \le \Delta MV_{i,j} \le \Delta MV_j^{\mathrm{U}} \quad (j = 1,2) \\ & \mathrm{MV}_j^{\mathrm{L}} \le \mathrm{MV}_{i,j} \le \mathrm{MV}_j^{\mathrm{U}} \quad (j = 1,2) \end{aligned} \tag{9.48}$$

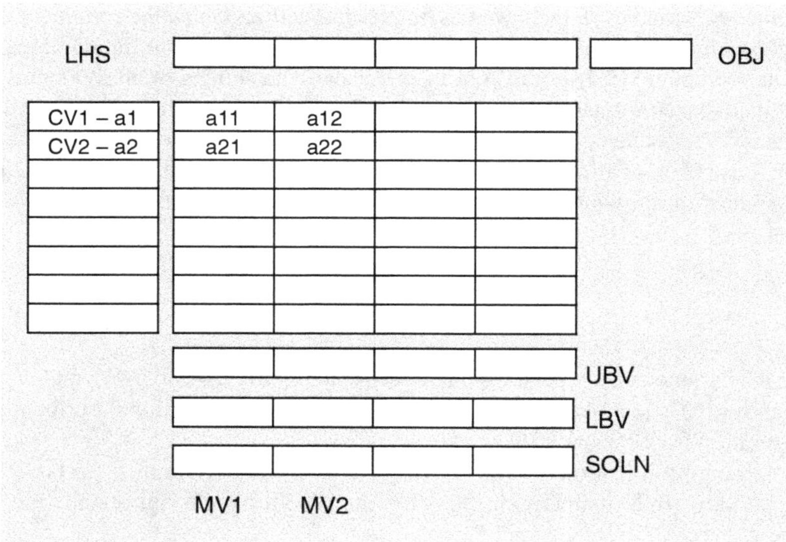

FIGURE 9.20. Linear model for multivariable control.

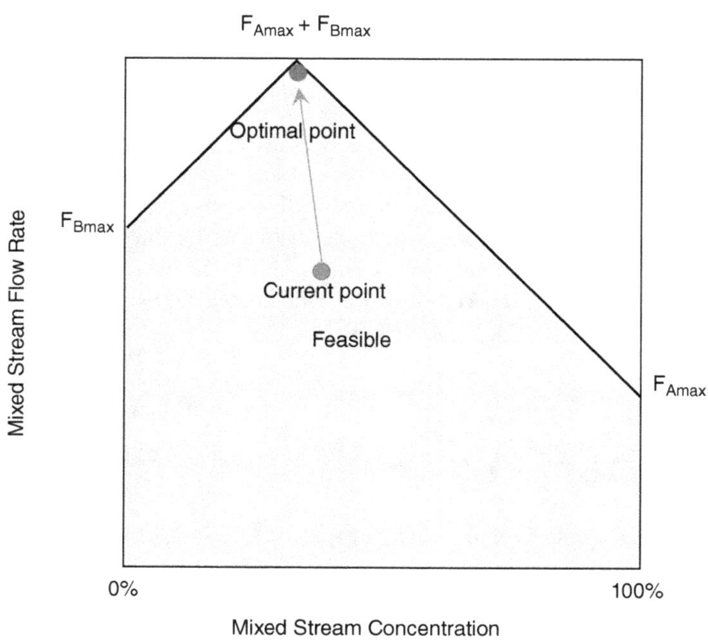

FIGURE 9.21. Operating window for the blending problem. APC improves probability by pushing operation closer to the true process constraints, without violating them.

where i = discretized time; H = total prediction horizon; m = control horizon; γ = CV weighting factors; λ = MV weighting factors that allow the tuning of the controller; CV^{sp} = CV target setpoint or reference trajectory; CV^{mp} = model prediction for CV; and ΔMV = MV moves.

The objective function J intends to find minimum delta moves of MVs (e.g. MV_1 and MV_2 in) order to achieve CV targets (Figure 9.22). All these potential moves can be made within possible operation windows defined by boundary conditions limited by MV_U, MV_L, CV_L, and CV_U.

Vector coefficients λ and γ determine what should be relative impact and with which direction of CV and MV in finding optimal solution. For example, we might decide that more important for us would be MV_1 (blending components is very expensive) rather than final composition of the blending product CV_2. Therefore, in such case in Equation 9.48, λ_1 will be greater than γ_2. Basically, coefficients λ and γ define optimization goal in respect to CVs and MVs.

In the optimization model, there are no dynamics considerations. Therefore, specification of CVs is set equal to the steady-state targets and is not dynamically optimum, which can be dynamically infeasible and would result in excessive MV moves without additional move suppression. This major shortcoming will be overcome by the MPC strategy.

This blending example explains the multivariable control concept and related algorithm, which is related to the APC control layer only. In the real design of the control system, it is more complex because it needs to include lower layer of the controllers (PID flow controllers for stream A and stream B) and some system-specific behaviors (delays, buffers, actuator and sensor dynamics, etc.).

9.5.2 Model Predictive Control

MPC is an advanced optimization-based control strategy applicable to a wide range of chemical process industries (CPI) applications. It is a multivariable strategy that relies on dynamic models and encompasses constraints, actuators, process states, process outputs, and other variables. Multivariable optimization problem described in the previous section can be extended for MPC.

To address control dynamics, every relationship is described by a series of parameters k_i, the step response at every time point. Summing up these parameters over time delivers the current value of the controlled variable (or process output) in question, if a step response y has settled after n samples:

$$y(t) = y(0) + \sum_{i=1}^{n} k_i \Delta u(t-i) \tag{9.49}$$

where k_i are the samples of the process unit step response and Δu or ΔMV is the step change in the process input.

As shown in Figure 9.22, a dynamic trajectory shows the path the process follows toward the new setpoint, which can be defined by the user or by the model. The control action is computed by finding a compromise in minimizing the deviations from the trajectory on one side and avoiding overactive movements of the MV(s) on the other.

By applying the concept of funnel design (Honeywell, 2010), the formulation below (Lu et al., 1996) based on range control algorithm (RCA) can determine the MV moves and the CV reference trajectories simultaneously:

$$\begin{aligned} &\underset{x,y}{\text{Minimize}} && W(Ax-y)^2 \\ &\text{Subject to} && \text{CV}_\text{L} \le y \le \text{CV}_\text{U} \\ &&& \Delta\text{MV}_i^\text{L} \le x_i \le \Delta\text{MV}_i^\text{U} \\ &&& \text{MV}_\text{L} \le Sx \le \text{MV}_\text{U} \end{aligned} \tag{9.50}$$

where A = process dynamic matrix as the state space model formulation

S = accumulating sum matrix for MV moves
x = control solution in terms of MV moves
y = CV response trajectories
W = weighting matrix including costs and prices

From every iteration of the optimization based on Equation 9.50, each future move of CVs can be predicted by knowing the process model Ax. If the predictions are in line with process CV targets, the RCA minimize the MV moves. Such solutions can achieve CV targets with use of minimal MVs (actuators)—flat curve of the white area on Figure 9.22.

9.5.3 Dynamic Operation Modeling

Let us consider with steady-state optimization first. When there are degrees of freedom available in a controller, the steady-state objective function could be generated

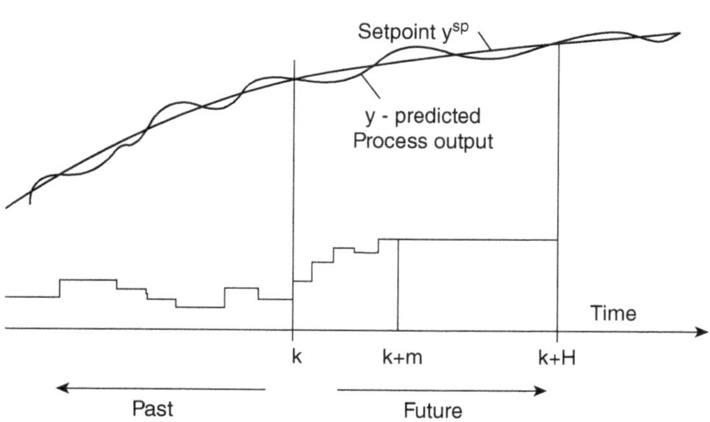

FIGURE 9.22. Dynamic trajectory for control variable.

as in Equation 9.48. However, the steady-state optimization does not consider the operation dynamics, and thus the optimal state (or setpoints) determined may be infeasible to reach from current state. In the dynamic optimization approach, optimization is integrated with process dynamic model so that optimization model becomes the integral part of MPC system.

To do so, Lu et al. (1996) proposed a dynamic optimization model extended from Equation 9.50 as

$$\min_{x,y} \left\| \begin{bmatrix} \mathbf{W} & \mathbf{0} \\ \mathbf{0} & \mathbf{W}_0 \end{bmatrix} \left(\begin{bmatrix} \mathbf{A} \\ \mathbf{S} \end{bmatrix} x - \begin{bmatrix} y \\ x_{ss}^{o} \end{bmatrix} \right) \right\|^2$$

$$\begin{aligned} \text{Subject to} \quad & \mathrm{CV_L} \le y \le \mathrm{CV_U} \\ & \mathrm{MV_L} \le Sx \le \mathrm{MV_U} \\ & \Delta \mathrm{MV}_i^{\mathrm{L}} \le x_i \le \Delta \mathrm{MV}_i^{\mathrm{U}} \end{aligned} \tag{9.51}$$

where

$\mathbf{A}$ = dynamic matrix as the state space formulation
$\mathbf{S}$ = accumulating sum matrix for MV moves
x = control solution in terms of MV moves
y = CV response trajectories
x_{ss}^{o} = the MV steady-state optimal value
$\mathbf{W}$ = weighting matrix including costs and prices
$\mathbf{W}_0$ = user-specified diagonal weighting factors for tuning optimization speed

The objective function is quadratic and can be solved effectively by SQP solver. The optimal solution is $\hat{x}$ (optimal MV moves) as well as the economically optimal dynamic path or the trajectory of "how to get there" from the current point to the optimal solution. To integrate with site-wide optimization, x_{ss}^{o} is introduced as it can be generated from a site-wide optimization model.

For practical applications, there usually are an infinite number of solutions to the model (Equation 9.51). In the advanced MPC approach (Honeywell, 2010), a minimum effort solution is sought, which means minimum MV movements (Figure 9.23).

This approach brings a significant robustness to the control solution. This treatment of robustness differs from the standard robust control design in that the control robustness is obtained from the freedom provided by CV upper and lower bounds but not by specification of model uncertainty. The integration of optimization and APC can be shown in a simplified block diagram in Figure 9.24.

At each execution of the optimizer, the optimizer gets process feedback. This frequent process feedback compensates for model mismatch and forms the basis for data reconciliation. Both controller and optimizer allow on-the-fly gain updating. For process areas where nonlinearity is significant such as the reaction system, frequent

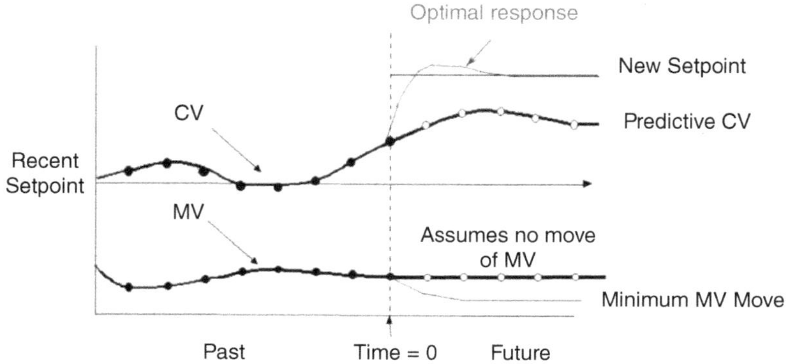

FIGURE 9.23. Minimum MV moves to reach the optimum.

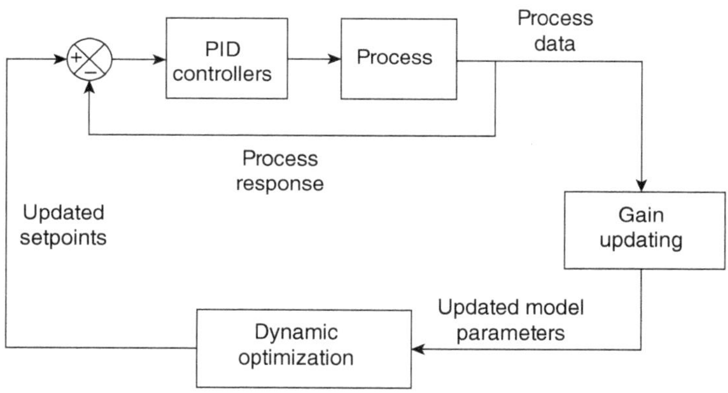

FIGURE 9.24. Integration of optimization and control.

gain updating is performed using empirical yield models. In this approach, there is a tight integration between optimization and controls. Not only does the optimizer pass down the optimized values to the controllers, but it also passes down the optimization speed factor to each of the controllers that takes into account the dynamic characteristics of each of the controllers. This dynamic optimizer moves the process toward the optimum along a minimum energy path that eliminates optimizer-induced dynamic violations of controlled variables. In addition, the dynamic optimizer also helps the process better cope with disturbances by applying corrective action to multiple controllers as and when needed. This approach is proven very successful in commercial applications (Nath and Alzein, 2000; Nath et al., 2011; Oleszczuk and Bozek, 2016).

If feedback control is like driving while looking in the rear-view mirror, MPC is like looking out the windshield. However, the accuracy of MPC is only as good as the model. Low-quality models can point you in the right direction, but high-quality models enable benefits such as enhanced optimization and better insight into the process. MPC in combination with quality models enables the controller to correct errors before they occur.

To handle possible model incorrectness and its impact on MPC performance, there special tuning factors introduced, that is, such FFWD/FBCK ratio. Such ratio allows implementation engineer decides if it is better to relay on the model prediction or feedforward (FFWD) and feedback (FBCK) measurements.

Generation of dynamic matrix. MPC employs dynamic modeling to capture and control complex interactions among PVs. The four general types of models used for simulating a process are explained in Figure 9.5. Models are selected based on step testing, which involves making a succession of step changes to an MV, measuring the process response, and generating a dynamic matrix as shown in Figure 9.25. The final dynamic models are the heart of an MPC application. As indicated in the MPC model in Equation 9.51, optimal changes to the MVs are generated based on the dynamic matrix A. These MV changes and CV setpoints are then passed along to the regulatory control system.

In Figure 9.25, multivariable dynamic model matrix developed for a pulp and paper digesting process is shown. By manipulating different types of liquor (MV_1, MV_2) and oxidation (MV_3 and MV_4) as well as by considering feedforward disturbances variable (property of the incoming pulp called Kappa), the MPC controller would try to maintain on the right trajectory CVs including residual alkali in the output pulp, pH, and Kappa. For example, by increasing NaOH, the model indicates that pH will increase after some time while Kappa (CEK) number will decrease.

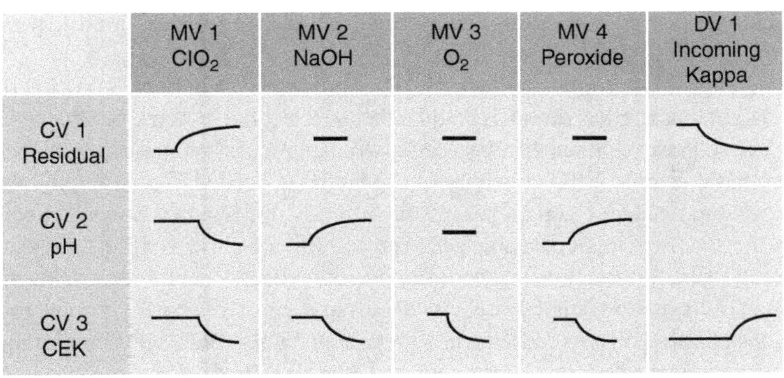

FIGURE 9.25. In this dynamic model control matrix for a pulp-mill bleaching unit, an increase in peroxide has no effect on the residual, a positive effect on pH, and a negative effect on the caustic extracted Kappa (CEK). Similarly, an increase in caustic (NaOH) has no effect on residual, but increases pH and reduces CEK. A model-based controller uses these relationships to predict future errors and calculate the optimum response. Source: From Nordh P (2016), reprinted with permissions of American Institute of Chemical Engineers.

9.5.4 Features of MPC-Based Optimization

The MPC-based optimization approach integrates optimization and MPC together into one model as shown in Equation 9.51. Thus, a process only needs one integrated process model for both control and optimization. In this approach, there is no duplication of modeling effort. Since the optimization part of the model is based on process dynamics for control, it generates not only the optimization solution but also a dynamic path from the current state to the optimal state.

Another implication of this unification is manifested in terms of a cooperative relationship that exists between the optimizer and the controllers. This approach can identify if dynamic violations exist or are imminent and if necessary attempts to solve the control problem from a global perspective. In this way, the approach unifies optimization and control problems.

Since the Honeywell MPC approach provides bulk of optimization modeling, initial development efforts as well as the long-term maintenance efforts for implementing the optimization part are greatly reduced.

Moreover, the above described functionality in reality allows the optimization part of the MPC controller to be switched off. In such case when optimization algorithm is turned off, the MPC controller will take care only about keeping CVs in the desired control ranges (not necessarily optimal ones). Process will be controlled and driven in the safety limits but without guarantees of its optimal path. Furthermore, the initial development effort for implementing this approach is minimal as the optimization capability is implemented on top of MPC without duplicated effort.

However, the linear dynamic model implemented in this approach is developed based on step testing at the most common operating point, and thus the linear model is valid in a small range around the operating point but invalid when operating point changes significantly (refer to Figure 9.3 and linearization around operation point x_0). That is the reason why control gains must be updated for MPC controllers when operating conditions change significantly.

New gains are determined by numerical perturbation of a nonlinear process model in simulation. Such process models could either be available in-house or obtained from third-party sources. Minimal effort is needed to adapt process models for gain updating since no basic change in functionality of the model is needed. Because only process derivatives are needed, rigorous parameter fitting is not required at each execution. Merely the source of input data changes from an offline file to an online database. The only additional effort required is that of filtering the process data for reasonable inputs. Execution of the process model could be triggered because of change in some process conditions or based on a fixed frequency for gain updating. Both MPC controllers and dynamic optimizer allow on-the-fly gain updating. For process areas where nonlinearity is significant such as the reaction system and feed change to the process, the frequent gain updating is performed using inferential soft sensors that describe process relationships or online measurements or simulation runs that work in parallel.

9.5.5 APC Benefits, Deployment, and Maintenance

Over the past 30 years, APC has been applied successfully in a variety of process industries, where it has:

- Increased throughput and improved yields;
- Reduced energy consumption and operating costs;
- Ensured consistent product quality;
- Provided operational flexibility;
- Improved process stability.

APC is mainly suitable for units that are operated as continuous processes, such as fluidized beds, boilers, and distillation columns. For example, in a pulp mill, APC could be implemented across the continuous digesters, thermomechanical pulping systems, brown stock washers, oxygen delignification units, bleach plants, evaporators, recovery boilers, lime kilns, and causticizing plants. However, APC could not be easily implemented for batch digesters, which require a different type of control strategy or pharmacy batch sequential processes where each batch differs significantly from each other.

A comprehensive plantwide performance audit can identify key optimization areas and provides an estimate of target values (e.g., savings, emissions) and thus reveals where APC can be deployed and where it would have the most immediate impact on plant operation. The project team then implements APC solutions in these areas of the plant to achieve the targets identified.

There are many potential APC applications. Some of the examples are listed below:

- Distillation column: APC applications would be designed to increase production/yields of the most desired fractions with minimum energy consumption.
- For fired heaters, APC will be responsible for maintaining desired outlet temperature (COT) with minimum fuel/oil gas consumption as well as CO_x and NO_X emissions.
- Hydrocracker reaction: APC will be searching for possible lower H_2 consumption, maintaining optimal reactor bed temperatures and keeping HC conversion at the desired value with desired sulfur and nitrogen content.
- For pulp digesters, MPC would be responsible for keeping maximum throughput in parallel keeping desired Kappa number and other quality parameters.
- For copper smelter fired heaters, APC will try to maintain stable oxygen/feed ratio with minimum slug and desired Cu purity.

- For ammonia plant, APC will try to maximize H_2 production with minimum usage of fuel gas, keeping appropriate steam/feed ratio and inserts in the synthesis loop.

A prerequisite to any APC deployment is to ensure that the field devices and the regulatory PID controllers are functioning optimally. Tuning the existing instrumentation, valves, controllers, and control infrastructure to their optimal performance setting can yield substantial performance and optimization gains even before APC is implemented.

Operator acceptance is often one of the main hurdles that APC implementation teams often encounter. Operators have to relinquish control to the intelligence embedded within the APC system and accept changes that may differ from established procedures. It is important to involve plant personnel early and ensure they receive the training necessary to effectively work with the new APC system. Operator skill level is a key factor in the long-term success of an APC system.

If not properly monitored and maintained, APC applications may degrade in a short period and yield only minimal benefits. Equipment modifications, operating strategy changes, feed rate and quality variations, instrumentation degradation, and fouling affect process units over time and can contribute to a decline in the economic performance of an APC system.

Issues related to how operators interact with the APC system may arise. Many plants fail to monitor APC controller uptime, which can reveal instances of operators turning off a controller during an abnormal situation that they do not know how to handle. Plants may also mistakenly assume that APC controller uptime indicates that the controller is running at its original design specification. If operators are not adequately trained to understand how a multivariable controller functions, they may reduce the available control range of the MVs, which results in reduced opportunities to be captured by the APC controller. Moreover, it requires multidisciplinary knowledge and skills for maintaining and tuning MPC controllers.

There are three common approaches to monitoring APC functionality and ensuring its benefits are sustained:

- Designate an on-site control engineer to manually monitor the APC system and how it is being used.
- Employ an external partner to provide monitoring and diagnostics and to help sustain the benefits of APC.
- For the highest degree of automation, purchase dedicated control performance monitoring software to handle the task of monitoring and sustaining the APC systems. The state-of-the-art APC software should include reinforcement learning algorithm that can conduct model parameter estimation and tuning automatically with a feedback loop.

9.5.6 Where We Go from APC?

At the top of process control hierarchy is site-wide production optimization and control as shown in Figure 9.1, which focuses on the entire site consisting of a good number of process units. The overall objective of production control is to coordinate all the process units to make the desired products and specified quality and then deliver to customers on schedule with minimum operating costs.

So far, optimization has been limited to individual unit operation, but it leaves greater benefits on table to capture. This is because the overall optimal solution cannot be achieved if only parts are optimized without optimizing the whole.

As APCs are implemented in various process units, the need to coordinate the controllers becomes more apparent. With the conventional approach, overall coordination of plant operation is typically accomplished using a planning and scheduling tool without integrated with control systems.

The problem of achieving the overall objective for production control via coordinating all the controllers in the plant is far from simple and can be further complicated by variations in the market prices, energy costs, and unit boundary conditions as well as by workforce competency within the organization.

The technology and software to solve this problem has been pursued for decades, and finally there is a light in the tunnel as new methods of solving this problem are emerging, which can provide dynamic coordination of multiple units to maximize overall profit, minimize energy costs, and manage soft and hard constraints. Detailed illustration of the new methods will be given in Chapter 13.

REFERENCES

Eder H (2007), Multivariable control, Chemical Engineering, **2007**, September, 46–54.

Erickson KT, Hedrick JL (1999), Plantwide Process Control, John Wiley & Sons, Inc.

Friedland B (1986), Control System Design—An Introduction to State-Space Methods, McGraw-Hill Inc.

Honeywell (2010), PROFIT OPTIMIZER (DQP): Concepts and Implementation Course Notes, Honeywell Training Document.

Lehman KA (2018), Implement Advanced Process Control, CEP, October, 60–66.

Lu ZJ, MacArthur JW, Horn B (1996), Method of multivariable predictive control utilizing range control, US Patent 5,351,184 granted to Honeywell.

Nath R, Alzein Z (2000), On-line dynamic optimization of olefins plants, Computers and Chemical Engineering, **24**, 533–538.

Nath R, Takinami A, Fujita A, Seki T (2011), Use of process simulation in ethylene plant advanced control and optimization, Paper No. 391, 2011 AIChE Spring Meeting, Chicago, March 13–17.

Nordh P (2016), Understand Advanced Process Control, CEP, June, 69–72.

Oleszczuk G, Bozek M (2016), Utilize APC solution to resolve hydrocracker conversion optimization challenges, Hydrocarbon Processing, **2016**, February, 69–74.

Smith CL (2016a), PID Explained for Process Engineers: Part 1 – The Basic Control Equation, CEP, January, 37–44.

Smith CL (2016b), PID Explained for Process Engineers: Part 2 – Tuning Coefficients, CEP, February, 27–33.

Ziegler JG, Nichols NB (1942), Optimum settings for automatic controllers, Transactions-American Society of Mechanical Engineers Journal of Dynamic Systems Measurement and Control, **64**, 759–768.

10

AI AND MACHINE LEARNING MODELING

AMIT GUPTA[1] AND FRANK (XIN X.) ZHU[2]

[1]*AIChE, USA*
[2]*Honeywell UOP, USA*

10.1 INTRODUCTION

In the context of Connected Plant for operation excellence, we aim to achieve collective intelligence via the digital system that integrates data with models based on first principle models, data analytics, and machine learning (ML). Thus, let computers do the repetitive, monotonous, and tedious tasks and complex calculations that is human's weakness. In contrast, let plant staffers focus on higher-value tasks: making more intelligent decisions and taking actions for implementation that is human's strength. The insights that the digital system provide act as the third eye. That is why the data and analysis become so important. As the result, the operation systems become so knowable and insightful. Engineers and managers can solve difficult problems and avoid human errors, and make changes to process with confidence. That is the result of human–machine work in synergy by integrating their distinctive strength and avoid their distinctive weakness. As a note, some of the discussions in this chapter largely comes from Gupta's articles (2018a, b) (reprint with permission of AIChE).

10.2 ARTIFICIAL NEURAL NETWORKS

The biggest breakthrough for AI research in recent years have been in the field of deep reinforcement learning (RL). This has been driven in part by the easy availability of data, but even more so by an explosion in parallel computing power in recent

Digitalization and Analytics for Smart Plant Performance: Theory and Applications, First Edition. Frank (Xin X.) Zhu.

years. From self-driving cars to robots, deep learning is fueling what many believe will be the biggest technology revolution the world has ever seen.

The fundamental element to ML is artificial neural networks (ANNs). These are brain-inspired networks of interconnected layers of algorithms, called neurons, which feed data into each other and which can be trained to carry out specific tasks by modifying the weighting factors attributed to input data as it passes between the layers.

The human brain is made up of about 85 billion neurons, resulting in a network capable of representing a tremendous amount of knowledge. As you might expect, this dwarfs the brains of other living creatures. For instance, a cat has roughly a billion neurons, a mouse has about 75 million neurons, and a cockroach has only about a million neurons. In contrast, many ANNs contain far fewer neurons, typically only several hundred, even a fruit fly brain with 100,000 neurons far exceeds the current state-of-the-art ANN (Lantz, 2015).

Though it may be unfeasible to completely model a cockroach's brain, a neural network may still provide an adequate heuristic model of its behavior. Suppose that we develop an algorithm that can mimic how a roach flees when discovered. If the behavior of the robot roach is convincing, does it matter whether its brain is as sophisticated as the living creatures? This question is the basis of the controversial Turing test (Turing, 1950), which grades a machine as intelligent if a human being cannot distinguish its behavior from a living creatures.

10.2.1 How Does Human Neuron Work?

Because ANNs were intentionally designed as conceptual models of human brain, it is helpful to first understand how biological neurons function. As illustrated in Figure 10.1, a human neuron uses dendrites to collect inputs from other neurons

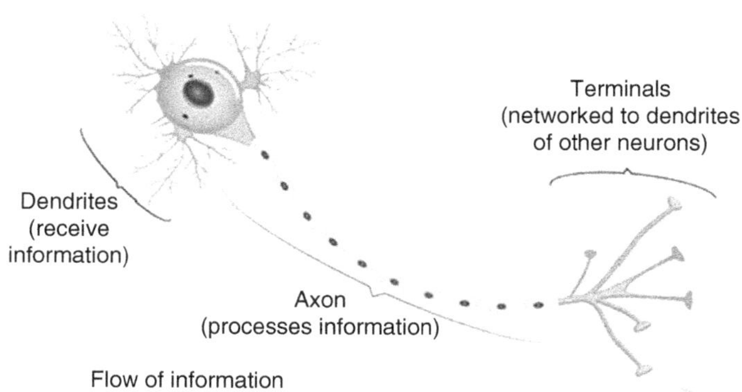

FIGURE 10.1. A human neuron collects inputs from other neurons using dendrites and sums all the inputs. If the total is greater than a threshold value, it produces an output. *Source*: From Gupta A (2018a). Reprinted with permissions of American Institute of Chemical Engineers.

and adds all the inputs, and if the resulting sum is greater than a threshold, it fires and produces an output. At the axon's terminals, the fired output in the form of electric signals is processed as a chemical signal to be passed to the neighboring neurons.

10.2.2 From Biological to Artificial Neurons

An ANN models the relationship between a set of input signals and an output signal using a model derived from our understanding of how a biological brain responds to stimuli from sensory inputs. Just as a brain uses a network of interconnected neurons to create a massive parallel processor, ANN uses a network of artificial neurons or nodes to predict output based on inputs.

As depicted in Figure 10.2, to mimic the human neuron mechanism of predicting output based on input, a perceptron is created in ML based on mathematical modeling of a biological neuron. Just like a real neuron, it receives inputs and computes an output. Each input has an associated weight. All the inputs are individually multiplied by their weights, added together, and passed into an activation function that determines whether the neuron should fire and produce an output.

Neural networks can be expanded into sprawling networks with a huge number of layers that are trained using massive amounts of data—it is called deep learning. It is these deep neural networks that have fueled the current leap forward in the ability of computers to carry out task like advanced learning and reasoning, speech recognition, and computer vision. For example, in 2016 the computer program AlphaGo equipped with deep learning software defeated the world champion of Go (AlphaGo, 2018). During the games, AlphaGo played very inventively and made smart moves.

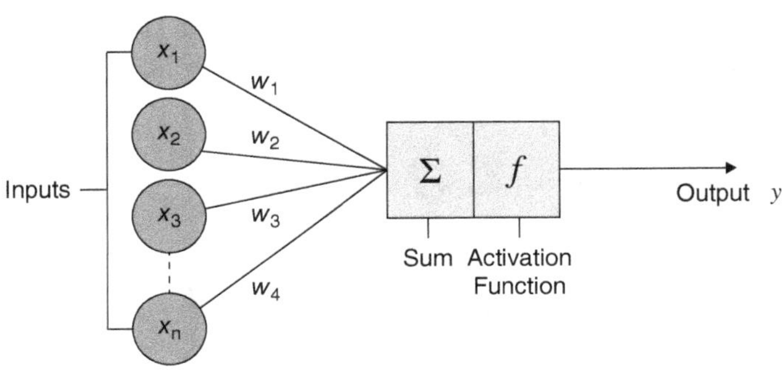

FIGURE 10.2. A perceptron is a mathematical model of a neuron. It receives weighted inputs, which are added together and passed to an activation function. The activation function decides whether it should produce an output. *Source*: From Gupta A (2018a). Reprinted with permissions of American Institute of Chemical Engineers.

10.3 KEY CONCEPT IN ML: PERCEPTRON

The building block of an ANN is perceptron, which is an algorithm inspired by the biological neuron (Rosenblatt, 1958). To mimic the human neuron mechanism of predicting output based on input, a perceptron is created in ML based on mathematical modeling of a biological neuron (Clabaugh et al., 2018). Just like a real neuron, it receives inputs and computes an output. Each input has an associated weight. All the inputs (x variables) are individually multiplied by their weights (w's), added together, and passed into an activation function f (threshold) that determines whether the neuron should fire and produce an output (y variable) (Figure 10.2).

There are many different types of activation functions with different properties, but one of the simplest is the step function (Sharma, 2017). A step function outputs a 1 if the input is higher than a certain threshold; otherwise it outputs a 0. For example, if a perceptron has two inputs (x_1 and x_2),

$$x_1 = 0.9$$

$$x_2 = 0.7,$$

which have weightings (w_1 and w_2) of

$$w_1 = 0.2$$

$$w_2 = 0.9$$

and the activation function threshold is equal to 0.75, then weighing the inputs and adding them together yields

$$y = x_1 w_1 + x_2 w_2 = (0.9 \times 0.7) + (0.2 \times 0.9) = 0.81$$

Because the total input (0.81) is higher than the threshold (0.75), the neuron will fire. Since we chose a simple step function, the output would be 1.

So how does all this lead to intelligence? It starts with the ability to learn something simple through training.

10.3.1 Training of Single Perceptron

ML algorithms describing perceptrons can be trained using sample data (called "training data") via making adaptive changes to the weights in order to make predictions close to the desired targets without being explicitly programmed to perform the task.

Training a perceptron involves feeding it with multiple training sample data and calculating the output for each of them. After each sample, the weights are adjusted to minimize the output error, usually defined as the difference between the desired (target) and the actual outputs (Figure 10.3). By following this simple training

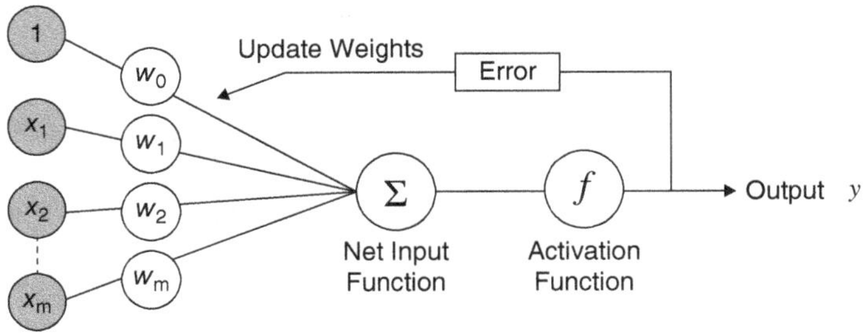

FIGURE 10.3. To train a perceptron, the weights are adjusted to minimize the output error. Output error is defined as the difference between the desired output and the actual output. *Source*: From Gupta A (2018a). Reprinted with permissions of American Institute of Chemical Engineers.

algorithm to update weights, a perceptron can learn to perform a certain calculation. For example, predicting a rotating equipment failure mode (output) based on rate of vibration (input).

The limitation for single layer of perceptrons is that they can only learn linearly classifiable patterns and are unable to handle nonlinear or more complicated patterns.

10.3.2 Multilayer Perceptrons

A single neuron is capable of learning simple patterns, but when many neurons are connected together, their abilities increase dramatically. Each of the 100 billion neurons in the human brain has, on average, 7,000 connections to other neurons. It has been estimated that the brain of a 3-year-old child has about one quadrillion connections between neurons. And, theoretically, there are more possible neural connections in the brain than there are atoms in the universe.

A multilayer perceptron (MLP) intends to mimic how connected neurons in human work. For instance, when you are watching your pet, each photoreceptor neuron in your retina generates signals about the color of light. Succeeding layers of neurons recognize even more complex patterns, such as edges, circles, squares, eyes, noses, and eventually your pet's overall face. Learning using MLP is called deep learning in which "sensed" data is processed through MLP; each layer uses the output of the previous layers as its input. MLP mimics human with multiple layers of neurons between input and output. MLPs are also called feedforward neural networks. Feedforward means that data flow in one direction from the input to the output layer. Typically, every neuron's output is connected to every neuron in the next layer. Layers that come between the input and output layers are referred to as hidden layers (Figure 10.4).

Such an MLP for recognizing digits would typically be trained by showing it images of digits and telling it whether it recognized them correctly or not. Initially, the MLP's output would be random, but as it is trained, it will adjust weights between the neurons and start classifying inputs correctly.

A typical real-life MLP for recognizing handwritten digits consists of 784 perceptrons that accept inputs from a 28×28 pixel bitmap representing a handwritten digit, 15 neurons in the hidden layer, and 10 output neurons (Zhang et al., 2017). Typically, such an MLP is trained using a pool of 50,000 labeled images of handwritten digits. It can learn to recognize previously unseen handwritten digits with 95% accuracy after a few minutes of training on a well-configured computer.

As an example, MLPs have been used for predictive control of chemical reactors (Spielberg et al., 2017). The typical setup trains a neural network to learn the forward dynamics of the plant. The prediction error between the plant output and the neural network output is used for training the neural network (Figure 10.5). The neural network learns from previous inputs and outputs to predict future values of the plant output. For example, a controller for a catalytic continuous stirred tank

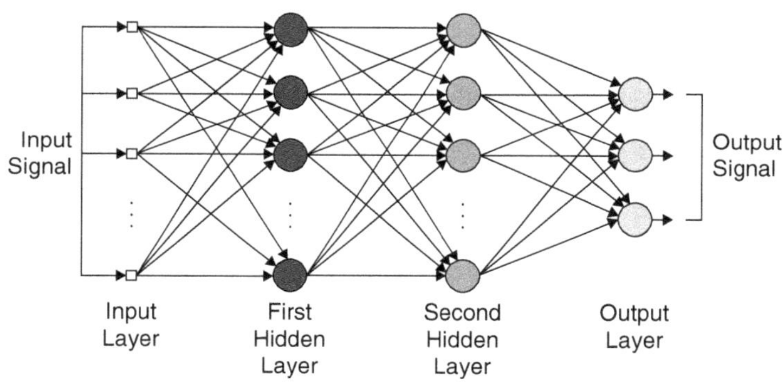

FIGURE 10.4. A multilayer perceptron has multiple layers of neurons between the input and output. Each neuron's output is connected to every neuron in the next layer. *Source*: From Gupta A (2018a). Reprinted with permissions of American Institute of Chemical Engineers.

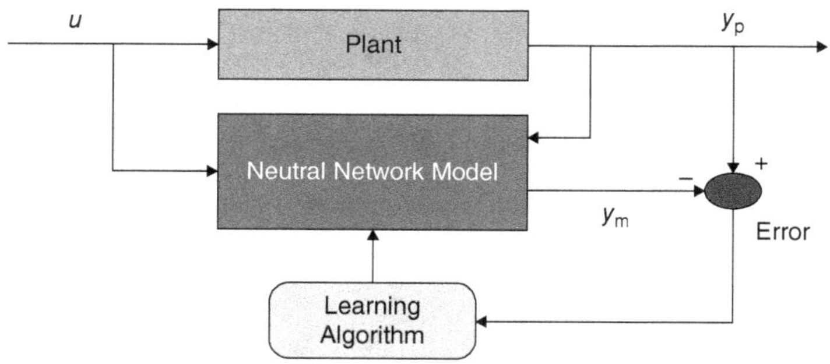

FIGURE 10.5. An MLP can learn the dynamics of the plant by evaluating the error between the actual plant output and the neural network output. *Source*: From Gupta A (2018a). Reprinted with permissions of American Institute of Chemical Engineers.

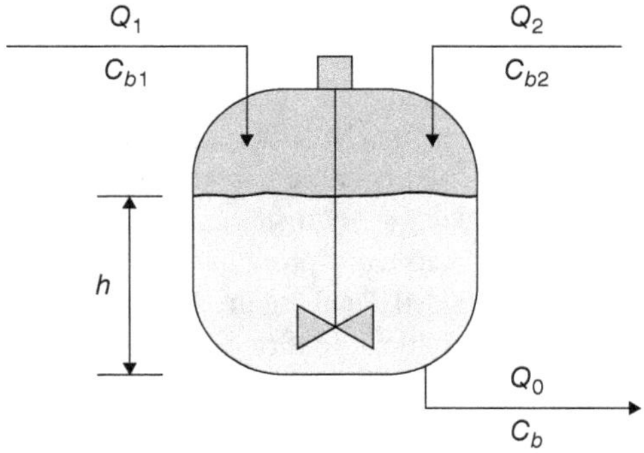

FIGURE 10.6. A continuous stirred tank reactor can be trained to maintain appropriate product concentration and flow by using past data about inflow, concentration, liquid level, and outflow. Source: From Gupta A (2018a). Reprinted with permissions of American Institute of Chemical Engineers.

reactor (Figure 10.6) can be trained to maintain appropriate product concentration and flow by using past data about inflow Q_1 and Q_2 at concentrations C_{b1} and C_{b2}, respectively, liquid level h, and outflow Q_0 at concentration C_b.

10.3.3 Training a Multilayer Perceptron

Training a single perceptron is easy—all weights are adjusted repeatedly until the output matches the expected value for all training data. For a single perceptron, weights can be adjusted using the following formulas:

$$\Delta w_i = \eta\left(t - o\right)x_i \tag{10.1}$$

$$w_i + \Delta w_i \rightarrow w_i \tag{10.2}$$

where w_i is the weight, Δw_i is the weight adjustment, t is the target output, x_i is the input and o is the actual output, and η is the learning rate—usually a small value used to moderate the rate of change of weights.

However, this approach of tweaking each weight independently does not work for an MLP because each neuron's output is an input for all neurons in the next layer. Tweaking the weight on one connection impacts not only the neuron it propagates to directly but also all of the neurons in the following layers as well and thus affects all the outputs. Therefore, you cannot obtain the best set of weights by optimizing one weight at a time. Instead, the entire space of possible weight combinations must be searched simultaneously. The primary method for doing this relies on a technique called gradient descent or the best first search.

Imagine you are at the top of a hill and you need to get to the bottom of the hill in the quickest way possible. One approach could be to look in every direction to see which way has the steepest grade and then step in that direction. If you repeat this process, you will gradually go farther and farther downhill. That is how gradient descent works. If you can define a function over all weights that reflects the difference between the desired output and calculated output, then the function will be lowest (i.e., the bottom of the hill) when the MLP's output matches the desired output. Moving toward this lowest value will become a matter of calculating the gradient (or derivative of the function) and taking a small step in the direction of the gradient.

Backpropagation, short for "backward propagation of errors," is the most commonly used algorithm for training MLPs using gradient descent. The backward part of the name stems from the fact that calculation of the gradient proceeds backward through the network. The gradient of the final layer of weights is calculated first, and the gradient of the first layer of weights is calculated last.

Before looking at how backpropagation works, recall that a perceptron calculates a weighted sum of its input and then decides whether it should fire. The decision about whether or not to fire is made by the activation function. In the perceptron example, we used a step function that outputted a 1 if the input was higher than a certain threshold; otherwise it would output a 0.

10.4 MACHINE LEARNING

ML consists of algorithms and statistical models that mimic ANNs to effectively perform prediction based on input. Today we think of ML as an intelligent modeling technique that can effectively perform prediction based on trained data sets. It is a sharp contrast with traditional computer programming. Standard algorithms like expert systems would follow a predetermined path set into motion by programmers' static instructions or code. An ML algorithm conversely can learn as it goes. With each new data set, it updates its models or weights and the way it "sees" the world. In this manner, machine can learn and change based on their experience and data; programmers become less like rule makers and more like teachers and trainers.

In general, there are three ML approaches: supervised, unsupervised, and reinforcement learning.

10.4.1 Supervised Learning

In supervised learning, the algorithm learns by being told for what is right and wrong. In achieving this, the algorithm builds a mathematical model from a set of data that contains both the inputs and the desired outputs. For instance, a child can group apples, oranges, and banana by being told what are they (been trained). In this case, the task is determining whether an image contains a certain object, the training data for a supervised learning algorithm would include images with and without this object (the input), and each image would have a label (the output) designating whether it contained the object. In overall, supervised learning needs manual intervention for the ML algorithm training, which requires a lot of data and effort. For example, it requires a lot of failure data to train the ML algorithms for prediction of equipment failure in real time.

This approach could lead to the rise of semi-supervised learning, where systems can learn how to carry out tasks using a far smaller amount of labeled data than is necessary for training systems using supervised learning today.

10.4.2 Unsupervised Learning

Unsupervised learning means learning without the need of a teacher or trainer. This is how we humans learn most of what we know about the world. Very simple example is that babies learn how gravity works without being explicitly taught. In achieving unsupervised learning, the algorithm for unsupervised learning builds a mathematical model from a set of data that contains only inputs and no desired output labels. Unsupervised learning algorithms are used to find structure or pattern in the data, like grouping or clustering of data points such as apples and oranges. Unsupervised learning can discover patterns in the data and can group the inputs into categories without prior knowledge. An example might be clustering together the equipment under good conditions. Thus the risk condition of an equipment can be predicted if it is outside of the health equipment cluster.

10.4.3 Reinforcement Learning (RL)

Besides supervised and unsupervised learning, RL involves learning by trial and error, solely from rewards or punishments. RL is a deep learning approach. Neural networks based on RL can start from a blank slate, and under the right conditions, they achieve superhuman performance based solely on knowledge about how to value the quality of outcomes (rewards). The RL approach combines intuition (make initial prediction based on general rules) and adjustment (based on the gap between the initial prediction and the goal or target). Over the procedure of self-learning, the RL algorithm will improve its prediction accuracy. RL is regarded as a game changer in AI.

RL involves an agent, a set of states {S}, a set of actions {A} per state, and the environment (Figure 10.7). By performing an action $a \in A$, the agent transitions from state to state. Executing an action in a specific state provides the agent with a reward (a numerical score). Figure 10.8 provides a similar but more visual explanation than Figure 10.7.

The goal of the agent is to maximize its total (future) reward. It does this by adding the maximum reward attainable from future states to the reward for achieving its current state, effectively influencing the current action by the potential future reward. This potential reward is a weighted sum of the expected values of the rewards of all future steps starting from the current state.

The most known learning method in RL is Q-function that estimates the impact of a certain action at a given state. Q-Learning is a model-free RL algorithm. The goal of Q-learning is to learn a policy, which tells an agent what action to take under what circumstances.

To help understand RL using Q-function, let us consider the process of boarding a train (https://en.wikipedia.org/wiki/Q-learning). The reward is measured by the

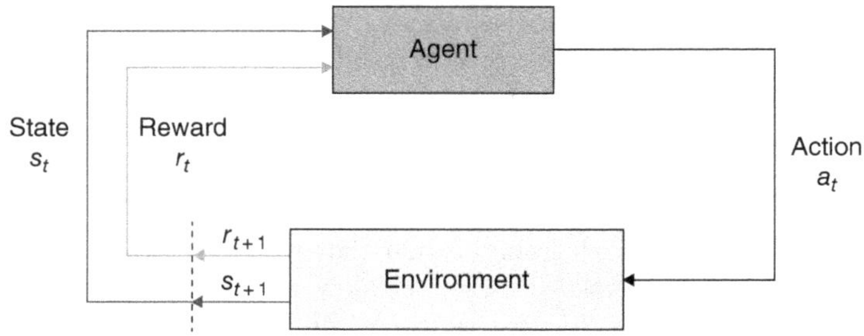

FIGURE 10.7. In reinforcement learning, the agent's goal is to maximize the total amount of reward it receives over time. The agent accepts the state and reward as input and outputs an action. Meanwhile, the environment sees the action, changes the state, and gives a reward. *Source*: From Gupta A (2018b). Reprinted with permissions of American Institute of Chemical Engineers.

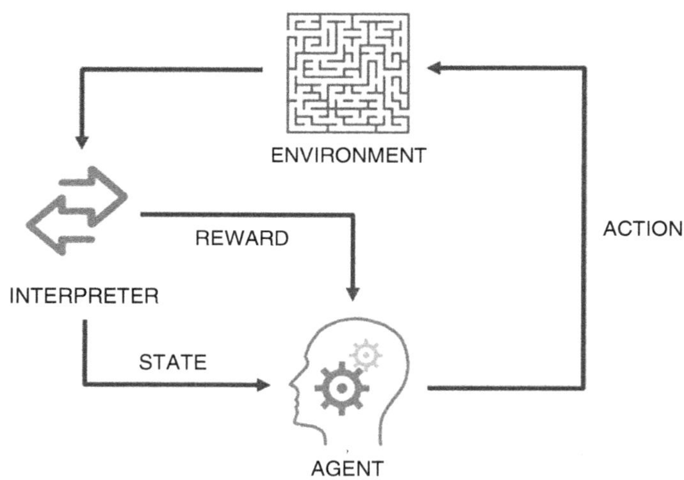

FIGURE 10.8. Reinforcement learning with reward/penalty function: an agent takes actions in an environment, which is interpreted into a reward and a representation of the state, which are fed back into the agent.

negative of the total time spent boarding (alternatively, the cost of boarding the train is equal to the boarding time). One strategy is to enter the train door as soon as they open, minimizing the initial wait time for yourself. If the train is crowded, however, then you will have a slow entry after the initial action of entering the door as people are fighting you to depart the train as you attempt to board. The total boarding time, or cost, is then

0-sec wait time + 1-sec fight time

On the next day, by random chance (exploration), you decide to wait and let other people depart first. This initially results in a longer wait time. However, time fighting other passengers is less. Overall, this path has a higher reward than that of the previous day, since the total boarding time is now

5-sec's wait time + 0-sec fight time

Through exploration, despite the initial action resulting in a larger cost (or negative reward) than in the forceful strategy, the overall cost is lower, thus revealing a more rewarding strategy.

RL, due to its generality, is studied in many other disciplines, such as game theory, control theory, operations research, information theory, simulation-based optimization, multi-agent systems, swarm intelligence, statistics, and genetic algorithms. The problems of interest in RL have also been studied in the theory of optimal control, which is concerned mostly with the existence and characterization of optimal solutions, and algorithms for their exact computation, and less with learning or approximation, particularly in the absence of a mathematical model of the environment. Using RL, various operating performance of equipment can be assumed, and the data can be used for RL training to generate different clusters of healthy, failure, and near failure.

10.4.4 *Q*-Learning in Reinforcement Learning

The central concept of Q-learning is that a function $Q(s, a)$ calculates the maximum future reward for performing action a in state s. It is called Q-function because it represents the quality of a certain action in a given state. Given a Q-function, the strategy for picking the best action (e.g., to open or close a valve) is simply to pick the action with the highest Q-value.

Obviously, a Q-function is an idealized concept—typically one cannot simply create a function that conclusively predicts the quality of future states. However, the Q-value of state s and action a can be stated in terms of the Q-value of the next state s':

$$Q(s,a) = r + \gamma \max Q(s',a') \tag{10.3}$$

where $Q(s, a)$ is the expected future reward of performing action a at state s; r is the reward, if any, of taking action a at state s; and $\max Q(s', a')$ is the maximum future reward. In other words, if taking action a at state s leads to state s' and there are multiple actions a' that can be taken in s', $\max Q(s', a')$ is the maximum among those. γ is called the discount factor. If set to a value near 0, the immediate reward is given preference, and, if set to near 1, future rewards are valued. Therefore,

Q-function calculates the maximum future rewards for performing different actions at state s and the action with the highest Q-value will be selected.

One area where RL can be applied is process control (Spielberg et al., 2017), where features describing the current state of a plant can represent states, controller inputs to the plant are actions, and the objective is to make the output reach the setpoint while meeting certain constraints. For example, in the manufacture of paper, the steam flow rate can be the control action, the current moisture content can represent states, and rewards can be associated with how close the current moisture is to the desired moisture content of a sheet of paper. Given such a set of rules, RL can come up with control policies without the need for handcrafted feature descriptors, controller tuning, deriving control laws, and developing mathematical models.

10.5 ML APPLICATIONS IN THE PROCESS INDUSTRY

ML has found applications tremendously in recent years. For example, ML becomes a major framework for artificial intelligence, with applications in information theory, signal processing, control theory, and optimization theory. Qin and Zhiang (2019) cover wide ranges of ML applications in the industries, while some of them are explained as below.

10.5.1 Straightforward ML Applications

There are straightforward ML applications to process monitoring, fault diagnosis, and inferential sensors. For example, supervised learning methods such as deep learning can be used as inferential sensors. Unsupervised learning methods, especially those that are nonlinear extensions of PCA (Principal Component Analysis), can be used for process monitoring and fault diagnosis.

Most work in process monitoring apply PCA and unsupervised learning methods. Although PLS (Partial Least Squares) has been used for process monitoring, the quality-relevant diagnosis has only been recently developed (Li et al., 2010; Qin and Zheng, 2013; Zhu et al., 2017). Unlike diagnosis based on unsupervised learning models where one can at most analyze abnormal changes in process variables, quality-relevant diagnosis refers to the use of supervised learning models for the diagnosis of a quality anomaly due to relevant abnormal process variations. Since this approach aims to monitor and diagnose quality anomalies before they are measured and confirmed, Zhu and Qin (2019) refer to it as supervised monitoring and diagnosis. Supervised process monitoring has received increased attention since the early work on total PLS and current PLS (Li et al., 2010). Liu et al. (2014) and Ge and Chen (2016) developed quality-relevant monitoring and diagnosis by taking into account of dynamics in the data. Huang and Yan (2017) proposed a two-block monitoring scheme based on mutual information (MI) and kernel PCA. Qin and Zhao (2018) designed a monitoring method based on quality information and process dynamics for closed-loop manufacturing processes. Supervised monitoring and diagnosis deserves much more attention since it is predictive and can reduce quality-irrelevant alarms significantly.

White (2016) reported an application case in which GE uses predictive analytics to monitor performance of over 1,500 turbines and generators in 58 countries. Each turbine typically has more than 100 physical sensors and over 300 virtual sensors that monitor temperature, pressure, vibration, speed, and so on. Data are routed over the Internet to a data center—central monitoring location. At the data center, predictive analytics algorithms check current and recent operating conditions against models of more than 150 failure scenarios to detect sign of impeding failures.

Routing data into central storage makes efficient use of highly skilled workers, allowing specialists to monitor all of the turbines and generator installations. This single-database method also helps to ensure that predictive models are continually refined and improved and enables best practices to be shared with customers simultaneously rather than on an incremental basis during system upgrades. GE estimated that this approach collectively saved US $70 MM last years for the GE customers.

10.5.2 Deep Learning Applications

There are two essential aspects of data analytics: prediction and interpretation. To achieve accurate prediction, often complex models such as deep neural networks are more capable. On the other hand, in order to achieve interpretability, simpler models are preferred. Interpretability is critical for industrial adoption, since operators and decision makers must establish a level of trust in the analytics algorithms. Deep learning has been applied to derive inferential sensors from data (Shang et al., 2014). Although the main goal for inferential sensors is prediction accuracy, it is necessary to provide a trade-off between model complexity and maintainability. The increased complexity of deep neural networks needs to be justified against the incremental accuracy gain comparing with other simpler models. The Google's GFT model for flu prediction is a classic example where the complex deep model fails during the test of time. The alternative autoregressive dynamic models perform much better (Yang et al., 2015). Deep neural networks have also been applied for image classification (Zhu et al., 2019) for a combustion furnace. This is directly applying image analysis methods to process operations.

From driving cars to finding welding defects by visual inspection (Khumaidi et al., 2017), vision represents a significant aspect of human intelligence. Because computers interpret images as a matrix of numbers, traditional computing approaches fail to scale due to size and variability of visual data. Deep neural networks, particularly convolutional neural networks, are very effective at image recognition where the data are voluminous but have structure and the relative location of features is important. For example, Chemception (Goh et al., 2017) use CNN to predict chemical properties of a molecule given 2D molecule structure in the form of an 80×80 pixel image as input. CNNs have also been applied to chemical process fault diagnosis (Wu and Zhao, 2018) and for high-throughput material and molecular discovery (Farimani, 2017).

REFERENCES

AlphaGo (2018), The story of AlphaGo so far, Deepmind, https://deepmind.com/research/alphago/, May.

Clabaugh C (2018), Neural Networks: The Perceptron, eRoberts course, Standford University, May.

Farimani BA (2017), Deep learning in chemical engineering. In: 2017 AIChE Annual Meeting, October 29.

Ge Z, Chen X (2016), Supervised linear dynamic system model for quality related fault detection in dynamic processes, Journal of Process Control **44**, 224–235.

Goh GB, Siegel C, Vishnu A, Hodas NO, Baker N (2017), Chemception: A deep neural network with minimal chemistry knowledge matches the performance of expert= developed QSAR/QSPR models, arXiv, https://arxiv.org/ftp/arxiv/papers/1706/1706.06689.pdf.

Gupta A (2018a), Introducing toe deep learning—Part 1, Chemical Engineer, 22–29, Prog (CEPe/AIChE), June.

Gupta A (2018b), Introducing to deep learning—Part 2, Chemical Engineer, 39–46, Prog (CEP/AIChE), October.

Huang J, Yan X (2017), Quality relevant and independent two block monitoring based on mutual information and kpca, IEEE Transactions on Industrial Electronics **64** (8), 6518–6527.

Khumaidi A, Yuniaro EM, Purnomo MH (2017), Welding effect classification based on convolution neural network (CNN) and Gaussian Kernel. In: 2017 International Seminar on Intelligence Technology and its Applications, ISITIA, August 28–29.

Lantz B (2015), Machine Learning with R, 2nd ed., Pack Publishing, Birmingham, Mumbai.

Li G, Qin S, Zhou D (2010), Output relevant fault reconstruction and fault subspace extraction in total projection to latent structures models, Industrial and Engineering Chemistry Research **49**, 9175–9183.

Liu Q, Qin S, Chai T (2014), Quality-relevant monitoring and diagnosis with dynamic concurrent projection to latent structures, IFAC Proceedings **47** (3), 2740–2745.

Qin S, Zheng Y (2013), Quality-relevant and process-relevant fault monitoring with concurrent projection to latent structures, AICHE Journal **59** (2), 496–504.

Qin Y, Zhao C (2018), A quality-relevant monitoring method for closed-loop industrial processes with dual consideration of static and dynamic analysis. In: 2018 10th International Conference on Modelling, Identification and Control (ICMIC). IEEE, pp. 1–6.

Qin SJ, Zhiang LH (2019), Advances and opportunities in machine learning for process data analytics, Computers and Chemical Engineering **126**, 465–473.

Rosenblatt F (1958), The perceptron: A probabilistic model for information storage and organization in the brain, Psychologial Review **65** (6), 386–408.

Shang C, Yang F, Huang D, Lyu W (2014), Data-driven soft sensor development based on deep learning technique, Journal of Process Control **24** (3), 223–233.

Sharma AV (2017), Understanding activation function in neural networks, Medium, https://medium.com/the-theoy-ofeverything/understanding-activation-functions-in-neural-networks, March.

Spielberg SPK, Gopaluni RB, Loewen PD (2017), Deep reinforcement learning approaches for process control. In: 6th International Symposium on Advanced Control of Industrial Processes, AdCONIP, May.

Turing AM (1950), Computing machinery and intelligence, Mind **LIX** (236), 433–460, October.

White D (2016), Big Data—What Is It?, pp. 32–35, CEP, March.

Wu H, Zhao J (2018), Deep convolutional neural network model based chemical process fault diagnosis, Computers & Chemical Engineering (115), 185–197, July.

Yang S, Santillana M, Kou SC 2015, Accurate estimation of influenza epidemics using Google search data via argo, Proceedings of the National Academy of Sciences **112**, 14473–14478.

Zhang J, Liu J, Luo Y, Fu Q, Bi J, Qiu S (2017), Chemical substance classification using long short-term memory recurrent neural network. In: IEEE 17th International Conference on Communication Technology (ICCT), October 27–30.

Zhu Q, Liu Q, Qin S (2017), Concurrent quality and process monitoring with canonical correlation analysis, Journal of Process Control **60**, 95–103.

Zhu Q, Qin S (2019), Supervised diagnosis of quality and process faults with statistical learning models, I&EC research revised.

Zhu W, Ma Y, Benton M, Romagnoli J, Zhan Y (2019), Deep learning for pyrolysis reactor monitoring: from thermal imaging toward smart monitoring system, AICHE Journal **65** (2), 582–591.

PART 3

CONNECTED PLANT FOR SMART OPERATIONS

Part 3 includes six chapters, which explains individual components of Connected Plant ranging from connected instruments to asset and safety management and from integrated planning and control to workforce enhancement and workflow management.

The Connected Plant is a suite of modern digital applications, in the form of digital twins, that delivers higher levels of safety, reliability, efficiency, and profitability. These proven industry solutions are based on decades of domain knowledge and controls experience. The Connected Plant method aims to connect **people**, **processes**, and **assets** within the organization much better than what has been doing with traditional digital solutions. The Connected Plant is more than just data collection and software tools. It is about applying model-based digital twin, located on cloud or at edge, to convert data to insights. It is about integrating new digital solutions with existing data infrastructure and work processes. It is about better management of asset, workflows, and knowledge. The Connected Plant creates value through real-time data analytics, models, and optimization to reveal new insights for opportunities leading to turnkey solutions. Use of the Connected Plant method can result in faster detection and resolution of operating problems as well as more effective best practice sharing and knowledge management.

The Connected Plant is about human-digital interactions. The method provides insights, and it requires people to make informed decisions and take smart actions for things to happen. One must understand the fact that new technology is the enabling tool; it is not the solution. If just expecting things will happen naturally after digital analytics are installed, it is doomed to disillusion and failure.

Digitalization and Analytics for Smart Plant Performance: Theory and Applications, First Edition. Frank (Xin X.) Zhu.

11

CONNECTED METERING AND MEASUREMENTS

MARTIN BRAGG

Honeywell Process Solutions, UK

11.1 INTRODUCTION

Process control requires accurate, responsive, and maintainable measurements. Within industry, and notably within oil and gas, the top four measurements are flow, pressure, temperature, and level. Each of these individual measurements may be connected in a digital transformation project that has the aim to bring new insight to the end user, realizing a valuable outcome not previously available. Each measurement can be connected to realize new outcomes; however, for this chapter, the focus is on flow measurement.

There is a fundamental requirement for flow metering accuracy and reliability where (a) the flow meter is the "cash register" of the business, (b) the typically associated flow computer is the "bureau de change" for the fiscal transfer of the product, or (c) the metering is utilized to ensure the "correct recipe" is achieved for the final product. The meter itself is often operating within a connected system with pressure and temperature transmitters and other measurement devices, depending on the liquid or gas and the meter's application.

After the initial oil and gas extraction, the industrial processes and plants must operate not only at desired capacity but also at optimal efficiency and safety: safety that enables the ability to predict undesirable process or plant conditions and equipment failures before they occur and efficiency that systematically addresses the metering as part of the continuous improvement process. The current approach to this asset (metering) management topic, at many facilities, is run-to-fail since

Digitalization and Analytics for Smart Plant Performance: Theory and Applications, First Edition. Frank (Xin X.) Zhu.

abnormal conditions and malfunctions are not identified until alarm warnings are evident or something reaches the point of failure.

A growing number of metering asset owners seek to unlock the power of the Industrial Internet of Things ("IIoT") to improve overall operational and business performance. This chapter examines the latest digital approach to the oil and gas industry that is the enabler to connected metering. Specifically, it describes how the digital economy, often referred to as the IIoT, can support continuous improvement and solve previously unsolved problems to increase meter availability and its operational performance, safety, and reliability while maintaining its desired accuracy. By taking advantage of streaming data from ubiquitous sensors and "things" to rapidly assess operational status, a digital system can identify potential warning signs, deliver alerts, and automatically trigger actions while preemptively servicing installed meters before problems arise.

Connected metering is digital transformation that is emerging for IIoT in which ubiquitous connectivity is coupled to cloud-borne data analytics and storage systems in enhancing metering accuracy and reliability. An approach encompasses three key connected topics of processes, assets, and people where each is considered to enable a positive-financial gain as a single-key customer outcome in reducing their operational expenditure while also enabling new learning within the essential metering elements of continuous meter diagnostics, condition-based monitoring ("CBM") analytics, near-real-time measurement uncertainty, digital twins, virtual metering, and technical auditing with a focus toward mismeasurements. Additionally, this chapter introduces the topic of positive-unexpected consequences of a digitally connected ecosystem in that it generates further subsets of data being readily formed from a larger master data set bringing about new insights—insights that are the positive consequence of connecting subsets of existing data in a "virtual world of digital connectivity." Finally, the chapter concludes with how to deliver ubiquitous connectivity in referencing the "Measurement-IQ™" and "Honeywell Connected Plant" products and solutions.

11.2 REVIEW OF METERING DEVICES

Metering has seen a tremendous growth since its early introduction in the 1800s in terms of both technology and market size. The global metering market at 2017 was estimated to be worth circa \$7B of which 28% was accounted for within the oil and gas industry—making it the largest consumer of meters.

11.2.1 The History of Metering: Flow Meter

Meters have existed, commercially, since circa 1830 with the positive displacement technology (particularly the diaphragm meter). The orifice meter appears to date back to 1904 with a published 1912 ASME paper by Thomas Weymouth on his previous orifice meter-based testing. The first turbine meter was drawn up in 1790 and became more commercially available after the 1940s. Vortex meters date back to the late 1960s, while the phenomena was first noted by Leonardo da Vinci in 1504. Ultrasonic meters have been in commercial use since the early 1960s and were also the technology springboard for "analytical diagnostics" that have existed since the 1980s. Thermal devices have been in use since the 1970s and gained momentum in the early 1900s with the hot-wire anemometers. Coriolis meters are based on an inertial force that was formulated by Gustave Coriolis in 1835 and commercially launched in the mid to late 1970s. Optical techniques have been shown to measure flow in research laboratories since the 1970s but have only really found commercial use in flare gas applications. It is sonar that is the latest technology to be deployed for metering with commercial usage recorded in 2003.

In April 2015, Springer published a book by Reader-Harris M (2015): Orifice Plates and Venturi Tubes. In the section "Introduction and History" of the Reader-Harris's book, it is noted that "Differential-Pressure flowmeters remain the most common type of meter for large and medium flowrates: their demise has often been predicted, and yet large numbers of them are still both used and bought." Mr. Reader-Harris continues: "It is estimated that at least 40% of industrial flowmeters in use at present are differential-pressure-based devices, with the Orifice plate being the most popular for accurate measurement of flow."

11.2.2 The History of Metering: Metering Supervisory Systems

As the flow meter technologies developed so too did the approach to "supervisory systems" where supervisory is defined as "that having or relating to the role of observing and directing an activity or a person." Prior to the evolution of the electronic computer, most of the installed meters' data was connected to chart recorders with local data collection by human operators. These paper charts and subsequent operatives' notes were dispatched to a central facility where the information was processed in terms of "measurement custody transfer" and "operations" reports. The main challenge here was that the centralized facility had to process reports based on inputs that saw discrepancies that could and were quite often some months old. This prompted a considerable change agenda within the industry that was to exist for some future decades.

There developed an overwhelming requirement to process real-time data from the metering and the metering operations personnel. Although there were telemetry systems in place, in the early years, getting the real-time data back to a centralized facility was not readily achieved within the control ecosystem. Most of the early remote terminal unit ("RTU") and supervisory control and data acquisition ("SCADA") systems had a limited capability in terms of signal input/output ("I/O") and subsequent processing. The early 1970s saw what was to become the ubiquitous microprocessor that quickly found applications within the oil and gas industry. By 1975 the first dedicated flow computer ("FC") was commercially available for which the primary purpose has changed little in the past decades: to accurately compute the amount of flowing fluid through an associated flow meter in terms of actual or standard engineering units. The essential inputs are flow (volume or mass), pressure, temperature, and often density or fluid composition (analyzer) as

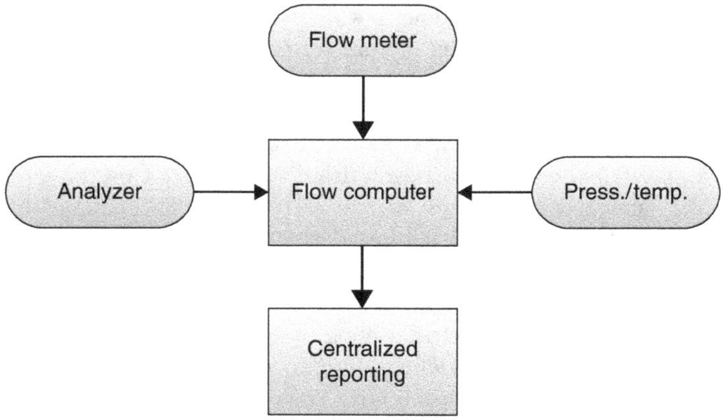

FIGURE 11.1. Early single-stream flow computer.

seen in Figure 11.1. The critical outputs are bulk flow rate and the integrated totalization over predefined periods of time from 1 to 24 hour, recording any notified alarms and any mismeasurement events.

Through the 1980s and 1990s, the universal multifunction controllers began to appear where there could be said to have been an overlap between the early programmable logic controller ("PLC") capabilities and the traditional FC, but the main issue was that the measurement engineers were not control systems engineers working with control programming languages or ladder logic. At this point in time, there grew a demarcation between the dedicated measurement systems engineers and the control systems engineers. This demarcation has continued to exist within the industry to this present day. In the measurement arena four main technological advances could be seen that were latterly embedded within the FC: redundancy, single-stream, multi-stream, and liquid provers. Traditionally, outside of Europe the multi-stream FC has gained popularity while being embedded within an RTU or PLC device.

The metering supervisory computer ("MSC") came into commercial operation as the number of FCs and the associated number of flow meters grew to see a typical metering system with multiples of tens of individual units. The MSC grew from the requirement to enable a greater communications network as well as computing power with its associated memory to implement advanced features across the total measurements system; ironically, these solutions have evolved into ever more powerful traditional SCADA control systems! Capturing this "metering ecosystem" data required a centralized facility that would connect to the individual FCs that would provide a similar data capture to the FC itself but now rolled up to address the total metering ecosystem. This MSC functionality has traditionally been provided by a SCADA system that latterly, as technology has developed, hands over flow rate data to the plant or processes distributed control system ("DCS") that subsequently records data in process historians within a control architecture that includes for operations data to be reported all the way to the management enterprise systems. As more and more metering technologies enter the market with onboard self-diagnostics as well as other advancing capabilities, the MSC is going to require

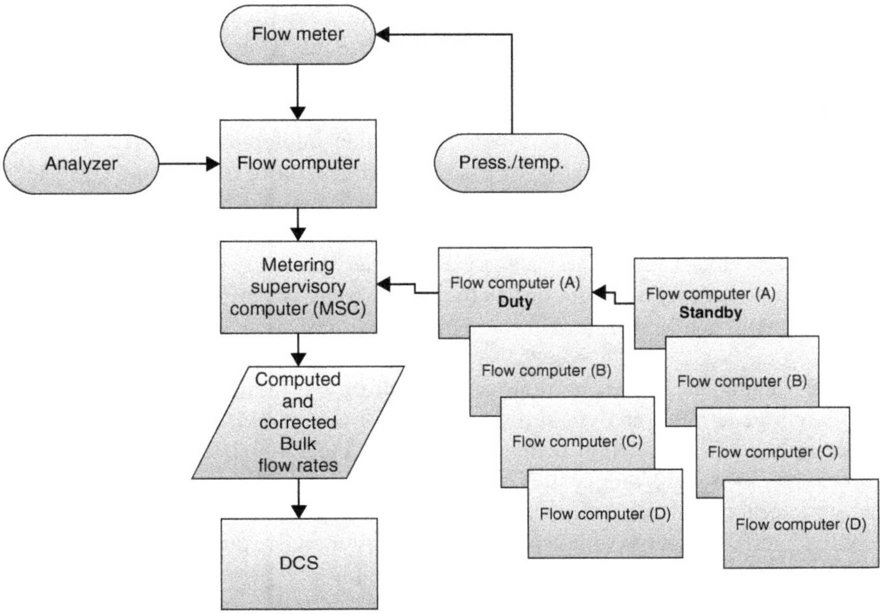

FIGURE 11.2. Traditionally connected flow meter and the metering supervisory computer.

more and more functionality. The MSC is a device that may be seen as the "chief financial officer for overall accounting services supporting the independent external auditors for the metering system" rendering the data acceptable to the governing bodies. A typical overview of an MSC system is provided in Figure 11.2.

In summary, meters of several different fundamental technologies have existed for centuries where the popularity of those with no moving parts has gained momentum within the oil and gas industry yielding the greatest accuracy with the highest reliability. The nonintrusive and noninvasive technologies are what we may cite as being "close to ideal" as they do not impact on the flowing fluid or they do not influence the flowing fluid due to any "restriction." The metering supervisory computer has existed for decades and may be regarded as a dedicated SCADA system for the metering engineers.

The key change being highlighted in this chapter is, therefore, in the connected digital transformation of these established technologies and not the fundamental metering technology itself. It is about the inherent digitization of a meter technology and its ability to relate to a larger system context to bring about new insightful outcomes.

11.2.3 The History of Metering Diagnostics and Condition-Based Monitoring ("CBM")

As the flow meters developed, so too did the ability to diagnose operational faults being largely based on the commercialization of the personal computer and ever-evolving generic software such as mathematical-based spreadsheets. As the software developed and the computing power grew exponentially, more complex analyses were performed on operational metering assets. Complexity encompassed not only the meter itself but also other associated devices that comprised the larger measurement ecosystem. Developments continued over several decades with a huge digital transformation enabler being the dual technologies of cloud computing and open-source frameworks for performing analysis on near-real-time collected, streamed data. However, while we have observed the growth of the various metering technologies and the associated standards and recommendations, there has been no published documentation to standardize either the industry definitions or the components of a specific solution for the digital transformation of meter diagnostics or CBM. It is for this reason the author is providing his definitions for clarity as follows.

Meter diagnostics and condition-based monitoring is digitization using physical computers generating analytical data to improve our knowledge and the maintenance of meter assets while better engaging people through the "automation of knowledge." A meter's diagnostics is a method that is used for reporting the original design tolerances against an operational meter's fault or failure enabling the end user to identify why it has failed: a mathematical approach to quantify first operating principles of the meter technology. A first principle is a basic, foundational, self-evident proposition or assumption that cannot be deduced from any other proposition or assumption.

A meter's CBM system is the process of monitoring parameters of condition within a meter, its associated devices (e.g. pressure transmitter, temperature transmitter, quality analyzer, etc.) and the local in situ environment for the meter in order to identify a significant change that is indicative of a developing fault. It is a major component of predictive maintenance and is, therefore, enabling an end user to identify what is going to fail at a future and predetermined time. Predictive maintenance is time series data that includes a timestamp, a set of device readings collected at the same time as timestamps, and device identifiers. The goal of predictive maintenance is to predict at the time "t," using the data up to that time, whether the equipment will fail in some specified period.

11.3 CONNECTED METERING

11.3.1 Why Connected Metering?

Connectivity in this setting is that enabled by the digital economy and its ubiquitous connectivity. The traditional installed-base meters within the oil and gas industry are great potentials for the automation-of-maintenance agenda. Automation through online self-diagnostics brings about "CBM" analytics that have been utilized to extend the traditional recalibration frequency of ultrasonic meters, a unified approach toward maintaining meter accuracy and reliability. Such an approach has been actively pursued within the industry for the past two decades and pursued with great success. Recalibration periods have been extended from an annual basis to a maximum of up to 7 years, while in the past decade software has been advancing to provide data that highlights the "operational health" of the meter by analyzing its in-situation performance with reference to the original "footprint" (or benchmark) when it was newly manufactured and subsequently calibrated at an independent third-party facility. The industry is governed by published standards and audited against these standards while being regulated by national legislation and legal metrology.

Not all meters are connected in such a way, and, indeed, the only metering technology to see considerable growth and acceptance in the past decade is the ultrasonic meter. With the advancement in other technologies, it is now possible to connect differential-pressure devices, Coriolis meters, and turbine meters to highlight only three other technologies and even analytical devices such as the gas chromatograph. The power of connected should embrace these other meter technologies as they represent in the case of mechanical meters (e.g. differential pressure) the industry's largest installed base. In other words, the existing connectivity toward diagnostics with CBM is only on the order of a few percentage points across the global meter ecosystem. Data leakage is evidently quite considerable and, if we apply the Pareto rule, is on the order of 80% and very likely higher, with leakage being defined as "an unintended loss of valuable data due to a lack of ubiquitous connectivity."

The ever-advancing digital economy is the IIoT that is simply the "Intelligent Interconnection for Operational Transformation (Bragg, 2017)" that starts with the "BIG WHY'S" and not the "What." Firstly, know why there is a clear need to connect and, subsequently, why is that connectivity going to transform and improve operations! Once the "Why's" are identified, the "How's" become a little more intuitive or, at least, more readily definable when it is clear "What" already exists. The digital economy toward connected metering is not a rip-and-replace strategy but one based on ubiquitous connectivity within an evolving digital ecosystem.

The first why (Yergin, 1991) is to connect the installed base of various metering types to leverage the technical advances historically seen with ultrasonic meters toward diagnostics and CBM. Advances that have reduced the traditional recalibration frequency while providing an online CBM facility to highlight when the meter is likely to drift outside its stated design performance or is trending toward failure. A strategic choice that moves away from the manual preventative maintenance "calendar" schedules toward online self-monitoring 24/7/365. The IIoT, the digital economy, must provide a positive-financial impact to enable its rollout, and it must deliver capability that was not previously possible. In other words, simply relocating an existing CBM package from a localized connection and quite likely stranded in terms of its connectivity across the measurement and control architecture to the cloud has limited positive-financial impact in terms of savings. The greatest saving shall arise from subsequent data analytics that are only enabled by the digital economy.

The second why is to enable data analytics not previously possible to bring about new capability to yield a significant positive-financial saving, the enabler being the IIoT's complex mesh framework with multiple data paths where subsequent advanced analytics shall be performed in multiple places in providing a technically auditable measurement. To further go beyond the traditional CBM approach, the advanced analytics will enable a near-real-time uncertainty analysis of a meter to determine its performance with respect to its uncertainty and the combined uncertainty of the overall metering system itself. The greatest negative-financial impact for any metering asset is its lack of online CBM with its associated near-real time uncertainty and not its extended recalibration frequency! A further example of such advancing data analytics is that which enables the physical meter to be complimented by its digital twin in what shall be an industrial metering paradigm shift.

Technology is developing toward CBM for meter types other than the traditional ultrasonic; however it is the advent of a near-real-time uncertainty analysis that is quite unique that became commercially available in April 2017. To be able to connect the meter and its associated CBM with a near-real-time uncertainty is a challenge for the industry for which the author proposes the solution to be the digital economy by way of the IIOT.

To highlight its impact on metering, the author has studied two real-life use cases that have shown a negative-financial annual impact of $25 and $27 million/year, respectively, in one metering instance on a natural gas export plant. One was an 18-in. ultrasonic meter (Lunde et al., 2007) and the other a 6-in. Venturi meter (Prognosis Brochure, 2015) for which the reporting is available within the public domain. How many more exist within the industry across its value chain? There are a great many millions of meters being sold globally every year by over 1,000 companies that are installed in hundreds of different applications in dozens of industry sectors (Drahm et al., 2013). If we further consider the history of metering, as provided within this chapter, it is evident that a significant number of meters are deployed within the historical installed base for which there is no common CBM capability nor any near-real-time uncertainty analyses. It is evident from within the industry that the number of CBM packages is increasing in terms of the various meter technologies for which there are commercially available solutions; however the approach to the individual meter analysis requires a number of disparate CBM packages for which there are no published standards. In other words, the end user is confronted with a barrage of diagnostics within a CBM package for which there is no standard toward data analytics, data presentation (visually), and the software program that is often the foundational basis of the CBM offering.

11.3.2 Condition-Based Monitoring and Meter Uncertainty

CBM is largely operating within the oil and gas industry as a stranded asset and typically for ultrasonic meters—as historically recorded. Stranded in its local connectivity, a separate industrial personal computer is running a software program physically close to the metering supervisory computer. The industry has seen what the author has termed phase 1 and phase 2 implementations. Phase 1 represents the first commercially available packaged solutions that had a focus toward, and a connection to, the flow meter itself. The phase 2 packaged solutions took a more holistic approach to include the other measurements as part of the metering system that directly contributed to the overall system health monitoring (i.e. CBM). There is often data that is provided without any engineering units and without a common scale rendering each instance of a CBM package somewhat unique. Indeed, interpreting the data collected and reported by a typical CBM package still requires an experienced metering engineer even with the advancements in reporting and advancements toward human factors for the user experience that culminates in a "traffic light" system of red, orange, and green for the status of the meter itself. The typical connectivity is shown in Figure 11.3.

The oil and gas industry has long debated the usefulness of CBM, but it has in this latter decade embraced the technology for ultrasonic meters. There remains a couple of challenges for the industry that are not resolved to date. Firstly, each instance of a CBM package requires some insight by an experienced metering engineer to be able to predict when the meter will tend to a failure state. The CBM packages can provide operating levels that once exceeded approach the set limits that will subsequently trigger an alarm, enabling the two-stage condition from

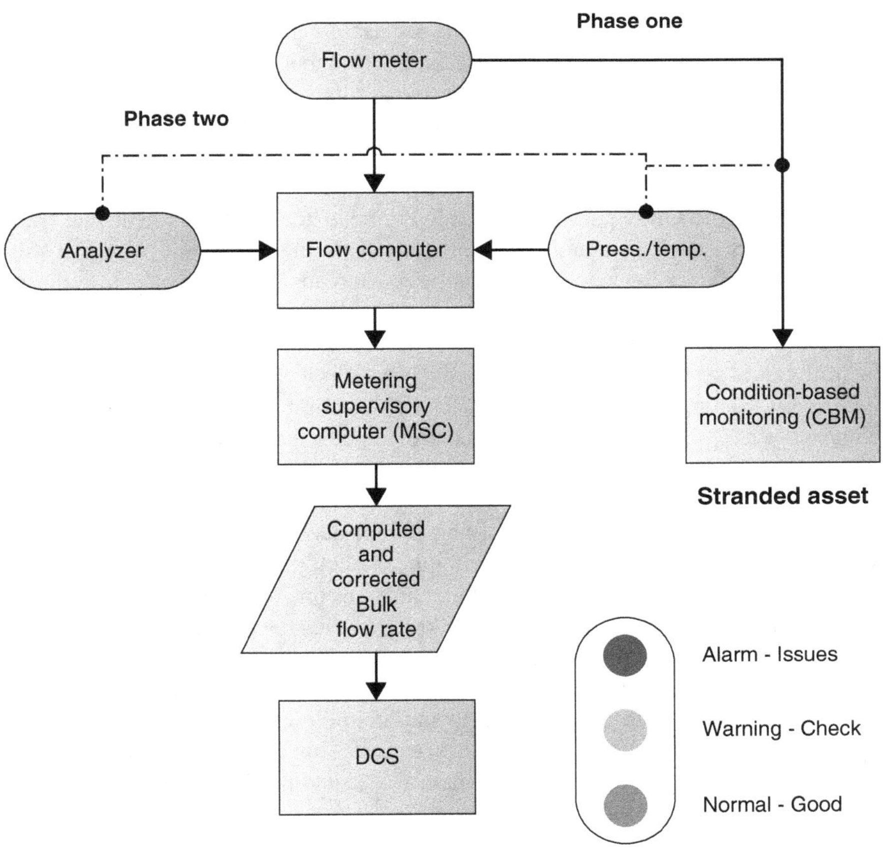

FIGURE 11.3. Typical condition-based monitoring packaged solution.

which the traffic light system is enabled. Secondly, no CBM package is providing an uncertainty analysis to indicate the change in the meter's uncertainty in deriving bulk flow rate due to a change in diagnostics. This is because there are many individual diagnostic parameters being analyzed within the CBM package for which the impact of each, or the sum of several diagnostics, is not directly attributable to the measurement uncertainty.

To be able to progress to the next higher level of diagnostics shall require advanced data connectivity for the meter as well as the process into which the meter is located. This is evidential as the ultrasonic meters' diagnostics are often influenced by the process into which they are located, and this influencing factor is not part of the diagnostics reporting. In other words, is it the true status that it is an influencing factor external to the meter itself that requires the attention of the maintenance team and not the meter, a cause-and-effect situation that would benefit from further analyses outside the meter's CBM package?

As stated, today's CBM packages do not provide any uncertainty data toward a meter's operating performance. Technology is reaching an enabling point where uncertainty can be derived from process data although this requires further connectivity outside the CBM package itself. The uncertainty analysis is based on process data that includes the meter's bulk flow rate reporting but does not resolve the issue of relating specific meter diagnostics to components of uncertainty. It is proposed that a connectivity model be established as shown in Figure 11.4 where the traditional metering system is enhanced with the addition of an uncertainty analysis package.

To connect the CBM package, for a given meter technology, a near-real time uncertainty analysis can attain the maximum performance of the meter. In this approach the traditional uncertainty analysis (Bureau International des Poids et Mesures (BIPM), 2008; ISO5168:2005, n.d.) is enhanced by the input of the relevant CBM package to further validate the uncertainty. Validation shall be tied out to the traffic light system within the CBM analytics. It is a challenge to connect where there is no existing connectivity, and it is a challenge to perform detailed uncertainty analyses within measurement and control systems that do not have the available capacity. It is for these reasons that the digital economy, in the form of the IIoT, is ideally suited. It is a challenge of connectivity, data analytics, and storage systems that is ideally suited to the ever-evolving digital ecosystem. The architecture to achieve this connectivity while addressing the challenge of the unique instances is shown in Figure 11.5. The choice of architecture was one founded upon the existing Honeywell Connected Plant providing the key elements toward connected metering in maintaining the essential metering attributes of accuracy and reliability within the utmost safe and secure architecture. An architecture with scalability beyond metering encompasses the requirement toward ubiquitous sensing being coupled to cloud-borne data analytics and storage systems. This is a proven and scalable architecture for connecting plant processes, assets, and people in an ever-evolving digital ecosystem.

Enabling a complete architecture as shown is quite challenging as a first step toward digitization; however having a view to the overall future architectural

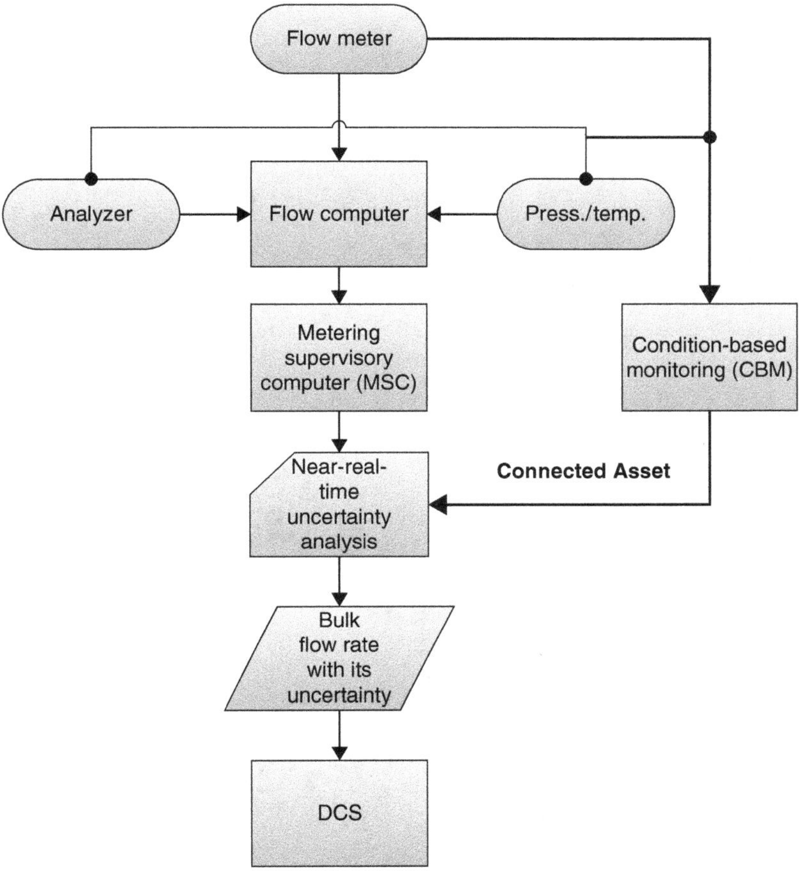

FIGURE 11.4. Condition-based monitoring with a near-real-time uncertainty analysis.

requirements can assist in the short-term decision-making process toward a smaller-scale rollout with the knowledge that this smaller rollout shall fit into the larger architecture as the digitization agenda is developed. Figure 11.6 provides an example of a potential smaller-scale initial step toward digitization with the first stage simplified connectivity to achieve a technically auditable flow rate measurement with its traditional uncertainty analysis. A technically auditable measurement is defined as "having special and usually practical knowledge toward metering that shall be authorized or sanctioned; it is justifiable." This simplified step takes the CBM monitoring to the cloud via a suitable edge gateway where the relevant flow meter data may be used to perform a near-real-time traditional uncertainty analysis. However, as data analytics develop in the cloud, a second stage becomes possible in that an analyses of the relationship between the actual process conditions and its impact on the various meter diagnostics can be performed to derive an actual near-real-time CBM uncertainty, an industry aim that

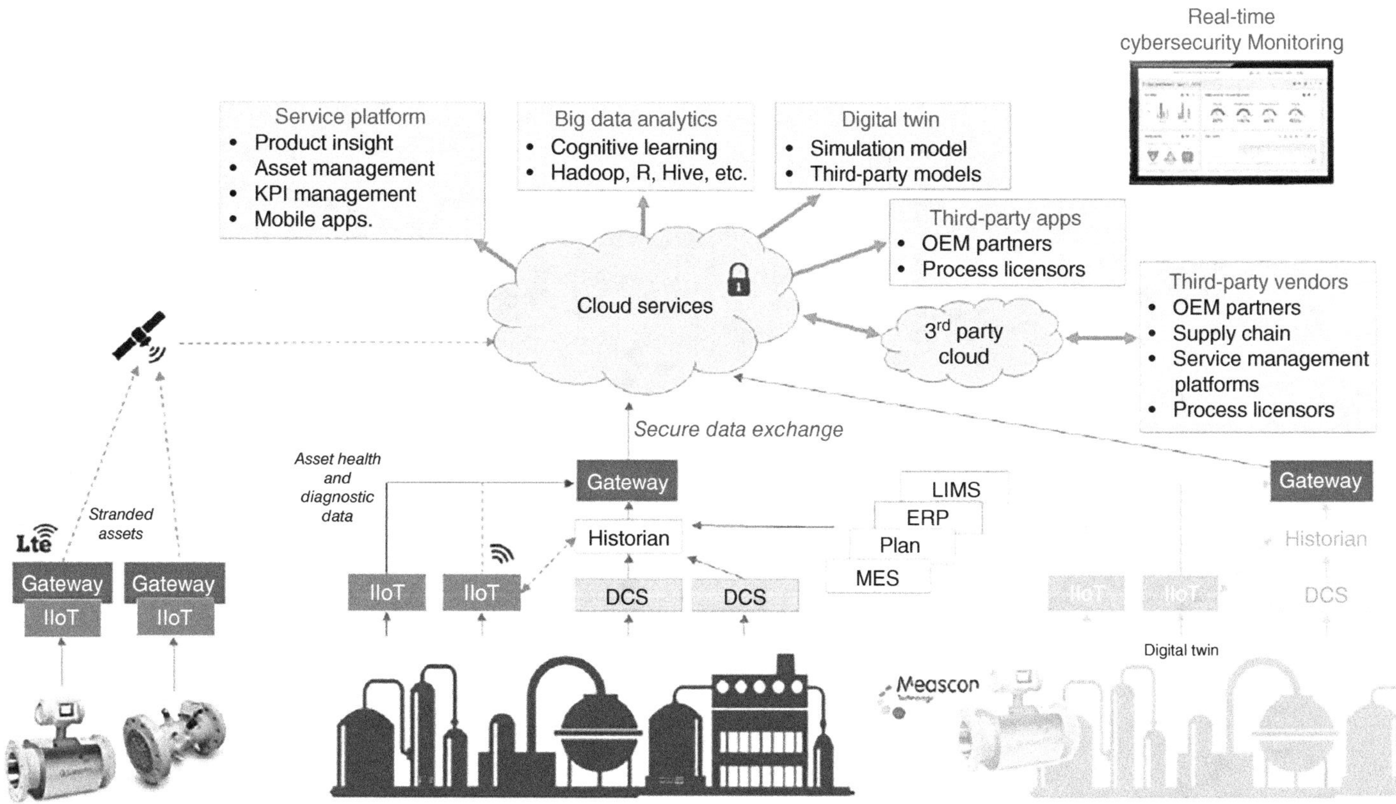

FIGURE 11.5. An IIoT connected architecture for connected metering.

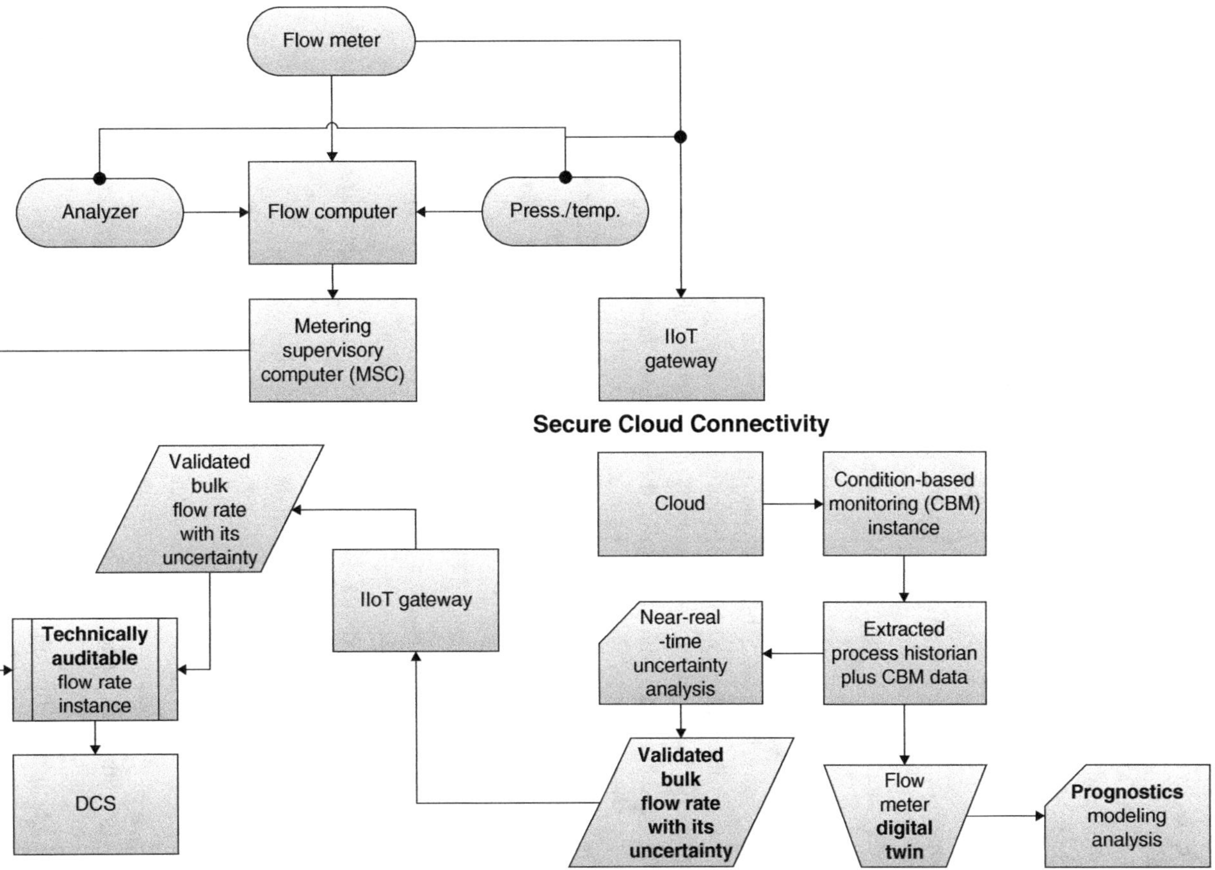

FIGURE 11.6. Connectivity toward a technically auditable measurement with its associated uncertainty.

is often requested by the end users but not deliverable with today's installed-base technology.

Even with this simplified connectivity, the power of connected is rapidly realized with the digital twin, a virtual twin with use cases toward conceptualization, comparison, and collaboration. The advantage of the digital twin within an automation-of-maintenance agenda is that the twin shall provide the platform to test changes within the physical meter without impacting upon the physical meter itself or the online resultant measurement. Simulating tests and applying changes to the twin negate the requirement to impact on the physical meter, yet the data and associated analytics are based on the physical meter's online data. It shall further assist in analyzing the physical meter's historical data sets within a complex analytics model to determine how the individual diagnostics potentially impact on the meter's uncertainty budget. It provides a platform to:

- A more effective assessment of a meter's current and future capabilities during its life cycle
  - Lowering costs while increasing uptime with a longer operational life span
- Optimize a physical meter's monitoring, diagnostics and prognostics leveraging models and simulations applied during the entire life cycle of the digital twin
  - Enabling actionable results through individual modeling of an on-site physical meter
- Ensure a continuous offline refinement of operational data embracing metering optimization to arrive at a sustainable lowest possible uncertainty that is readily cross-referenced to the physical meter's real-time performance
- Marry sensors from the physical meter's ecosystem—both immediate or within a larger process footprint—and simulation modeling in the digital twin
  - Learning how the meter is responding within its measurement environment or ecosystem that is itself subject to change
  - Connecting the digital silos and analyzing hereto unused readily available unstructured data.
- Structure inherent unstructured data collection
  - Reconfigure data collection and refining the process after an event is a static read and react model
  - The digital environment enables a living model to be achieved through pre-event scenario modeling of the digital twin
- Bringing the benefit of continuous improvement engineering to a meter throughout its life cycle
  - Substantially reducing the timeframe as the duplicate runs with "virtual real-time data" bringing it to life
  - Closer to real-time corrective actions for optimum performance
- Meeting and going beyond the industry trending for flow measurement
- Performing "What-if" analyses in the duplicate twin without any impact on the physical in situ meter

Once the digital economy is active toward the IIoT, the possibilities for connectivity within the digital realm increase exponentially. An increase toward connectivity can bring about a positive-financial impact for the asset owners.

11.4 POSITIVE-UNEXPECTED CONSEQUENCES OF THE DIGITAL ECONOMY

It was evidently noted by the author that the digital economy shall result in "positive-unexpected consequences" in terms of its data through advancing connectivity. The unintended consequences (sometimes unanticipated consequences or unforeseen consequences) are outcomes that are not the ones originally foreseen and intended by a purposeful action—as adopted from the Lean Six Sigma methodology. A positive-unexpected consequence of collecting data within a digitization strategy is that further subsets of data may be readily formed from the larger master data sets bringing about new insights. Insights are the positive consequence of connecting subsets of existing data in a "virtual world of digital connectivity."

With the power of digital connectivity, the connected metering concept becomes a reality. A concept that brings about online CBM with near-real-time system uncertainty that is not available within today's physical metering systems. A concept that shall provide savings well beyond that of traditionally extending recalibration frequency or preventative maintenance schedules with estimated savings on the order of up to $250 million/year across a metering ecosystem, depending on (a) the total number of meters and (b) the "uncertainty creepage" for the given metering assets. A positive-unexpected consequence would be that where there is enough data to implement "virtual metering." A virtual meter being a software model that uses ubiquitous connectivity to generate data that enables one to estimate or model flow rates instead of using a physical meter. The benefits of both a physical flow meter with the enhancement of further process validation can leverage an "autonomously self-maintaining virtual meter" model, one that can be checked based on process historian data, ubiquitous physical sensors, and physical meters.

This positive-unexpected consequence for the physical meter complimented by virtual flow metering is further enhanced by a physical meter's digital twin. It was the original design intent of connected metering to enhance CBM with the addition of a near-real-time uncertainty analysis but with the larger connected data set and subsequent analytics the digital economy quickly enables the virtual ecosystem in terms of metering. The ecosystem would, therefore, include for (a) the virtual metering being made possible through software modeling based on the ubiquitous physical sensors within the overall process (i.e. the object's environment) with (b) the physical meter's digital twin in representing the actual physical asset as seen in Figure 11.7.

A unique set of consequences can be seen within this metering ecosystem, as shown in Figure 11.7, in that the physical meter is technically audited by its CBM and uncertainty analysis with "what-if" scenarios being implemented via its digital twin, while the process is further enhanced by being provided with virtual meters

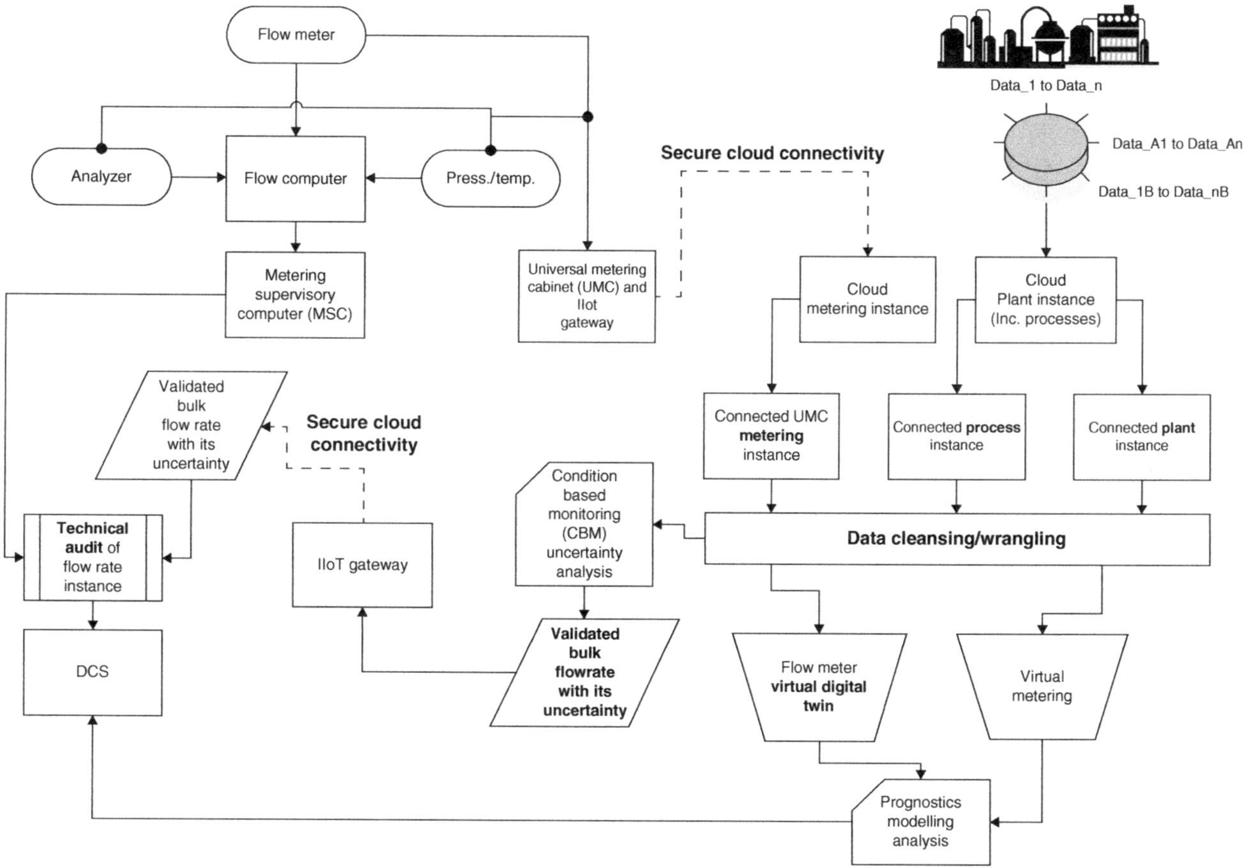

FIGURE 11.7. Connectivity within a connected plant's metering ecosystem.

where there are no installed physical meters. Should the physical meter itself fail without warning, the digital model can continue to provide a valid and auditable flow rate to the plant's DCS avoiding a potentially costly interruption to plant operations. An auditable flow rate shall be one that is performed by an accredited auditor, engineer, or subject matter expert who evaluates failures or deficiencies or areas of improvement in metering systems where the outcome-based technical audit addresses the technological aspects of the metering technologies being installed within the localized process or overall plant.

As the digital economy is further embraced, it is quite possible for the topic of augmented reality ("AR") to be introduced as another positive-unexpected consequence (see Chapter 16 for details). Leveraging the power of AR to show digital data upon the physical asset in terms of its condition and overall performance while providing 3-Dimensional design drawings and traditional services manual (e.g. Users' Manual) can assist a field technician in maintaining the asset. Maintenance is now reduced to essential field-based intervention, at a prescribed time, with the advanced information at a detailed parts level (provided to the technician in advance), all being achieved prior to the asset's failure. This approach may be directly compared with that of a traditional calendar-prescribed regular preventative maintenance schedule where the technician discovers the failure, during or after the event (responding to an alarm status), and then has the challenge of bringing the asset back online (over a period of calendar time) as quickly as possible. Meanwhile the asset is not available and is, as such, a financial loss, while the cost for the corrective measures increases as time becomes critical and "critical time" adds unnecessary costs!

Of note is that the overall master architecture of a connected metering system, as shown in Figure 11.5, remains the same because it is the digital connectivity with its associated data analytics that is the differentiator between the various use cases. Data analytics is a virtual approach based on software that is enabled due to the connectivity of the physical ecosystem. The greater the "depth and breadth of connectivity," the relatively smaller the data leakage (i.e. data not utilized) and the greater the opportunities for ever advancing data analytics. We know what we know, we often think we know what we do not know, but we do not know what we do not know without exploring the possibilities and opportunities afforded by advanced data connectivity within a digital ecosystem!

In summarizing, it can be stated that the positive-unexpected consequences exhibit a greater potential for savings as a result of the original connected metering strategy. A strategy goes beyond the traditional approach to meter maintenance that was originally founded upon an exercise to extend the recalibration frequency of the meter asset.

11.5 THE OUTLOOK FOR CONNECTED METERING

In considering the possibilities what is the outlook for flow metering? What is the next generation of human resource going to be driving toward?

The future of metering is all about ubiquitous connectivity and enhanced collaboration within an ever-advancing digital economy. It is about tighter integration between measurements and controls where the historical approach to systems architecture shall be simplified while removing the overlap in functionality. It is about building upon the Honeywell MeterSuite™ solution that already embraces the traditional metering system within the plant DCS and that has received legal metrology approvals having been commercially available within the market for several years.

As the technologies develop and the digital economy is embraced within the IIoT arena, there shall be a time when the physical process or plant is connected to nothing more than Universal I/O, while the subsequent measurements and controls hierarchy shall be simplified into one-single-mesh instance, wired, wireless, or other communications removing the overlap between the traditional RTU, PLC, SCADA, and DCS in considering the topic of metering. Substantial costs shall be removed, while simplification shall be evidential, and collaboration shall span the entire organization without the traditional silos toward pieces of the overall architecture, an approach that shall enable the highly mobile and connected workforce of the future. Such an approach leads to a simplified cloud-based metering ecosystem as shown in Figure 11.8.

If we now consider the positive-unexpected consequences of the digital economy, whether as a simplified future state as in Figure 11.8 or in embracing an initial step toward CBM and near-real-time uncertainty as in Figure 11.6, we can readily deploy a meter's digital twin within the virtual world while leveraging the vast

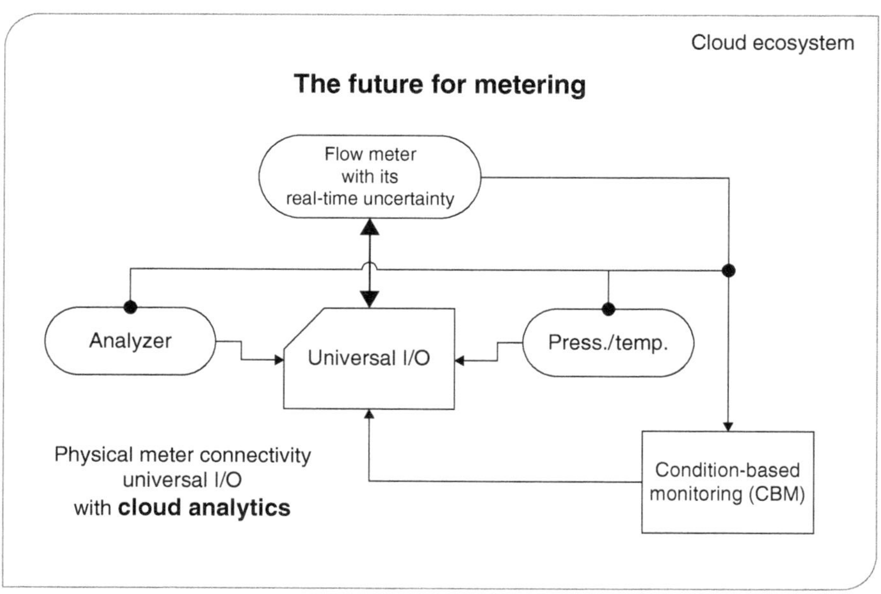

FIGURE 11.8. The future for connected metering.

mesh of connectivity in which the meter is installed and even further across the entire plant, operation or enterprise.

With reference to Figure 11.9 and the future connected "4D-Ecosystem," the abbreviated form "DC" represents the "data connectivity" within a digital economy leveraging the IIoT. The four-dimensional element is one based on connectivity that has width, height and depth along with a time dimension that is, in this case, one of near-real time being supported with an associated historian for further analytics. The various layers of data connectivity shall be extracted from the complex mesh that shall evolve within the digitization agenda. A digital mesh of multiple data-points shall be realized within a framework with multiple data-paths where subsequent analytics shall be performed in multiple places. With enough connectivity the virtual agenda toward virtual flow metering readily evolves. A virtual meter is a software model that uses digital data to estimate flow rates instead of using a physical meter. The benefits of both a physical flow meter with the enhancement of further process validation leveraging an "autonomously self-maintaining

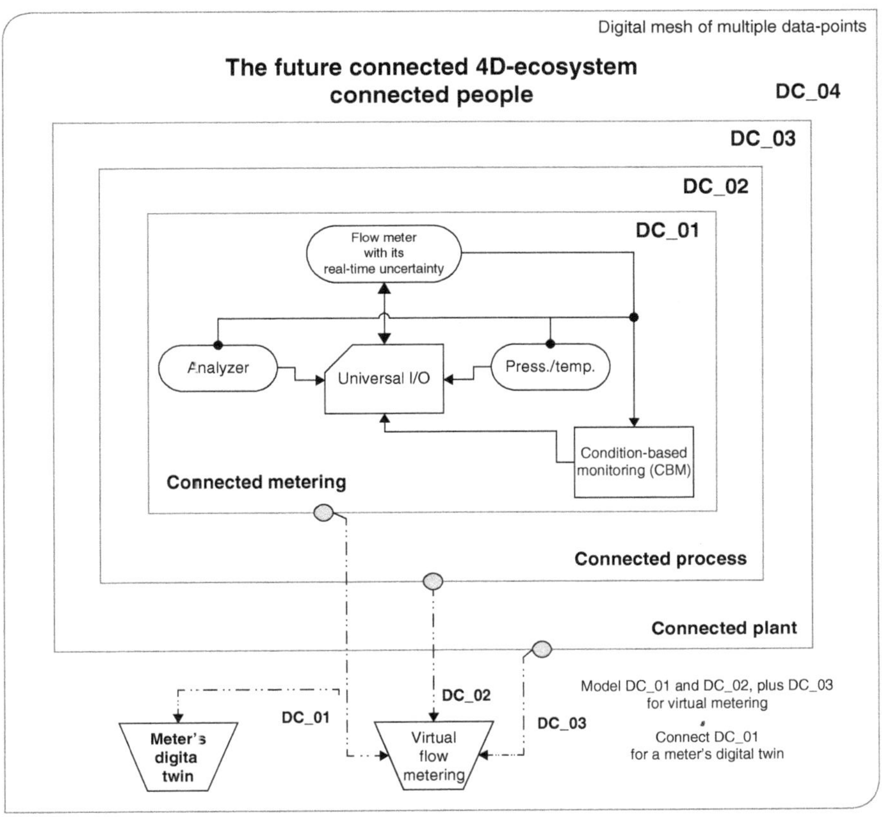

FIGURE 11.9. The future connected 4D ecosystem: a metering perspective.

virtual meter" model, one that can be checked based on process historian data, ubiquitous physical sensors, and physical meters.

The data connectivity to DC_01 provides validation toward the virtual flow metering resultant from the model derived from primarily DC_01 and DC_02 plus further modeling via DC_03. Such an approach could enable an in-line calibration of a physical meter negating the requirement to remove it and ship it to an independent third-party facility, a process that includes for a great many man-hours in terms of process isolation, meter removal, onward shipping to the recalibration facility, and its subsequent return into process. Man-hours include the metering technicians, health and safety personnel and process operators to name but a few.

The future of connected metering within a digital mesh of multiple data points shall quickly enable:

- Simpler measurements and controls architecture with less overlap in products/ solutions with a highly integrated agenda.
- The considerable reduction of data leakage generating a positive-financial impact.
- Advanced data analytics developing with augmented reality solutions.
- Physical meters analyzed with an online CBM and near-real-time uncertainty analysis increasing measurement confidence.
- Digital twins with use cases toward conceptualization, comparison, and collaboration.
- Virtual metering across a process, plant, or enterprise.
- In situ calibration of online physical meters.
- Flow assurance (technically auditable) with both physical and virtual metering.
- Enhanced metering for process control, management systems, and well optimization are a digital mesh of connectivity.
- Enhanced pipeline systems leak detection integrity.
- Enhanced data capture toward advanced analytics bringing about new learning.
- Enhanced flow modeling, visualization, and fluid mechanics in an ecosystem that can readily deploy advanced computational fluid dynamics to resolve challenging applications.
- Automatic generation of reporting for regulatory, compliance, and auditing purposes.
- Connecting people in a metering ecosystem where future faults are predicted before an alarm condition occurs and the pre-ordering of the correct spares is implemented to reduce the number of service visits to the asset.

11.6 CONCLUSIONS

It has been shown how the metering systems within the industry have developed, historically, to embrace not only the meter itself but the flow computer and the metering supervisory system that is a dedicated SCADA solution. Embracing early digital technologies to provide Condition Based Monitoring software that is a data analytics package can be founded upon the meter's operational diagnostics. An ecosystem is an evolving digital solution to the challenges of metering and its maintenance agenda in considering the legal metrology, national, international, and global standards that have been developed over several decades. Standards have been developed based upon experiential learning (i.e. learning by doing) and quite often written by subject matter experts with a deep-domain knowledge of the specific subject. A subject should now embrace the digitally connected ecosystem, often referred to as the IIoT, as offering the promise and potential to be the most influential and disruptive influence on metering systems since the advent of the first meter in the early 1800s.

The meter quite typically is not a stranded and isolated measurement device but one that is installed within a measurement ecosystem bringing about some form of process control. Metering has been approached from the perspective of the meter itself and its immediate localized piping diameters often referred to as the "meter run" being essential for an accurate and reliable measurement. Therefore, the operation of these meters is one based upon limited connectivity within disparate systems that are a mixture of partial automation and manual processes. In this regard it was essential in innovating with connected metering not to have made strategic choices based on a retrospective analysis but to have dreamt of the possibilities instead of the limitations while being willing to re-examine the status quo. Furthermore, connected metering can turn the possibilities into deliverable products or solutions in bringing about innovative technologies that can generate a positive-financial gain (i.e. cost savings) for our customers.

Connected metering therefore goes beyond the meter itself to consider a greater connected ecosystem in three key areas being (a) its connected processes, (b) further connected assets, and (c) in connecting people. This innovative approach to connected metering is therefore encompassing these three key topics where each was considered to enable a positive-financial gain as a single-key customer outcome in reducing their operational expenditure. An approach has been shown to be deliverable when considering not only the overall architecture but also a smaller number of steps or stages within an ever-evolving digital ecosystem. Steps or stages that bring about new learning within the essential Metering elements of: Continuous Meter Diagnostics; Condition Based Monitoring Analytics, Near-Real Time Measurement Uncertainty; Digital Twins; Virtual Metering; and Technical Auditing with a focus towards Mismeasurements. A connected metering choice is founded upon a more ubiquitous approach to connectivity that diminishes the present impact of data leakage that exists because there is either limited or no connectivity within the installed-base toward the evolving digital economy that is resulting in "an unintended loss of valuable data."

It has further been noted that there are what the author has termed positive-unexpected consequences from ubiquitous connectivity. It has been evidentially noted that the digital economy shall result in positive-unexpected consequences in terms of its connected data and subsequent analytics. The unintended consequences (sometimes unanticipated consequences or unforeseen consequences) are outcomes that are not the ones originally foreseen and intended by a purposeful action—as adopted from the Lean Six Sigma methodology. A positive-unexpected consequence of collecting data within a digitization strategy is that further subsets of data may be readily formed from the larger master data sets bringing about new insights. Insights are the positive consequence of connecting subsets of existing data in a "virtual world of digital connectivity."

In closing, it should be noted that the connected metering digital transformation is an ever-evolving subject matter due to the ubiquitous connectivity being enabled by the IIoT encompassing cloud-borne data analytics and its storage systems.

REFERENCES

Bragg M (2017), Intelligent Interconnection for Operational Transformation, V.11, Issue 2, Exponent, Honeywell.

Bureau International des Poids et Mesures (BIPM) (2008), JCGM 100:2008 GUM 1995 with minor corrections, Evaluation of measurement data—guide to the expression of uncertainty in measurement.

Drahm W, Popp O, Touzin M (2013), Future Challenges of Flow Measurement: Towards Smart and Compact Sensor, Endress+Hauser GmbH & Co.. DOI: 10.5162/sensor2013/D2.1.

ISO5168:2005 (n.d.), Measurement of fluid flow—procedures for the evaluation of uncertainties.

Lunde P, Frøysa K-E, Folkestad T (2007), Pressure and temperature effects for Ormen Lange ultrasonic gas flow meters. In: 25th International North Sea Flow Measurement Workshop, Gardermoen, Norway, October 16–19.

Prognosis Brochure (2015), Prognosis Case Studies, Case 07 Venturi (Wet Gas), Swinton Technology, Yorkshire, June 2015.

Reader-Harris M (2015), Orifice Plates and Venturi Tubes. Experimental Fluid Mechanics, Springer, Cham, ISBN-13: 978-3319168791.

Yergin D (1991), The Prize: The Epic Quest for Oil, Money, and Power, Simon & Schuster, New York, ISBN: 0-671-50248-4.

12

CONNECTED ASSET AND SAFETY MANAGEMENT

FRANK (XIN X.) ZHU[1] AND TONY DOWNES[2]

[1]*Honeywell UOP, USA*
[2]*Honeywell Process Solutions, USA*

12.1 INTRODUCTION

Asset management is to maintain the ability of an asset to perform its required function effectively and efficiently while safeguarding people's life and the environment. Effective asset management integrates people, asset and process performance, with digital twins to make asset operation in good shape over the whole life cycle of the asset. Maintenance of these assets has evolved dramatically from the earlier viewpoint of repair upon failure to routine repair for prevention and then to fix it before failure occurs (predictive) and to today's proactive strategy of detecting failure at the earliest opportunity to minimize repair costs and at the same time increase equipment utilization (Figure 12.1).

The IT technology is changing industries especially with pivot technologies such as wireless and smart sensing, Internet of Things (IoT), advanced data analytics, augmented reality (AR), and virtual reality (VR). These modern technologies offer differentiated opportunities for companies to further improve the ways for asset management and retain knowledge. By understanding and effectively using these modern technologies, companies can develop proactive maintenance program to improve plant reliability and utilization, operational efficiency, safety, and cost-effectiveness by working with technology vendors.

Digitalization and Analytics for Smart Plant Performance: Theory and Applications, First Edition. Frank (Xin X.) Zhu.

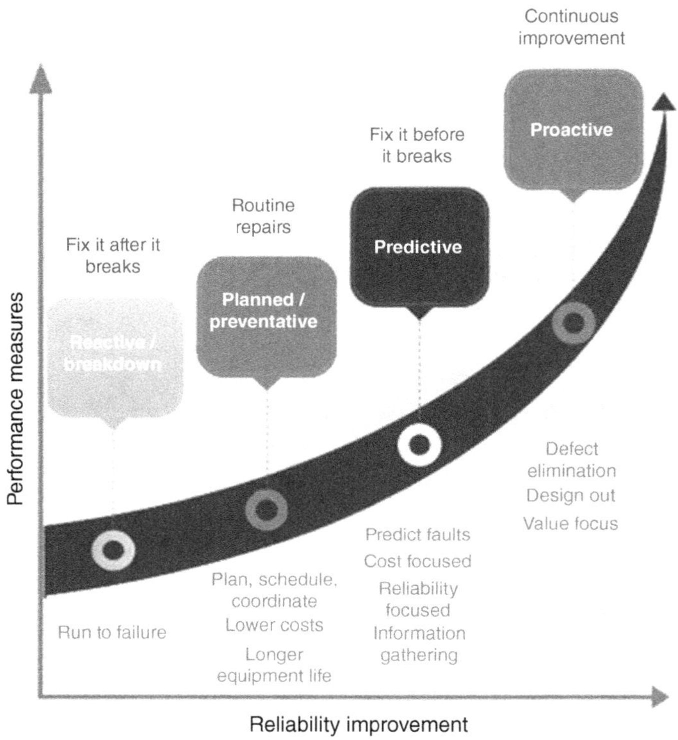

FIGURE 12.1. Evolving maintenance strategies over time.

12.2 REVIEW OF DIFFERENT MAINTENANCE STRATEGIES

To gain some basic understanding of the past, present, and future in the field of maintenance, a brief review is provided below.

12.2.1 Preventive Maintenance

In the past, preventive maintenance programs had been the bread and butter for the industries, which is about checks and services for inspections and updates that are scheduled based on calendar time, run time, or cycle count depending on specific equipment and related parts. The assumption for this approach is that the probability of equipment failure increases with use and time. However, data on failure patterns compiled by NASA and the US Navy show that only 18% of failures are age related and 82% have a random pattern. Therefore, preventive maintenance provides a quantitative benefit for just 18% of assets. The balance of faults that cannot be quantified can be avoided using preventive maintenance.

12.2.2 Predictive Maintenance

Recently, predictive maintenance has replaced or augmented the preventive approach. The predictive maintenance approach uses condition monitoring to provide advanced notice of a failure so that appropriate maintenance can be scheduled and performed to prevent the unplanned downtime. This approach leads to increased equipment availability as well as reduced maintenance costs. The fundamental idea is to transform the traditional "fail and fix" maintenance practice to a "predict and prevent" approach.

In many cases, the underlying analytics modeling for prediction is often based on a single variable. Consider heat exchanger as a use case: when the pressure difference between the inlet and outlet increases beyond an upper limit, the tubes need to be cleaned. This form of predictive maintenance is an improvement over periodic preventive maintenance where the process is shut down and the exchanger opened for inspection, often resulting in unneeded downtime and maintenance costs. While an improvement over preventive maintenance, predictive maintenance uses basic analytics and typically applies to correspondingly simple assets.

12.2.3 Proactive Asset Management

Predictive fault detection starts with the identification of failure modes and then develops a failure detection model, which is a reactive approach. In contrast, a proactive approach does not start with a failure problem already identified; instead it starts discovering the root causes of failures before they occur and identify failure mode target. Thus, the key is to determine how to detect those failure modes. For example, bearings can be replaced before failure, based on condition monitoring when unacceptable vibrations occur. But if the vibration was caused by misalignment, this state will be frequently repeated. Once the machines are aligned, increased vibrations will take a much longer time to occur as the operational conditions of the machine have been improved. Another example of this can be lubrication; an increased amount of abrasive particles in oils may speed up the wear of gears. The gears may be replaced frequently before they actually fail, but improved oil filtration can help to reduce the wear of gears. Again, the improved operational conditions result in the extended life of machine components.

There are three approaches for identifying the proactive failure mode targets. The first one is advanced pattern recognition (APR). The second is based on the failure rate specification from design failure modes and effects analysis (FMEA) to identify the most likely future problems. This method is called fault injection. The third approach is based on fundamental physics-based condition indicators (CIs), which can respond as symptoms of a broad range of failure modes. A CI is a measure of detectable phenomena, derived from sensors, that show a change in physical properties related to a failure mode or fault. CI values are trended over time for monitoring equipment health.

12.2.3.1 Advanced Pattern Recognition One approach for proactive asset management involves engineering and designing condition monitoring using customized algorithms. This approach works well for equipment OEMs that can spread the cost of developing the algorithm over many machines. For operators, the limited availability of engineering resources often makes this approach difficult to implement and sustain. An emerging, more automated approach uses APR.

APR technology generates an empirical model by "learning" from an asset's unique operating history during various stable and dynamic process conditions. This model becomes the baseline for normal operations for a specific equipment. The APR system automatically compares an asset's model with real-time operating data to detect meaningful changes—some of which could be subtle, interrelated, and beyond normal human recognition using visualization tools. APR provides advanced warning of an impending equipment failure before they reach alarm levels or an unplanned shutdown.

Current applications of APR typically use data from existing sensors from which higher fidelity model enhanced by statistic and ML techniques can improve condition monitoring. The combination of these two emerging trends—IIoT with structured and connected data with APR—offers an opportunity to take proactive maintenance to a new level with broad adoption across critical assets.

After an APR system learns and has a proven track record with its associated equipment, it becomes a virtual model for a certain type of equipment. A key benefit of this "virtual equipment" is the ability to replicate the model (somewhat like a template) across many similar devices. Examples could include pumps in a refinery, transformers for power transmission lines, or stamping machines at an automotive supplier. This approach helps make the best use of the often-limited technical resources available and offers the opportunity to have condition monitoring coverage for a much larger portion of critical and complex assets.

12.2.3.2 Fault-Injection Method Since there typically is no field data to train fault detection algorithms for proactive targets, a good approach is to develop a fault-injection model of the component/system. A fault-injection model ideally consists of a high-fidelity model of normal behavior and the addition of fault-injection handles, which allow the user to inject failure modes at various severities. Deciding where and how to add these handles in the model requires domain experience and engineering judgment to determine the most realistic and valuable fault-injection techniques and severity levels.

Once development of the fault-injection model is complete, one can synthesize a variety of potential fault signatures. With an understanding of the various fault signatures, it is possible to develop a strategy for the failure detection model. Next comes coding and validation of the algorithms to a preliminary maturity level.

Then the model is deployed in a pseudo-production environment to identify potential fault cases as they occur in the field. This testing step is so-called the model introduction phase. This step has the benefit of allowing the model developers to collect feedback from operators as the component failure is evolving and to collect detailed information on the exact nature of the failure mode. But one challenge with this type of model is to keep them up to date with changing processes or asset conditions.

12.2.3.3 Condition Indicator (CI) Method This approach is to develop a fundamental physics-based CI that is likely to respond to multiple failure modes. With this approach, the developer initially does not have specific cases from which to train a detection algorithm. Rather, the CI comes first, and its trends are evaluated over time to identify various fault signatures as faults occur in the field. Again, this requires some investment in the form of exploratory investigations to determine if the various patterns are indicative of normal behavior or faults. This approach has the advantage of having a rough indicator already in place when field issues arise. Even though there may not be a precise detection algorithm available, just having some CI that can be trended and evaluated enables the revelation of deeper insights. The algorithm developer can use these insights to design the detection algorithm. Section 12.7 discusses an example of how we used a fundamental indicator for fault detection and to discover very valuable insights. These insights enabled the development of algorithms that would have been much more difficult or impossible to envision without first having some CI trends to analyze.

The comparison among various asset management approaches is given in Table 12.1.

TABLE 12.1. Comparison of Different Maintenance Strategies

Approach	Method	Applicability	Comparison
Reactive	Run to failure and then repair	Noncritical assets with low impact from failure	10X plus when failure occurs
Preventive	Service in a fixed time or cycle interval	Failures increase with age or usage	2X maintenance costs
Predictive (single variable based)	Monitor process data, identify bad trends, and alert prior to failure	Simple systems where single variable model predicts a failure	1X maintenance costs
Proactive (multiple variables)	Equipment-specific data acquisition with multi-variable based analytics modeling	Complex systems where multiple variables and analytics can predict failure	Unscheduled downtime approaches zero

12.3 THE CONCEPT OF OPERATING WINDOWS

The operating window is an important concept in proactive asset management. As defined by the American Petroleum Institute's (API) RP 584, integrity operating windows (IOW) are established limits for process variables (parameters) that can affect the integrity of the equipment if the process operation deviates from the established limits for a predetermined length of time. In short, IOWs establish acceptable limits of process variations before an asset begins to incur damage.

The purpose of defining an IOW is to establish, implement, and maintain a program to identify any potential damage mechanisms that may adversely affect the process and then use that information to create a system where parameters can be modified as equipment and processes evolve over time. IOWs identify safe operating limits with the purpose of avoiding equipment degradation. The goal of IOWs is to make sure healthy operation and lower the risk of operating equipment. Integrity operating windows are typically defined by their criticality and API RP 584 divides IOWs into three categories, namely, IOW critical limit, IOW standard limit, and IOW informational limit.

12.3.1 The IOW Critical Limit

An IOW critical limit would be one that, if exceeded, could lead to rapid deterioration of process equipment. A critical limit requires immediate operator action to return the process variable to a predetermined value in order to prevent potential equipment damage in a short time frame. An example of a critical limit could be a flow rate for centrifugal compressor as the mass flow rate drops to a critical value at which discernible flow instabilities take place. This could lead to severe surge causing damages to the compressor.

12.3.2 The IOW Standard Limit

An IOW standard limit is an integrity parameter, which, if exceeded over a specific time frame, could cause increased corrosion rates or eventually lead to cracking or other damage to materials of construction. Standard limits are typically time based, in that the time required for equipment to be adversely impacted generally defines the response to a standard limit IOW. The consequence of damage to the equipment associated with the IOW will also influence an SME's response to the parameter being exceeded. An example of a standard limit IOW could be the flooding percent in a fractionation column. When the operating flooding factor is larger than the design limit, it gives an indication how close the column downcomer could have liquid backup, which could cause unstable column operation.

12.3.3 The IOW Information Limit

IOW informational limits are generally used by SMEs to predict the long-term integrity of equipment, or to analyze impact on the asset-integrity program. Informational IOWs are parameters that may or may not be affected by operations. These IOWs do not typically require operational responses but may be utilized to assess future repairs of turnaround and shutdown frequencies. Informational IOWs can be used to evaluate the process assumptions used to establish risk-based inspection (RBI) programs.

These terms vary across companies and industries. Critical limit IOWs may also be referred to as safe operating or safety critical limits. Standard limit IOWs have been referred to as key operating limits or reliability limits, and informational limit IOWs may also be referred to as corrosion and fouling control limits and so on, depending on the industry in which they are used.

12.3.4 Modeling of Dynamic IOW Limits

An IOW program is to identify all potential damage or degradation mechanisms that may adversely impact the process equipment. A diverse team of engineers will evaluate the operating parameters of each process condition to understand their potential impacts. The subject matter expert (SME) team will develop a damage map involving indicators to detect exceedance of an IOW limit.

Most operating parameters with IOW limits can be measured via sensors or meters, e.g. temperature sensor for measuring fired heater tube skin temperature, flow meter for a centrifugal compressor. In other cases, it requires lab testing, for example, for sulfur content in the feed stream, or pH excursion in a process stream, or magnetic particle testing for metal surface cracks. In some cases, an IOW limit involves several parameters that interact with each other, and it cannot be directly measured by sensors. It requires complex modeling (statistic model, first principle model) to determine the combined contribution to the limit. These parameters may change whenever process conditions vary, and thus the dynamic IOW limits must be calculated continuously in operation so that these limits can be observed and not violated. To do so, inferential sensors can be applied effectively.

As an example, consider the feasible operation window for a fractionation column, which is bounded by tray capability limits. Although every tray has a specific operation window based on distinctive vapor and liquid equilibrium, it is the operating window for the most constrained tray that should be monitored. These limits are determined based on rigorous hydraulics modeling of vapor and liquid equilibrium inside of the column. A specific feasible operating scenario represents an operating point within this operating window (the point C in Figure 12.2). There are two important observations for the column operating window:

- The operating point is very sensitive to variation in feed rate, product rates, and operating conditions such as feed temperature, tower pressure, and reflux rate. In contrast, the column operating window is relative insensitive to changes

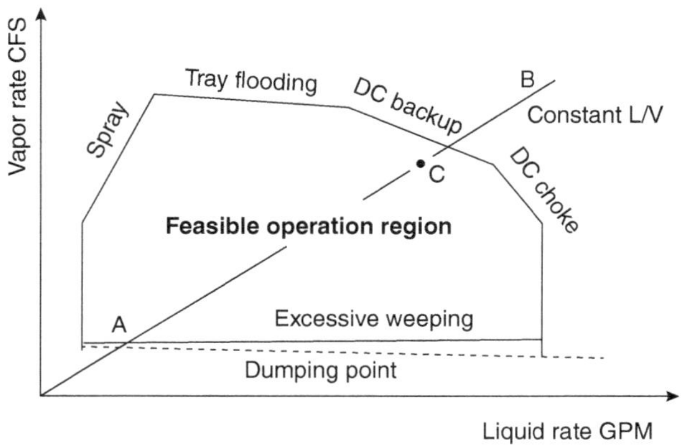

FIGURE 12.2. Capacity limits for distillation tower. *Source*: From Zhu (2014). © 2014 John Wiley & Sons. Reprinted with permissions of John Wiley & Sons.

in operating conditions but sensitive to column tray design (diameter, the number of trays, tray type, and configuration). Thus, different tray designs provide different capacity limits resulting in different operating windows. Consequently, distillation efficiency will be different as reboiling and condensing duties vary.

- Vapor and liquid flooding are the most important indicators to monitor in operation because when column is flooding over the limit, effective separation stops as heavies are entrained to top of the column.

The first question that the operating window can answer is: how close is the tower operation to the unstable operation? The current operating point located in the operating window could give you a good understanding if the column is operating under healthy conditions. If an operating point is outside of this window, the desirable separation cannot be achieved. When the column is under severe flooding, no separation happens at all.

If the answer is: the operation is healthy, then the follow-up question is: is the operation optimal? The minimum energy operating policy is achieved via increased feed preheating and reduced reflux ratio to obtain minimum reboiling duty. However, if product specification cannot be met under increased feed rate, the reboiler duty must be increased for obtaining higher vapor flow. In this case, both the operating window and the operating point conditions change at the same time.

Thus, it is challenging to design a tower with the lowest overall cost (capital and utility) and simultaneously provide sufficient operation flexibility to deal with process variations. In contrast, the challenge for operation is to determine the optimal operating conditions within the feasible region under which the column is operated in the most efficient manner. Detailed discussions for column operating window and guidelines for tower design and operation optimization can be seen in Zhu (2014).

12.3.5 Integration of IOW Limits

The IOW program must be fully integrated into the plant's operations and maintenance, reliability and mechanical integrity programs, and engineering processes. Changes in feedstock, operating conditions, and flow characteristics can influence IOW limits for individual assets. It is critically important for a facility to fully understand the IOW limits and monitor the operating point in association with operating windows. It must be recognized that the effects that changes in one process parameter can have on other parameters.

This step can be time-consuming and meticulous. In order to maximize its effectiveness, an IOW program must be integrated into plant operations. The most effective way is to develop a digital twin embedding a comprehensive mechanical-integrity data-management software program, in which IOW limits, inspection, and maintenance data and more are stored, organized, trended, and analyzed; furthermore, dynamic operating windows are calculated—helps to ensure that IOW data are centrally located, consistently and automatically updated, and continuously monitored.

12.4 THE MAJOR GAPS IN CURRENT ASSET MANAGEMENT

It must be said that the asset management has gone through several major phases of improvements over time and the value created is clearly noticeable and the success is worth celebrating. However, there are still several gaps in current practice that can be observed across process industries. It can be demonstrated that use of digital solutions to close these gaps would be a perfect fit. Let us discuss the gaps first and then explain the digital solution.

12.4.1 No Sufficient Consideration of Process Conditions

Traditionally, asset management programs are based on a given set of operational parameters that were established during the implementation of the programs. To develop these programs, process conditions—such as the various chemical components present in the process streams, temperature, pressure, and flowrates—are analyzed by mechanical integrity (MI) SMEs. These experts identify the potential damage mechanisms associated with anticipated operating conditions and help to establish inspection methods and testing frequencies. In the subsequent management of the program, reliability and MI groups rely on these parameters and testing methodologies that were established during the program's development.

But the issue is that operational parameters change over time, and the rate of equipment degradation changes as well. Extrapolations of operating conditions based on experience in the form of rules of thumb usually lead to either unnecessarily conservative operation or too aggressive operation. This approach could cause the asset program to become ineffective. This can shorten equipment life spans, raise costs, and greatly enhance potential risk to personnel, assets, and the environment.

For example, rotating equipment reliability is a major source of throughput constraint, maintenance expense, and energy cost, but there is a trade-off that deserves much consideration. When an overly conservative operating approach is taken, a unit is run light, resulting in missed profit opportunities. If such a conservative approach is taken, reliability suffers causing lost throughput, excess maintenance cost, wasted energy and potentially downtime to make necessary repairs. One of the effective solutions to tackle the issues is the definition of operating windows, which is discussed in Section 12.3.

12.4.2 Ineffectiveness in Communication and Execution

A common observation is that many plants are not equipped with an effective digital system (or digital twin) with common access to asset information for various stakeholders—maintenance, operations, and plant management. Thus, the plant reliability team spend many hours in numerous meetings in order to communicate with stakeholders and influence the decision-makers to explain the implications for reliability. They try to monitor the high-risk equipment and feed back to the organization on performance, using whatever influence they can gather to make reliability a priority with many competing requests. This approach requires a lot of effort with little assurance of delivering the desired outcomes. The success of the plant reliability team depends on how well it can change behaviors by wielding influence and imparting knowledge within itself and in other functions.

The consequence is a major loss of effectiveness in how plant reliability is managed. Gaining cross-functional buy-in can get political very quickly, which can take a long time. Making your priority someone else's priority is often met with resistance.

12.4.3 Knowledge and Skill Gap

Like everything else, asset management requires people with sufficient knowledge and skills to make things happen. However, due to the aging of the workforce worldwide, experienced personnel are set to retire and will pass the baton to young generation who are good at IT but with much less know-how and experience. Thus, industries face the challenge in retaining knowledge and skills as young and inexperienced professionals require training and practice to acquire all the related knowledge, skills and abilities to perform the job satisfactory. At the same time, resources continue to be cut, how can companies achieve the goals for safe, secure, and reliable and economic operation?

12.5 DIGITALIZED ASSET MANAGEMENT

To close the gaps as discussed above, the answer is in digital transformation. IIoT and advanced analytics provide the digital solutions to enable such a shift from predictive maintenance to proactive asset management smoothly with the goal for enhancing production and business performance via increased asset utilization. With IIoT and scalable cloud computing and storage, high volumes of data from different parts of the business can be securely gathered and stored from the direct source in real time or batch (bulk transfer). Not all analytics need be done in real time or with real-time streamed data—lots of examples of bulk collection at the source and then non-real-time transfer to the data lake. But for some data, it requires data analytics or models (statistic, first principle, machine learning, optimization), which can convert data to new insights. These models can help identify anticipated failures earlier and provide more accurate diagnosis of the pending issue with fewer false predictions. Beyond prediction using models, optimization for asset management can then determine the optimal process conditions, resource allocation, and maintenance scheduling to fulfill both asset management and production requirements. The set points from optimized conditions control the automation of machines and processes.

With proper integration, proactive asset management can be digitalized to make a digital twin, which can help companies to face the twin challenges of maximizing productivity and opportunities for growth while driving down operational costs and striving for sustainability.

As demonstrated through applications, digital twin can generate significant values for asset management through integration of online data, process simulation, models for fault diagnosis and prediction, monitoring IOW limits, and operation systems (procedure, best practices, control). The integration of these enables the reliability team to push the digital twin for reliability as the single point of the truth to be used more effectively across the OT domain supported by IT systems. Thus, the rigor, accuracy, and consistency of the equipment reliability analysis are increased to gain the trust of those who receive it.

12.5.1 Digital Twin for Asset Performance Management

The digital twin is the replicate of physical assets in real-world conditions, which mimics the design, conditions, and limits of actual processes and equipment via simulation based on real-time operation and lab data. The digital twin must be an accurate representation of an asset over its full range of operation and its full life cycle to be useful. It works in the present, mirroring the actual plant in simulated mode, but with full knowledge of its historical performance and accurate understanding of its future potential.

The foundation for digital twin is the process operation simulation that is converted from process design simulation. The simulation represents both process and equipment conditions and inter-relationships. Data gathered from the plant operation is validated and reconciled to ensure that all physical and chemical laws are respected, and gross measurement errors and dynamic effects are eliminated through filtering, thereby ensuring that data quality issues are identified and mitigated. The treated data becomes the input for simulation.

On top of process and equipment simulation that provides the replicate of physical operation, analytic models are used to predict future events for equipment health.

12.5.2 Intelligent Models: Cornerstone of the Digital Twin

The cornerstone for the digital twin for asset management is process simulation together with a set of prediction model with high accuracy. Good models can help us to categorize, explain, and most importantly predict. Indeed, experience and information can be good teachers, but many times in asset management, we simply cannot afford to learn on the job. One cannot go through several incidents to learn how to protect critical equipment. Our experience shows that failure forecasting can be inaccurate up to half the time. Suboptimal service providers are unable to offer high-fidelity prediction models and thus cannot provide coordinated, integrated, and timely maintenance.

This is why process simulation together with analytics models become so valuable in digital twin methodology: accurately predict what will happen before it occurs and interpret the status of the installed base, thus enabling operators to quickly initiate maintenance measures when irregularities occur. An overview of modeling techniques used in digital twin for equipment monitoring can be shown in Figure 12.3 and discussed in details below.

12.5.2.1 Process Operation Simulation (Mimic Physical Assets) It is converted from process design simulation that consists of hydrocarbon management, proprietary rigorous models for reaction molecular-based kinetics, product fractionation and impurity separation, rotating equipment, fired heaters, heat exchangers, operating limits, and many more. In this manner, process design is integrated with operation requirement and business optimization. Process changes and technology improvement often provide business opportunities that enable more effective maintenance schedule and deployment of resources in the context of production needs.

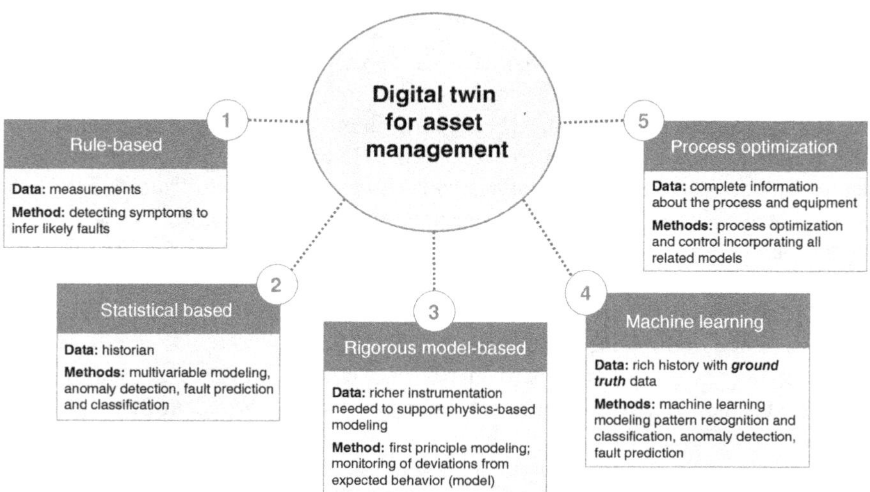

FIGURE 12.3. Model thinking in the APM digital twin.

12.5.2.2 Rule-Based Models or Expert System (Mimic SMEs Know-How) Rule-based models are used as the first natural opportunity for fault diagnosis analytics. Due to relatively small number of sensors and limited availability of historical data, it has been recognized that the rule-based approach is the most easily deployable. Generalized rule sets were specified based on industry expertise and domain knowledge and can be easily incorporated and used. Although keeping rule-based systems configured for a specific equipment up to date might represent a challenge, this technology is widely used today. Generalized rules can overcome the silos among SMEs who did not naturally grow to work as collectively as they should—at least within the level of harmony that we now know is required to accelerate continuous improvement to next-level productivity gain.

12.5.2.3 Statistic Models (Descriptive: What Went Wrong?) Statistic analytics provide a newer means for deriving greater asset insight through statistical analysis. In general, they identified a problem (what go wrong now?), and then predictive models can determine possible damage and maintenance options. The descriptive models can be the models for defining the operating windows for fired heaters, distillation columns, compressors, pumps, heat exchangers, etc.

Multivariate statistical models for fault detection based on historical operating data have been found to be useful with processes having a large number of measured variables and when rigorous models are unavailable. Multivariable characterizations are derived from simple third-order polynomial building blocks. Such functions can be used to build real-time operational condition diagnostic and prognostic, i.e. fault classification and time to failure estimates. For fault isolation or diagnosis, they have been less powerful because of the non-causal nature of the data on which they are based. To improve the fault isolation with these methods, additional data on past faults have been used to supplement the models.

12.5.2.4 First Principle Models (Descriptive and Predictive) Rigorous models based on first principles (i.e. equations derived from the laws of physics and chemistry) have and will continue to be successful in machinery performance assessment as they can describe the nature laws of physics and chemistry. However, such modeling techniques require extensive experience and skills with appropriate calibration techniques.

12.5.2.5 Machine Learning Models (Predictive and Prescriptive) Machine learning techniques are used to predict the unknown part of the process due to operation disturbances. Supervised machine learning is the most developed and widely used analytics technique. Working with labeled historical data, one can develop classification model for future event predictions and regression model to predict key process and product properties. APR model can also be developed for anomaly detection for machinery reliability and process safety improvement.

One can also build clustering model using unlabeled data by using unsupervised machine learning to uncover patterns and trends in the data that were not previously distinguishable. This creates group of similar events (or clusters) and explicitly express what feature is decisive in these results. Machine learning models can be used to accomplish the following:

- **Predictive maintenance** (what is likely to go wrong and what will be the consequences): The predictive analytics models can be developed based on machine learning such as principal component analysis (PCA) and partial least square (PLS). In situations when there is enough ground truth data about various equipment faults, it should be possible to apply machine learning methods to train models for fault prediction. Predictive maintenance can leverage rule-based algorithms to continuously monitor individual control and equipment performance. Predictive modeling is then applied to a selected condition indicator that deteriorates in time.
- **Prescriptive maintenance** (what is the right solution): This is an emerging analytic technology that started to be embraced by oil and gas companies. Prescriptive analytics help to determine what is the best maintenance option and when it should be performed based on actual conditions rather than a time-based maintenance schedule. The prescriptive analytics moves beyond descriptive and predictive analysis. Once a problem is identified via descriptive analytics, then possible damage, failure, or maintenance options can be predicted by predictive analytics, followed by prescriptive analytics to select the right solution based on the company's capital, operational, and other requirements. The objective for prescriptive maintenance is to "fix it on the first visit." High fidelity of failure forecasting models is critical for maintenance in a timely fashion, extends the useful life of critical machinery, keeps maintenance costs down, develops optimal service centers, remotely fixes troubled equipment, and maximizes the cost efficiency.

12.5.2.6 Optimization Models (Proactive: Predict the Optimum) Optimization uses real-time data to determine optimal operation performance (be proactive). The optimization models can determine if operation is too conservative (e.g. tray flooding far away from the design flooding limit) or too aggressive (e.g. tray flooding larger than maximum allowable flooding limit). In the former case, process condition can turn up a couple of notches (e.g. increasing the feed rate for the column), while in the latter case, process condition must be turn down a little. When a critical asset deteriorates either faster or slower than expected toward a planned maintenance event, there is an opportunity to adjust operation conditions such as production rate, temperature, recycle, furnace duty, reboiling duty, and so on. The recommended operating range is fed to the operator in a simple, easily understood form—a dynamic operating window for each key operating parameters. The dynamic operating window changes over time. Visualization of this operating window provides insights into equipment for determining the economic and safe operating policy: run the assets close toward the hard limits without sacrifice of reliability and safety based on the reliability engineers' knowledge and experience.

Furthermore, optimization uses real-time data to coordinate multivariable controllers by adjusting set-points and schedules for optimal operation. This approach can provide cost savings in the operation of equipment and reduce performance drifts. The optimized set-points and schedules can be communicated directly to the underlying controllers, which allow an automated closing of the loop from analytics to control.

In this manner, the digital twin has its teeth—it is not just a technological nicety but is insightful and trustable with the prediction and is pointing to the right directions for actions.

The digital twin runs on-site (e.g. edge) or in a remote data center (e.g. cloud) featuring fast computing and large data storage. Some digital twins are better established on-site, in situ, unless cloud scale is desired. With cloud-based technology, there is no need for expensive hardware, making the digital twin an affordable, accessible, and scalable option. Multiple web-connectivity options and the ability to access diverse data in one place mean that continuous monitoring, asset management, failure detection, and troubleshooting can become one seamless operation. The ability to employ wireless technology across multiple sites may be particularly pertinent to bigger organizations.

Digital twin for asset management is an embodiment of integrated IT (data centric), ET (model centric), and OT (control centric), which is designed to overcome the gaps described above and provide enhanced prediction capability and asset reliability. The overall goal is to improve asset uptime. Significant value can be achieved by using digital twins through the integration of process simulation, SME's know-how, operating windows, and data analytics techniques for asset management and manufacturing execution systems.

12.5.2.7 Advantages of Digital Twin for Asset Management Digital twin has following advantages over traditional asset management methods:

- Data treatment: All data-based solutions are subject to the quality of incoming data, i.e. error prone, unreliable data often means error prone, unreliable analyses. To mitigate these effects, data cleansing methods are applied to make sure the treated data will be correct for fit-in-purpose applications for process optimization, control, and monitoring of equipment reliability.
- The digital twin provides a single source of the truth for what's going on inside the plant at an asset level via trustable models. It enables everyone to see inside and perceive things that are not being directly measured. It is wired up so that the insights are instantly available that everyone can understand and agree on. It drives convergence in comprehension and unity in decision-making. Thus, it is a single point of responsibility for onboarding, provisioning, sustaining, and improving of asset management.

- Reporting equipment healthy conditions, potential failures trending, and the associated value lost if they occur: This sets the common platform for communication and the basis for decision making. Detecting potential failures before they occur and allowing operators and maintenance engineers to generate work orders automatically based on recommendations made from predictive models.
- Digital twin can also be used to assess "what happened," "what if," and "what is optimal" scenarios automatically to determine available strategies that maximize profitability and reliability. SMEs can review the recommended strategies to assess the impact of each recommended actions.
- To mitigate the risk of lacking sufficient skills at customer side, a SME technical service can be provided by technology vendors on top of digital twin software. The intention is to provide an enhanced service on an ongoing basis for troubleshooting, catalyst performance assessment, and more, to ensure the customer can get the most out of the digital solution as well as make the customer hassle free to focus on running your units to peak performance.
- Digital twin can be used for training purpose.

The concept of APM digital twin can be extended to cover process safety, which is discussed below.

12.6 PROCESS SAFETY MANAGEMENT

Process safety is about managing the integrity of operating systems by identifying hazards, estimating their risk, and applying engineering safeguards and disciplined operating practices. It deals with the prevention and mitigation of incidents that have the potential for severe, even catastrophic damage. Poor process safety performance is very bad for business. Closing the loop on the safety life cycle is more important than ever—but it is also more achievable than ever before.

Over recent decades, improved understanding of process safety risk and decreasing risk tolerance have led to many more safety barriers, particularly safety integrity level (SIL) rated interlocks, being installed in process industry facilities. Unfortunately, the barriers an organization have to prevent major accident hazards from occurring can become less robust without continual checks that they are solidly in place and truly achieving the company's risk targets. Closing the loop "manually" is possible, though painful and expensive. Keeping a half-dozen documents—managed by four or five different departments—in perfect harmony might be theoretically possible (with a lot of effort), but in practice, it does not work. With critical process information held in silos in different domains, it is just not "operational." As SMEs in the safety field quoted, "our problem is with 'the last mile'—what really happens in the field."

A growing number of companies are looking for a way to compare their facility's actual performance against its intended design. They need solutions, making it possible to monitor performance across their operations, identify bad actors wherever they exist, and sustain minimal risk throughout the life of the plant. The goal is to ensure operations run for the next 30+ years with suitably reliable independent protection layers (IPLs) to safeguard assets from potential hazards. Advanced analytics—programmed from the expectations in the HAZOP/LOPA/SIL—hold the key.

12.6.1 Traditional Safety Strategy

Historically, the way that process safety has been approached requires many resources with varying expertise (internal and external) and different tools for gathering relevant data and conducting hazard-related studies. This has been, and continues to be, an expensive, inefficient, and costly process. In this traditional process, unit information switches back and forth many times, and manual data entry, revision control, and normal human behavior create havoc and lead to systematic errors. The common elements in this process safety strategy in the process industries include the following.

12.6.1.1 HAZOP Industrial facilities seeking to optimize their process safety performance typically start by conducting a hazard and operability (HAZOP) study to find hidden hazards in their processes and give a preliminary risk ranking. This process should be deliberately conservative. If the HAZOP team overestimates the risk of a particular scenario, this can be revisited during the expert review in the LOPA step.

12.6.1.2 LOPA The second step in modern risk management systems is a layer of protection analysis (LOPA), which builds on the information developed during the HAZOP. The primary purpose of LOPA is to determine if there are sufficient layers of protection against the consequences of an accident scenario (i.e. can the risk be tolerated?). Because the LOPA process has more parameters available, it allows a more precise estimate of the risk—on both the consequence and the likelihood side. Many companies require that LOPAs be led by more experienced risk analysts. They generally establish tables for things like initiating event frequency (IEF) and probability of failure on demand (PFD) for active safeguards, as well as conditional modifiers, probability of ignition, and so on.

The LOPA also establishes how effective each of the protection layers is supposed to be. Let us illustrate with a real-world example. Operating a gasoline storage tank (Figure 12.4) has a specific risk. A HAZOP team might look on it as follows:

There are two main potential hazards: an internal explosion in the headspace and an overfill and spill. Let us focus on the overfill hazard. Most analysts would expect the worst consequence of a spill of a flammable liquid. Figure 12.4 shows a

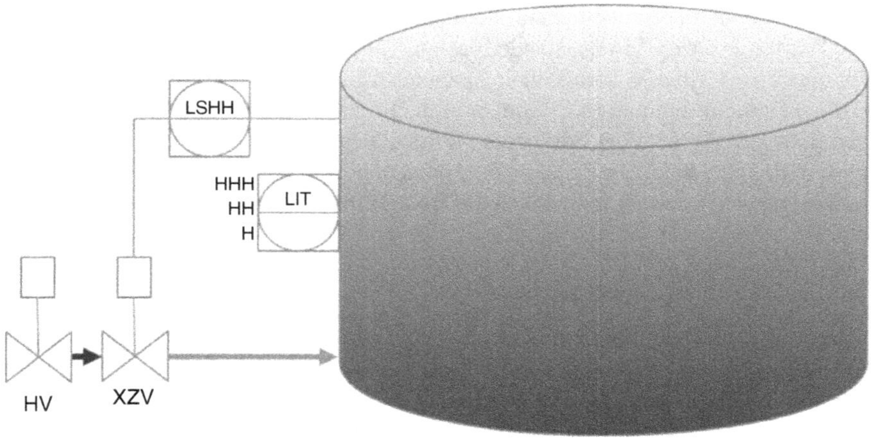

FIGURE 12.4. Gasoline storage tank. LSHH = level switch high–high; XZV = emergency block valve; HV = level control valve.

simplified view of the situation at a storage tank at Buncefield in the United Kingdom several years ago. Although reality is always more complex, this simple example illustrates the issues and opportunities.

Here is how the LOPA might have looked for a situation like this:

- Initiating event: LIT[1] Error. IEF[2] = ~1/10 years or 0.1/year (Typical failure rate)
- Safety interlock: LSHH (Level Switch High-High) → XZV (Emergency Block Valve) SIL[3] (Assumed RRF[4] of 100 or PFD[5] = .01)
- Conditional modifier: Probability of ignition ~0.99 (heptane at 60 °F)

Thus, combined probability of fire = $0.1 \times .01 \times 0.99$ = ~.001/year.

If a layer is an active safeguard like a high–high level interlock, then it has a certain reliability—and thus a probability that it will be in a fault state when a demand (like a high level) occurs. Better hardware and more testing can help—as can analytic software. And analytic software can confirm the demand rates the HAZOP team assumed or flag when demands occur much more often than expected.

Much has been written about the explosion at Buncefield in December 2005. Much of it focuses on the extraordinary pressure waves generated and thus the damage done. But the causes of the overflow that led to the vapor cloud that

1 LIT = level indicating transmitter.
2 IEF = initiating event frequency.
3 SIL = safety integrity level.
4 RRF = risk reduction factor.
5 PFD = probability of failure on demand.

eventually exploded are less exotic. Here's what happened. A storage tank was being filled. Its level indicating transmitter (LIT) was stuck, so it was not giving an accurate indication of the level. Nor was it providing any high level alarms. This particular transmitter had stuck a number of times—about once a week for several months. It had been repaired, but continued to stick. The independent high–high level switch (LSHH) had been inadvertently left in "bypass" after its last test. By the way, instrument engineers have favored analog transmitters with a continuous level signal for years rather than the point switches. The reason is that it is possible to see when a transmitter is not moving with the process conditions in the expected way. You have to either look at the signal, or configure a deviation alarm or (today's solution) configure an analytics engine to monitor your safety-critical devices against the digital twin. More about digital twins for safety is in Section 12.6.2. So there were no safeguards in place, beyond unaided operator diligence, as the tank was being filled. It is not surprising that the tank was overfilled.

12.6.1.3 Safety Integrity Level Analysis Once the reliability of the active safeguards has been established such as Table 12.2, the functional safety experts can design the safety instrumented systems (SIS) to achieve that level of reliability.

Recent events have propelled industrial companies to work toward an evergreen functional safety management plan, including monitoring the performance of protection layers (Figure 12.5). The focus is on understanding how issues affect their risk profile, as well as managing risks and demonstrating adherence to policies. Work has been done to synchronize:

- PHA studies
- LOPA findings
- SIL calculations and verification
- Safety requirements specifications (SRS)
- Daily operating data

TABLE 12.2. Safety Integrity Requirements

Safety Integrity Requirements		
Demand Mode of Operation		
Safety Integrity Level (SIL)	PFD_{avg}	Required Risk Reduction
4	10^{-5} to 10^{-4}	>10,000 to ≤100,000
3	10^{-4} to 10^{-3}	>1,000 to ≤10,000
2	10^{-3} to 10^{-2}	>100 to ≤1,000
1	10^{-2} to 10^{-1}	>10 to ≤100
A	More than 10^{-1}	>1 to ≤10

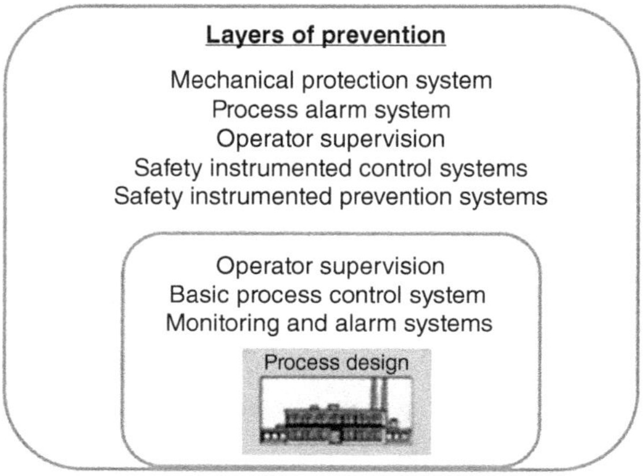

FIGURE 12.5. Layers of prevention.

There is a growing emphasis on developing interrelated processes for addressing safety hazards and risks throughout all levels and functions of an industrial organization. In this way, safety becomes part of everyone's job. By connecting all of the life cycle elements and the expectations that accompany them, functional safety experts can analyze data to validate their safety strategy in real time. That is the concept of digital twin for safety management, which is extended based on the digital twin for asset management.

12.6.2 Creates Digital Twin for Process Safety Management

If you think about it, the Hazard and Operability Study—and particularly the LOPA—is a good model of how you expect the plant to operate. You expect that someone might make an error loading a storage tank about once every hundred times they do it. You add an alarm or a high–high interlock to reduce the chances of an overflow. If this "data" is deliberately collected in a form that can be used for comparison, then it can act as a digital twin. This enables you to use process historian (alarm data in this case) to see how often that high level alarm goes off. If it is more often than you expected in the LOPA, there may be some problem that you should work on before you have a real overflow.

In most cases, however, the traditional approach to safety is ineffective, and the documents produced are just documents without any connection to each other or to the real-world operation. It is a shame, because the HAZOP, LOPA, and SIL calculations have to be done anyway. Disparate reports and tools are commonly employed to support the steps of the safety lifecycle, requiring information to be manually moved from step to step. For example, users have had to retype information from the LOPA into the specialized software used to calculate the expected availability of the safety shutdown systems. Contractors on major projects have reported that transcription errors show up in about 10% of these cases. Imagine mistyping a tag name for a level alarm (LAL-001 instead of LAL-002). In today's modern control systems, LAL-002 is likely to be on a different screen and warning about an unrelated problem.

Technology advancements enable industrial organizations to collect hazard review data into a single enterprise-wide digital system. The enterprise system makes it easier for corporate experts to oversee and manage the company's risk management process. And these safety life cycle systems enable continuous analysis of intended safety performance (from the HAZOP and LOPA) versus actual safety performance (from the process historian) and fix problems before they lead to serious incidents. While doing so, they can reduce man-hours, not just for design but also for investigating issues as called for in IEC-61511. They can also drive out errors, ensure process compliance, and save money on functional safety engineering. And most importantly, they can explain risk to their leadership in a way that lets leaders make better risk-based business decisions.

The digital twin for safety management can utilize enterprise tools to enable safety personnel at all levels of the organization to monitor process conditions by comparing actual performance from the plant historian with pre-defined hazard conditions from the risk analysis and take immediate action to minimize risk. These tools also facilitate analyses such as PHA, LOPA, SRS, and SIL calculations and classifications without excessive engineering costs.

The digital twin can be utilized to validate the safety system design before the system is physically manufactured. It can be used in an "offline" mode to run "what if" scenarios, determining potential elevated risk levels. Used in "online" mode, the digital twin can provide a dynamic model of the operating asset to prevent problems before they occur or to identify the root cause of an issue. In addition, the digital twin can use cloud-based data capture and analytics, together with off-site resources, to analyze the performance of SIFs against expected or assumed behaviors for trip events. This capability also helps operating companies understand the range of trips in similar process units across sites.

12.6.3 Advantages of the Safety Digital Twin

The cost of collecting process history data has come down dramatically. And most often, this information is now readily available to the engineers who understand what it means. In fact, the process industries today are practically swimming in process data or perhaps drowning in it. But only a small fraction of that lake of data is important to process safety. As always, the key is to focus on the critical information. So, what is critical? The HAZOP/LOPA team has already decided what information is really critical and has to be watched.

All successful organizations seek out and use feedback on minor issues to identify and fix them before they can become major issues—or even catastrophes. The safety digital twin—built up from the HAZOP/LOPA—enables companies to identify very small issues like an alarm going off more often than the team expected or an interlock being bypassed or perhaps not being tested per the schedule the designers intended.

Process facilities can now program the digital twin to alert them to possible issues. The other side of the coin is that it can also alert them to possible opportunities to reduce testing costs when a system is performing *better* than expected.

12.6.3.1 Expanded View of the Process Based on the Data At many industrial facilities, it is not uncommon for operating metrics to suggest that everything is under control, when digging below the surface reveals signs of significant risks to the safety layers and barriers meant to prevent major accidents. Generally speaking, safety personnel only look at historians after a process upset has occurred.

Plants can now utilize a methodology for managing process safety information that connects design assumptions to real-time operating data and proactively removes risk. This strategy relies on modern computerized analytic systems to optimize safety functions that have been handled manually for years. It is intended to identify potential vulnerabilities and address them before an upset or incident occurs.

New process safety life cycle management tools are penetrating the marketplace as IIoT/digitalization revolution takes off in the process industries. These tools can cost-effectively remove risk from business. Their premise is to consolidate process safety data into a single data repository. This data is fed to an analysis engine, where it can be efficiently and automatically leveraged to unlock hidden process safety information.

At the same time, the centralized database approach brings together key work-groups such as process and functional safety experts, automation specialists, and operations and maintenance personnel to interact with the same information and stay in-sync within a shared system.

Process safety life cycle management tools look for a host of crucial issues:

- Is the SIS prevention layer ready?
- Is the SIS prevention layer degraded in any way?
- Are any SIFs in bypass today? Or over the past year?
- How did this layer perform during the last demand (real or spurious)?
- What are the risk reduction factor (RRF) and PFD averages over the past year or so?
- How steady is the basic process control system (BPCS) layer?
- Are any critical control loops degraded?
- What is the real initiating event frequency?

For any plant, the ideal operating state is when production processes follow control strategies designed, built, and checked during the HAZOP stage for safe operations.

Emerging process safety techniques are innovative in the way they enable plant operators to compare what should happen with process safety to what actually does happen. They also ensure a single source of truth for process data.

An inherited data model linking risk rankings down to tagged devices makes it easier to support daily operations and maintenance activities. Previously invisible risk become identifiable by analyzing deviations in operations compared with the original design assumptions.

Making all the "what should happen" information available in a database means plant personnel can now configure analytic tools to do automated tracking, recording, and validation on the condition of the SIS and final elements. They can take into account events from all operations (planned and unplanned) to track and automatically validate them versus the expected behavior. This makes it easy to analyze SIL-related key performance indicators (KPIs) to track failed operations, demand rate on SIFs, failure rate on final elements, and test intervals.

For plants that are already up and running, the safety digital twin can help PHA revalidation teams move smoothly through their work of identifying issues using data from their own unit's performance. The safety digital twin can also improve the speed and accuracy of numerous technical activities needed to properly analyze risk during revalidations.

Process safety specialists can use the safety digital twin to look for gaps between plant design and historian data and thus help make operations safer, reduce spurious trip rates, and minimize engineering costs. Information from historians as well as computerized maintenance management systems (CMMS) is collected and supplied to the safety digital twin, which examines time-stamped event journal data and analog historian data for field devices and other equipment.

The safety digital twin can continuously executes performance, health, efficiency, and safety-related calculations and compares the results of current actual operation to an expected performance model. Predicted or detected deviations from these models are used to generate notifications to facilitate investigation and intervention to minimize the cost and frequency of an event.

12.6.3.2 Valuable for Plant Operators No matter where they are in the process safety life cycle, plant owners can utilize new safety methodologies implemented in the safety digital twin to reduce risk and save money while complying with best engineering practices throughout the operating life of their facility. The latest process safety tools use a plant's test and operating history to minimize risk and optimize technology investments.

Manufacturers can do more with data via a connected, intelligent digital twin to process safety. This includes turning process historian data into valuable safety insights. Employing a single data system for all process safety risk management information and analysis, they gain the means to improve monitoring of operational hazards and streamline work to enhance functional safety:

- Integrate process hazards and risk analysis as part of a cohesive safety strategy
- Find and fix safety problems that previously went undetected
- Realize the benefits of drawing PHA and LOPA processes closer together

- Do away with enormous 10X safety factors through a better understanding of actual risk
- Eliminate the difficulty of keeping many separate documents in sync
- Implement template-based SIFs to streamline safety design efforts
- Simplify preparation of SRS and SIL calculation documents
- Improve methods for passing outputs to the SIF implementation team

In addition, continued progress in the area of process safety enables plants to take steps to ensure reliable operations and maintenance:

- Comply with best practices like IEC 61511 requiring robust testing as well as investigation of trips to ensure safeguards stay in place and work properly every time
- Ensure data can demonstrate that the site is in compliance with risk management requirements, or indicate when there's an issue
- Provide reliability information as an input to the next revalidation cycle of the PHA

The ability to leverage a digitized process safety life cycle during capital project execution has the potential to drive significant efficiency and cost improvements. This includes a reduction in time and effort due to the typical copy factor between plant equipment and SIFs and savings in functional safety engineering services. Furthermore, data generated in the design phase can be used in operations to validate assumptions and make risk-based business decisions.

12.6.4 Remarks for Safety Digital Twin

Experience has shown that having the HAZOP, LOPA, cause-and-effect matrix, and historian data in a safety digital twin enables a sustainable analytics solution. Thanks to innovations in process safety management, industrial organizations can take advantage of integrated enterprise systems and effective performance metrics that truly reflect the state of risk controls and allow plant managers to target resources toward the areas of greatest concern. This approach connects design assumptions with actual performance and helps facility operators to be confident with the protection layers that will work.

The addition of the safety digital twin makes process safety monitoring and analysis practical and effective all the time. And today's new enterprise tools that integrate PHA/LOPA and SIL calculation systems along with analytics systems that continually check the real-world performance of the process systems (using Historian data) and the maintenance systems against the assumptions take this approach to the next level.

As a concluding remark, simply having technology connect various data sources and data analytics models does not guarantee results. It takes much more than that. These systems need to be implemented and controlled with a deep understanding of the business and a cross-functional approach to bring together a variety of necessary skill sets. The safety digital twin serves as a rigorous and consistent platform on top of which asset management strategies can be implemented as well as the necessary feedback loop to keep the content relevant. But it must demonstrate how informed decisions add strategic value to the business and thus have a full buy-in from all stakeholders and motivate people to adopt the systems and use them to their full capacity to deliver the value for increasing equipment availability. At the same time, the customer must also be aware of the functionality, power, and limitations of the digital twin technology as no technology can be like penicillin—a cure-all for all pains. The data flows, task structure, and execution and architecture as well as cybersecurity in digital twin will be discussed in Chapter 17.

12.7 CASE STUDY: APM DRIVES CAPACITY IMPROVEMENT

12.7.1 Background

It is a Honeywell Performance Materials and Technology (PMT) Specialty Material plant which produces low-density polyethylene (LDPE) thermoplastic in Orange, Texas. PMT collaborated with Honeywell Connected Industry (HCI) business, who provides industrial asset monitoring and predictive analytic solutions from the Honeywell Connected industry software suite—a family of cloud-connected solutions geared to maximize productivity from process, people, and assets.

12.7.2 Challenges

The plant at Orange had incurred several costly unplanned shutdowns, and the metric for unplanned downtime has been rising rapidly. There is a need to reduce the degree of unplanned downtime by transforming from schedule-based or highly reactive maintenance to predictive maintenance. At the same time, the plant management wanted a way to extend service intervals without increasing the risk of an unplanned event. Costly unplanned shutdowns combined with operational challenges, long-lived assets, and a worn plant infrastructure made it an ideal target for the Honeywell APM solution. The desire from the customer is to implement the Honeywell Connected industry software quickly to uncover hidden inefficiencies, within existing plant resources and without interrupting operations in order to ensure on-time delivery given that the plant has been running at full capacity. The plant sought to increase uptime yet needed an improved means to understand existing plant data and derive actionable insights.

12.7.3 Solution

The kick-off initiative involved an assessment of the plant's assets, processes, and operational needs. Then HCI and PMT engineers collaborated to solve the problem through configuration of 28 assets with embedded site implicit operational knowledge, implementation of 58 key performance indicators (KPIs) and predictive models across selected assets using Honeywell Asset Performance Management (APM) and data analytics engine of Connected Plant.

Of the selected assets, four were recognized as opportunities to reduce downtime, prolong maintenance intervals, and capture untapped throughput potential—cumulatively reducing both planned and unplanned downtime of plant machinery and increasing overall plant efficiency.

Implementation included APM training for reliability, maintenance, and process engineering personnel, and operational procedures were updated to include roles and responsibilities of the system's users. Likewise, a documented chain of command was created for APM alarm recognition and an approval workflow established for implementing corrective actions.

12.7.4 Benefits

The pilot began proving its worth shortly after going online in the summer of 2018. The first identified benefit is heat exchanger performance monitoring. APM model alerted heat exchanger fouling issue earlier than scheduled maintenance. With this information, heat exchanger cleaning was scheduled during routine planned maintenance of the major equipment that the exchanger is serving, so reduced throughput due to fouled exchanger and a secondary outage is avoided.

The most significant contributors to the derived benefits were found directly from descriptive analytics. HCI-PMT engineers created a unified historical dataset from the data of machines and their associated processes. Using the unified dataset as the basis, the engineers developed enhanced predictive models that both recognized impending machine/process deterioration and provided accurate estimates for prolonging time to required maintenance.

The most significant finding involved three critical machines where engineers ran descriptive analytics and found unique "shift" patterns in the machines' bearing vibration and temperature signatures. Interestingly, the shifts revealed the ability to predict failure of a nearby packing gland (i.e. seal) rather than the bearing itself. Previously the glands were a preventative maintenance item and replaced early, but with the new insight the plant engineers realized they could confidently prolong the change-out interval. By doing so the plant avoided 3 days of unnecessary downtime. Instead, the plant was able to run uninterrupted through the end of the year and plan the packing replacement in conjunction with the plant's planned 2019 turnaround.

12.7.5 Results

By the closing of its fiscal year, the plant had captured additional revenue as a direct result of the APM implementation. Income on the incremental revenue is setting a recovery rate of project costs to within 1 year.

12.7.6 Summary

This case demonstrates the value that the asset digital twin can provide toward recognizing hidden asset performance degradation and magnifies how assets interact with the processes they serve. Leveraging the capabilities of Honeywell APM, insight was derived from asset monitoring that revealed hidden improvement opportunities in the plant's LDPE production process—improvement opportunity that would otherwise have gone unnoticed.

REFERENCE

Zhu FXX (2014), Energy and Process Optimization for the Process Industries, Wiley/AIChE.

13

INTEGRATED PRODUCTION PLANNING AND PROCESS CONTROL

13.1 INTRODUCTION

We discussed production planning optimization in Chapter 7, which aims to determine the optimal feed selection and key operating targets for making desirable product slates based on market demand and economics. It is observed that the planning optimization is to optimize the whole site (which is like a bird's-eye view), but with a very coarse process representation based on yield vectors and without addressing nonlinear and dynamic behavior of process operations and nitty-gritty operating constraints. Thus, planning optimization can provide operating targets directionally, which can not be directly implemented in operation. On the other hand, we illustrated process control in Chapter 9 and understood that control optimization considers nitty-gritty details such as operating conditions and limits (analogy to a worm's view). In current practice, there is inconsistency between planning and process control, which is a significant gap existing in industry. Thus, extensive manual intervention based on experience is needed to determine control setpoints in order to implement planning recommendations.

Therefore, the challenge is how to integrate planning with process and control optimization in a coherent and automatic manner so that planning optimization can coordinate process optimization and control while respecting nitty-gritty operation constraints at the same time. This is the focus of discussions in this chapter.

Digitalization and Analytics for Smart Plant Performance: Theory and Applications, First Edition. Frank (Xin X.) Zhu.

13.2 CURRENT PRACTICE IN SITE-WIDE OPTIMIZATION AND CONTROL

A production planning model looks at the entire site with a bird's-eye view based on overall material and energy balances among all process units and then optimizes them as a whole to achieve maximal overall profit. Thus, production planning is also called site-wide optimization. This planning model focuses on interunit material and energy balances based on steady-state relationships. The model variables include production rates, product yields and quality, key operating conditions that affect yield and quality, intermediate and final product inventories, energy usage, and so on. This planning model represents each individual process unit on a very coarse basis on yield vectors, typically involving three to seven key variables, and excludes non-production- and nonprofit-related constraints such as hydraulics (e.g., column flooding), reliability (e.g., compressor surge), and safety (e.g., flux limit in fired heater).

On the other hand, an MPC (model predictive control) model represents a process unit on a much more detailed scale. It includes many operating variables such as temperature, pressure, flow rates, levels, valve openings, vessel levels, product properties, etc. as well as reliability- and safety-related parameters. For example, an MPC model may contain 100 control variables (CVs) (outputs) and 40 manipulated variables (MVs) (inputs) for a fluid catalytic cracking (FCC) unit. However, a planning model for the FCC unit includes only the causal relationships between the feed rate, riser temperature, and product yield. As a result, the planning model may have as few as 10 outputs and two to three inputs for the FCC unit. In general, the scale difference in terms of the number of variables for a process unit can be one order of magnitude.

In addition to the difference in modeling process details, the time scales of the planning and control models are also drastically different. An MPC's time horizon typically ranges in minutes so that it may capture the nature of operation dynamics and describe changes in a timely manner, whereas a planning model ranges from weeks to months and thus describes changes based on averages within the planning period.

The differences in the process details and time scale are a formidable barrier that prevents proper integration of planning and MPC optimization (Lu, 2015; Shobrys and White, 2000). Table 13.1 highlights the large inconsistency between planning optimization and control models (and thus illustrates the challenges to unifying them). The profit derived from planning optimization cannot be fully captured in control, and as a result, the opportunity loss can be significant: in the order of US \$30–50 M/year for a typical refinery. Although the opportunity for real-time site-wide optimization is huge, there exist paramount challenges for integrating planning and control into one optimization system.

13.3 SIMULTANEOUS APPROACH FOR SITE-WIDE OPTIMIZATION AND CONTROL

The reason for the solution inconsistency lies in the fact that the planning model and MPC models are built and run in isolation without proper integration. One may think: why cannot a single NLP model be formulated by lumping the planning

TABLE 13.1. Comparison of Planning Model and MPC Model

Model Types	Users	Purpose and Advantages (Also Spatial Scale)	Typical Tasks	Typical Decision Variables	Time Scale
Yield model	Planning personnel	• Plant-wide or multisite sharp focus on key economic factors and interunit material and energy balances • Bird's-eye (compact) view of plant-wide operations • Free of unnecessary or obscuring details	• Produce the right amounts at the right specs • Coordinate the productions among different units • Satisfy the inventory (and possibly other logistic) interunit constraints	• Plant/unit material flow configurations • Unit feed rates, operating modes, and/or severities • Feed choices • Product mix • Buy vs. sell decision	• Longer time horizon • Less frequently executed • Open loop
MPC model	APC and process engineers	• Intraunit or multiunit • Sharp focus on the safety, operability, and product quality of a process unit • Unit-level efficiency is a secondary goal	• Control all safety- and operability-related CVs • Control me product quality within the specs • Operate the unit at a desirable operating point	• Flows, including the unit feed rates • Temperatures • Pressures • Levels • Pump-arounds	• Shorter time horizon • More frequently executed • Closed loop

Source: From Lu (2015).

model and rigorous process models together so that both planning and process optimization can be conducted simultaneously?

Suppose that we had such an overall NLP model built. We would have great difficulties in solving such a complex model computationally in a timely manner for control purposes. Even if we could solve the model quickly, operators do not trust such black-box solution as there are so many moving parts, and it could lead to non-optimum; furthermore it would have serious reliability issues due to likely occurrence of instances such as data error, model malcalibration, equipment malfunction, and changes in operating limits, which may not be detected and addressed in the model.

From a practical perspective, this monolithic MPC solution would be too complex for users to interpret any given part of the solution. It would have issues in practice and thus with low user acceptance, not to mention difficulties associated with run-error diagnosis and model maintenance. In fact, no such single-model MPC solution has been successfully implemented anywhere in the process industries. The intent of the discussions here is to demonstrate the feasibility of using a decomposition approach based on centralized planning optimization but decentralized controls at the process unit level. Decentralized MPC solutions are much desired because of their autonomy, better operability, and flexibility in dealing with unit upsets, equipment failures, and maintenance. These individual MPC controls are coordinated and directed by planning optimization while any inconsistency between the two levels can be overcome by means of a feedback loop, which is discussed below.

Let us start with the mathematical structure of such a single overall model including planning decisions and process details. We can build a block angular structure for a general problem involving planning and n number of processes (Figure 13.1).

A_i and B_i are blocks or matrices of coefficients. b_i are columns of coefficients forming the right-hand side. A_i represent the common rows, which usually involve distribution of resources among processes and describe interconnections between processes and mixers/splitters. The objective function is also included in the common rows. B_i are submodels describing process i. The major feature of this angular structure is that the common rows describe the interactions between processes and there are no common variables between submodels for processes.

This mathematical structure naturally makes us ask the following question: since all the submodels virtually do not have common detailed operating variables, is it possible to separate these submodels while retaining the overall optimality? This question is valid if and only if the allocation of common elements (e.g., intermediate products) is determined optimally from the total site point of view.

By analyzing the above angular structure of the total site model, we can use the common rows to form a master model, by which we optimize the selection of feeds and products, connectivity between processes, process flows and their associated properties, and allocation of common resources. Relatively simpler models (e.g., yield models) are used in the master model to account for processes,

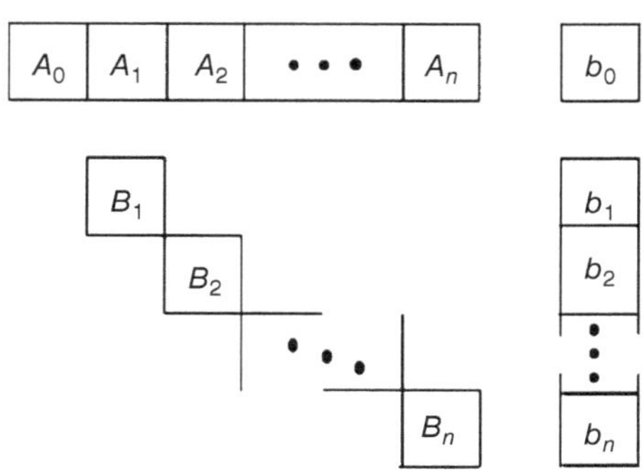

FIGURE 13.1. General angular structure for multiple process problems.

which are also part of the total site model. These results can then be used as the basis for optimizing the submodels for individual processes. This logical thinking leads to a general decomposition idea for solving large-scale problems of site-wide optimization. The master model makes decisions on site-wide aspects, while submodels determine process operating issues that help to achieve the targets and requirements set by the master model. Effectively, the master model coordinates individual submodels. The consequence of this decomposition is that we can eventually obtain an optimal solution to the large problem of site-wide optimization together with process optimization without ever building the total site model and solving it directly.

Example 13.1

We use the following simple example to illustrate the rationale and feasibility of the above decomposition (Zhang and Zhu, 2006). For illustration, a nonlinear problem is deliberately formed that may not be physically meaningful. A company has two processes, I and II, to produce the same two products, A and B, with net profits of \$12 and \$15/t (ton), respectively. Each process has two production sections, PS1 and PS2. The capacities of PS1 and PS2, together with the processing time required for making products A and B, are given in Table 13.2. In addition, making each ton of each product consumes 4 t of raw material, and the availability of raw material for the company is 125 t/hr.

Solution

Based on the above information, a company model can be built to maximize the overall profit as follows:

$$
\begin{aligned}
\text{Maximize} \quad & f = 12x_1 + 15x_2 + 12x_3 + 15x_4 \\
\text{s.t.} \quad & 4\left(x_1 + x_2 + x_3 + x_4\right) \le 125 \\
& \frac{40}{60}x_1^2 + \frac{15}{60}x_2^2 \le 80 \\
& \frac{20}{60}x_1^2 + \frac{45}{60}x_2^2 \le 60 \\
& \frac{30}{60}x_3^2 + \frac{20}{60}x_4^2 \le 60 \\
& \frac{30}{60}x_3^2 + \frac{45}{60}x_4^2 \le 75 \\
& x_1, x_2, x_3, x_4 \ge 0
\end{aligned} \tag{13.1}
$$

where x_1 and x_2 are the quantities of A and B produced by process I, while x_3 and x_4 are the quantities of A and B produced by process II. One can observe that the above model features an angular structure. Divided by 60 is to convert processing time from minutes to hours.

TABLE 13.2. Company Information

	Process I		Process II	
	PS1	PS2	PS1	PS2
Capacity (hr)	80	60	60	75
Processing time (min/t^2)				
Product A	40	20	30	30
Product B	15	45	20	40

By solving the company model, the optimal production plan is that process I produces 9.41 t/hr of product A and 6.38 t/hr of product B while utilizing 63.14 t/hr of raw material and process II produces 7.67 t/hr of product A and 7.80 t/hr of product B while utilizing 61.86 t/hr of raw material. The optimal company profit is $417.52/hr.

Alternatively, if we allocate 63.14 t/hr of raw material to process I and the remaining 61.86 t/hr of raw material to process II beforehand, we can build models for the two sites separately.

- Process I's model:

$$
\begin{aligned}
\text{Maximize} \quad & f_1 = 12x_1 + 15x_2 \\
\text{s.t.} \quad & 4(x_1 + x_2) \le 63.14 \\
& \frac{40}{60}x_1^2 + \frac{15}{60}x_2^2 \le 80 \\
& \frac{20}{60}x_1^2 + \frac{45}{60}x_2^2 \le 60 \\
& x_1, x_2 \ge 0
\end{aligned}
\tag{13.2}
$$

- Process II's model:

$$
\begin{aligned}
\text{Maximize} \quad & f_2 = 12x_3 + 15x_4 \\
\text{s.t.} \quad & 4(x_3 + x_4) \le 61.86 \\
& \frac{30}{60}x_3^2 + \frac{20}{60}x_4^2 \le 60 \\
& \frac{30}{60}x_3^2 + \frac{40}{60}x_4^2 \le 75 \\
& x_3, x_4 \ge 0
\end{aligned}
\tag{13.3}
$$

By solving the two process models separately, we obtain the same production plans for processes I and II as that from the overall company model. From this example, we can observe that if the allocation of raw materials is determined optimally, the overall solution from separate process optimizations is the same as that from total site optimization. In other words, a total site model can be correctly decomposed into individual process models if common aspects are determined optimally.

13.4 GENERAL DECOMPOSITION STRATEGY

By applying the above mathematical analysis, a site-wide optimization problem is decomposed into two levels, namely, a site level (master model) and a process level (submodels), which is proposed by Zhang and Zhu (2006). The relationship between the master model and submodels is shown in Figure 13.2. The master model is built to find the best allocation of resources for each process based on their current performance, while the submodels try to improve process performances based on their allocated resources. A feedback loop is developed to address the convergence between the master optimization and process optimizations.

At the site level, the objective is to maximize the total site profit by taking into account major aspects associated with site-wide operation, which are basically about how to manage resource distributions among processes. These considerations, which have significant impacts on overall economic performance, include selection of feeds and products, distribution of intermediate products, connections between different processes, and allocation of utilities. These aspects feature nonlinear and/or discrete elements. At this level of optimization, the existing planning model can be used in which processes are modeled via sequential linear yield correlations (see Figure 7.3). The nonlinear problem is solved using sequential linear programming (SLP) techniques (see Figure 7.4).

The objective of process-level (submodel) optimization is to optimize detailed process operating conditions (intermediate flows, temperatures, pressures, etc.) to fulfill the targets determined from site-level (or planning). The main feature of process-level optimization is that feed conditions are fixed to be the same as those determined from the current site-level optimization.

Depending on the optimization method chosen, the convergence criteria for site and process levels can be the relative difference of the objective values between the current and the previous site-level results. The iteration procedure and the convergence criteria for these two results will be explained below.

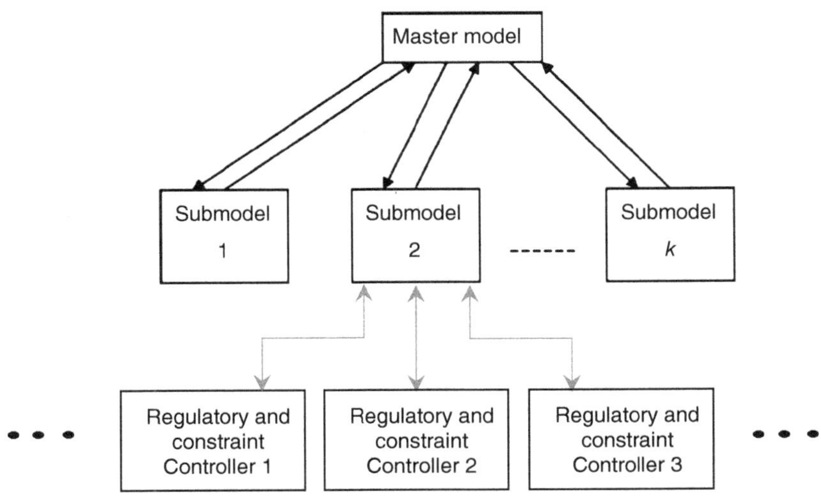

FIGURE 13.2. Model decomposition.

13.4.1 Demonstrating the Decomposition Strategy

Example 13.2

From Example 13.1, a master model could have been built as follows (Zhang and Zhu, 2006):

$$\begin{aligned}
&\text{Maximize} && f = f_1(x_5) + f_2(x_6) \\
&\text{s.t.} && x_5 + x_6 \leq 125 \\
& && x_5, x_6 \geq 0
\end{aligned} \tag{13.4}$$

where x_5 and x_6 are the allocation of raw material to processes I and II, respectively, and $f_1(x_5)$ and $f_2(x_6)$ are the profit generated by processes I and II, respectively.

Process models are modified as:

- Process I:

$$\begin{aligned}
&\text{Maximize} && f_1(x_5) = 12x_1 + 15x_2 \\
&\text{s.t.} && 4(x_1 + x_2) \leq x_5 \\
& && \frac{40}{60}x_1^2 + \frac{15}{60}x_2^2 \leq 80 \\
& && \frac{20}{60}x_1^2 + \frac{45}{60}x_2^2 \leq 60 \\
& && x_1, x_2 \geq 0
\end{aligned} \tag{13.5}$$

- Process II:

$$\begin{aligned}
&\text{Maximize} && f_2(x_6) = 12x_3 + 15x_4 \\
&\text{s.t.} && 4(x_3 + x_4) \leq x_6 \\
& && \frac{30}{60}x_3^2 + \frac{20}{60}x_4^2 \leq 60 \\
& && \frac{30}{60}x_3^2 + \frac{40}{60}x_4^2 \leq 75 \\
& && x_3, x_4 \geq 0
\end{aligned} \tag{13.6}$$

As a result of optimization, $x_5 = 63.14$ t/hr of raw material is allocated to process I and $x_6 = 61.86$ t/hr to process II, and the total company profit is \$417.52/hr. With this decomposition scheme, the company achieves the maximum profit without the need to solve the total site model described by equation set (13.1).

13.4.2 About Intermediate Products

Example 13.2 shows that there are two levels of optimization. The master model is to optimize the allocation of resources for each process while the process optimization determines the best process operations and resource utilization for a given allocated resources In Example 13.1, there are no intermediate flows involved. However, when a process produces intermediate products, as often happens in the chemical industry, their economic values may not be well defined. In this scenario, the objectives of individual process optimization become unclear and may lead to suboptimal conditions by optimization of individual components without proper coordination of each other.

In general, intermediate products are derived from upstream resources and act as direct resources for downstream processes. Their economic contributions are measured by using marginal values in traditional approaches. However, marginal values cannot be applied to the objective functions of process optimizations without limitations, since they are only valid in a very narrow range. Thus, when applying marginal values to process optimizations, the valid limits must be specified.

Example 13.3

Let us convert Example 13.2 as follows: add process III into the flowsheet to provide feed to processes I and II (Figure 13.3) (Zhang and Zhu, 2006). The total amount of raw material is given as 125 t/hr. Process III is actually a splitter. However, here we treat it as an independent process, and it can be modeled as below:

- Process III:

$$\begin{aligned} \text{Maximize} \quad & f = a_5 x_5 + a_6 x_6 \\ \text{s.t.} \quad & x_5 + x_6 = 125 \\ & x_5, x_6 \geq 0 \end{aligned} \tag{13.7}$$

where a_5 and a_6 are marginal values for x_5 and x_6, respectively. Marginal values are defined as the ratio of incremental profit and incremental change in an independent variable x. For example, a_5, the marginal value for x_5, is defined as $\Delta f/\Delta x_5$.

The models for processes I and II remain the same as equation sets (13.5) and (13.6).

Let us start assuming an equal distribution of raw material and initial flow rates for x_5 and x_6 as

$$\begin{bmatrix} x_5^0 \\ x_6^0 \end{bmatrix} = \begin{bmatrix} 62.5 \\ 62.5 \end{bmatrix}$$

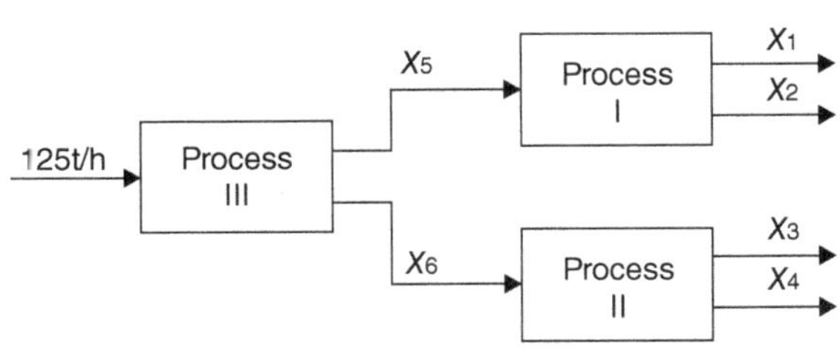

FIGURE 13.3. Example 13.3—with intermediate flows as variables.

By optimizing processes I and II, we have marginal values for x_5 and x_6 as

$$\begin{bmatrix} a_5^0 \\ a_6^0 \end{bmatrix} = \begin{bmatrix} 1.873 \\ 0.928 \end{bmatrix}$$

Now, if we directly apply these marginal values to optimize process III without any new limitations, all the raw material will be allocated to x_5, since it has a larger marginal value than x_6.

It is obvious that more raw material should be given to x_5. Once we have done that, the marginal value of x_5 is decreased, and that of x_6 is increased. If the two are equal, the best balance between x_5 and x_6 is achieved. Based on this analysis, the following approach is used.

After obtaining the marginal values, the next step is to determine how far the process optimization can go by using those values. Theoretically, at the optimal point, a_5 should be at least as much as a_6^0, and a_6 is unlikely to be greater than a_5^0. When $a_5 = a_6^0 = 0.928$, the optimization of process I gives x_5 = 63.782. Therefore, the maximum allowable change in $x_5 = 63.782 - 62.5 = 1.282$. On the other hand, if $a_6 = a_5^0 = 1.873$, the optimization of process II gives $x_6 = 61.248$. Therefore, the maximum allowed change in $x_6 = 62.5 - 61.248 = 1.202$. Since the optimal marginal values are between

$$\begin{bmatrix} 1.873 \\ 0.928 \end{bmatrix}$$

and

$$\begin{bmatrix} 0.928 \\ 1.873 \end{bmatrix},$$

we then compromise so that the maximum allowed change of x_5 and x_6 is in the middle:

$$\lambda = \frac{1}{2} \times \frac{1.282 + 1.202}{2} = 0.621 \tag{13.8}$$

Then we set the overall maximum change of x_5 and x_6 to be less than or equal to λ. Thus, the optimization model in Equation 13.7 for process III can be modified:

- Process III:

$$\begin{aligned} \text{Maximize} \quad & f = a_5 x_5 + a_6 x_6 \\ \text{s.t.} \quad & x_5 + x_6 = 125 \\ & x_5 - x_5^0 \le \lambda \\ & x_6^0 - x_6 \le \lambda \\ & x_5, x_6 \ge 0 \end{aligned} \tag{13.9}$$

The next step is just a repeat of the above procedure. The iteration continues until the convergence criteria are met. It should be noted that at the optimal point, we do not always have the same marginal values for the split streams, due to other constraints such as throughput limits. For instance, in the above example, if the maximum rate of x_5 is 60 t/hr, the marginal values for x_5 and x_6 at the optimal point are not the same. This example demonstrates that with given resources, optimal process operations can be determined with intermediate flows involved. The optimal resource utilization can then be obtained by comparing the results of different resource usage and allocations, which leads to the overall optimum solution.

13.4.3 About Recycles

Decomposition becomes difficult when recycles exist among processes, since processes are linked more closely to each other. One way to handle recycles in the decomposition algorithm is to merge those processes linked by recycles into one process (Figure 13.4). However, this approach makes the subproblems more difficult to solve especially when a large number of processes are involved, which conflicts with the decomposition objective. Alternatively, recycles can be handled in a sequential way. As shown in Figure 13.5, recycles are initially cut off for decomposition. After a search direction and a step size are determined with an initial x_1, which means new process operating conditions are found, Equation 13.10 is added to the site-level model. Then, the recycles are determined accurately by the iteration of the site-level optimization and process simulation as discussed below:

$$x_1 = x_2 \tag{13.10}$$

By applying this decomposition strategy for integrating site-wide and process optimization, two different approaches are developed: one is the integrated MPC approach, and the other is based on rigorous models, which are discussed in detail below.

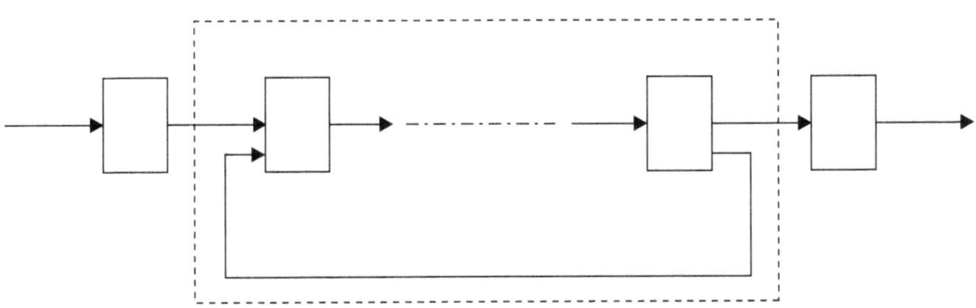

FIGURE 13.4. Modeling of recycles.

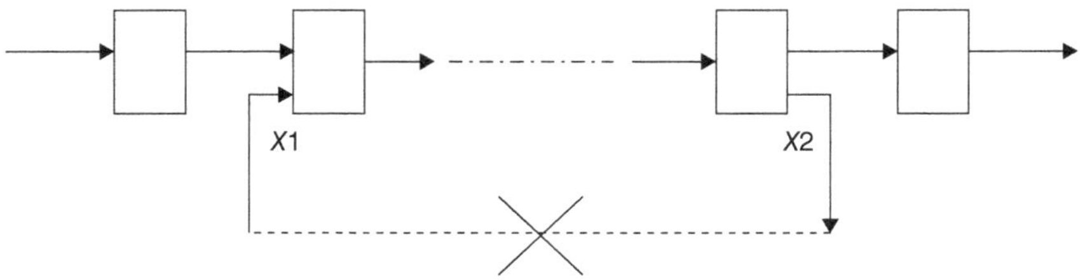

FIGURE 13.5. Decomposition with recycles.

13.5 MPC-BASED INTEGRATION APPROACH

The integrated MPC-based approach, so-called *1-to-n* MPC cascade approach (Figure 13.6), was proposed by Lu (2001, 2015). This approach aims to fill the gap between the planning optimization and process MPC controllers and to address their contradictive needs in an integrated manner. To do so, the master model mimics the planning model, which depends on MPC controllers to provide actual constraints within every process unit to create a feasible region for site-wide optimization. At the same time, planning optimization provides an indication of sensitivity associated with each constraint, reveals which constraints are true site-wide bottlenecks, and assigns values to push them toward hard limits.

More specifically, the master model for economic optimization uses a preexisting linear yield model as the initial steady-state gain matrix of its MPC model, while process dynamics can be determined from operating data. The master model controls the interunit variables such as feed rates, product yields, and quality within the entire site. Thus the optimization of the master model can be furnished with the same planning model structure and economic expressions.

The site-wide master MPC controller (*1: single master model*) for planning optimization sits on top of multiple slave MPC controllers (*n* slave models) at the process unit level. Note that this MPC cascade needs a two-way connection between the master and the slave models. It is this two-way connection that also allows the slaves to provide the master with the operating constraints inside every unit (see discussions on proxy limits in Section 13.5.2.4 for details). With this supplemental information, the real-time master planning solution of the MPC cascade eliminates the aforementioned handicap and will respect all the unit-level operating constraints. Jointly, the master and slave MPC controllers provide simultaneously decentralized controls at the low level with fine-scale MPC models and centralized real-time planning optimization at the high level with a coarse-scale yield model—all in one consistent cascade control system.

13.5.1 *1-to-n* Site-Wide Optimization Model Formulation

In this section, the model formulation is provided based on three-tiered structure (Lu, 2001). The top tier is for site-wide optimization (master model), while the third tier is for MPRC (model predictive range control) controllers. The middle tier is for slave MPC controllers and is the coordination "collar" for preventing each controller from receiving a locally infeasible command.

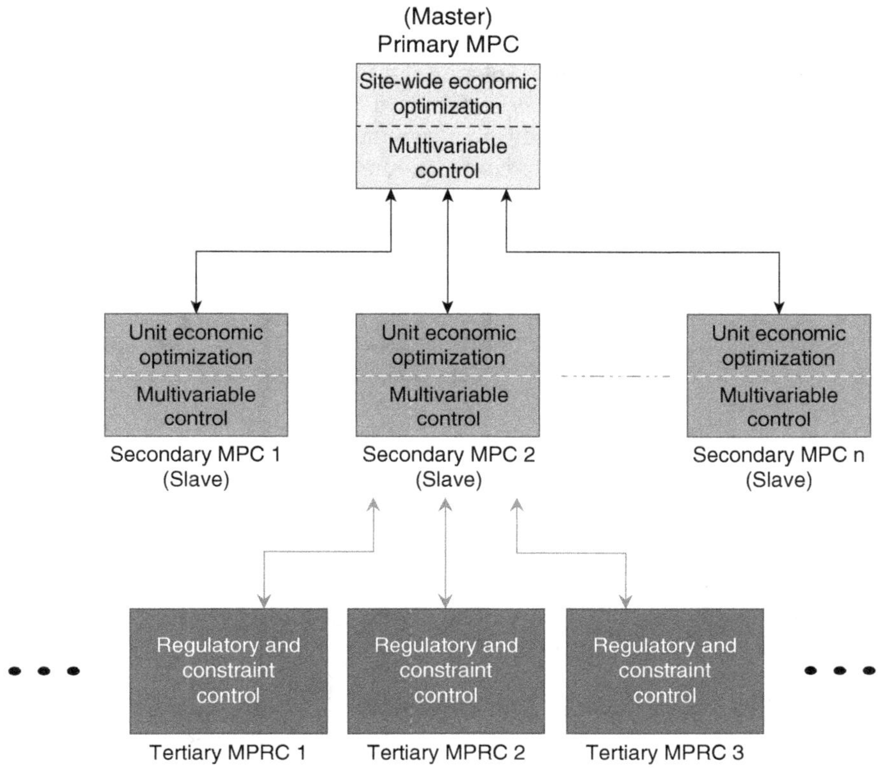

FIGURE 13.6. Three-tiered structure of multiscale MPC cascade.

13.5.1.1 Tier 1 Formulation Master controller—Perform site-wide optimization based on the predicted future constraints transferred from Tier 2 MPC models:

$$\begin{aligned} &\text{Maxmize} \quad f(x_{ss}) \\ \text{s.t.} \quad &\mathrm{CV}_{\min} \le G \cdot x_{ss} \le \mathrm{CV}_{\max} \\ &\mathrm{MV}_{\min} \le x_{ss} \le \mathrm{MV}_{\max} \end{aligned} \tag{13.11}$$

where

$f(x_{ss})$ = the economic objective function, which can be nonlinear
G = the steady-state gain matrix
x_{ss} = the steady-state control move
CV = control variable
MV = manipulated variable

For the master controller, a typical single-period planning problem can be readily described by Equation 13.11. The constraints in Equation 13.11 represent all the actual process constraints including proxy limits that are transferred from each of the MPC controllers in Tier 2 via inquiry calls from the master model to individual

MPC controllers. With this constraint transfer, the master optimizer will honor the same set of constraints as in MPRC controllers.

Once the economically optimal solution, x_{ss}^*, to Equation 13.11 is obtained, various algorithmic approaches can drive the slave MPC controllers to the economically optimal point. One approach is to assign the optimal CV solution, $y_{ss}^* = Gx_{ss}^*$, as the internal targets of the MPC controllers. Another approach is to use x_{ss}^* directly in a manner that takes advantage of the inherent extra degrees of freedom in the MPRC formulation, which is used in Tier 2 formulation.

13.5.1.2 Tier 2 Formulation Slave MPC controllers—For each of the MPC controllers, find the closest locally feasible point to the global optimum:

$$
\begin{aligned}
&\underset{x_{ss}}{\text{Minimize}} && \sum_i \left(x_{ss,i} - x_{ss,i}^*\right)^2 \\
&\text{s.t.} && \text{CV}_{\min,i} \le G_i x_{ss,i} \le \text{CV}_{\max,i} \\
& && \text{MV}_{\min,i} \le x_{ss,i} \le \text{MV}_{\max,i}
\end{aligned}
\tag{13.12}
$$

In MPC optimization as shown in Equation 13.12, the economic optimal solution x_{ss}^* from a site-wide optimization becomes recommended "setpoints" for MPC controllers in each process unit. The MPC controller serves as a dynamic enforcer of optimization setpoints. This is exactly how MPC controllers integrate with a site-wide master optimizer.

13.5.1.3 Tier 3 Formulation MPRC controllers—Pass the solution of Tier 2 to the corresponding *i*th controller; use MPRC to accomplish the regulatory and constraint control based on tight and dynamic constraints. In many aspects, MPRC is similar to other MPC algorithms used in the industry. The key difference lies in the control formulation. In the open-loop control solution step, constraint control is woven into the formulation by adding a slack variable for each pair of high/low constraints (or a range). Thus, the MPRC can be described as Equation 13.13:

$$
\begin{aligned}
&\underset{x,y}{\text{Minimize}} && \left\|\mathrm{W}(\mathbf{A}x - y)\right\|^2 + x^{\mathrm{T}}\Lambda x \\
&\text{s.t.} && \text{CV}_{\min} \le y \le \text{CV}_{\max} \\
& && \Delta\text{MV}_{\min} \le x \le \Delta\text{MV}_{\max} \\
& && \text{MV}_{\min} \le \mathrm{S}x \le \text{MV}_{\max}
\end{aligned}
\tag{13.13}
$$

where

$\mathbf{A}$ = the dynamic (step-response) matrix of the model
$x = \Delta$MV move
y = control variable (CV)
S = accumulating sum matrix
$\mathbf{W}$ = the diagonal weighting matrix
$\left\|\mathrm{W}(\mathbf{A}x - y)\right\|^2$ = the CV error penalty function
$x^{\mathrm{T}}\Lambda x$ = the MV move-penalty term (Λ is semi-positive definite)

Although many forms of MPC offerings are described in various publications, only the ones that provide dynamic constraint control are suitable for this task. For the process industries, a majority of the CVs in a typical MPC application are constraint CVs, meaning that they need to be constrained within their high/low process limits but that they should not be controlled (i.e., regulated) to an internal setpoint that is placed within high and low limits. The importance of dynamic constraint control here can hardly be overstressed, particularly for the master controller, because more than 95% of its CVs in a typical plant-wide production planning problem are constraint CVs.

Moreover, the number of CVs in a typical application is about twice the number of MVs. Any artificial internal setpoint can potentially introduce undesirable behavior, particularly, for example, when parallel or near-parallel CVs exist in the controller. MPRC (Lu, 2001) is a form of MPC that unifies both regulatory and constraint control. An augmented form of MPRC Equation 13.13 is developed by Lu (2015), and it is called RMPCT (robust MPC technology).

13.5.1.4 Discussions Constraint consistency from Tiers 1–3 is essential to a successful integration.

In Tier 1, the first two sets of constraints, $CV_{min} \le Gx_{ss} \le CV_{max}$ and $MV_{min} \le x_{ss} \le MV_{max}$, represent all the constraints that are transferred from each of the MPC controllers. G in the first set of constraints is the global gain matrix. With this constraint transfer, the optimizer is bound by the same set of constraints as the MPRC controllers.

In Tier 2, the two sets of constraints, $CV_{min,i} \le G_i\, x_{ss,i} \le CV_{max,i}$ and $MV_{min,i} \le x_{ss,i} \le MV_{max,i}$, represent the constraints in the ith controller at the end of its prediction horizon (i.e., a time period of interest in a model that considers process dynamics, often a time to reasonably approach steady state). Matrix G_i is the model gain matrix for the ith controller, which also equals the ith diagonal matrix block in the global gain matrix G. Tier 2 is a coordination collar, through which one can protect the MPC controller from receiving a locally infeasible maneuvering destination. It is important to note that a globally feasible solution may not be feasible to a local controller. The reason is that, in Tiers 1 and 2, different prediction horizons, different scopes, and different gains, G and G_i, are used.

In a plant, a local infeasibility in a process can often occur if it is connected to its upstream unit via an intermediate storage tank or other processing units. The coordination collar (Equation 13.12) provides a way to protect the "local interest" and, in fact, favors the local interest the most. If one prefers to balance the local and global interests, one could construct an objective function that would trade off the local interest for the global considerations or even skip the collar Equation 13.12 completely. Notice that the local and global interests always converge at the end of the optimization horizon and that the conflict arises only in transience.

In Tier 3, the MPRC layer adopts an RMPCT model, which unifies all three common tasks of advanced process control (APC), namely, regulatory PID control, constraint control, and local economic optimization, in a single range control formulation, that is, Equation 13.13. This formulation provides a way to protect the "local interest" via identifying actual constraints for conjoint variables (or proxy limits) at the process unit level and sending them back to the master optimizer for iterative optimization with newly generated proxy limits. This feedback mechanism enforces consistency between master MPC and slave MPC controllers via use of proxy limits. Honeywell's RMPCT has been successfully implemented in more than one thousand applications to date in various process industries.

13.5.2 Features of the MPC-Based Site-Wide Optimization Approach

There are several major improvements from this *1-to-n* approach compared with the traditional approach, which are discussed below.

13.5.2.1 Integration of Planning Optimization via Multiscale Modeling In the traditional approach, a large gap exists between a planning model and APC models. The gap originates from different scales in terms of process details and run time used in the planning model and APC models. To close the gap, laborious manual interactions based on simulation are required in conjunction with experience to ensure the feasibility of planning optimization results implemented in APC.

In contrast, the one-to-many MPC cascade approach can integrate planning and MPC models seamlessly and close the gap automatically without sacrificing the optimality and feasibility between planning and process control. Multiscale modeling is the key to achieve seamless integration in an automatic manner. With multiscale modeling, a site-wide master MPC controller cascades to multiple (n) slave MPC controllers.

The master model uses coarse-scale yield vectors from the planning model as the gain matrix to represent process units in the master dynamic model. The master model performs overall economic optimization of total site operations, which, depending on the market needs, may include feed mixing, key operating conditions for individual process units, utility allocation among these units, product blending, etc. The master MPC model, like a planning model, has a very coarse scale as it excludes nonproduction or noneconomically related variables and hence represents a process with several relevant yield vectors and property relationships.

On the other hand, slave MPCs adopt finer-scale process models involving all the necessary operating variables for control purposes to ensure simultaneously safe and economic operation. Thus, the scale difference in terms of the number of variables modeled for a process unit in the master mode and a slave MPC model could be one order of magnitude.

Each slave MPC model focuses on intraunit dynamic relationships between CVs and MVs and performs two functions: (a) MPC for the process unit and (b) computation of proxy limits that represent the current constraints inside the slave model. The master's economic optimizer amends the planning optimization in real time with the slave's proxy limits, and the embedded planning model is thus reconciled with the MPC models for process units in the sense that the master's optimal solution now satisfies the slave's constraints.

In this manner, this approach can run the planning optimization in real time and determine the targets for key operating conditions, while slave MPC models determine detailed operating setpoints for individual PID controllers and provide a dynamic path to move each MV from its current state to the optimal state. Let us focus on three noteworthy features, namely, dynamic control, feedback mechanism, and proxy limits as below.

13.5.2.2 Dynamic Path from Current to Optimal Point Dynamic control is about determining the path to be taken from the current state to the optimum state. In general, a multitude of paths exist, and the "best" path (from steady-state optimization) may not necessarily be feasible for a process with dynamic interactions. The *1-to-n* MPC cascade approach can achieve plant-wide optimization by providing dynamic paths for the MV movements or the trajectory of "how to get there" from the current point to the optimal solution.

The *1-to-n* MPC-based site-wide optimization has a mathematical property that will move the process toward the optimum along the minimum energy path. The minimum energy path is the path of minimum MV moves. What can happen for nonlinear processes with significant gain changes is that predictions of the optimal state may shift from execution to execution. Practically speaking this iterative approach to the optimal state is not a problem, as the optimizer will settle out at the ultimate optimum within the optimization time horizon. In addition, the path will be robust in the sense of featuring the least amount of MV moves and always moving in the direction of improvement.

To accomplish the above claim, the master MPC model contains dynamic gain information. For a single-period planning model, the single gain matrix in the planning model is used as the gain matrix for the master MPC model. For a multi-period planning model, there are multiple gain matrices, and one gain matrix corresponds to one-period planning. The gain matrix used in the master MPC is consistent with those gain matrices used in slave MPC models.

13.5.2.3 Feedback Loop for Solution Consistency If a large gap in production rates exists between the master MPC controller and slave MPC controllers, the effective way to solve this problem is to extend the feedback control to the master optimization layer. To accomplish this, an MPC scheme is used to establish the basis for constraint optimization. At each execution, the constraints for optimization are predicted out to a future point, utilizing dynamic models. The predictions, available in each MPC controller, can be used as a starting point. The future point is referred to as the optimization point at the slave MPC level, and the distance between it and the origin (time zero) is referred as the optimization horizon. The slave optimization point can be a user-specified parameter, commonly selected as the nearest point after all the dynamics settle out. The site-wide optimization is then performed at the slave optimization point, as opposed to the original steady state.

In this *1-to-n* MPC cascade approach, the economic optimization is performed at the optimization horizon inside the model predictive controller. The global economic optimization is thus defined in a future feasible space, bounded by the predicted constraints. The optimization can be executed in synchronization with its downstream MPC controllers, eliminating the need for detecting and waiting for a steady state.

There is two-way connection between the master model and multiple (n) slave MPC controllers at the process level, unlike a traditional one-way PID-to-PID cascade. It is this two-way connection that allows the slave models to provide the master model with operating constraints, which could restrict the planning optimization (Figure 13.7). With the operating constraints generated online from slave models, the next-round planning optimization will honor such constraints. With this two-way communication, the master model jointly with the slave models provides simultaneously decentralized controls at the lower level with fine-scale MPC models and centralized real-time planning optimization at the high level with a coarse-scale yield model: an all-in-one consistent cascade control system. Therefore, the planning model and control models are coupled via a closed-loop feedback mechanism.

13.5.2.4 Proxy Limits: A Conduit for Reconciling Multiscale Models A proxy limit is an alternative representation of slave MPC controller's constraints in the master's space, and it can be viewed as a conduit for adding the slave's constraints to the master MPC. Proxy limits from all slave MPCs are combined and then included in the master's multivariable control and economic optimization formulations. Inherent relationships exist between the master and slave models—while the CVs of the master do not intermix with the CVs of any slave, the MV/DVs (disturbance variables) are interconnected. To illustrate these interweaving relationships, the master model matrix is placed on top of the slave model matrices in Figure 13.7. The top half of the diagram represents the master model matrix that is partitioned into multiple column groups to align its MV index with the slave MV indices. The lower half contains the slave model matrices that are arranged block diagonally.

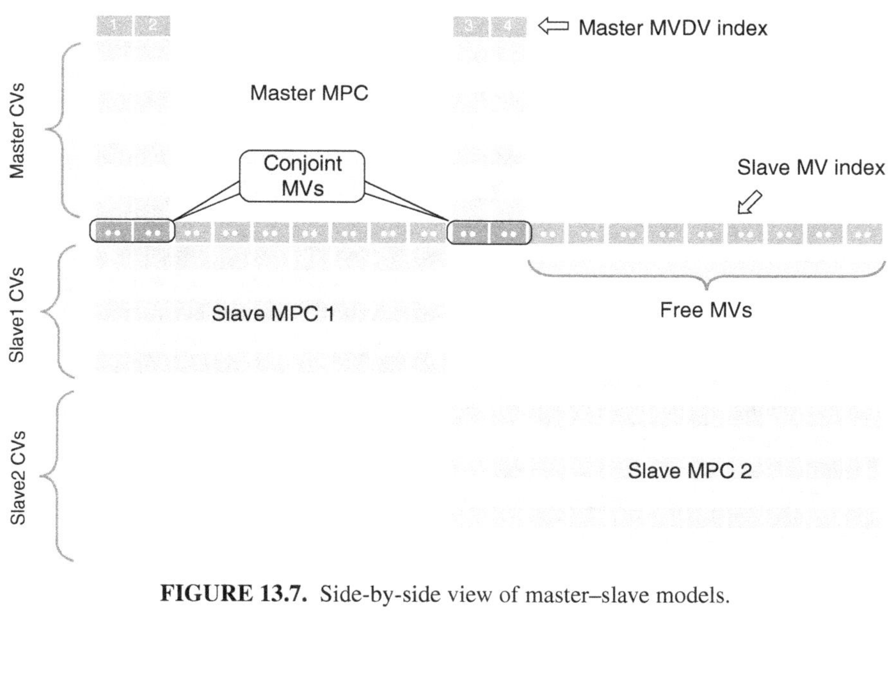

FIGURE 13.7. Side-by-side view of master–slave models.

Each cell in the diagram represents a potential dynamic model between its corresponding input and output.

As a matter of fact, an MV of the master may also be an MV of a slave, and vice versa. In the specific case depicted in Figure 13.7, the third MV of the master controller is connected to the first MV of slave MPC 2; they are the same MV. For illustration purposes, suppose this MV is the feed rate of Unit 2. In this case, the master controller needs to manipulate the feed rate for interunit production inventory and quality control, whereas the slave controller also needs it for local control objectives within Unit 2. These connected MVs are denoted as conjoint variables, for the obvious reason that these variables conjoin the master and slaves together. Wherever there is a conjoint variable, either the slave or the master loses one degree of freedom, since neither of them can manipulate the conjoint MV independently of the other. The non-conjoint MVs are denoted as free variables. They can be manipulated unilaterally by the slave for unit-level optimization. Whenever extra degrees of freedom remain, they can also be used by the slave to make more room for the conjoint variables, which can then be used by the master to further optimize the entire site.

As the simplest case, in which a given slave controller has only one conjoint MV, the proxy limit can be expressed simply in terms of an MV limit in the master controller. This limit value can be determined by computing the maximum distance that the master can move the conjoint MV before one or more slave variables are pushed against their corresponding CV and MV limits.

Take a hydrocracking unit (HCU) in an oil refinery for example. Suppose that the HCU feed is configured as MV3 in the master controller and as MV1 in the second slave MPC. The current feed rate is at 33.5 MBPD. The master can then make an inquiry call to the slave and request the slave to maximize the feed rate in its steady-state optimizer. The inquiry call's result is returned to the master, which indicates that the maximum feed rate of 38.1 MBPD can be requested before one or multiple slave CVs and MVs hit their limits. The returned maximum value for the conjoint MV3, 38.1 in this example, is then used as the proxy limit in the master for all of the constraints in the second slave MPC. The low proxy limit can be derived similarly by the master making an inquiry call to minimize the conjoint MV (or the feed rate in the above example). The returned minimum value is then used as the low proxy limit in the master.

In this simplest case of one conjoint MV per slave, regardless of how many slave constraints can limit the master's MV, the master MPC only needs to know the point where it must stop pushing the conjoint MV (otherwise some low-level constraint violations will result). This predicted stopping point is a proxy limit, which represents the entire set of potentially active slave constraints that could limit the master's MV. In this specific HCU example above, no more than two proxy limits are required for representing the entirety of slave constraints!

As a result, the high and low conjoint MV limits of the master, unlike the MV limits of a regular MPC controller, are not set by the end users. Instead, they are set by the slave MPC controllers based on the current operating constraints in the process unit.

When a slave MPC controller has two or more conjoint variables with the master, proxy limits for this slave can be multivariate in nature. In this general case, the proxy limit concept will be extended and expressed in either the MV or CV space

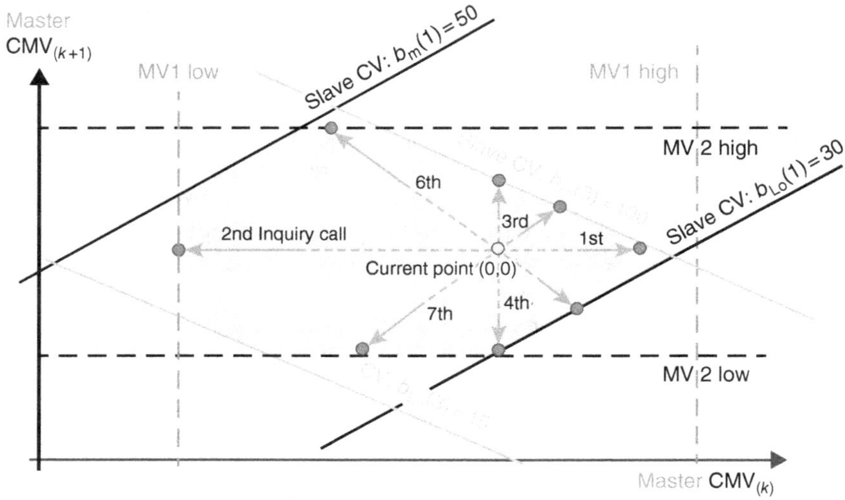

FIGURE 13.8. Slave constraints as plotted in the master MV space and projections of inquiry calls.

of the master controller. To do so, more inquiry calls in different directions will be made to determine the feasible region defined by the slave constraints.

To explain the general concept of proxy limits, the relationship between the master and slave constraints for two conjoint MVs is illustrated in Figure 13.8.

If the free variables of the slave controller are fixed at their current operating values, the (current) MV and (predicted) CV constraints of the slave can then be expressed in the MV space of the master. For a linear, time-invariant slave controller, its MV and CV constraints form a convex region in the master MV space (the gray polygon in Figure 13.8).

To estimate the feasible region of the conjoint MVs in the master, multiple inquiry calls in different directions are required. The turquoise dashed arrows inside the convex region illustrate how these inquiry calls work in a two-dimensional space. The first inquiry call is made with an objective function to maximize the first conjoint MV. This inquiry call "pushes" all the movable edges to the right as depicted by the horizontal dashed arrow labeled as "1st." The black point at the end of this arrow denotes the maximum limit to which the CV constraints can be pushed. The mathematic formulation for the inquiry calls are given in Lu (2015).

13.6 RIGOROUS MODEL-BASED INTEGRATION APPROACH

Similar to the MPC approach, there are two levels of optimization for this model-based approach, namely, master and process optimization (Zhang and Zhu, 2006). In the master optimization, the site-wide model is based on a planning model. In the process optimization layer, rigorous models are used to describe processes. Optimization of process rigorous models provides the setpoints for the regulatory controllers. In contrast, in the MPC approach as discussed previously, the second level consists of slave or MPC controllers for which process modeling is based on linear gains.

Depending on the optimization method chosen for the site level, the convergence criteria can be the relative difference of the objective function values between the current and the previous site-level results. The iteration procedure and the convergence criteria are discussed below.

13.6.1 Integration of Site-Wide Planning Model and Process Rigorous Model

The site-level optimization captures major operating aspects for an overall plant by finding the best feed distribution and resource allocation under fixed process operations. Then, the optimal ways of operating each process are determined in the process-level optimization by exploiting additional degrees of freedom associated with process operating details. For a given feed distribution and resource allocation, the process-level optimization determines the setpoints for individual controllers. The combination of the two (site-wide and process systems) gives the overall optimal solution.

Functionally, process optimization is a supervisory control layer, which continuously strives to maintain the plant optimal operating conditions. In order to achieve this functionality, the process optimization must have the following four capabilities, namely, state estimation, model calibration, optimum state, and dynamic path. These are discussed below:

- **State estimation**: This refers to the determination of the current steady state of the system. The plant is a dynamic entity with a wide mix of process dynamic characteristics (fast, slow, inverse, and delayed responses) and all manners of process disturbances. State estimation defines both where the plant is right now and where it can be expected to be in the future.

  In this approach state estimation is resolved by only optimizing when the process is steady. Steady-state detection however is problematic for a typical process plant that is seldom at perfect steady state. In practical terms this becomes a tuning issue as to what constitutes a steady state. There are two extremes for steady-state detection: (a) strict criteria for steady state that could mean a long wait before conducting the following steps and (b) relaxed criteria for steady state to reduce the wait. Both approaches have problems. In the first approach, all the waiting means lost opportunity for optimization, although when steady state is detected the results are more meaningful. In the second approach, optimization can run more frequently, but there is less certainty of the current state that in turn questions the integrity of model calibration and data reconciliation. In fact, this uncertainty compromises the credibility of the whole optimization.
- **Model calibration**: This refers to a calibration of the process models based on real-time feed conditions (flow rate, composition), operating conditions (temperature, pressure), reaction kinetics, product yields, catalyst deterioration, and equipment performance. Because the quality of calibration relies on consistent and accurate operating data, the model calibration starts with gross error detection for bad operating data followed by model tuning.

- **Optimization**: This refers to identification of the steady-state optimum point. For a system described by a set of algebraic equations, there are powerful mathematical optimization tools commercially available.

  Optimum state determination has two issues. First, the optimizer takes significant execution time, typically 40–60 min (Nath et al., 1994; Sowers et al., 1998) for a complex plant. Second, intermediate disturbances during program execution can invalidate results of an optimization solution. Because of the lengthy wait for steady state, long execution time, and intermediate disturbances, as an example, a typical ethylene plant optimization runs fairly infrequently. It is typical to run a plant optimization model once per every 4–8 hr (Nath and Alzein, 2000).
- **Consistency check**: This refers to validation before downloading results to the underlying MPC controllers. This is to ensure that the state of the process after the completion of optimization is not significantly different than the initial state established in the first step.

13.7 COMPARISON BETWEEN THE MPC AND RIGOROUS MODEL-BASED APPROACHES

The major difference between these two approaches is in the complexity of process models used. In the rigorous model-based approach, rigorous models featuring high nonlinearity are used, while in the MPC approach, linear process models (or empirical models) are adopted although nonlinearity can be taking into account by using successive linear programming (SLP). Rigorous models can predict process variations directly and precisely. In contrast, with linear models, the control gains often vary and must be updated through SLP if the process conditions are significantly different from the base conditions; otherwise, the optimized results can be misleading.

So a critical question can be asked naturally at this point: why is the rigorous model-based approach not adopted in commercial applications more widely than the MPC approach? The answer is largely to do with model fidelity, dynamic implementation path, and model development and maintenance effort.

13.7.1 Process Nonlinearity

Nonlinear-based rigorous models can provide high-quality prediction for process performance. In general, all processes are nonlinear within normal operating region, some obviously more than others. However, MPC-based approach employs a linear gain matrix in the master planning model and linear dynamic models in slave MPC process models. Thus, the MPC approach requires gain updating to improve controller performance over a wider operating range.

However, in current practice, it is a manual procedure for gain updating, which requires plant performance test to determine gains under changing conditions. This procedure is very tedious, and it could take a couple of weeks. The length of time, along with associated costs, is a significant barrier for updating control gains on regular basis. Therefore, an automated procedure for gain updating needs to be developed. Rigorous process models can be used as inferential sensors. New gains are determined by numerical perturbation of a nonlinear process model. It is important to get the first derivatives correct for gain updating, while rigorous parameter fitting for the nonlinear process model is not required at each execution. Execution of the rigorous process model can either be triggered by a major change in some operation condition or can be on a fixed frequency.

13.7.2 Dynamic Path

For the rigorous model-based optimization approach, determination of a dynamic path for control implementation is problematic. Typically this problem is resolved by restricting optimization step size. Since the optimization model is steady state in nature, typically small optimization steps are taken to minimize the risk of dynamic CV violations. Subsequent small steps will slowly move the process to optimum conditions, which partly explains the reason why this approach takes a relatively longer time to reach the optimal solution. Another reason is the long wait to reach steady state.

In contrast, in the MPC-based approach, the complete unit dynamic optimization is solved together using the MPRC control algorithm. It not only solves for the steady-state optimum but also determines the trajectory of "how to get there" while handling the control tasks currently at hand. This trajectory is a combination of control response and economic optimization paths. Thus this approach unifies the control and optimization problems.

13.7.3 Model Maintenance

The benefit of using rigorous models for better prediction comes at the expense of a high initial cost for model development and high model maintenance effort over time. Maintaining model fidelity in real-time applications requires handling data errors, equipment malfunction, performance deterioration, etc., which could occur on an irregular basis. There are no sufficiently good techniques available for detecting such instances and revising models effectively and automatically. When feed type, product specifications, and process conditions change significantly, model tuning and verification are required. In some cases, the model must be changed accordingly. The effort requires modeling experts to tackle these challenges and maintain model fidelity for online applications. Lack of proper modeling skills is a major roadblock to apply the rigorous model-based optimization approach.

REFERENCES

Lu J (2001), Challenging control problems and emerging technologies in enterprise optimization, IFAC Proc. **34** (25), 23–24.

Lu J (2015), Closing the gap between planning and control: a multiscale MPC cascade approach, Annu. Rev. Control **40**, 3–13.

Nath R, Alzein Z (2000), On-line dynamic optimization of olefins plants, Comput. Chem. Eng. **24**, 533–538.

Nath R, Laiseca M, Poje JB (1994), Operations optimization of cold end of an ethylene plant using OPTCOM/RT-OPT. In: Paper Presented at ASPENWORLD, Boston, MA.

Shobrys DE, White DC (2000), Planning, scheduling and control systems: why can they not work together? Comput. Chem. Eng. **24**, 163–173.

Sowers G, Reed C, Corripio B, Starks D, Mainan M (1998), Multivariable control and online optimization of an ethylene plant. In: Paper Presented at the AIChE 10th Ethylene Producers Conference, Houston, TX.

Zhang N, Zhu XX (2006), Novel modelling and decomposition strategy for total site optimization, Comput. Chem. Eng. **30**, 765–777.

14

DIGITALIZING THE ENERGY MANAGEMENT

14.1 INTRODUCTION

In many organizations, cutting energy operating costs has been an ongoing effort for a long time. However, "cherry-picking" has been the dominant approach. To achieve greater benefits, we need a systematic approach covering a solution methodology and supporting structure and activities. The objective of applying this approach is to identify energy-saving projects, capturing hidden opportunities in process operation, closing of various loose ends in steam system and off-site utility, and optimizing utility system operation.

But where to start and how to make it happen? You may ask more specific questions:

- What metrics are applied to measure the process energy efficiency?
- What energy indicators and their targets are defined for key operating parameters?
- What are the methodologies to follow through to implement the opportunities identified?
- How to digitalize the energy management for sustaining the benefits achieved?

These questions will be answered in this chapter.

Digitalization and Analytics for Smart Plant Performance: Theory and Applications,
First Edition. Frank (Xin X.) Zhu.

14.2 THE CONCEPT OF ENERGY INTENSITY

Let us start with the specific question: how to define energy performance for a process? People might think of energy efficiency first. Although energy efficiency is a good measure as everyone knows what it is about, traditionally it does not relate energy use to process feed rate and yields, and thus it is hard to connect the concept of energy efficiency to plant managers and engineers.

To overcome this shortcoming, the concept of energy intensity is adopted, which connects process energy use and production activity. The energy intensity was originated from Schipper et al. (1992), who attempt to address energy intensity through historic energy use and economic activity in five nations: the United States, Norway, Denmark, West Germany, and Japan. The concept of energy intensity allows them to better examine the trends that prevailed during both increasing and decreasing energy prices.

By definition, energy intensity (I) is described by

$$I = \frac{\text{Energy use}}{\text{Activity}} = \frac{E}{A} \tag{14.1}$$

Total energy use (E) becomes the numerator, while common measure of activity (A) is the denominator. For example, commonly used measures of activity are vehicle-miles for passenger cars in transportation, kWh of electricity produced in power industry, and unit of production for the process industry, respectively.

Physical unit of production can be barrel/day or tons/hr or m^3/hr of total feed (or product). Thus, industrial energy intensity can be defined as

$$I = \frac{\text{Quantity of energy}}{\text{Quantity of feed or product}} \tag{14.2}$$

Energy intensity defined in Equation 14.2 directly connects energy use to production as it puts production as the basis (denominator). In this way, energy use is measured on the basis of production that is in the right direction of thought: a process is meant to produce products supported by energy. For a given process, energy intensity has a strong correlation with energy efficiency. Directionally, efficiency improvements in processes and equipment can contribute to observed changes in energy intensity.

Therefore, we can come to agree that energy intensity is a more general concept for measuring of process energy efficiency indirectly.

Before adopting the concept of energy intensity, the measure of activity must be defined by either feed rate or product rate. For plants with a single most desirable product, the measure of activity should be product rate. For plants making multiple products, it is better to use feed rate as the measure of activity. The explanation is that a process may produce multiple products and some products are more desirable than other in terms of market value. Furthermore, some products require more energy to make than other. Thus, it could be very difficult to differentiate products for energy use. If we simply add all products together for the sum to appear in the denominator in Equation 14.2, we will encounter with a problem that is the dissimilarity in product as discussed. However, if feed is used in the denominator, the dissimilarity problem is nonexistent for cases with single feed. The dissimilarity is much less a concern for multiple-feed cases than for multiple products because, in general, multiple feeds are more similar in compositions to each other than multiple products are similar to each other.

The above discussions lead us to define the process energy intensity on the feed basis as

$$I_{\text{process}} = \frac{\text{Quantity of energy}}{\text{Quantity of feed}} = \frac{E}{F} \tag{14.3}$$

In the case where there a process unit has multiple feeds but single product, the energy intensity is defined on the product basis as

$$I_{\text{process}} = \frac{\text{Quantity of energy}}{\text{Quantity of product}} = \frac{E}{P} \tag{14.4}$$

It is straightforward to calculate the energy intensity for a process using Equation 14.3 or Equation 14.4 where E is the total net energy use and F or P is the total feed consumed or product produced from the process. Net energy use is the difference of total process energy use and total process energy generation from the process. Process energy use mainly includes fuel fired in furnaces, steam consumed in column stripping and reboiling, and steam turbines as process drivers and electricity for motors. In addition, it also includes other energy usage such as boiler feed water generation, cooling water, and condensate return from steam heaters and turbines. Process energy generation mainly comes from process steam generation, and power generation from process pressure reduction.

14.2.1 The Concept of Fuel Equivalent for Steam and Power (FE)

There is an issue yet to be resolved for the energy intensity defined in Equations 14.3 and 14.4. The energy use (E) for a process consists of fuel, steam, and electricity. They are not additive because they are different in energy forms and quality. However, if these energy forms can be traced back to fuel fired at the source of generation, which is the meaning of fuel equivalent, they can be compared on the same basis, which is fuel. In other words, they can be added or subtracted after converted to fuel equivalent. For simplicity of discussions, definitions of fuel equivalent (FE) for different energy forms are given in Zhu (2014), while examples of FE calculations are provided below.

In general, fuel equivalent (FE) can be defined as amount of fuel fired (Q_{fuel}) at the source to make a certain amount of utility (G_i):

$$\text{FE}_i = \frac{Q_{\text{fuel}}}{G_i} \quad i \in (\text{fuel}, \text{steam}, \text{power}) \tag{14.5}$$

In most cases, Q_{fuel} is calculated based on the lower heating value of fuel. G_i is quantified in different units according to specifications in the marketplace, namely, Btu/hr for fuel, lb/hr for steam, and kWh for power. Thus specific FE factors can be developed as follows based on this general definition of fuel equivalent.

14.2.1.1 FE Factors for Fuel By default, fuel is the energy source. No matter what different fuels are used, tracing back to itself makes "fuel equivalent for fuel" equal to unity, that is,

$$\mathrm{FE}_{\mathrm{fuel}} = \frac{Q_{\mathrm{fuel@source}}}{G_{\mathrm{fuel}}} \equiv 1\ \mathrm{Btu/Btu} \tag{14.6}$$

14.2.1.2 FE Factors for Steam A typical process plant has multiple steam headers, typically designated as high pressure, medium pressure, and low pressure. In some cases, very high pressure steam is generated in boilers, which is mainly used for power generation. For calculating fuel equivalent of steam, a top-down approach is adopted starting from steam generators. The total FE for each steam header is the summation of all FEs entering the steam header via different steam flow paths, which include steam generated from on-purpose boilers and waste heat boilers, steam from turbine exhaust, steam from pressure letdown valves, etc. The FE for each steam header is the total FE divided by the amount of steam generated from this header, that is,

$$\mathrm{FE}_i = \frac{Q_{\mathrm{fuel}}}{G_i} = \left.\frac{\text{Total FE consumed}}{\text{Total steam generated}}\right|_i \text{header} \quad i \in (\mathrm{HP}, \mathrm{MP}, \mathrm{LP})\ \mathrm{kBtu/lb} \tag{14.7}$$

14.2.1.3 FE Factors for Power For power, $\mathrm{FE}_{\mathrm{power}}$ is expressed as

$$\mathrm{FE}_{\mathrm{power}} = \frac{Q_{\mathrm{fuel}}(\mathrm{Btu/hr})}{Q_{\mathrm{power}}(\mathrm{Btu/hr})} = \frac{1}{\eta_{\mathrm{cycle}}}\ \mathrm{Btu/Btu} \tag{14.8}$$

where η_{cycle} is the cycle efficiency of power generation and Q_{power} represents the amount of energy associated with power in unit of Btu/hr.

By using the conversion factor of 1 kW = 3,414 Btu/hr, Equation 14.8 is converted to

$$\mathrm{FE}_{\mathrm{power}} = \frac{1}{\eta_{\mathrm{cycle}}}(\mathrm{Btu/Btu}) \times 3{,}414(\mathrm{Btu/kWh}) = \frac{3{,}414}{\eta_{\mathrm{cycle}}}\ \mathrm{Btu/kWh} \tag{14.9}$$

Equation 14.9 can be generally applied to different scenarios for power supply such as power import and on-site power generation from backpressure and condensing steam turbines as well as from gas turbines.

14.2.1.4 Energy Intensity Based on FE By converting different energy forms to fuel equivalent, process energy intensity on feed basis in Equation 14.3 can be revised to give

$$I_{\text{process}} = \frac{\text{FE}}{F} \text{ Btu/unit of feed} \tag{14.10}$$

Or process energy intensity on product basis in Equation 14.4 can be revised to give

$$I_{\text{process}} = \frac{\text{FE}}{P} \text{ Btu/unit of product} \tag{14.11}$$

where FE is the total fuel equivalent as summation of individual fuel equivalent for different energy forms across the process battery limit and P is the most desirable product generated from the process.

Let us walk through calculation of process energy intensity via an example.

14.2.2 Calculate Energy Intensity for a Process

In industry, steam is measured in mass flow, while fuel in volumetric flow and electricity in electrical current. To compare them on the same basis, all the energy use and generation need to be traced back to fuel fired at the source of energy generation in order to obtain fuel equivalent, which is a cardinal rule for energy benchmarking calculations.

For energy benchmarking of a process unit, the important thing is to identify the main energy consumers and provide a reasonable estimate for missing data. Going overboard to collect miniature details and chase utmost precision should be avoided. Doing so may be wasted effort because such fine details are most likely not needed in the benchmarking calculations and will not make meaningful impact on energy optimization.

Example 14.1 Calculate process energy intensity for the existing aromatics complex design where paraxylene is the desirable product. The complex consists of six process sections, namely, aromatics extraction, benzene–toluene fractionation, xylene fractionation, transalkylation, xylene isomerization, and p-xylene selective adsorption. The last row of the table shows energy use on fuel equivalent (FE) basis. We will explain how FE usages are calculated below.

Table 14.1 gives an example for the relevant data needed at this stage for establishing the process energy balance and calculating energy performance. As a general guideline, the fuel produced from a process unit, in the forms of fuel gas, vent gas, and fuel oil, should not be included in the energy balance for the unit. The value of vent gas and liquid slop streams is kept separate from the energy balance because these streams may be processed further to valuable byproducts instead of burned for heating value only.

Assumptions: Firstly, related fuel equivalent factors need to be obtained as below, and the basis for deriving these assumptions will be explained later. Assumed fuel equivalent (FE) factors for this example are as:

- FE for purchased power = 9.09 MMBtu·FE/MWe
- FE for MP steam = 1,350 Btu·FE/lb
- FE for LP steam = 1,200 Btu·FE/lb
- FE for condensate = 150 Btu·FE/lb
- FE for BFW at 221 °F (105 °C) = 177 Btu·FE/lb
- FE for CW at 90 °F (32 °C) = 5 Btu·FE/lb
- FE for standard (std.) oil = 39.7 MMBtu·FE/Metric Ton std. oil

Convert energy inputs and outputs to fuel equivalent:

- FE for power usage = −20.5 MW × 9.09 MMBtu·FE/MWh = −186.8 MMBtu·FE/hr
- FE for MP steam export = 5.2 klb/hr × 1.35 MMBtu·FE/klb = 7.0 MMBtu·FE/hr
- FE for LP stem usage = −26.4 klb/hr × 1.2 MMBtu·FE/klb = −31.6 MMBtu·FE/hr
- FE for BFW usage = −16.3 klb/hr × 0.177 MMBtu·FE/klb = −2.9 MMBtu·FE/hr
- FE for condensate return = 23.4 klb/hr × 0.15 MMBtu·FE/klb = 3.5 MMBtu·FE/hr
- FE for cooling water usage = −3,811.1 klb/hr × 0.005 MMBtu·FE/klb = −19.1 MMBtu·FE/hr
- FE for fuel = −832.7 MMBtu/hr × 1.0 MMBtu FE/MMBtu-fuel = −832.7 MMBtu·FE/hr
- Net total FE for the complex = summation of all the above = −1,062.5 MMBtu·FE/hr

Let us define specific energy use based on the energy intensity:

$$\text{Specific energy} = \frac{\text{Net energy input}(\text{FE})}{\text{Product rate}} \tag{14.12}$$

Assume on-stream utilization as 95%; thus hours on-stream per year is 8,322 and metric ton (Ton) p-xylene production on-stream per year is equal to 1.2 MM·Ton/year/8,322 = 144.2 Ton/hr.

Applying Equation 14.12 yields

$$\text{Specific energy use for the complex} = 1{,}062.5\ \text{MMBtu} \cdot \text{FE/hr}/144.2\ \text{Ton}\cdot\text{p-xylene/hr}$$
$$= 7.37\ \text{MMBtu}\cdot\text{FE/Ton}\cdot\text{p-xylene}$$

Alternatively, specific energy use can be expressed based on standard oil equivalent (SOE) as below:

$$\text{Specific energy usage for the complex} = 1{,}062.5\ \text{MMBtu} \cdot \text{FE/hr}$$
$$/\,[39.7\ \text{MMBtu} \cdot \text{FE/Ton std} - \text{oil} \times 1{,}000\ \text{kg}\cdot\text{SOE}$$
$$/\ \text{Ton std} - \text{oil}]/144.2\ \text{Ton} \cdot \text{p-xylene/hr}$$
$$= 185.6\ \text{kg SOE/Ton} \cdot \text{p-xylene}$$

Specific energy use is a very insightful concept as it represents the *energy intensity of production* indicated by the amount of energy required for producing one unit of product. Let us see how to apply this concept that is discussed below.

Example 14.2 As an exercise, you may calculate the operational specific energy usage for the above process based on the FE data given above, and the operation data are given in Table 14.2. For reference, specific energy usage for this example = 7.72 MMBtu/Ton product (equivalent to 194.3 kg SOE/Ton product).

14.2.3 Energy Intensity for Total Site

The structure of energy intensity indicators can be organized in a hierarchical manner. That is, intensity indicators are developed for processes first and then toward a total site. One may question why the concept of energy intensity does not apply to process sections (e.g., reaction section, product fractionation section) of a process. The reason is that a process section could have high energy intensity if this section exports energy to other sections in a process. In contrast, a section could have very low energy intensity if it receives energy from other sections. Thus, when there is strong heat integration between sections of a process unit, energy intensity for sections cannot fairly represent section energy performance. In this case, section energy intensity loses its meaning of measuring. Energy transfer across process units could occur, but the chance is much slim compared with between-process sections. In case of heat transfer between processes, some adjustment must be made to account for it.

TABLE 14.1. Example Data Set for Energy Use

Process Sections (Total pX Product = 1.2 MM·Ton/Year)	Electric Duty (MW)	MP 150 psig (klb/hr)	LP 60 psig (klb/hr)	BFW at 221 °F (klb/hr)	Condensate 3 psia (klb/hr)	Cooling Water at 90 °F (klb/hr)	Fuel Fired Duty (MMBtu/hr)	
ED Sulfolane™	−0.4	0.0	−0.3	0.0	0.3	−2,310.1	0.0	
BT Frac	−0.6	0.0	−1.0	0.0	1.2	−68.3	0.0	
Tatoray™	−2.4	0.0	−1.6	0.0	1.8	−524.2	−35.4	
Xylene Frac	−3.2	−13.1	−4.2	−0.5	0.0	−185.4	−190.3	
Parex™	−12.3	18.2	−17.4	−16.2	18.6	−714.4	−532.9	
Isomar™	−1.8	0.0	−1.9	0.4	1.5	−8.5	−74.0	
Net total energy usage	−20.5	5.2	−26.4	−16.3	23.4	−3,811.1	−832.7	
Multiply FE factors	9.09 MMBtu/ MWe	1.35 MMBtu/ klb MP	1.2 MMBtu/ klb LP	0.177 MMBtu/ klb BFW	0.15 MMBtu/ klb Cond.	0.005 MMBtu/ klb CW	1.0 MMBtu/ MMBtu Fuel	Grand total
Total FE use, MMBtu/hr	−186.8	7.0	−31.6	−2.9	3.5	−19.1	−832.7	−1,062.5

Note: A positive value indicates quantity produced. A negative value (−) indicates quantity consumed.

TABLE 14.2. Process Energy Use Data for Example 14.2

Process Sections (Total pX Product = 1.2 MM·Ton/Year)	Electric Duty (MW)	MP 150 psig (klb/hr)	LP 60 psig (klb/hr)	BFW at 221 °F (klb/hr)	Condensate 3 psia (klb/hr)	Cooling Water at 90 °F (klb/hr)	Fuel Fired Duty (MMBtu/hr)
ED Sulfolane	−0.4	0.0	−0.3	0.0	0.3	−2,310.1	0.0
BT Frac	−0.6	0.0	−1.0	0.0	1.2	−68.3	0.0
Tatoray	−2.9	0.0	−1.6	0.0	1.8	−524.2	−41.0
Xylene Frac	−3.7	−13.1	−4.2	−0.5	0.0	−185.4	−196.2
Parex	−12.0	18.2	−17.4	−16.2	18.6	−714.4	−544.1
Isomar	−1.9	0.0	−1.9	0.4	1.5	−8.5	−93.5
Net total energy usage	−21.4	5.2	−26.4	−16.3	23.4	−3,811.1	−874.8

Note: A positive value indicates quantity produced. A negative value (−) indicates quantity consumed.

To calculate the energy intensity index for the whole site, *aggregate* energy intensity could be defined simply as the ratio of total energy in fuel equivalent divided by total feed rate, which is

$$I_{\text{site}} = \frac{\sum_i \text{FE}_i}{\sum_i F_i} \tag{14.13}$$

where FE_i is the fuel equivalent across the site battery limit.

However, there is a problem here with this simple aggregate approach: although energy in fuel equivalent is additive, feeds (F) are not because processes usually have different feeds with very different compositions. In other words, the problem with Equation 14.13 is the dissimilarity in feeds that cannot be added without adjustment.

To overcome this dissimilarity problem in feeds, we could think of a reference site with energy intensity indicators for each process known in prior. Thus, the total amount energy use could be calculated for the reference site, which can be compared with the actual energy use for the site of interest. Let us derive the mathematic expressions along the line of this thought.

When the energy intensity for reference process is known or can be calculated, applying Equation 14.10 gives the energy use for a reference process as

$$\text{FE}_{i,\text{ref}} = I_{i,\text{ref}} F_{i,\text{ref}} \tag{14.14}$$

Since FE is additive, the total energy use for the reference site is

$$\text{FE}_{\text{site,ref}} = \sum_i \text{FE}_{i,\text{ref}} = \sum_i I_{i,\text{ref}} F_{i,\text{ref}} \tag{14.15}$$

Then, an intensity index can be defined as the ratio of actual and reference energy use:

$$I_{\text{site}} = \frac{\text{FE}_{\text{site,actual}}}{\text{FE}_{\text{site,ref}}} \tag{14.16}$$

The total energy use for a site is the net energy input and output, which can be readily calculated from the battery limit of the site. It must be noted that a total site energy use consists of all processes, boiler house(s), off-site utility for material transportation, and tanks farms heating.

You may ask a critical question: a real process could differ from the reference process in terms of feed rates and process conditions. How can we deal with these differences in the energy intensity index calculations? This question can be addressed by defining the intensity as a function of three major factors as

$$I_{i,\text{ref}} = f_{i,\text{ref}}(\text{design,conditions,weather}) \tag{14.17}$$

where design, conditions, and weather reflect the actual process. In this way, Equation 14.17 describes the energy performance for the reference processes with the same attributes as the actual processes, but the energy intensity could be different. This is because the energy intensity in Equation 14.17 for a reference processes is developed based on peers' performance, while the energy intensity for actual processes is calculated based on real data.

The simplest form is linear expression. For example, if two operating parameters are considered, the linear form becomes

$$I_{i,\text{ref}} = a + bx_1 + cx_2 + d\left(T_{\text{ambient}}\right) \tag{14.18}$$

where a is a structural term that reflects the design performance while b and c are the sensitivity factors for x_1 (process condition 1) and x_2 (process condition 2), respectively; d is the correction factor for weather; and T_{ambient} is the ambient temperature in local area.

14.2.4 Remarks for Energy Intensity

Decline in energy intensity is a proxy for efficiency improvement, but energy intensity reflects production and hence is much more universal and communicable across the process industry.

Clearly, structural and operational changes for efficiency improvements in processes and equipment can contribute to reduction in process energy intensity in a big way. A state-of-the-art process gives low energy intensity by design. However, it could end up with high operating energy intensity if the process is poorly operated. On the other hand, a poorly designed process could achieve its best potential if it is operated with diligence. But diligent operation could reach the limit since the performance is handicapped due to inherently inefficient design. To improve the process beyond this design limitation, structural changes must be made.

14.3 ENERGY BENCHMARKING FOR PROCESSES

The most important result of energy benchmarking is the indication of energy intensity for individual processes in the context of the best alternative. The energy intensity is then used with a performance target as defined by a corporate goal, or industrial peer performance, to determine the process energy performance in comparison with targeted performance. In general, benchmarking assessment can give several indications:

- The need for an overall energy optimization effort: If large gaps are available for majority of the process units, this could imply there are many opportunities available and require coordinated effort across the plant. A dedicated energy team may need to be established to identify and capture the opportunities.

- Areas for focus: Process units with large performance gaps can be selected as focus areas. This allows us to effectively concentrate efforts on the areas with the greatest potential for improvement. Specialists may need to be assembled to form a project team for individual process units.
- Update targets: If all major process units are performing well relative to the targets, the plant may concentrate efforts on continuous improvements via monitoring and control.

The use of the best alternative as the basis for assessing process energy efficiency is discussed in the following.

14.3.1 Benchmarking: Based on Energy Performance Index (EPI)

This FE calculation method is built on the concept of guideline energy performance, which is used as a benchmark against which actual energy performance is compared. The rationale of using this concept as the basis for assessing process energy efficiency is revealed in the following.

Let a ratio of actual energy performance (AEP) and guideline energy performance (GEP) be defined as below. This ratio shall be labeled as the energy performance index (EPI):

$$\text{EPI} = \frac{\text{Actual energy performance}}{\text{Guideline energy performance}} = \frac{\text{AEP}}{\text{GEP}} \tag{14.19}$$

By definition, EPI represents the energy efficiency for the process unit on the basis of GEP. In this way, any improvements in operation, design, equipment, and technology upgrade can be measured using EPI.

Generally speaking, an EPI gap of less than 5% between AEP and GEP belongs to an operational gap. In other words, better operating practices and control could close this gap. An EPI gap of 5-10% may require small energy retrofit projects, which can feature a quick payback, for the gap to be closed. If the EPI gap is on the order of 15+%, it may require significant energy and process retrofit projects to close the gap.

14.3.2 Benchmarking: Based on the Best-in-Operation Energy Performance (OEP)

By applying the method for calculating specific energy use, you can obtain a plot of specific energy versus time based on the historic data. This plot can pinpoint the best-in-operation energy performance (OEP) that your process unit has achieved at a time when there was institutionally dedicated effort for operation performance and with technical know-how available. You could confirm this by talking to engineers and operators who have worked in the plant during this period. As a result, you will be able to determine the OEP as the energy guideline performance representing the best-in-plant performance.

Assume the specific energy use based on OEP is the same as the design performance of 185.6 kg SOE/Ton p-xylene as calculated in Example 14.1. With the actual energy use of 194.3 kg·SOE/Ton product calculated from Example 14.2, EPI for the process unit can be calculated via

$$\text{EPI} = \frac{\text{AEP}}{\text{OEP}} \times 100\% = \frac{194.3}{185.6} \times 100\% = 105\% \tag{14.20}$$

Equation 14.20 indicates that AEP has a total energy consumption of 5% higher than the OEP, which indicates poor operation performance. Such a gap is significant, which should alert you to initiate investigations for root causes. The mere factor of determining EPI gives you an immediate indicator as to where your process unit stands in energy performance so that you can quickly spot problematic areas.

14.3.3 Benchmarking: Based on Industrial Peers' Energy Performance (PEP)

In industry, there are peer survey groups organized based on industrial sectors and process technology. Organizers for the survey groups send questionnaires to survey members to gather sample data on yearly and conduct performance calculations. Consequently, the peer performance results are shared among survey members. If your plant belongs to the survey group, you could obtain the best peer energy performance (PEP) via the representative in your organization. For certain large companies, there are CoP (community of practice) networks based on process technology. You should seek out the best PEP for your process unit via the CoP in your company.

Assume the operation is improved for the process and the energy use is reduced by 5% from 194.3 to 185.6 kg·SOE/Ton product. Also assume that the specific energy for PEP is 170 kg·SOE/Ton product for this example. Based on the actual energy use of 185.6 kg·SOE/Ton product, we can calculate EPI for the process unit as

$$\text{EPI} = \frac{\text{AEP}}{\text{PEP}} \times 100\% = \frac{185.6}{170} \times 100\% = 109.2\% \tag{14.21}$$

This indicate that the current energy consumption in the plant is 9.2% higher than the best peer performance, which is significant and requires effort to get it down in order to stay in the same peer group. Usually, the survey group is divided into tiered performance structure such as first, second, third, and fourth quartiles. Based on the energy performance index calculated above, you can find out which performance quartile your process unit belongs to. This indicates where your process unit stands among your peers, which may motivate your management to consider to revamp the process.

14.3.4 Benchmarking: Based on the Best Technology Energy Performance (TEP)

With technology advancement in catalyst, equipment, process design, and control, process energy efficiency could improve. It is not difficult to gather the performance data for state-of-the-art technology. In some cases, the data is published in public by government offices, and you could find them via web search. If not available in public, you can contact technology companies—they are often eager to provide the data to customers.

The plant management is interested to know the scope of further energy improvement by applying better process technology. Assume the TEP is 155 kg·SOE/Ton product. Thus the EPI for the process unit can be calculated by

$$\text{EPI} = \frac{\text{AEP}}{\text{TEP}} \times 100\% = \frac{185.6}{155} \times 100\% = 120\% \tag{14.22}$$

Technology updates can improve the energy performance, further by 20% on top of 5% operation improvement, but technology improvement usually requires high capital costs and long implementation periods. Justification of technology improvement may mainly come from combined benefits from throughput increase, yield, and energy improvement.

14.3.5 Remarks for Energy Benchmarking

The energy benchmarking methods discussed above make it a much simpler yet effective task for assessing the energy performance for a process unit, which requires minimal data. At this point, we have a very good starting point. You know three essential facts: the energy intensity for your process unit, the performance target (GEP), and the gap against the target. Your mind may be racing with questions like: What has gone wrong with my process unit? What areas to focus and what takes to reach the GEP target? These questions will be answered below via applying key indicators.

14.4 THE CONCEPT OF KEY INDICATORS

The rationale of introducing key energy indicators (KEI) is to seek answers to this critical question: "How can engineers characterize energy use in a process unit with an emphasis on *major energy users* in terms of *their needs and the reasons and practical ways to minimize energy use for the needs*?" Application of the KEI in reality follows a methodology based on three steps: defining key indicators, setting targets, and identifying actions to close gaps. Using this methodology, a process unit can be described by a small number of key indicators to measure energy performance, which can be developed based on process knowledge and experience. Application of key indicators will allow focus on important issues and avoid falling into trap of details.

14.4.1 Define Key Indicators

The above discussions reveal the fact that major improvement opportunities can be captured by paying attention to key operating parameters. This is the rational of introducing key indicators. The method for determining key indicators follows a basic thought process: understand the process objectives ➔ understand energy needs ➔ develop measures for the needs ➔ define the key indicators representing the measures. As an example of how to define key indicators for process, let us consider an ethylbenzene dealkylation type of xylene isomerization unit in an aromatics complex.

The flow scheme as shown in Figure 14.1 is a commonly used xylene isomerization unit (such as the UOP Isomar process), converting a raffinate stream from an upstream paraxylene separation unit that is depleted in paraxylene to a product that is a near equilibrium mixture of xylene isomers. In the most common style of this process unit, ethylbenzene in the feed is converted to benzene and ethane.

To reduce the charge heater duty, the reactor effluent stream is heat-exchanged against both feed and recycle gas to recover process heat in a combined feed exchanger (CFE). Afterward, reactor effluent is sent to product separator where the non-condensable vapors that remain after the stream passes through the CFE and condenser are rejected to a recycle gas stream and liquid is sent onward to fractionation. In more recent designs, two separators may be used—the first is a hot separator that is placed upstream of the condenser, and the second is a cold separator placed downstream of the condenser. This reduces the duty requirement of the condenser, as already condensed liquid from the CFE does not need to pass through the condenser itself. The fractionator, a deheptanizer column in this case, is partially condensing. Off-gases are vented, as the feed to the column will still contain dissolved light gases such as hydrogen and ethane (and small amounts of other light gases that will have

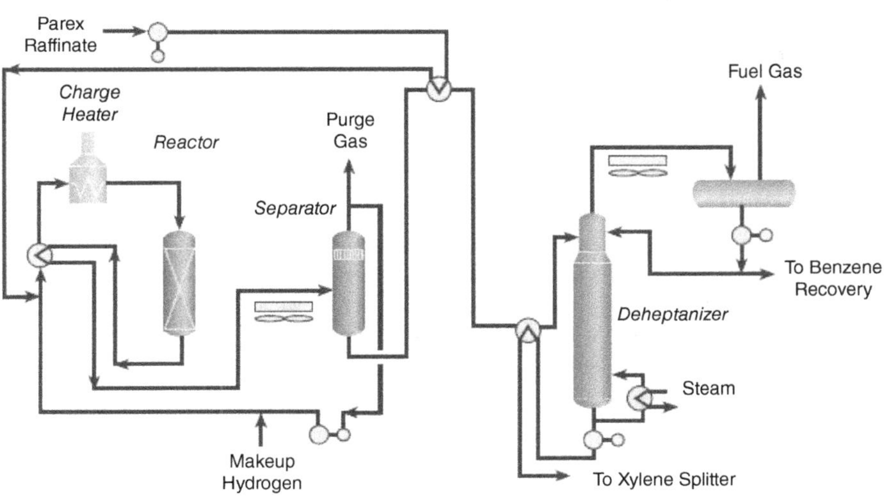

FIGURE 14.1. Typical xylene isomerization unit, example from a UOP aromatics complex.

entered with the reactor section makeup gas such as methane and propane). The overhead goes to a light aromatics recovery system, first typically a stripper column to remove light non-aromatics and any remaining dissolved light gases and then an aromatics extraction unit, with benzene as the main product from a downstream benzene–toluene fractionation unit. The fractionation column bottoms liquid is recycled back to the xylene fractionation unit upstream of the paraxylene separation unit.

14.4.2 Simplifying the Problem

To simplify the overall task of defining key indicators for the whole unit, the unit can be divided into two main sections, namely, reactor section and fractionation section. The goal is to define a set of key indicators for each section, which can be used for monitoring and optimization. The explanation below may be a little wordy, but it is necessary to illustrate the thinking process behind the key indicator (KI) concept. You may apply this thinking process for defining KI's for other process of interest.

14.4.3 Developing Key Indicators for the Reaction Section

14.4.3.1 Understanding the Process In this example, the primary process objective is actually twofold—conversion of feedstocks to an equilibrium mixture of xylenes and conversion of most of the ethylbenzene in the feed to benzene and ethane. The equilibrium mixture of product xylenes will typically contain around 23.5–24.0 wt% paraxylene, compared with a feed that contains less than 1 wt%. The ethylbenzene conversion is typically around 60–80 wt%, with the feed ethylbenzene content typically around 2–15 wt%. The target ethylbenzene conversion and concentration in the feed are heavily dependent on the complex processing objectives and feed sources, respectively. The reactor contains a catalyst that is intended to maximize production of the desired products, with minimal side reactions. The energy efficiency for this unit will largely depend on how effectively the reaction effluent heat is recovered and how well the side reactions can be minimized.

14.4.3.2 Understanding the Energy Needs The following items are identified as major energy users in the reactor circuit and their distinct roles and significances will be discussed.

The feed heater is used to increase the feed temperature and control the reactor inlet temperature. Although the heater efficiency depends on how it is designed and operated, the heater duty is determined by feed preheating. For a given feed preheat, the heater duty is mainly a function of the heat of reaction and heat recovery. Xylene isomerization processes typically have a low exotherm (5–10 °C is typical). A process engineer can determine the ways to maintain process heat recovery, heater efficiency, and heat flux.

The compressor for recycle gas is a large power user. The role of a recycle gas compressor is to provide required amount of hydrogen to the reaction and to provide catalyst stability and selectivity in the presence of hydrogen. The recycle compression work depends on the gas flow and its molecular weight. The gas flow depends on the hydrogen purity in the recycle gas and pressure of the reactor section. The role of a process engineer is to optimize the recycle gas rate, purity, and hydrogen to hydrocarbon (H_2/HC) ratio. The optimization may indicate to operate the compressor at the lowest possible recycle gas rate, purity, and H_2/HC that will still provide good catalyst stability and selectivity.

14.4.3.3 Effective Measures for the Energy Needs Based on the above understanding of the energy needs, we can go one step further to develop efficiency measures in providing these needs:

- Reactor activity: Operating at a higher reactor temperature will result in higher ethylbenzene conversion. As the catalyst deactivates, the temperature requirement to maintain ethylbenzene conversion will increase. The xylene isomerization function is relatively constant (though there is a minor shift in the equilibrium concentration as temperatures change) over the typical temperature window and life of the catalyst. Operating at higher severity will result in shorter catalyst life, higher hydrogen consumption, and higher side reactions. Ethylbenzene conversion is the standard measure of severity for these units.
- Heater reliability: Heat flux (for large heaters) or tube wall temperature (for small heaters) is the key reliability parameter for a heater. When heater operation is higher than the limit of heat flux or TWT, a heater is under risk of reliability failure because the tube life is shortened at higher fluxes. On the other hand, operating a heater at much lower than the limit makes the heater underutilized, which represents an opportunity for increased feed rate or higher process severity. Besides these two operating limits, flame impingement is another reliability measure that is usually caused by too low O_2 content. The combustion flame becomes longer with too little O_2 and could reach the tube and pose a serious reliability risk.
- Heater efficiency: Efficiency can be affected by the excess O_2 content or extra air for combustion and a high stack temperature. Inappropriate O_2 content could be caused by lack of control, air leaks, and poor burner performance. A high stack temperature corresponds to high heat loss in the flue gas. A high heater approach temperature, defined as the temperature difference between flue gas to the stack and heater feed inlet, could be caused by heater fouling in operation and by poor or outdated heater design. Xylene isomerization is generally considered a non-fouling service, as such heater fouling here is unlikely.
- Feed exchange: The feed is mainly heated by the reactor effluent in feed exchangers before the charge heater. The hot-end approach temperature on feed exchangers is a good indication of heat recovery performance by the feed preheating system.
- Reactor product condenser inlet temperature: After the reactor effluent transfers its heat to the feed and recycle gas, it goes to an air condenser. Thus, the condenser inlet temperature on the reactor effluent side is a good indication of how effectively the reactor effluent heat has been recovered.

- H_2/HC ratio: This ratio defines the amount of hydrogen in the recycle gas compared with the amount of hydrocarbon in the feed on a molar basis. Too high of a H_2/HC ratio could cause greater power usage due to the higher recycle gas rate, while too low of a ratio could impact yield and shorten the catalyst life. The allowable H_2/HC for xylene isomerization reactors is getting progressively lower throughout the industry, as catalyst manufactures are able to design more stable catalysts. This reduces the required compressor power requirement for newer units. The lower hydrogen requirement of new catalysts is particularly helpful for a unit when higher throughput is desired and the compressor is thought to be a key equipment limitation.

*14.4.3.4 **Developing Key Indicators for the Energy Needs*** Through the above exercise of simplifying the problem and developing understanding of major energy needs and measures of efficiency in providing the needs, we can define the following key indicators for the reaction section:

- Ethylbenzene conversion
- H_2/HC ratio
- CFE hot-end approach temperature
- Product condenser inlet temperature

Specific indicators for heaters could include:

- Heater O_2 content
- Heater stack temperature
- Heat flux
- Flame impingement

14.4.4 Developing Key Indicators for the Product Fractionation Section

*14.4.4.1 **Understand Process Characteristics*** The deheptanizer is intended to remove toluene and lighter components from the bottoms product of the column. The bottoms product should be primarily C_8 aromatics and heavier, which will be further reprocessed in a column upstream of the paraxylene separation unit to remove heavies. It is okay to let small amounts of toluene in the C_{8A+} fraction, but most of it should be recovered overhead as it will increase the load on the paraxylene separation unit.

The overhead aromatics product, mostly benzene and toluene, is typically taken as a distillate product, but may be also taken as a side draw near the top of the column. This product should be mostly devoid of C_8 aromatics. This is important to maintain proper function of downstream units, in particular the aromatics extraction unit. For sake of storage tanks, the RVP of this stream needs to be controlled. As such, a stabilizer column is often used as well, though it is not always within the scope of the xylene isomerization unit itself.

It is essential to avoid flooding and/or slumping this column, which would severely impair its separation capability and thus energy efficiency. Fractionation efficiency can be monitored by the column internal vapor to liquid ratio (V/L). The ideal V/L can be achieved jointly by optimizing the reboiler heat input and overhead composition control scheme. This should be done to maintain good vapor–liquid contact. Operating at slightly below the flood point is a good practice, though this is not always straightforward to measure. An optimal reflux to feed ratio should be determined, typically the lowest possible that achieves the desired splits. This will generally reduce heat input per amount of feed processed.

Column pressure also has an impact on fractionation efficiency. While lower pressure can allow for better separation of the key components, reducing column pressure will bring the column closer to its flood point. Higher pressure will increase the column temperatures, reducing the LMTD in the reboiler, making heat exchange more difficult. There may also be considerations for the destination of the column off-gas that set a minimum allowable column pressure.

14.4.4.2 Understand the Energy Needs Heat input is provided to the deheptanizer column to facilitate separation of benzene and toluene overhead, from the main bottoms product of C_{8+} aromatics. The main role of the process engineer is to determine the appropriate heat input and R/F ratio that will allow for good rejection of C_8 aromatics from the overhead and toluene from the bottoms product. The overhead product is typically controlled by a temperature controller tied to a column tray temperature. Increasing or decreasing this temperature will increase or decrease the amount of xylenes allowed into the distillate product, respectively. Higher heat input will generally result in a better separation, but the separation typically does not need to be perfect. Further, very high input can result in column flooding.

If an additional stabilizer exists, its purpose is to remove light non-aromatics from the light aromatics product stream from the deheptanizer overhead. This is frequently required to suit vapor pressure requirements for downstream destinations of this stream or to reduce the likelihood of foaming in a downstream extraction unit.

14.4.4.3 Effective Measure for the Energy Needs Fractionation efficiency can be determined by the minimum R/F ratio that allows for the target values of the key component splits in the deheptanizer distillate and bottoms liquid product streams.

Heat input for each column should be measured as heat transferred in the reboilers for the deheptanizer and stabilizer (if applicable) columns. It is easier to strictly measure the flow rate of the reboiler heating medium, for instance, a condensate return flow rate, but this is not always a suitable proxy for heat input. The conditions of the heating medium (i.e., temperature and pressure) may be different, changing the heat available for exchange. For this reason, the actual heat input should be measured based on inlet and outlet conditions of the hot and cold sides of the reboiler exchangers.

14.4.4.4 Developing Key Indicators for the Energy Needs in the Fractionation System Based on the understanding of major energy needs and measures of efficiency in providing the needs, we can define the following key indicators for the fractionation section:

- Fractionator column reflux ratio(s)
- Reboiler heat input(s)
- Column pressure(s)

14.4.5 Remarks for the Key Indicators Developed

For a typical xylene isomerization unit, there could be several thousands of data points measured and collected. The key indicators above only account for a very small fraction of the overall data available but capture the key performance parameters that ultimately reflect directly on operating costs. If these indicators can be monitored and optimized, the unit can operate closer to optimal performance.

In many cases, parameters related to feed and product yields are measured and controlled using basic control systems or advanced process control (APC) systems. However, process indicators are typically not integrated with energy use. Furthermore, many energy parameters are not even measured, much less recorded. By identifying the key process and energy indicators and optimizing them together, the optimization does not only reduce energy cost but may also allow for an increase in throughput, improve product quality, and minimize product specification giveaway.

14.5 SET UP TARGETS FOR KEY INDICATORS

To improve from current performance, targets must be established for the key indicators. These targets provide standard against which existing facilities are measured and equipment improvements are evaluated. The difference between a target and the current performance for each key indicator defines the performance gap. Each performance gap should be associated with dollar value that represents opportunity to be captured. Each indicator is correlated to a number of parameters, including process and equipment conditions together with equipment limits. In this way, energy optimization is connected with process conditions and constraints.

How would one make the concept of KEI work for a process unit or the complex as a whole? Consider the example of the deheptanizer for the xylene isomerization unit previously discussed.

14.5.1 Problem

The deheptanizer column sketch is shown in Figure 14.2. The reboiling at the bottom of the tower is to provide sufficient vapor flow on the trays for separating toluene and lighter components from the C_8 aromatic and heavier components in the feed. Most of the toluene and lighter boiling components will be withdrawn at the overhead, while the C_8 aromatics and heavier components leave at the bottom. A certain amount of toluene in the bottom product is allowed as a component of energy optimization for the complex. Too much toluene will limit the paraxylene separation unit, but

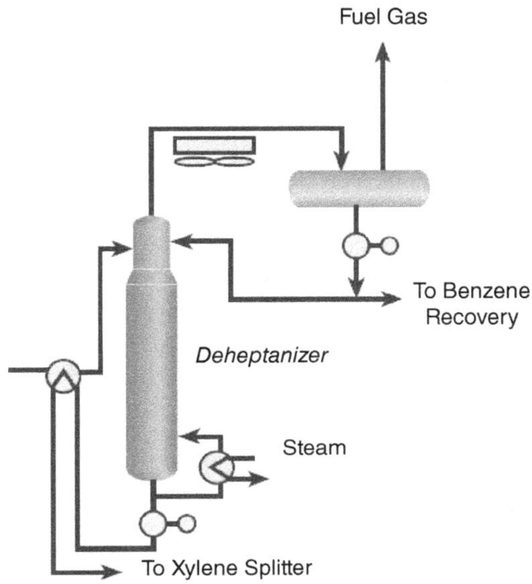

FIGURE 14.2. Deheptanizer column in a xylene isomerization unit.

trying to remove too much will require more heat input to the deheptanizer than it is worth. Reboiling duty is the main variable in controlling the toluene amount in the bottoms product. In other words, the reboiling duty must increase when toluene amount in the bottom exceeds the target. The operating objective is to minimize the reboiling duty while achieving the maximum amount of toluene that the downstream units can process. This typically optimizes to less than 1 wt% in the bottoms product.

14.5.2 Rationale

The task at hand is to develop a relationship between reboiling duty and toluene in the deheptanizer bottoms product so that the toluene content in weight percent can be controlled by adjusting the reboiling duty. However, other operating parameters also affect the reboiling duty, which include column feed conditions, feed enthalpy, and tower conditions. If a correlation of reboiling duty against the above influencing parameters could be generated, reboiling duty can be adjusted according to any of the changes in the related parameters and thus avoid the need of trial and error.

14.5.3 Solution

There are several ways to develop such a correlation. The simplest way is by use of a data historian. This method can be applied if three conditions are met:

(a) The related parameters are measured and data available in the historian.
(b) The measured data reflects the operation at the time the toluene content was measured.
(c) The historian data covers all possible operating scenarios. After all, online data is the true representation of real simulation!

Development of a correlation using the historian data can be conducted readily in a spreadsheet using regression techniques. After gathering the data from the historian, multiple-variable regression can be applied to develop such a correlation. The overall correlation coefficient must be higher than 85% for sufficient regression fidelity.

The second option is to use the step-test method usually used for developing a parametric relationship for control systems. By making a small step change to the independent variable of interest, a response from the dependent control variable (in this case, reboiling duty) can be recorded after reaching to the steady-state condition. This response can be called an energy response. Finally the regression method is applied to derive the correlation of reboiling duty against all related variables.

In many cases, the conditions above for using the data historian are difficult to satisfy. Also, it could be labor intensive and inconvenient in operation to adopt the step-test method. Thus, the most common method is to use the simulation method for developing relationship correlations. To do this, a simulation model for the tower can be developed readily based on the feed (rate and compositions) and tower conditions (temperature, pressure, theoretical trays), with product specifications (toluene weight % in the bottom and C_8 aromatic weight % in the overhead) established as setpoints in the simulation. Operating parameters such as reflux rate and reboiling duty can be adjusted to meet the product specifications. The simulation model is verified and revised against high-quality performance test data.

For evaluating the effect of individual parameters, simulation cases can be developed by prespecifying the values for independent variables of interest, and the energy response (reboiling duty) will be recorded automatically. For example, to evaluate the effect of feed preheating, the UA value of the feed preheating exchanger is varied with prespecified values (U = overall heat transfer coefficient; A = heat exchanger surface area). During simulation runs, the feed temperature before the tower will change according to the UA values, which will cause the reboiling duty to vary automatically in the simulation. A set of four curves for such one-to-one relationships can be obtained as shown in Figures 14.3a–d. Brief explanations are given below for each figure.

Reducing the column overhead pressure will reduce reboiling duty with the trend as shown in Figure 14.3a. Toluene in the bottom is the specification that the ultimate C_8 aromatic feed to the paraxylene separation unit should meet. But too low toluene content is not necessary; there are diminishing returns for sake of the paraxylene separation unit performance compared with the cost of extra reboiling duty due to the steeper part of the curve, shown in Figure 14.3b. C_8 aromatics in the overhead product, shown in Figure 14.3c, is the indication of C_8 aromatics lost to the aromatics extraction unit, which should be avoided. This loss may affect the operation of the extraction unit and also results in unnecessary recycle of C_8 aromatics through the complex. Adding feed preheat also reduces the reboiling duty but raises condensing duty (Figure 14.3d).

One must be aware of the capacity limit at the existing condenser when increasing feed preheat, as this will require extra condensing. On the other hand, when the feed toluene concentration changes, the reboiling duty will need to be adjusted according. There is no control for the feed quality for the deheptanizer operation because the

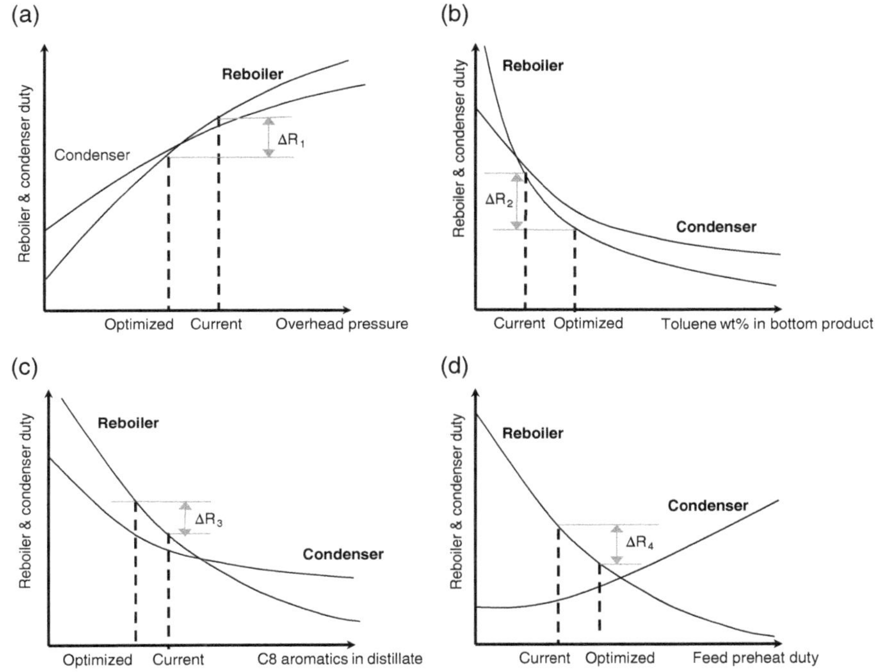

FIGURE 14.3. Correlations of deheptanizer reboiler and condenser duty with other parameters. (a) column overhead pressure (b) toluene weight % in the bottom product (c) C_8 aromatics in the distillate (d) feed preheat duty.

feed quality is the consequence of side reactions in the xylene isomerization unit. For the other four parameters above, operators can make changes to the column overhead pressure, toluene content in the bottom product, C_8 aromatics in the overhead product, and feed preheating. Optimizing these parameters could give around 5% reduction in reboiling duty, compared with operating based on experience and meeting target key component splits. This can be very significant, even more so if this methodology is applied to every column in a complex.

Reboiling and condensing duty can be described based on the relationship with individual parameters as shown in Figures 14.3a–d. If assuming a polynomial form of correlations with order three, each correlation as shown in Figures 14.3a-d can be developed as

$$\begin{aligned} R_i &= a + bx_i + cx_i^2 + dx_i^3 \\ x_i &= (\text{Toluene\% in bottom, C8A\% in ovhd, preheat, ovhd pressure}) \end{aligned} \tag{14.23}$$

where

- x_i is the particular operating parameters defined in the variable set in equation 14.23
- R_i is the reboiling duty calculated for a given x_i value

The incremental effect (ΔR_i) from each individual parameter (Δx_i) can be determined as

$$\begin{aligned}\Delta R_i &= R_{i,\text{new}} - R_{i,\text{base}} \\ &= a + b\left(x_{i,\text{new}} - x_{i,\text{base}}\right) + c\left(x_{i,\text{new}} - x_{i,\text{base}}\right)^2 + d\left(x_{i,\text{new}} - x_{i,\text{base}}\right)^3\end{aligned} \tag{14.24}$$

If there are no interactions among these four operating parameters, the total effect of changes in these parameters on the required reboiling duty would be simply the summation of the four individual effects:

$$\Delta R = R_{\text{new}} - R_{\text{base}} = \sum_{i=1}^{4} \Delta R_i \tag{14.25}$$

However, in some cases, there could be strong interactions among operating parameters. If this were the case, two or more parameters could appear together in one term. To develop a relationship of parameters with interactions, several parameters need to vary at the same time in the plant test or simulation, and the effect on reboiling duty can be seen as the statistically significant result of the interaction parameter. A set of data with changes to the operating parameters and the energy response can be obtained, and regression can be applied to derive a correlation that accounts for these interactions.

When dealing with correlations involving multiple variables, economic sensitivity analysis is essential to determine the most influential parameters. For example, the reboiling duty requirement is very sensitive to feed preheat and column overhead pressure, more so than other operating parameters for the deheptanizer. Getting the most sensitive parameters right in operation can get the greatest economic and technical response.

The correlation developed can be implemented into the control system so that reboiling duty can be controlled automatically to achieve the minimum energy requirement at all times. Alternatively, the correlation can be used as a supervisorial tool. Whenever a variation is expected, adjustments to operating parameters should be made to optimize the reboiling duty. This reboiling duty is the minimum with major independent variables considered and is the target for the conditions at hand. This target and operating cost value for closing the gap must be communicated with board operators in each shift so that actions will be taken for achieving the targets and the actual cost savings can give operators a sense of pride in their direct contribution.

14.6 ECONOMIC EVALUATION FOR KEY INDICATORS

Operation variability is a major cause of operation inefficiency. In general, there are two kinds of variability that can be observed in reality: inconsistent operation and consistent but nonoptimal operation. Figure 14.4 represents the operating data of a reboiler heat input as measured by condensate rate in the deheptanizer in a xylene isomerization unit. In Figure 14.4a, the heat input data points appear to be randomly scattered showing an example of inconsistent operation. This is usually caused by

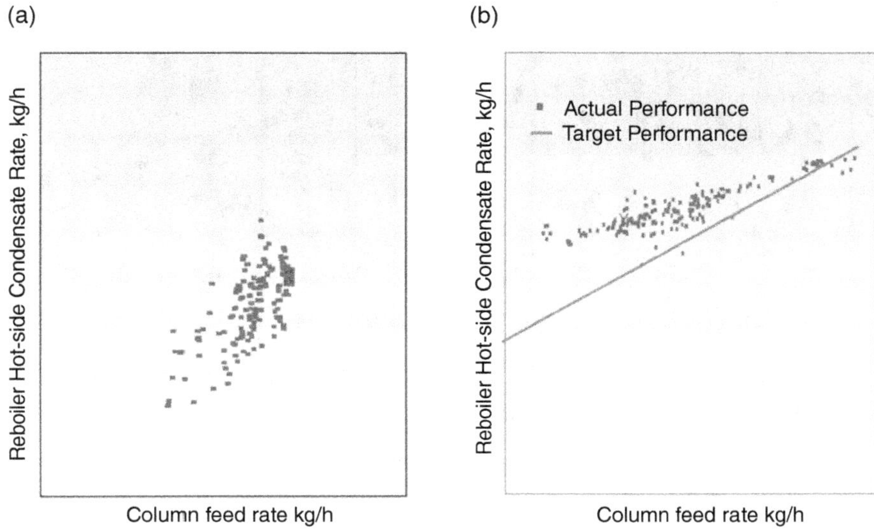

FIGURE 14.4. Two common operating patterns: (a) inconsistent operation and (b) consistent operation but nonoptimal.

either poor control strategy or different operating philosophies used by operators for running the column. In contrast, Figure 14.4b shows a consistent operation, but one that is nonoptimal. In this case, a consistent operating strategy was adopted, but it was far away from the target for adjusting reboiling duty against column feed rate. The target operation represents the minimum reboiling duty required to achieve product specification.

The variability of any operating parameter occurs due to various reasons. The question is how to identify variability in operation and the economic value of minimizing this variability.

Variability assessment starts with simple statistical analysis of operating data. For example, the operating data for C_8 aromatics in the deheptanizer column overhead product under normal conditions can be extracted from the historian as shown in Figure 14.5a, with the specification limit provided. To understand the variability, data in Figure 14.5a is converted to a normal distribution curve, which represents frequency of observations as shown in Figure 14.5b. Frequently the operating data will exhibit a normal distribution.

Two parameters describe the normal distribution, namely, the mean or average (μ) and the variance or variability (σ). σ defines the shape of the normal distribution. The larger the σ value, the broader the curve peak and the larger the variability. Conversely, the smaller the σ value, the thinner the curve peak and the smaller the variability. μ and σ can be calculated by

$$\mu = \frac{\sum x}{N} \tag{14.26}$$

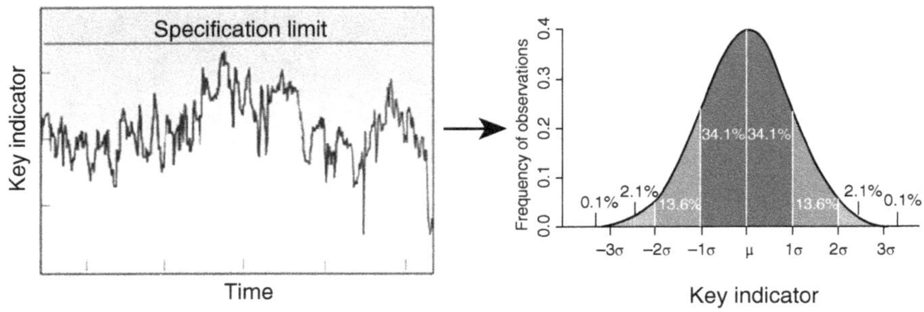

FIGURE 14.5. Operating data: (a) historian and (b) frequency distribution.

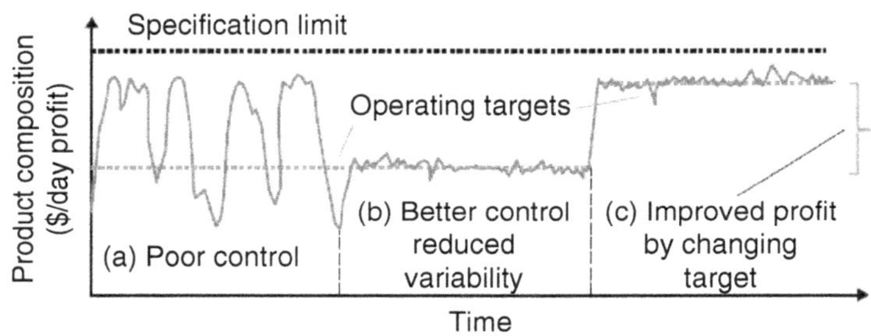

FIGURE 14.6. Operation performance: (a) current operation, (b) reduced variability, and (c) increased profit. *Source*: From White DC (2012). Reprinted with permissions of American Institute of Chemical Engineers.

$$\sigma = \frac{\sum (x-\mu)^2}{N} \tag{14.27}$$

where

- x is the value of key indicator obtained from the historian
- N is the number of sample data points for the key indicator

Refer to the example discussed in White (2012) shown in Figure 14.5. There are two shortcomings in the operation performance shown in Figure 14.6a. The first is the large variability, and the second is too much conservatism from reaching the specification limit. If the operation or control strategy improves, the variability could be minimized to achieve more consistent operation (Figure 14.6b), but performance still remains far away from the specification limit. The limit is usually set by a physical limit such as product purity specification, maximum temperature or pressure, maximum valve opening, maximum vapor loading in a separation column, maximum space velocity in a reactor, and so on. The operation can be improved further (Figure 14.6c) by moving the average closer to the limit by adopting a better

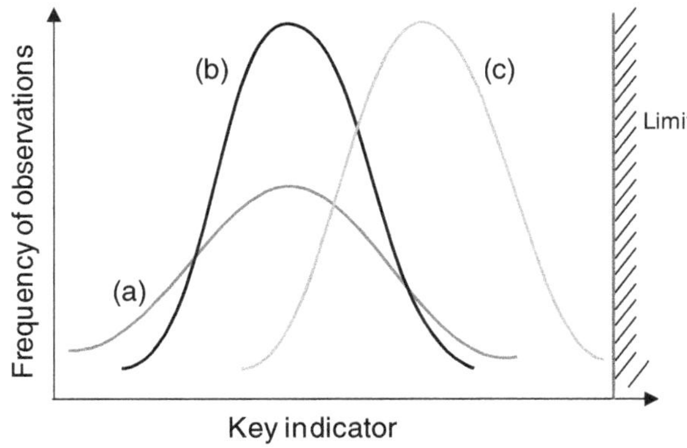

FIGURE 14.7. Convert time series data in Figure 14.6 into normal distribution curves: (a) current, (b) reduced variability, and (c) increased profit.

control strategy. Time series data in Figure 14.6 can be converted to normal distribution curves as shown in Figure 14.7.

The economic value, V_i, for a given frequency that a particular key indicator is observed is given by

$$V_i = C_i \cdot x_i \cdot f_i \tag{14.28}$$

where

- C_i is the economic value for the key indicator
- x_i is the numerical value of the key indicator of interest
- f_i is the frequency of observations for the key indicator

For example, C_8 aromatic content in the deheptanizer overhead product represents the high-value component C_8 aromatics lost in the deheptanizer overhead, representing an additional cost for reprocessing. The key indicator could be defined the difference of actual C_8 aromatics in the deheptanizer overhead and the C_8 aromatics operating target. As an example, if the distillate rate produced from the column is 1,000 Ton/day with C_8 aromatics at 10% higher than the specification, or x_i = 10% with frequency of occurrence as f_i = 30%, and the cost of reprocessing these C_8 aromatics is \$10/Ton, the economic value to avoid this occurrence is

$$V_i = C_i \cdot x_i \cdot f_i = \$10/\text{Ton} \times 10\% \times 1{,}000\ \text{Ton/day} \times 30\% = 300\$/\text{d} \tag{14.29}$$

Similarly for other occurrences, C_8 aromatics in the distillate could be lower or higher than 10%, and the economic values can be calculated accordingly.

In this way, the normal distribution curve can be converted to an economical curve as shown in Figure 14.8. The conversion is calculated to an economic difference.

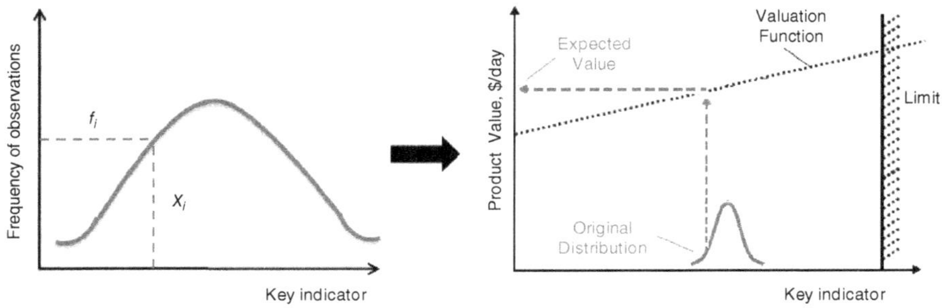

FIGURE 14.8. Converting the normal distribution curve to an economic curve. (a) normal distribution to represent the variation for the key indicator (b) incorporate the normal distribution for the key indicator into the economic curve of the product value per day. *Source*: From White DC (2012). Reprinted with permissions of American Institute of Chemical Engineers.

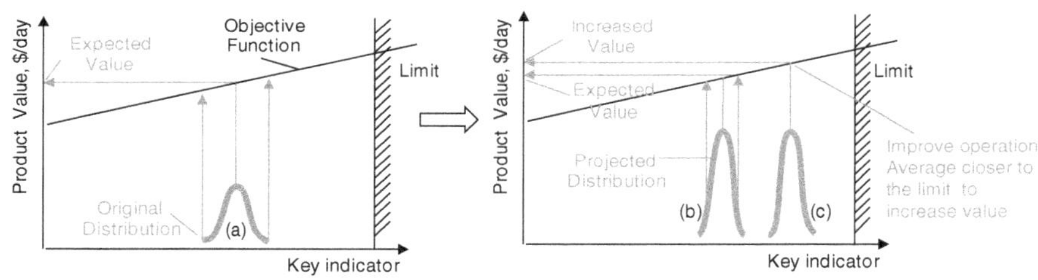

FIGURE 14.9. Economic curves generated based on normal distributions: (a) the original distribution with a relative large variability in the key indicator resulting in lower product value per day, (b) the improved distribution with reduced variability of the key indicator, (c) the key indicator has the similar low variability but has been optimized to provide higher product value per day than (b). *Source*: From White DC (2012). Reprinted with permissions of American Institute of Chemical Engineers.

With the statistic-based economic evaluation method mentioned above, improved operation can be quantified with economic values based on statistical distribution of operating data. The current operation with large variance, such as the case shown in Figure 14.9a, is improved by more consistent operation and/or control strategy to reduce variability as shown in Figure 14.9b, while optimized operation shown in Figure 14.9c utilizes the potential capability available in the process and equipment and pushes the economic value even higher.

14.7 SITE-WIDE ENERGY MANAGEMENT STRATEGY

There is no single solution for energy improvements. As a matter of fact, improving energy efficiency within the process can be achieved in a variety of ways ranging from little or no capital cost operational improvements to optimize the process, recovering more heat by improving the heat integration of the process or adopting new process technology that fundamentally improves the process efficiency (Zhu, 2009). To manage and sustain the improvement benefits, a continuous improvement culture must be established supported by enablers such as performance matrices, key indicators, roles, and responsibility. The strategy consisting of five steps to achieve the best energy efficiency are discussed below (Sheehan and Zhu, 2009).

14.7.1 The First Step: Minimize Waste and Losses via Diligence

In reality, steam generated in the boiler house is distributed through an extensive network of steam pipelines to end users. The losses in steam distribution can be 10–30% of fuel fired in boilers. Hence the net boiler efficiency could be 10–30% lower from the user's point of view.

The losses do not necessary attribute to a single cause but are the result of a combination of various causes. It is common to observe the major steam loss caused by steam trap failure and condensate discharge problems. Steam loss could also occur due to poor insulation of steam pipes, leaks through flanges and valve seals, opened bypass and/or bleeder valves, etc. For example, a ¼ in. leak in a 600 psig line loses over 1,000 lb/hr steam. In addition to noise and safety risks, this leak could cost $104 k/year for the steam priced at $11.9/klb. Simple measures such as maintenance of steam traps and monitoring of steam distribution to determine if steam generated is in accordance with steam consumed can lead to significant cost saving benefits.

Apart from distribution losses, other forms of energy losses could occur due to poor insulation, condensate loss to drainage, pressure loss from steam letdown through valves, pump spill backs, and so on. To detect losses, you must know how much energy is generated versus how much are used in individual processes. The benchmarking method could be used to determine the overall gap of the energy performance and individual losses are identified using different methods. Process energy losses can be detected using the energy loss assessment methods discussed in chapter 8 in Zhu (2014), while identification of steam losses and the ways to overcome the losses in the steam system are discussed in details in chapter 18 in Zhu (2014). Although these solutions can vary greatly, typical improvements to energy efficiency via minimizing losses are in the range of 3–5%, and it can be even higher in some cases.

14.7.2 The Second Step: Effective Energy Use via Optimizing Operation

Early action should be taken in closing the overall energy performance gap in order to optimize operation. This is because operation improvement can be implemented quickly with little investment. Optimizing operation is to operate a process in the most energy-efficient manner via exploring synergetic interactions and running key equipment tight and close to their design limits. Simply put, optimizing operation is to get the most you can out of existing assets. At the same time, equipment must be operated within their reliability limits to avoid "equipment abuse." Otherwise, it could cause mechanical failure unpredictably. More detailed discussions on technical limits are discussed in Chapter 12 in this book.

By implementing operation opportunities at the early stage, it can boost engineers' confidence for implementing other improvement ideas. More importantly, the benefit captured creates a financial budget to pay off other investment projects during the energy optimization cycle.

To determine how well a plant or a unit is doing, it is necessary to build energy monitoring capability so that current energy use can be compared with a consumption target that reflects the current operations. Only then is it possible to do some analysis to determine the cause of deviations from target and take appropriate remedial action. A good energy monitoring solution can be a digital twin for energy management, which allows the user to quickly **observe** the situation and assess the relative performance of multiple units, **orient** themselves by being able to drill down to get more details on KEI of the most critical areas, **decide** on a set of possible actions based upon the determination of possible causes for deviation from target, and **act** quickly and decisively based upon well-informed decisions. Actions can be best practices or control based depending on the resources and control tools available. The key to the success in optimizing operation is the concept of key indicators.

The KEI and targets can be readily implemented into an energy dashboard that can show the performance gaps between current and targets on the computer screen as a snapshot summary. The level of a gap indicates the severity of deviations that forms the basis to assign traffic light for each KEI, that is, green implying the current performance is acceptable as it is within the target range; yellow, a warning sign indicating that a gap occurs and requires attentions; and red, an alarm sign urging for taking actions at the earliest time possible. An example of monitoring for key indicators is Honeywell's Energy Dashboard (Sheehan and Zhu, 2009). This tool could be tremendously valuable to operators and engineers as to what to watch, what to focus on, and what actions to take.

For each key indicator, technical targets are established based on online data and simulation as the basis to compare with current performance. The difference between a target and the current performance for each key indicator defines the performance gap. Different gap levels indicate the severity and level of urgency.

An example of a large multivariable control strategy was applied to an ethylene complex (Sheehan and Zhu, 2009). This involves a total of 17 multivariable controllers that were linked together by an overarching optimization strategy that included the use of nonlinear steam cracking model to predict product yields. The result of this control project was to enable the customer to increase feed rate by 3% over the previous best rate by being able to operate the process up against multiple constraints simultaneously. In addition, the application was able to reduce energy consumption by 3.25% by reducing steam consumption in the fractionators and minimizing excess O_2 in the furnaces. This resulted in a payback of less than 5 months.

Opportunities to operate process units more efficiently exist in most process plants. Little or no capital operational solutions can improve energy efficiency by 2–5%, which can be higher in some cases.

14.7.3 Third Step: Low Capital Investment via Increased Heat Recovery

Monitoring and controlling to improve process energy efficiency usually results in pushing the process up against multiple equipment constraints. To get to the next level of energy efficiency requires capital cost modifications to increase heat recovery within and across process units. Indeed, one of the key values of implementing operational solutions is that it highlights where the equipment constraints to the process are more clearly. Once specific units have been identified for improved heat integration, heat recovery opportunities can be identified via applying the network pinch method as discussed in chapter 10 in Zhu (2014). This method not only considers value and cost of improved heat recovery but also the impact in terms of product yields, operating flexibility, especially with respect to start-up, shutdown, maintenance, and control.

A typical example of redesigning for improved heat recovery involved an older 1970s vintage diesel hydrotreating unit that had a CFE, charge heater, one reactor, and a stripper. After assessing this unit, the modifications of adding four heat exchangers were identified to recover more heat from the process and also generate steam. This scheme reduced the product rundown temperature by 107 °C and the temperature to the product condenser by 150 °C, which reduced the amount of heat lost in the air coolers. The capital cost for this project was estimated to be \$3 MM but resulted in energy savings of \$4.5 MM/year.

Projects to improve process unit heat recovery can typically improve energy efficiency by 5–10%.

14.7.4 Fourth Step: High Capital Investment via Process Changes

Improved heat recovery is the most common type of capital project implemented to improve energy efficiency. However, recent work has identified other areas less commonly explored that may provide significant opportunities. These opportunities include process design changes and use of advanced process technology such as enhanced heat exchangers, high capacity fractionator internals, new reactor internals, new power recovery turbines (PRT), improved catalysts, and other design features and more.

One example of process changes comes from process flowsheet optimization for a hydrocracking process. This example was discussed in detail in chapter 9 in Zhu et al (2017). The original design features one common stripper that receives two feeds containing very different compositions, one from the cold flash drum and the other from the hot flash drum; these two feeds are originated from the reaction effluent and go through a series of separations before mixing. The inefficiency caused from this process sequence can be described very briefly as separation, mixing, and then separation again. This design feature can be observed in many other process plants.

To avoid this inefficient design feature, it was proposed to use two strippers, namely, a hot stripper that receives the hot flash drum liquid as the feed and a cold stripper that is used for the cold flash drum liquid. Furthermore, the cold stripper bottom does not pass through the main fractionator feed heater but goes directly to the main fractionators. Only the hot stripper bottom goes to the main fractionator feed heater. The net savings by the proposed two stripper design is 110 MMBtu/hr, which is 42% reduction of the fractionator heater duty or 23% of total energy reduction for the hydrocracking unit. This improvement is significant. In implementing this design change in reality, a stacked column is used in which the cold stripper is stacked over the top of he hot stripper to form a common overhead system. As the result, capital cost is nearly neutral comparing with the base case but energy saving is 110 MMBtu/hr. The details for this design can be seen in Zhu et al (2017).

Another example is power recovery for an existing plant (Zhu, 2014). Power recovery often represents a good opportunity for economic energy optimization. For an fluid catalytic cracking (FCC) unit with 60,000 bbl/day throughput, the FCC catalyst regeneration flue gas at high temperature was being used for steam generation alone via a waste heat boiler. A power recovery system was quickly identified as a method for significant energy efficiency improvement, as the flue gas could be used for both steam and power generation simultaneously. Further improvement could be achieved by installing a PRT combined with a steam turbine. The objective was to not only generate electricity from the regenerator flue gas but also produce electricity from high pressure steam letdown to make the medium pressure and low pressure steam required in the FCC unit. Compared with the base case that does not include a PRT and uses a condensing steam turbine to drive the main air blower, this new scheme could achieve a net energy benefit of $14 MM/year with payback in less than 2 years.

There are a variety of advanced technologies that can be applied, which vary in terms of cost to implement and return on investment. Careful evaluation of each of these solutions is required as capital is always limited, so it is necessary to select only the best opportunities that can be best integrated with the existing processes that provide the highest return on capital employed. Although these solutions focus on throughput increase or product yield optimization, typical improvements to energy efficiency are in the range of 3–7%.

14.7.5 Fifth Step: Utility Optimization for Reduced Energy Generation Cost

In addition to using energy more efficiently in the process, another common strategy is to produce energy more efficiently. In this step, all the energy users and producers as well as commercial options are combined in a single system, with the objective to maximize overall energy efficiency. Many manufacturing sites have their own on-site industrial power plants that primarily exist to provide steam and power to the process units but may also supply electricity to the grid at times of excess capacity. Supply and demand optimization is achieved by a simulation of the utility plant that can take a set of process forecasted demands from the production schedule and determine the configuration and operating profile of the boilers and turbines to meet demand while taking into account of tiered pricing for natural gas and power contracts. Detailed discussions on utility assessment and optimization are provided in chapters 16–19 in Zhu (2014).

The opportunity for utility system optimization is evident when the electricity price varies from season to season, in some cases from daytime to evening. Optimizing the choice of on-site power generation and purchasing from the grid could reduce operating cost on the order of millions of dollars. Secondly, the opportunity is relevant when there is a large amount of steam going through the letdown valve. It could reduce generation cost by minimizing the letdown rates. Another key opportunity to reducing energy costs in utility plants is to balance changing energy demands from the processes with adequate supply from the utility plant without wasting energy by keeping spare capacity on hot-standby.

Therefore, a utility optimizer could determine the most efficient and economic operation to supply energy to satisfy process energy demands. There are several major options to achieve this overall goal including driver selection, utility equipment loading allocation, steam head pressure control, steam letdown rate minimization, power generation versus purchasing, and contract management.

For load allocations of boilers and turbines, boiler efficiency curves are used as the basis to distribute the total fuel requirement among all the boilers in the lowest-cost manner. Combined with the pressure control, this strategy allows for the fastest dynamic response while always trending to the most economic steady-state position. Turbines are modeled based on performance curves that can operate in either pressure control mode where the set of turbines aim to maximize power generated while maintaining steam pressure or in power generation mode where steam consumption is minimized while maintaining a total generated power target.

When it comes to natural gas and electricity purchasing, the plant could face contract management issue, and better contract management could save millions dollars per year. This is because the plant could set up a contract with a choice of a fixed amount with a flat cost or a tied structure with different prices. For a tied structure, the base amount is fixed with a flat price, while additional amount can be provided with price(s) higher than the base price. With better forecasting, the plant will predict the amount of deficit in fuel and electricity more accurately and thus can make the purchase order well in advance. This could dramatically reduce purchasing costs.

An example of where the utility optimization solution was applied was at a petrochemical site in Korea (Sheehan and Zhu, 2009). The plant had three oil-fired boilers and three backpressure steam turbines that provide steam and power to the process units and also supplied excess power to the national grid. The solution used the combustion controls, the steam header pressure controls for three headers, and equipment load allocation across the boilers and the turbines. The results from the implementation led to significant reduction (>10%) in CO_2 and NO_x emissions and improved boiler efficiency, leading to overall benefits of more than \$1 MM/year.

Optimization of the energy supply system can typically improve energy efficiency by 2–3%, while the main driver of utility optimization is in cost reduction.

14.7.6 Summary of Potential Savings from Five Steps

Table 14.3 combines all of the potential saving opportunities from the five steps to provide a perspective on the level of benefits for a typical refinery of 100,000 BPSD that could be achieved by adopting a comprehensive energy management program. Recent studies indicate that typical benefits of 15% energy reduction are expected, rising to 30% for refineries plants. The carbon credits associated with the reduction in greenhouse gas emissions range from 180,000 to over 360,000 Ton/year.

14.8 DIGITAL TWIN FOR ENERGY MANAGEMENT

A digital twin for KEI and their targets can be readily implemented, which can quickly show the performance gaps between current performance and targets on a computer screen. The magnitude of a gap indicates the severity of deviations and forms the basis to assign a "traffic light" or similar indication for each KEI–that is, a green light indicates the current performance is acceptable as it is within the target range; a yellow light is a warning sign indicating that a gap occurs which continuously drifts away from the target and requires attentions; or a red light is an alarm urging for taking actions at the earliest time possible. An example tool of monitoring key indicators is Honeywell's Energy Dashboard (Sheehan and Zhu, 2009) and has been further improved upon in Honeywell's Connected Plant applications. Tools such as these could be tremendously valuable to operators and engineers, as they indicate the most important variables to watch, taking actions to bring them back to the targets.

KEIs applied to a system can be defined in a hierarchical structure, from the overall complex to each process unit, down to major equipment, and finally to individual operating parameters. The sum of all incentives (the opportunity gap between current performance and optimal targets) from all major equipment represents the total opportunity for the entire process and the overall site. This hierarchical structure allows engineers to drill down from overall performance to specific

TABLE 14.3. Typical Energy Saving From Different Categories of Energy Savings

Saving Opportunities	Energy Improvement (%)	Energy Saving MM$/year	CO_2 Reduction kTon/year
Step 1: Get basics right	3–5	2.2–3.7	36–60
Step 2: Improved operation and control	2–5	1.5–3.7	24–60
Step 3: Improved heat recovery	5–10	3.7–7.5	60–120
Step 4: Advanced process technology	3–7	2.2–5.2	36–84
Step 5: Utility optimization	2–3	1.5–2.2	24–36
Total	15–30	11–23	180–360

operating parameters and thus identify specific actions. Listed below is an example of what can be visualized in the digital twin:

- **Overall site view** shows the site-wide energy consumption and emissions versus overall targets. On the same screen, the overall site view shows the relative amount of energy consumption and emissions from each process unit. A traffic light color is assigned to indicate which processes are furthest away from the targets.
- **Process unit view** indicates the process performance that can be measured by 10–20 KEIs (key energy indicators). These KEIs and their targets are developed from a combination of design, process simulation, and historical data. These predicated energy targets are automatically adjusted to reflect current operating conditions such as feed rate and compositions, product yields and quality, etc. Color coding is assigned to each KEI, which could suggest the need to drill down in the next level of key indicators to identify root causes and logical actions.
- **Equipment view** shows equipment performance via several key operating parameters with indications of current values versus their corresponding targets. The operators may decide to perform more detailed investigations for the root causes if the gap between these values is large.
- **Deviation trend view** allows engineers and operators to review the time periods during which the KEIs deviated significantly from the targets and to determine the major causes of the deviation. By building up a history of causes, operators are able to look back over time and see the most common causes of deviations. This can lead to recommendations about remedial actions for improving equipment performance.

For each key indicator, a target is established as the basis to compare with current performance. The difference between the target and the current performance for each key indicator defines the performance gap. Different gap levels indicate the severity and level of urgency for actions.

Gap analysis is then used to identify root causes. Potential causes may include inefficient process operation, insufficient maintenance, inadequate operating procedures, poor operating practices, poor process control, inefficient energy system design/heat integration, and outdated technology. Gap analyses are translated into specific corrective actions to achieve targets via either manual adjustments or by automatic control systems. Finally, the results are tracked so the improvements and benefits achieved can be quantified.

14.9 ESTABLISHING ENERGY MANAGEMENT SYSTEM

It is necessary to have an effective energy management system in place to drive measures for energy efficiency. It is not enough to just have good intentions from management and motivation from personnel. Some years ago, in a chemical plant, an energy program kicked off with enthusiasm, slogans, revival meetings, and pledges. An audit showed visible improvements from month to month. Management expected the path of improvement to continue. Instead, improvement ground to a halt. The trend leveled off after a couple of years and gradually turned upward.

What had happened? The rapid improvements seen at first came from correcting simple matters such as leaks, steam traps, insulation, and so on, which were detected by common sense. But after the obvious improvements were accomplished, the improvement curve leveled off and become stable at an unacceptable level. After a while, despondence kicked in and the program arrived at the dead end since there was no preparation to keep digging deeper for hidden opportunities and thus no resources were assigned.

The lesson learned is that intention and motivation can produce good results only if right directions are developed, sufficient resources are given, and enabling tools are provided. Therefore, it is necessary to put into place a digital twin for the energy management with clearly defined roles and accountability, metrics and measurement, review, and actions. Simply put, a digital twin based energy management system is about operating targets and gaps, accountability, work process, metrics, audit, and communications. For sustainability, the digital twin based energy management system must be integrated with existing technical management structure for continuous improvement, which is very important. Let us walk through these components one by one.

14.9.1 Accountability

The key element in the energy management system is a motivated energy coordinator or energy manager who has the passion for driving energy cost down, technical know-how, and communication skills. This person should come from within the plant so that he knows the people in the organization, processes, and existing management system. The role of the energy coordinator is to manage the energy management work process, facilitate implementation of improvement projects, handle interunit and off-site issues, and drive an optimization culture in the organization.

A large site may require an energy team consisting of a process engineer, an experienced operator, and/or a control engineer, all of whom are assigned to multiple responsibilities. The role of the operator is to go out to identify any inefficient operations visible on field such as furnace air register opening, bypass valve opening, flaring, steam leaks, hot flange, etc. The role of process engineer is to conduct opportunity evaluations and formulate business cases for improvement options in order to push them through gate review process. The control engineer is to develop control strategy for improvement ideas and assist unit control engineer to implement actions in the DCS.

14.9.2 Work Processes

There are two major fronts for an energy management system, which are energy capital projects and energy operation improvement. The work processes for these two fronts are different due to the nature of the work. For example, energy projects should follow techno-economic review process: define (process changes)–evaluate (cost/benefit assessment)–select (option prioritization)–execute (option implementation).

In contrast, for operation improvements, the work process should be based on the operation process: define (key indicators, targets, and limits)–monitoring–gap analysis (current versus targets)–actions/control.

The work process will define schedule, milestones, deliverables, and responsibilities. A well-developed work process can avoid cracks under which good opportunities could fall through, while bad ideas are selected. The work processes should be put on test in trial for revision before it is integrated with the existing technical management structure.

With well-defined work processes for energy management which can be implemented into a digital twin, it is like providing a perpetual engine. Often we observe situations an energy management program driven by some strong personality. When this person leaves, the campaign suffers as this would cause a major disruption to the energy program.

14.9.3 Scorecard and Tracking

Success in operating a plant depends on awareness and alignment of objectives together with enabling tools. The digital twin based energy system must be developed as part of overall business and operating plan. This starts with sharing annual target for energy cost reduction including milestones and benchmarks. Tracking must be in place for reviewing progress toward goals that could include monthly review with plant management and reporting scorecard of process energy performance.

Using a scorecard to show energy performance toward goals enhances alignment and awareness, while tracking progress promotes motivation. Drilling the performance scorecard from plant level to process areas level until equipment level is the key for empowerment. This is similar to sports where the team will keep doing what they are doing well when winning. But when losing, the team will change the game plan. The essence of the scorecard is to tell if the team is winning or losing.

14.9.4 Documentation

Beyond the scorecard and tracking, it is critical to have a corporate-wide integrated documentation system for all management systems across the organization. When a new subsystem within a management system is created, it gets implemented throughout the corporation via the digital twin. The integrated document system, combined with the company intranet, can allow the company to integrate all management system processes throughout the company.

14.9.5 Technical Target System

Key indicators define the basis for continuous monitoring. Technical targets for key indicators must be developed and applied across the organization. Many industrial facilities have the potential to increase the efficiency of their systems but have difficulty doing so because there are no metrics available to measure energy efficiency improvement. The absence of energy performance metrics makes it difficult to establish value structure for the improvements. Practically, a review of KEIs and targets should be part of process performance review, which is a key part of integrating energy management system with existing management structure. Since development of key indicators and targets is a big ticket for energy management system, it requires management support to acquire resources and technical capability.

14.9.6 Corporate Energy Audit

The energy audit is intended to assist plant personnel in identifying both operational and capital projects that have short payback periods. The audit can be conducted in stages. The first phase is *benchmarking* of overall site and each major process to determine where they stand. The result of this assessment indicates the actual performance, which can be compared with either cooperate target or with peer performance to give the overall performance gaps. If the benchmarking is conducted from corporate, it can be done off-site with the plant sending required data. The effort for conducting benchmarking should be minimal as the digital twin provides the majority of the data required.

If the overall gap is large enough, it warrants the second phase of audit or specific audit, which is designed to *identify root causes* and process improvement opportunities. The audit methods are explained in the book by Zhu (2014). The methods for energy audit can be put into two categories, equipment based and system based.

If the specific audits are conducted by a team organized by corporate, the company should provide audit standards and work processes that describe assessment topics and the requirements for conducting assessments, reporting, and documentation.

The third stage of the audit is to *review* the energy projects selected by the plant and recommend project prioritization although ultimate approval of projects will be made by management. Successful or failed results will be selected for sharing as experience and lessons learned throughout the company. Specific tasks can be developed, which could include heat exchanger cleaning optimization, steam trap repairing and maintenance, product fractionation system optimization, control strategy for optimizing column energy-yield trade-off, utility contract management, flare gas reduction, key indicator system, and so on.

The key for a successful corporate audit is to have good preparation from plant personnel and from the lead auditor. Lead auditor needs to make plant engineers aware of what will be audited as well as the auditing process. At the same time, the lead auditor needs to get the audit team on the same page for the agenda and schedule. On the other hand, plant process engineers need to prepare and make the plant's performance data available in organized manner. The effort for process engineers should be minimal as they can find the majority of the data needed.

REFERENCES

Schipper L, Howarth RB, Carlassare E (1992), Energy intensity, sector activity, and structural changes in the Norwegian economy, Energy – The International Journal **17**, 215–233.

Sheehan, BP, Zhu, XX (2009), Improving energy efficiency, PTQ **Q2**, p29–p37.

White DC (2012), Optimize energy use in distillation, Chemical Engineering progress (CEP), AIChE, March.

Zhu XX (2009), Challenges and solutions for reducing industrial energy and CO2 emissions. *Exponent*, Vol. 5, December Issue, Honeywell.

Zhu XX (2014), Energy and Process Optimization for the Process Industries, Wiley/AIChE, Hoboken, NJ.

Zhu XX, Hoehn R, Thakkar, V., Yuh E. (2017), Hydroprocessing for Clean Energy, Wiley/AIChE, Hoboken, NJ.

15

INTEGRATING THE WORKFLOWS

Frank (Xin X.) Zhu[1] and Joe Ritchie[2]

[1] *Honeywell UOP, USA*
[2] *Formerly in Honeywell UOP (Retired), USA*

15.1 INTRODUCTION

Process companies today are faced with unprecedented supply chain complexity. Regardless of the industry, supply chain optimization is a critical factor ultimately defining business success by effectively reducing the cost of production. To that end, product demand needs to be accurately estimated, raw materials procured, production planned and executed as planned, and distribution costs minimized. For maximum efficiency, this needs to be accomplished using an integrated platform to provide transparency, visibility to stakeholders, and continuity. Legacy work processes and systems addressed different aspects of supply chain optimization, but traditionally the weak link was at interface handoffs that lacked continuity in assumptions and data from the start to end of the supply chain. Modern supply chain optimization models the entire supply chain ensuring consistency and validation of assumptions from raw material purchase to delivered product. These models are also powerful tools for scenario analysis and timely update of plans to incorporate the latest intelligence on product demand and prices and reflect the inevitable unplanned supply, production, and distribution events that occur in a manufacturing environment. This provides a level of economic agility simply not possible with legacy systems.

In this chapter, we will focus on the petroleum industry as an example of workflow integration and automation via digitalization. The supply chain cycle for petroleum is briefly described in Figure 15.1 with end-to-end needs to be addressed.

Digitalization and Analytics for Smart Plant Performance: Theory and Applications, First Edition. Frank (Xin X.) Zhu.

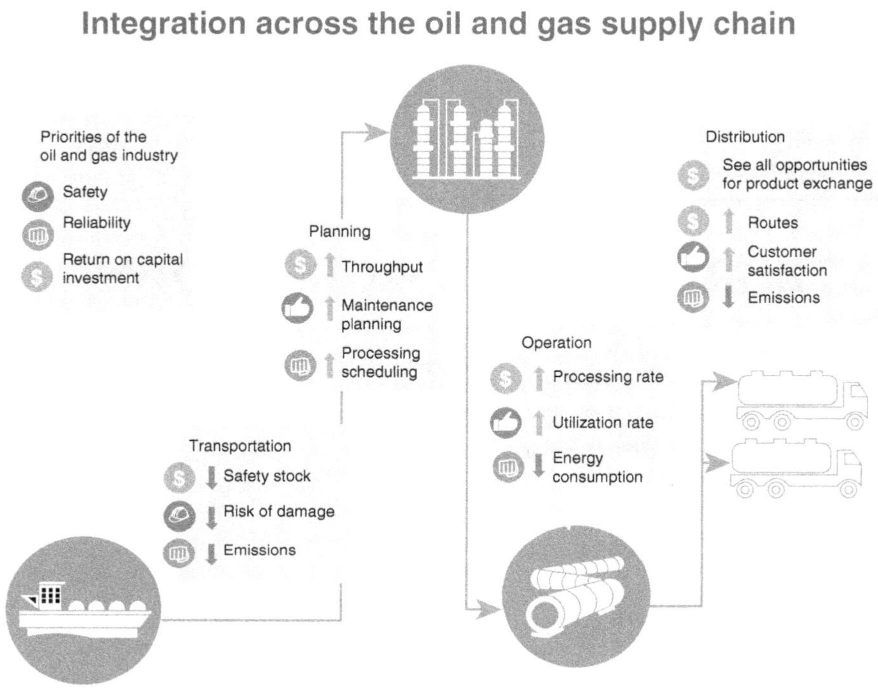

FIGURE 15.1. Manufacturing supply chain.

Essentially, the key elements of a supply chain are planning, purchasing, scheduling, operation, control, sales, transportation, and distribution, which are fundamental business processes that exist in every production environment, and the financial incentive to perform these activities in an integrated, uniform, and automated fashion is substantial. Pace-setting organizations have shown that the vision can be realized by deploying integrated platforms for bringing crude scheduling and product and dock scheduling with optimization capability for blending and rudimentary unit optimization. But on the whole, many organizations still approach the problem in a siloed fashion, dividing responsibilities to different parts of the organization to users using their own spreadsheets and systems with limited communication. In this chapter, key gaps and the best practices for integrating supply chain management are discussed.

15.2 KEY ELEMENTS OF INDUSTRIAL SUPPLY CHAIN

Industrial supply chains start with raw material acquisition and end with product delivered to the customer. Depending on the industry, there is a myriad of steps from start to finish, but the fundamental steps include planning, purchasing, scheduling, operation, control, sales, transportation, and distribution. The steps are orchestrated in concert to minimize production cost, thereby maximizing margin. Purchasing and sales will often fall under the umbrella of planning. Many decisions will be made throughout the supply chain including which materials to procure at what time, which products to make, when and how to deliver the products, how to make the products, whether to make or buy the products, and how to maximize use of assets. Supply chain optimization decisions are not static—they are time dependent and refreshed frequently. These decisions will be reexamined when product demand or crude supply changes, production upsets occur, transportation logistics change, or market price differentials shift.

The objective of supply chain planning is to maximize net cash margin across all time horizons. Supply chain planning is complex, and due to the inherent complexity, the work process is hierarchical, operating at several tiers with increased detail at each level as the problem becomes better defined and preparations are made to move from plan to execution. The process is inherently iterative to reflect updates on inventories, receipt and lifting schedules, and a host of other data that move from notional to transactional as time progresses.

At the first tier, strategic planning structures the business model and determines the optimal assets needed to support that model. As an example, a refinery could be supply driven emphasizing making the products needed to fulfill marketing needs. Alternatively the refinery model can simply be to monetize crude or pursue a "merchant refinery" model that buys opportune crudes and responds in an agile fashion to product market opportunities.

At the second tier, once the assets are in place, a centralized supply and distribution (S&D) function focuses on economically supplying corporate product demand via conducting "make or buy" decisions, purchasing crude that is the best fit for assets and product goals, and continually identifying opportunities to be responsive to a dynamic market. Movement and freight cost logistics, particularly with multisite refineries, is a key optimization handled by S&D. S&D takes the supply chain lead for long- and mid-term operations but collaborates closely with on-site planning to ensure high-level feasibility. As purchase and sales decisions are finalized, the supply chain transitions the leading role from S&D to planning. S&D remains responsible for movement logistics and receipt and lifting updates.

At the third tier, the planning goal is to define the operating plan with sufficient granularity so it can be executed by operations. A linear program (LP) tool used by planning is primarily an economic model, given robustness by representative and technically sound, though simplified, process models that emphasize variables that impact economics. The LP will be used in conjunction with scheduling tools to provide timing for operating campaign changes, process unit schedules, blending optimization, blend scheduling, and liftings. As the granularity increases, potential problems with the plan, or problems arising during execution, are resolved in the most economic fashion involving all parties of the supply chain team as needed.

Consistent with the organizational goal, the goal of the supply chain is to maximize margin, and to do so, the different tiers must operate in unison, not in isolation. Depending on the stage of execution, those leading the effort change based upon their specific expertise. The continuity and collaboration between the different functions is essential to execution and to capture margin. The benefits of an integrated platform for supply chain optimization for visibility, continuity, and accuracy are clear given the level of collaboration needed for success.

The following sections elaborate on the key supply chain elements using the petroleum industry as an example. Given the inherent flexibility in both raw material and a multitude of products, this supply chain is very complex.

15.2.1 Strategic Planning

A strategic plan is a high-level business model that articulates the core purpose of the business and major operating parameters, forecasts financial performance, and defines production goals and capital planning. Each refinery typically generates a 5-year strategic plan annually. More granularity will be applied for the first year of the strategic plan since the horizon is much closer, giving more confidence in underlying assumptions, so this typically becomes the annual operating plan or budget. The financial and production results from the strategic plan are integrated with comparable results from the organization's other operating assets and for financial projections for shareholders. The strategic plan will incorporate the following considerations during its development:

- The impact of scheduled and mandatory turnarounds is incorporated for the entire plan horizon. Major unit turnarounds have 3–5 years' cycles with downtimes ranging from 30 to 60 days. Due to the interconnectivity of the process units' turnaround, scheduling often brings down units in entire blocks, for example, if an FCC is taken down, all the downstream units such as alkylation and gasoline hydrotreating no longer have feed. The impact of turnaround planning is twofold from a financial standpoint: the physical cost to rejuvenate the units and more importantly the margin loss associated with lost production. The plan will evaluate the most cost-effective timing to conduct scheduled maintenance considering the lost production cost and the actual turnaround cost.
- Catalyst changeouts for units with cycles that do not coincide necessarily with turnarounds, such as hydrotreaters and hydrocrackers, that may require catalyst replacement anywhere between 1 and 3 years.
- The capital plan for payout and regulatory projects needs to be superimposed, including both potential downtime affects and capital cost. These may end up being linked with scheduled maintenance to minimize downtime.
- Onstream time assumptions for critical units are included, often based on past measured performance.
- Known and anticipated contractual commitments for raw materials and finished products to set the minimum requirements. An LP will be used to optimize the gap between baseload, long-term crude contracts, and optimal crude mix using crude availabilities and perhaps some judgment based on crudes that have generally proven economic in the past. Likewise, products marketed above the contractual minimum will be based on optimization with imposed market constraints based on historical perspective or judgment. In the end the plan should be practical and achievable to ensure credibility of financial projections.

- Forecasted prices can be developed either internally or by market consultants if additional insight or objectivity is desired. It is not unusual to evaluate multiple price sets, including perhaps historical low and high demand/price scenarios, to better understand the potential envelope of operation and determine if the plan is sufficiently robust. Understanding price points, or key differentials, that materially change the operation and product mix should be explored so the plan can serve more so as a roadmap to identify different possible scenarios than as a point solution.
- Key performance indicators are developed to provide a framework to understand deviations from plan. As an example, "crack spreads" that represent the price difference between a finished product such as diesel and a marker crude are used to analyze market fluctuations and the impact on margin performance. Reliability or utilization metrics are used to distinguish asset performance from market variability.

15.2.2 Product Demand and Price Forecasting

Production planning starts with a set of assumptions on product demands, raw material availability, and a price set. The assumptions will all be tested in time and replaced with decisions on which products to make, which products to buy, how to best supply each of the regions served, and which raw materials to buy—all with the aim of maximizing the overall business margin. The assumptions are a starting point, or a "straw man," to hone and optimize the plan prior to handing it over to execution. Annual plans, monthly plans, or historical context is often used as the starting point since those operations reflect what had been optimal in the past.

Demands from each of the distribution centers are estimated considering historical internal marketing demands and third-party contracts on a ratable basis. This is typically done by a centralized S&D function. Seasonality, contractual changes, inventory positions, product exchange agreements, pipeline schedules, and pricing strategy will be considered in setting the forecast. If trading has identified and committed to opportunities, these will be consolidated in the estimate. At about 30 days prior to execution, the product demand will become more granular as initial waterborne lifting schedules and pipeline schedules are needed to ensure product is delivered on a timely basis. The schedules will be updated yet again when vessels are secured and product parcel sizes are finalized to accommodate vessel capability and onshore storage. At this point the demand moves from an initial ratable estimate to known vessels and parcel sizes for shipment. With this information inventory positions at both the source and destination will be reassessed to ensure working supplies are adequate and there is minimal risk of either product runout or product containment issues leading to ship demurrage.

Product demand will directly impact the amount and timing of raw material purchases, and it may also set the minimum amount of production, or baseload, considered for a given time period since it reflects near-term business commitments. The baseload is typically below peak production capacity, so there is discretion on both rate and product mix that can be exercised to adapt to market conditions with agility. Operation above the baseload is typically economic given the assets are in place and economies of scale.

Price forecasting is equally important in developing the operating plan. Responsibility for this typically falls with the trading group within S&D. The price set will be the key driver in optimizing the discretionary product mix as well as the crudes selected. Futures prices are often used for the more widely traded products like gasoline and diesel. For smaller markets differentials pegged to crude, gasoline, or diesel are often used. Differentials can be derived from publications or market intelligence based on inquiries or past trades.

15.2.3 Crude Oil Evaluation

An effective program for crude oil evaluation is an integral part of optimizing economic margin. There have been many alternative approaches to evaluating crude over the years. Earlier approaches were simple spreadsheets that cut the crude into naphtha, jet, diesel, VGO, and residue fractions. Simplified process unit models were used to calculate downstream yields and ultimately generate LPG, gasoline, diesel, and other refinery yields. The crude margin was calculated using the plan prices for crude and finished products. Crude options would then be ranked for purchase based on margin. The downside to this approach is that the specific crude yields were looked at in isolation and downstream capacity was assumed to be sufficient. In reality, the incremental value for a given crude will depend not only on the base crude mix that it supplements but also on unit capacity and operating constraints, constraining finished product specifications, and schedule considerations. As a general rule, margin declines as more constraints are imposed.

LP models are now routinely used to determine crude values given the complexity of the problem. LP models are business/economic models as opposed to rigorous process unit models. The LP will have submodels for each of the process units that are representations of process unit yield and product quality behavior over an operating window range of potential feed properties and operating parameters. The submodels are often derived from rigorous models, but simplified, with pertinent feed properties and unit operating parameters selected by the modeler based on how influential they are to overall economics. The submodels will also reflect major unit operating limits such as capacity, product draw rates, and key equipment limits based on the frequency that they limit and if the limit is economically material.

When an LP is used for crude evaluation, a reference crude slate is often defined based on an historical mix, annual plan or monthly plan, or past purchases. The crude to be evaluated is then substituted for a given amount of reference crude, meant to represent the last decision on filling up the crude unit. The amount of the subject crude substituted typically reflects a cargo-size parcel over the month. The amount is arbitrary but should be meaningful enough, so constraints resulting from the specific characteristics of crude are evident and quantifiable from an economic perspective.

The trading organization will have indicative prices for each of the crudes at this stage based on recent differentials versus markers such as Brent or Dubai. This crude price will be used at this stage to determine the margin, and the process will be repeated with different crude to generate a ranking based on margin. For the most promising candidates, traders will initiate negotiations to progress the indicative price into a real transaction cost and then finalize the purchase. Margin priority may change once indicative prices are refreshed based on initial negotiation, so the crude priority may shift influencing the final decision.

Once the crude cargo is purchased, the reference crude mix will be updated to reflect this, and this process continues with subsequent purchases until all the crude needed for the month is secured. Proceeding in this manner gives a continually improving picture of the plan for the upcoming month and provides the consistency and granularity needed for execution and scheduling when the torch is passed.

In certain situations the LP solution may not run the crude unit to maximum capacity. This is often due to the specific characteristics of the crude that result in downstream constraints earlier than the reference mix. Frequent and quantitative assessments of constraints are critical to obtaining valid LP solutions. The LP in effect should be a digital twin of the operation, showing consistent process unit and blending constraints as the facility for the same crude mix; otherwise the results are not economically sound.

While price often dictates the best margin crude, it is also often the case that a crude providing the best "fit" to the facility is optimal because its yields are such that multiple downstream capacities can simultaneously be satisfied giving higher overall asset utilization. Along similar lines, it is possible that crudes that intrinsically yield products that have higher crack spreads are favored, for example, summer operation may favor crudes with more naphtha content, and winter may favor crudes with a higher distillate content.

Scheduling considerations may preempt the choice of the highest margin crude. Crude suppliers publish production schedules that include time windows when each of the cargoes for the month can load based on the crude production rate and tankage. For maritime refineries, the voyage to the refinery from the production site can range from several days to 60 days, and the receipt timing may not work from a crude containment standpoint. Alternatively, certain crudes may need to be blended with others in order to run at prescribed rates, so receipt timing and inventory considerations need to be part of the decision. Close collaboration between trading, planning, and scheduling functions is required to ensure the logistics are feasible as part of the purchase decision; otherwise the intended margins will not be realized.

Purchasing the optimal crude mix is a collaboration between the trading, planning, and scheduling functions. While trading brings insight on crude market dynamics and pricing, the planning function ensures that facility constraints reflect current operation and the LP submodels reflect performance over a realistic operating window.

15.2.4 Planning

Planning, scheduling, and blending act in a hierarchical, iterative, and collaborative fashion. Demand forecasts provide the minimum required products that need to be supplied. Demand forecasts and price forecasts are used to determine which products should be made versus purchased and are used to identify opportunities. Planning will start at a high level based on an annual plan but continually refine the plan to reflect decisions taken such as crude purchases or spot sales. The objective of planning is margin improvement, but to be realistic high-level constraints are imposed to reflect scheduling realities. There will be scheduling input to planning throughout the supply chain cycle, particularly in an integrated platform.

Production or operation planning is typically conducted on a 90-day rolling basis, with major refreshes as the current month ends. The immediate or current month will be well defined, whereas as the time horizon goes out further, more assumptions are needed to fill in the gaps. The plan is continually updated to reflect decisions taken, information that is finalized, and routinely refreshed such as inventories. The 90-day planning horizon is based primarily on the lead time to procure and receive crude, which can often be more than two months. In the plan, crude slate and rates are typically defined before product liftings. Crudes are typically purchased globally, whereas product distribution is regional.

For products that will go to outlying company-owned terminals for subsequent sale, S&D provides a date window (nominally 5 days) along with a parcel size for each lifting. Third-party long-term contracts will also nominate lifting date windows, product grade, and product parcel size for the upcoming month, at least 10 days prior to commencement. Windows will be narrowed down once the vessels are nominated. Pipeline scheduling is more precise, while waterborne moves have more uncertainty in timing due to weather and port traffic. S&D will coordinate the overall lifting schedule, with a high-level objective to ensure offtake is close to ratable for each product over the month to minimize the likelihood of product runouts and product containment issues. The scheduling function will provide initial commentary if certain product inventories are unworkable, and some adjustment to the lifting schedule may be made.

The trading department makes a call on prices over the 90-day planning horizon. This is often linked to futures prices as spot products can be sold forward against these markers. With this information and the lifting schedule, the planning function puts together 3 LP programs for each of the months in the plan. If optimal production exceeds liftings, then spot sales will be arranged to close the gap, choosing to the extent possible the most profitable opportunities, for example, diesel may be favored over gasoline so site-wide cutpoints shift to optimize accordingly.

If optimal production falls short of required liftings, then S&D will need to buy products to make up the shortfall, and there may be an optimization regarding what products to purchase. Specific opportunities that may present are evaluated at this point, for example, selling alkylate may improve margin rather than blending it to premium. Much of this exercise reduces to "make or buy" decisions for each of the products with the market giving you the price to buy and the LP providing the cost of production.

It is important to point out that an LP solution is an "average" solution, assuming all crudes are purchased at once, all products sold at once, and capacities are available throughout. This is not a scheduling tool; it is an economic optimization tool.

Based on the LP results, the planner will provide scheduling and operations with unit throughput targets and operating modes such as maximum distillate or gasoline, crude unit cutpoint targets, reformer RON, and FCC riser temperature as well as expected limitations. Planning, scheduling, and blending functions are typically under one roof, so blending recipes are communicated as initial targets.

Scheduling is about executing the plan, understanding the production objectives, and optimization direction and opportunities communicated by planning. Scheduling systems will also refine the schedule as transactions are completed and provide input to planning in the decision-making process on crude and product transactions in a highly collaborative environment. Early in the process planning has the primary role; this is gradually passed on to scheduling as the time horizon for execution nears. Planning relies on scheduling for an aggregated view of the starting point for the next planning cycle for considerations such as opening inventories and inventory qualities. Resolutions of potential issues prior to execution are communicated from scheduling to planning, which are worked collaboratively to ensure the best economic outcome.

15.2.5 Scheduling

Due to the complexity in refinery scheduling, the problem is often broken into three activities as shown in Figure 15.2 and highlighted below:

- Crude oil scheduling covers unloading from vessels or pipelines, crude blending, and the charge and yield schedule for each crude oil blend.
- Process unit scheduling includes the charge and yields from each of the downstream units as well as mode of operation or severity changes that are part of the overall plan. Component rundown rates and properties to the blending component tanks are estimated.
- Finished product blending covers the scheduling of the blends and the recipe and schedule optimization to meet the lifting schedule.

The three scheduling tasks have much in common. Crude storage is typically 15–30 days of crude unit capacity, and product storage is also about 15–30 days of product. That said, the net horizon where scheduling is critical is this same window from receipt to product lift, nominally 30–60 days. Scheduling takes place in several tranches, each with increased level of detail and confidence, until the plan is fully executed. Simple heuristics and daily balances suffice as the time period to receipt is long, whereas more detail is needed as execution approaches.

Scheduling has a more immediate focus than planning, dealing typically in the 1–10 days range, with attention more toward feasibility than optimization. In this immediate window, crude and raw material decisions have already been made and the receipt schedule is fixed. Commitments have been made for term and spot sales, and a product lifting schedule has been fixed. Crude blending, unit throughput targets, unit modes of operation, and specific operating parameters are determined,

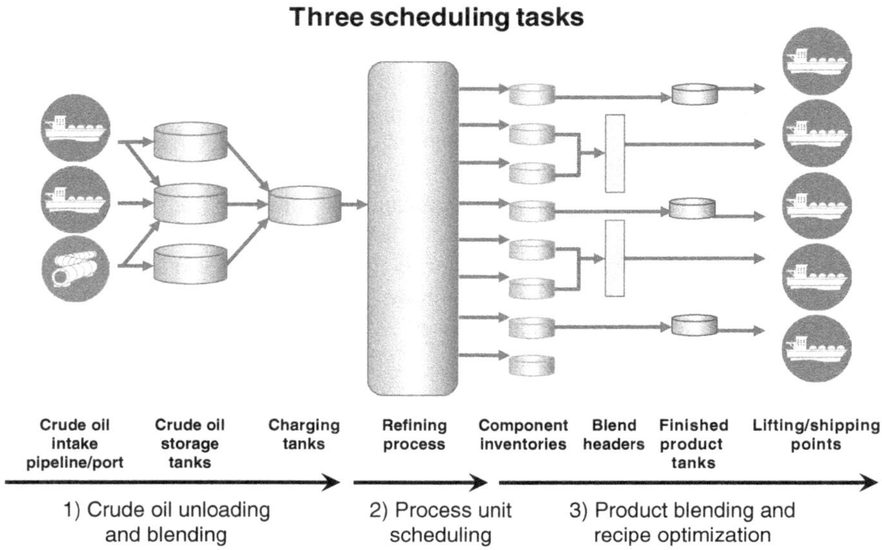

FIGURE 15.2. Three scheduling tasks.

and product blending is planned to ensure the solution stays within storage limitations. At this stage, several passes have been made to foster feasibility, and clear infeasibilities uncovered in initial schedules have been resolved. In general, the earlier that schedule adjustments can be made to ensure feasibility, the lower the cost will be. Unanticipated events such as delays in receipts or liftings or under performance of production can still throw a schedule into disarray, and if this occurs, the scheduler will compensate, involving planning if there is a cost or tradeoff for remedying the problem.

Scheduling uses the monthly or weekly operating plan as a starting point but examines operations on a more granular basis, considering the impact of crude shifts on downstream operations, the need to change operating campaigns to suit specific product requirements, tank availability, and the lifting schedule. Scheduling will work with daily or often shorter-term balances dictated by specific events such as ship unloading to ensure feasibility and to provide guidance to operations on anticipated unit throughputs and timing of campaign changes to meet product specifications.

Receipt and lifting projections are updated routinely to determine if schedule slippage requires changes to the execution plan. Monthly planning often considers heuristics such as ratable lifting offtake and initial crude scheduling to minimize future possible infeasibilities before detailed scheduling. Waterborne receipts and liftings can be problematic due to weather and tides and therefore frequently deviate from plan, so the scheduler needs to adjust accordingly. Scheduling focuses on establishing feasibility through detailed inventory projections and solving potential infeasibilities by modifying operations, blending schedule, lifting schedule, or spot feed or product transactions.

Updating past projections with actual data is critical, so scheduling starts with the most recent validated information of initial inventories and properties as well as unit performance. Rundowns from the process units will be predicted based on current operation and anticipated changes to unit operations due to unit feed quality changes and operating mode. Scheduling will minimize working capital by minimizing oil inventories to the extent possible while ensuring lifting schedules are met. Ship demurrage on crude and product vessels and target blend giveaway are other important scheduling considerations that significantly affect profitability.

15.2.5.1 ***Crude Scheduling*** Most refiners run a mixture of crudes to each of their crude units. The mixture is determined based on maximizing the crude unit rate and ensuring that major downstream conversion units such as FCC, hydrocracker, and delayed coker remain full since these units have by far the largest impact on margin and hence overall economics. The general objective is to simultaneously hit as many constraints as possible with the crude mix by tailoring the selection and ratio of crudes.

There are two types of crude blending. With an online crude blender, the operator can pull from several tanks with different crude grades to meet the target synthetic blend. Likewise, this configuration gives more flexibility to adjust composition on the run to ensure multiple constraints are met. Older systems use tank-to-tank blending of crude to meet the target mix. This requires more overall storage and more upfront planning on crude receipts since the preference would be, where possible, to discharge from the ship to the final charge tank. Once the charge tank is online, the crude mix cannot be adjusted as with online crude blending. Crude scheduling needs to allot time for tank dewatered and testing to certify for charge. In a tank blending configuration, there is a sequential rotation of charge tanks with each tank online perhaps 3–5 days. Each charge tank swing will result in a swing in crude unit operations as the unit equilibrates to new yields based on the next mix, so there is some inefficiency during transition.

Scheduling crudes from pipeline are easier than waterborne receipts. Depending on the crude origin and supplier, it is not unusual for a waterborne receipt to be several days late, and this often requires reformulating the crude blend with stocks on hand.

Segregation of crudes for an online blending system is typically based on the major differences in quality and the impact of those qualities on optimizing the downstream. Heavy and light segregations are typical as well as low and high sulfur. If high TAN crudes are run, this could be another logical segregation. The objective here is to have as much leverage as possible between crude properties in the segregations to give more leverage on the final blend.

15.2.5.2 ***Process Unit Scheduling*** The purpose of process unit scheduling is to run the refinery in the most economic fashion with a known crude mix and set product liftings. Process unit scheduling will communicate expected unit capacities throughout the plan horizon, typically 7–10 days, along with operating campaigns, modes of unit operation with specific targets, and an overall refinery balance. As an example, an operating campaign could be a 3-day run where the reformer operates at high octane to produce premium gasoline or a campaign when crude unit diesel endpoints are reduced to produce a low cloud point diesel cargo. Operating targets such as FCC riser temperature and guidance on unit optimization will be provided to ensure the philosophy is understood. Detailed operating instructions are provided daily to operations and terminal personnel to ensure effective coordination of activities. Average blend recipes for finished products are typically provided to blending personnel for guidance. These recipes will be based on assumed stream rundown properties.

Scheduling works backward from the lifting schedule to ensure blends are available just in time for the scheduled lift. Likewise, if unit operating conditions need to change to enable certain products, such as premium gasoline, a campaign will be devised. Unit rates and possibly target conditions will change if the crude mix changes, so operations are advised of the yields so they can plan for the transition.

Scheduling communicates key targets for unit throughputs, modes of operations, and changes associated with operating campaigns to operations. Blending is typically part of the scheduling organization, and their input is part of developing the detailed schedule. Operations will work with technical support to modify control setpoints as needed to accomplish the plan.

Production rates and product quality targets from the scheduling system are sent to the control system, while process optimization and control updates scheduling on the status of the production process and the parameters used in process models. The updates on production are synchronized with updates on material availability, customer demand, and transport availability. Monitoring, analysis, and accounting functions provide feedback on actual plant operation.

15.2.5.3 Product Blending and Lifting Schedule Product blending is addressed in several tranches, with detail and accuracy improving as the plan prepares for execution. As part of the monthly plan, nominal blend recipes for all products are provided. These recipes are indicative only since the LP represents monthly average operation and does not reflect the impact of schedule constraints on recipes and rundown properties are estimated. A weekly plan will provide updated recipes typically based on a more recent assessment of unit performance and rundown properties but also averaged over time. From a conceptual standpoint, the blender should be able to come close to the averaged recipes if product rundown properties are accurate, unless scheduling conflicts require adjustment.

There are two key considerations for the blender: first, minimize product quality giveaway for properties that have economic consequence, such as octane, and second, keep the components close to property balance and contained in tankage. The second consideration has priority since if components go out of balance, for example, large buildup of low-octane components due to insufficient octane generation, this puts the next period in jeopardy and there will be a cost to correct. As an example, if all the high-octane components available were blended to meet premium requirements, there may be insufficient octane in the next period to produce regular gasoline so the pool is unsalable. In this situation either the premium liftings should have been reduced, or the reformer octane should have been increased well in advance. In general, for gasoline, the rundowns should come close to octane balance with octane demand implied from sales. The thought process and balances need to go out beyond the short period currently in focus, that is, do not solve today's problem and create tomorrow's nightmare.

A blender typically has component tank storage and finished product storage. Initial detailed scheduling will designate the start and end of each blend and use earlier provided blend recipes to determine daily finished product tank inventories and component tank inventories. Blends will essentially work back from when the product is needed for lifting, with a 1- to 3-day window for certification and blend touch-up, if needed. Moving the product into the blend tanks can take up to several days depending on blend size and in-line blender capacity. In any event, the blend scheduler will use the daily balance to determine and address any pinch points in either component or finished product storage. Adjustment may include starting a gasoline blend early because FCC naphtha storage is topping.

Blend scheduling has traditionally been done with spreadsheets but has transitioned into production programs to facilitate and expedite the process. Linkage of these systems directly to inventory and lab LIMS systems gives quicker and more accurate blend scheduling than past manual processes. Blending optimization programs can be anywhere from sophisticated nonlinear blending to fixed recipe blends.

If online blending is used rather than tank batch blending, the system will use ratio control for component rundowns and finished product property control. The components will be blended into the finished tank based on a specified blend recipe, but certain designated component compositions will be modified online by the system in order to minimize key property giveaways such as octane. The blender will typically specify a high-octane and low-octane trim component to be used to adjust online measured octanes as needed to achieve minimal giveaway. Likewise, butane will be used for trim RVP control, fine-tuning the original recipe rate as required to reach the target blend RVP. Though a target recipe was designated, the final blend will differ somewhat as the system strives to minimize property giveaway and because the component rundown properties will differ slightly from assumptions used in the initial formulation. The final recipes are typically sent back to planning so LP blending algorithms can be improved. This practice reinforces continual improvement.

Blending is typically on a just-in-time basis to directionally minimize inventory and to maintain product flexibility by keeping the inventory in component tanks versus finished blend tanks. Initial blend recipes are provided by planning and later adjusted by scheduling to account for changes in unit rundown properties or other constraint changes. Maintaining close to the recipes in the long run is critical to keeping components in balance and economically used given the fixed slate of product liftings.

The effectiveness of blending is measured by the giveaway of key properties that have direct economic significance. For gasoline, this is typically octane and RVP. If a finished blend has octane giveaway, this implies that either more premium gasoline could have been made by blending the components better or if additional premium cannot be sold then the severity of the reformer was too high, consuming more fuel and reducing gasoline yield. Likewise, if gasoline RVP is below specification, it means less butane was blended and more will need to be downgraded to LPG. Each refiner will be aware of what mechanisms they use for trim octane and RVP control and that understanding will be the basis for determining the cost of product giveaway. The cost associated with giveaway should be rooted in the tangible economic benefit based on what mechanism is the primary control at the facility. In this way the profit or loss naturally flows to the financial bottom line. As an example, if more premium gasoline cannot be sold, then valuing octane based on the market premium–regular differential overstates the real octane cost that should be based on the reformer cost of octane generation. Planning will typically provide daily guidance to scheduling on the cost of various quality giveaways so potential trade-offs can be evaluated.

Target property margins above specification and the actual giveaway may be influenced by scheduling considerations. If production is behind schedule and the ship is at the dock, blends may target more giveaway to ensure the blend is on specification on the first pass to minimize demurrage cost or supply impact to the customer.

15.2.6 Process Operation and Control

Day-to-Day Operation Optimization The business process provides economic guidance to refinery operations on a day-to-day basis as well as a 7- to 10-day look ahead. A review of the economic options for common operating difficulties may also be provided in order to facilitate an expedited, economic response. Larger upsets will require review and evaluation by planning, scheduling, and S&D to determine the best path forward.

Control Advanced control and online optimization translate the goals set by planning and scheduling to real-time process targets or setpoints, which are executed by regulatory control. The production process involves real-time manipulation of selected production variables to reject short-term process disturbances and hold key product qualities and production rates near their desired target values while ensuring that equipment operating limits are not violated (Shobrys and White, 2000). Where degrees of freedom exist, it also includes incorporating economic criteria into determination of the current target values for selected variables. Control is generally implemented in a hierarchy with individual equipment control at the lowest level, process unit control at the next level that coordinates multiple items of equipment, and site-wide control at the top linking multiple process units (Joseph Lu, 2015).

15.3 LITTLE INTEGRATION OF SUPPLY CHAIN WORK PROCESSES

Historically, planning and scheduling have been treated as separate problems, often being solved in sequence but largely independently. Planning has been long rooted, going back decades, in LP optimization for guidance setting key plan parameters to optimize margin. Scheduling evolved first from manual estimates to spreadsheets designed to separately address aspects such as crude, product, and blending scheduling. The work process was laborious and often involved multiple assumptions that were not only inconsistent between schedulers responsible for different product lines but also inconsistent with planning assumptions (Hartmann, 1998). Given the complexity and lack of integration, scheduling concentrated more so on daily volume balances with little emphasis on quality balancing and little if any optimization. Skill sets and experience for scheduling and planning have also been quite different. Given how time consuming scheduling has been, there has been little time to recast projections or look at different scenarios to solve scheduling conflicts or to respond to unplanned outages.

A comparison and analysis of monthly performance versus plan will often show where there have been disconnects between planning and scheduling. This exercise also tends to accentuate the need for a stronger feedback loop from scheduling, which is closest to the data, back to planning in order to improve LP models on issues such as blend recipes and even unit yields.

Scheduling and blending are often consolidated in a single group since the activities are both part of execution and highly interdependent. For the most part lifting schedules set blend schedules, and in turn blend schedules influence unit scheduling. The scheduling process by nature is iterative involving trial and error, examining a tentative schedule for product runouts and storage containment, adjusting the schedule as needed to mitigate. The advent of personal computers facilitated a basic level of automation, but with multiple schedulers responsible for different product lines, there remained major consistency gaps between crude unit and gasoline scheduling due to assumption variations as well as between scheduling and planning on unit performance expectations and blending recipes. Integrated scheduling systems consolidating receipts, liftings, and very basic unit and blending representations first appeared in the late 1990s and have been widely adopted. At a minimum these systems allowed all schedulers to work from a common platform, eliminating artificial boundaries, so the crude schedulers estimate for yields would propagate to the scheduler responsible for diesel and gasoline blending. With automation scheduling resolution was improved by changing the architecture to be event driven rather than simply by daily balances. This was a significant improvement since infeasibilities are caused by certain events and may occur at any point in time, not necessarily at the daily close. Though faster due to automation, scheduling remained iterative.

When scheduling problems arise, they tend to be highly visible and urgent, and the cost to remedy is tangible, for example, demurrage costs may be incurred. When

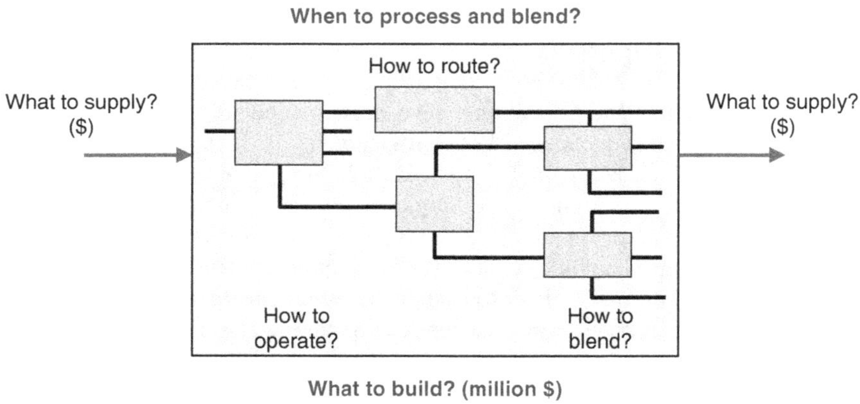

Decisions on	Model for	Objective
What?	Planning	Integrated economics
How?	Operation	Detailed economics • Plant optimization • Blend optimization
When?	Scheduling	Check on feasibility • Time (arrival/liftings) • Volume (empty/full)

FIGURE 15.3. Difference among planning, operation, and scheduling (Hartmann, 1998).

planning issues arise, the options are not always as clear-cut or tangible. When planning and scheduling advocate different paths, it is important that that planning decisions and efforts not be subordinate to scheduling considerations. Of course, the best path forward is for planning and scheduling to resolve differences. Planning, scheduling, and control activities and tools are separate from each other as shown in Figure 15.3, which compares objectives and activities.

15.4 GAPS EXISTING IN CURRENT SUPPLY CHAIN MANAGEMENT

As identified above, a major gap for current supply chain management exist in that the activities are conducted by different departments, based on different data, tools, and people.

This negatively impacts consistency, continuity, and visibility and is a barrier to collaboration and margin optimization. A common work process with assigned responsibilities, working from a common automated digital platform, is vital to improving supply chain performance. This integration needs to span from the onset to the end of the supply chain cycle.

Forecasting is a precursor to planning that is often neglected in the process industries. Forecasts in many companies are still based on access to and interpretation of historical data, and this approach often leads to multiple forecasting functions, both formal and informal, across different business areas in a company, as each area interprets history based on local information. Another reason for the existence of multiple forecasting functions is that different areas of the business work with forecasts at different levels of aggregation, that is, no "standard format" has been established. This often spawns multiple forecast functions using tools with different levels of sophistication (Shobrys and White, 2000).

15.5 INTEGRATED WORK PROCESS FOR SUPPLY CHAIN MANAGEMENT

The best practice of supply chain management should address workflow integration and feedback loops between lower-level and higher-level activities and should eliminate barriers as needed to automate data transfer and execute planning, scheduling, and blending tasks. Data should only be input once, and owners responsible for data input and integrity need to be nominated.

A common platform with integrated workflows allows multiple planners and schedulers to enter the input to a common platform for visibility—including crude schedulers, gasoline schedulers, and diesel/HSFO schedulers—each doing their part to balance the system. This common platform tool lets the scheduler view and analyze their production schedule and project inventory levels and the distribution terminals they replenish. The timing and volume of production activities go electronically to process control. Process control in turn provides feedback on the status of the production process and production volumes. S&D can query the scheduling tools as part of order commitment to verify the availability of capacity and materials. Orders are seen by the forecasting tool as they are accepted.

With this best practice, the planning and scheduling functions dynamically update the blend formulations passed down to control, and feedback of actual component properties is critical in predicting blend properties.

With the planning function on-site, engineers are much closer to the data and interact more with refinery operations personnel. They are better positioned to reconcile data and have realistic and comprehensive constraints based on operations feedback built into the LP model. Since short-term volume balances are much more critical to the scheduler than the planner, the scheduling system excels at highlighting imbalances early so metering issues can be resolved in a timely fashion. This is much more difficult for the planner working with monthly balances for process units that most times are not consistent with more reliable tank gauges. What goes in needs to come out, and the scheduling system forces the units to balance over a small enough time period to identify and ultimately correct balance issues. Feedback from schedulers to planners on unit yields and blend recipes is very important in keeping both planning and scheduling systems in optimal shape. Decisions resulting from poor planning tools can easily cause scheduling conflicts and margin loss. Poor planning tools will incorrectly assess potential margin performance and adversely impact crude selection and planning choices.

A simple example by Shobrys and White (2000) as shown in Figure 15.4 is used to explain the concept of workflow integration and potential benefits. A single, collaborative forecasting function serves all of the business areas. Online analytical processing (OLAP) statistical methods are used to analyze history, create and adjust the forecast, and provide the views of the forecast required by the different areas of the business. The forecast incorporates all open orders and goes to production planning and scheduling. The forecasting function monitors actual orders through the course of the month, flagging "out of pattern" demand and adjusting the forecast accordingly. A single production planning function uses a multi-period LP model. It provides feedback to forecasting on the feasibility of current demand (if demand exceeds capacity, customers may be placed on allocation or external capacity may be obtained).

The scheduler creates a base production plan for each plant, requirements for long lead time materials, inventory targets, transportation budgets and working capital, and expected replenishments to the bulk distribution terminals. Commitments are made based on the plan and passed to scheduling in the form of material and transport constraints for feasibility check on demand/production.

The automation of the overall scheduling process has also facilitated a wider time horizon for scheduling activities, essentially back to crude purchases. These systems give a more enlightened view of the schedule and margin impact of decisions as they are made and the impact of more definitive timing of receipts and liftings as the plan crystallizes. These systems also highlight well in advance major potential problems so schedulers and planners can resolve these issues earlier at a much lower cost. This includes dock optimization to mitigation issues such as dock congestion, which can escalate demurrage costs.

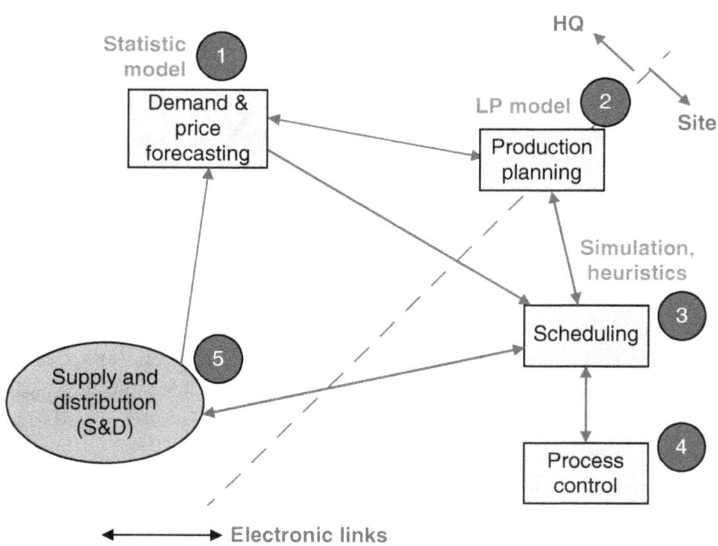

FIGURE 15.4. An example of the best practice of supply chain management.

Companies aggressively pursuing integration report substantial financial incentives for better integration. Exxon Chemicals estimated that a major integration initiative reduced annual operating costs by 2% and operating inventory by 20% (Shobrys and White, 2000). DuPont aggressively pursues the improvement and integration of these functions. An indication of the resulting financial impact is that a single polymer business reduced working capital tied up in inventory from $160 million to $95 million. Although no numbers have been published, oil companies internally estimate incentives of up to $1 per product barrel for better integration of planning, scheduling, and control for gasoline blending.

15.6 SUPPLY CHAIN DIGITAL TWIN: ONE PLATFORM FOR WORKFLOW INTEGRATION AND AUTOMATION

To implement the supply chain workflow integration and automation, a digital twin system can be developed, which involves common data sources, data cleansing, feed and product analysis, model validation, information transfer, workflow automation, and results visualization. As an example, the workflows for different activities in supply chain management and their connections together with operation execution are described in Figure 15.5 (AspenTech, 2020; Coker et al., 2014). Better coordination of these activities through such a digital system can increase throughput, reduce quality giveaway and demurrage, and improve the refinery's agility and ability to evaluate special opportunities that may arise.

This supply chain digital twin should have the following features (AspenTech, 2020):

- **For consistency**: The common platform allows multiple planners and schedulers to enter the input to a common platform for visibility—including crude schedulers, gasoline schedulers, and diesel/HSFO schedulers—each doing their part to balance off the system.
- **For accuracy**:
    - *Accurately predict properties for better crude purchase decisions*: Accurate characterization data for enhanced property prediction leading to better crude purchase decisions.
    - *Forecast accuracy*: Improvements in forecast accuracy lead to better demand forecast, increase service levels, reduce excess inventory, and provide better input to purchasing decisions.
- **For transparency**:
    - *Visualize planning and scheduling results at the same time*: React faster to discrepancies between the plan and schedule and effectively exploit optionality to make smarter, more profitable supply chain decisions.
    - *Schedule all activities from one platform*: Aspen Petroleum Scheduler is an event-based, single blend optimization tool that supports scheduling of refinery activities for global refineries.

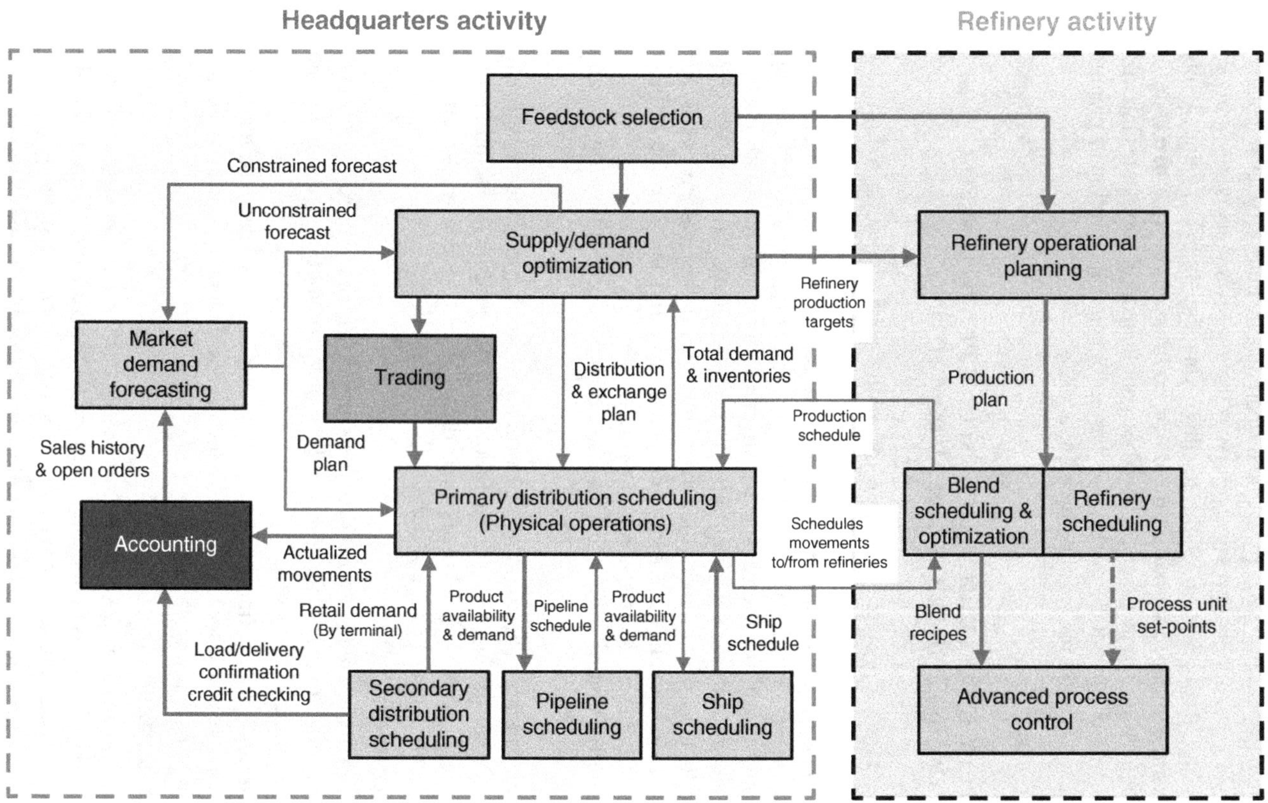

FIGURE 15.5. Vertical integration of workflows for supply chain management. (AspenTech, 2020)

- **For optimization**:
  - *Optimize economics of supply distribution network*: Aspen Petroleum Supply Chain Planner is a powerful economic tool that solves multi-period transport optimization problems related to allocating resources, helping you balance them to maximize profitability.
  - *Optimize operation decisions*: Accurately and effectively manage hour-to-hour, day-to-day operational activities while respecting tanks and inventory constraints and more easily meet product requirements.
  - **For self-calibration**: Models used in planning, scheduling, blending, and control can be calibrated regularly when feeds, catalyst performance, and process conditions vary beyond the range of model prediction. Furthermore, incorrect data used in planning can be identified.

15.7 INTEGRATION OF ENGINEERING MODELS WITH SUPPLY CHAIN DIGITAL TWIN

Major gaps include lack of routine calibration to process models and updating of unit constraints, insufficient equipment constraints, and incomplete unit material balances. While it varies among sites, the quality and completeness of process unit data is usually poor, especially for older facilities. This is a major impediment to constructing robust LP models.

There is no prediction of crude properties per se; these come from either purchased crude assays, supplier assays, or internally generated assays. Best practice is to check all crude receipt key properties against assays, with updates required when deviation is pronounced. If key receipt properties deviate significantly from assay, crude yields and properties, and for the matter crude value, may be significantly off.

For each of the process units, robust tools are needed to estimate unit yields with different feeds and operating conditions, which in turn is to build LP models. LP vectors will be calibrated by plant data or preferably test runs if possible. Test runs may also be needed to adequately calibrate process models. If possible, it is valuable to confirm plant behavior is reasonably consistent with model predictions, perhaps by examining data on the outside of the envelope for feed properties and operating conditions.

To maximize refining margin, it is important to incorporate engineering models into the key business decisions in a more automatic manner (Figure 15.6). It is essential that these organizations communicate and that they have common business processes, preferably supported by a common IT system. As Figure 15.6 illustrates, this integration ranges from planning, scheduling, blending, control, and process engineering (Mulick, 2012).

The key for this integration is to develop an automated procedure for gain updating where new gains are determined by numerical perturbation of a nonlinear process model. It is important to get the first derivatives correct for gain updating,

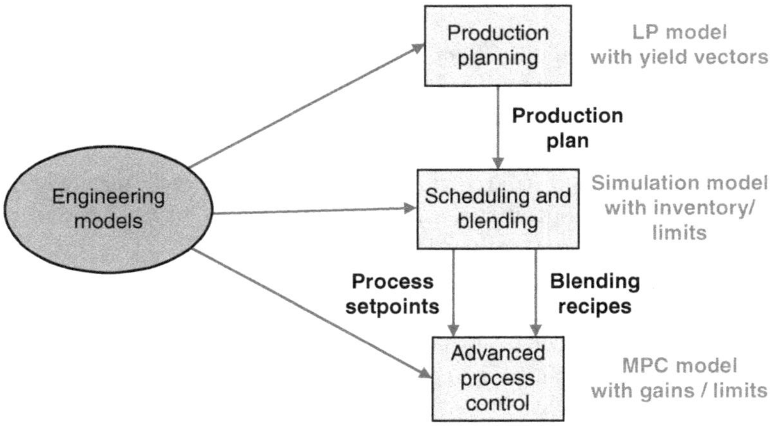

FIGURE 15.6. Integrated engineering models for accuracy and consistency.

while rigorous parameter fitting for the nonlinear process model is not required at each execution. Execution of gain updating can be triggered by change in some process conditions or can be on a fixed frequency.

REFERENCES

AspenTech (2020), aspenONE Petroleum Supply Chain, https://www.aspentech.com/en/products/pages/aspenone-petroleum-supply-chain, accessed on May 7.

Coker R, Shah A, Khan A (2014), Advanced process control collaboration with planning and engineering, Galveston Bay Refinery, Marathon Petroleum Company in collaboration with AspenTech, May.

Hartmann JCM (1998), Distinguish between scheduling and planning models, Hydrocarbon Processing, July, pp93–100.

Lu J (2015), Closing the gap between plannin and control: A multiscale MPC cascade approach, Annu. Rev. Control. **40**, 3–13.

Mulick S (2012), Benefits of integrating process models with planning and scheduling in refining operations, An Industry White Paper, Aspen Tech.

Shobrys DE, White DC (2000), Planning, scheduling and control systems: Why can they not work together, Comput. Chem. Eng. **24**, 163–173.

16

DIGITALIZING THE WORKFORCE

ROHAN MCADAM

Honeywell Process Solutions, Australia

16.1 INTRODUCTION

There are many ways in which digitalization can enable and empower a workforce to work more efficiently, reliably, securely, and safely. The roots of this lie in the astonishing development in human–computer interaction technology over the past two decades. The consumer market is now awash with personal information devices in many different forms, including smartphones, tablets, smartwatches, virtual and augmented reality headsets, biophysical monitors, and so on. These devices are backed by increasingly powerful distributed information systems that allow users to perform a huge range of tasks from the mundane to the extraordinary. A current smartphone and its applications would likely be "indistinguishable from magic" to someone from the 1980s, according to Arthur C. Clarke's famous adage. Combining current and emerging human–computer interaction technologies with other digitalization enablers such as analytics, empirical and first principles modeling, machine learning, and artificial intelligence provides many opportunities for enabling and empowering an industrial workforce. More specifically, digitalization can help enable the workforce through more effective competency development and management. It can help empower the workforce through the transformation of the on-the-job performance of workers.

Competency development involves equipping workers with the knowledge and skills needed to do their jobs. Digitalization provides new ways to develop and deliver training material to build competency that are superior to traditional methods in terms of learning outcomes, accessibility, and cost-effectiveness. Competency development can be further enhanced by providing more direct ways of transferring

Digitalization and Analytics for Smart Plant Performance: Theory and Applications,
First Edition. Frank (Xin X.) Zhu.

knowledge from experienced to less experienced workers without the formality of a training course, something that is particularly important in a climate of high staff turnover due to demographic change. On-the-job performance can be dramatically enhanced in the key areas of access to information, communication, decision making, and worker augmentation. Digitalization can make access to information essentially "frictionless," satisfying the information needs of workers that are often deprived of information they need to work as efficiently, reliably, and safely as possible. Communication is core aspect of any enterprise and especially in an industrial enterprise where timely, effective collaboration between roles is essential. With digitalization, communication between workers moves from low bandwidth, high latency methods to rich real-time communication that allows people to work more effectively together, even when separated by large distances. Decision making benefits greatly from improved information access and communication together with digitalization enablers such as analytics, modeling, machine learning, and artificial intelligence. The result can be much faster and effective decision making that closes the loop on problem identification, analysis of causes, evaluation of potential courses of action, and actual problem resolution.

In addition to considering the benefits and opportunities that stem from digitalizing the workforce, it is also important to identify and mitigate several challenges that can prevent the benefits of digitalization being realized. These include dealing with the rapid churn in technology that can make it difficult to realize a return on the investment in a technology before it becomes obsolete, understanding issues arising from the relationship between digital tools and human workers that may prevent technology from being allowed to deliver its benefits, addressing security and privacy issues related to digitalization, and channeling workers' everyday relationship with technology into a productive engagement with technology at work without the unnecessary intrusion of "digital distractions."

This chapter explores these opportunities and challenges in enabling and empowering the workforce through the considered adoption of digital technology. The first section of the chapter describes the ways in which technology can help build and maintain a competent workforce. The second section examines how the key benefits of digitalization in the areas of information access, communication, and decision making can transform key roles in an industrial enterprise to improve the efficiency, reliability, and safety of the workforce. The final section surveys key challenges in digitalizing the industrial workplace and how these problems might be mitigated.

16.2 ENABLING THE WORKFORCE

There has always been a need to enable the workforce to perform their duties by providing them with the knowledge and skills needed to do their jobs on an ongoing basis. Increasingly, this needs to be done more quickly, efficiently, and cost-effectively. This need is even greater now with increased levels of staff turnover as industry rides through the so-called Silver Tsunami of Baby Boomer retirements with replacements drawn from the Millennials and Generation Z. The need for more effective approaches to providing workers with the knowledge and skills they need is further exacerbated by change in the workplace such as more agile operating and maintenance philosophies and the movement of people between locations and roles.

Traditional approaches to skilling the workforce based on structured training courses and informal knowledge sharing fall short in face of increased demand and high staff turnover. Sending personnel off-site to an equipment vendor's office for hands-on training is expensive and time consuming. Alternatives such as paper, electronic, or video-based training lack the experiential aspects of hands-on training that is important for achieving the best learning outcomes. Less formal knowledge transfer is also a key problem in complex industrial settings where it can take an operator or engineer many months to become fully familiar with the structure and operation of a process unit. Knowledge transfer through one-on-one interaction with more experienced personnel is disruptive and slow and difficult in the face of high staff turnover where corporate knowledge often walks out the door.

Digitalization can help solve these problems. Firstly, digital technology can provide more effective forms of training delivery using new ways of preparing and presenting training material. Secondly, the rise of knowledge sharing platforms and technology flowing from social media platforms can provide much simpler ways to capture and transfer knowledge between workers. Finally, mobile technologies can help to reduce the need for "off-the-job" training by providing guidance to workers at the point where they need it, when they need it.

16.2.1 Training Delivery Using VR and MR

Digitalization can help with more effective training delivery in several ways. Firstly, the presentation of training material can be made more effective and accessible. Operator training simulations of plant production processes and their associated control systems have been used for many years to train console operators. However, combining simulation of processes, equipment, and other aspects of an industrial plant with new and emerging human–computer interaction technology can provide training experiences that are demonstrably more effective than other forms of training delivery for a wide range of training scenarios. For example, virtual reality (VR) is a maturing technology that makes it relatively easy to construct training scenarios in which students learn and hone skills needed to perform a task in a simulation of the actual physical environment in which the task takes place. VR provides a user with an immersive visual and auditory experience of a three-dimensional physical environment and is commonplace in the consumer market in computer gaming. VR has also been used extensively in architecture, archaeology, medicine, data analysis, and so on. In the industrial context, VR can be used to provide a trainee with a simulated physical experience of the work situation, for example, a piece of equipment to be serviced, driven by a simulation of the behavior of the equipment. The trainee can learn to "virtually" service the equipment in an environment that provides detailed guidance and the opportunity to perform the actions required to complete the task. Figure 16.1 illustrates the use of VR for training technicians

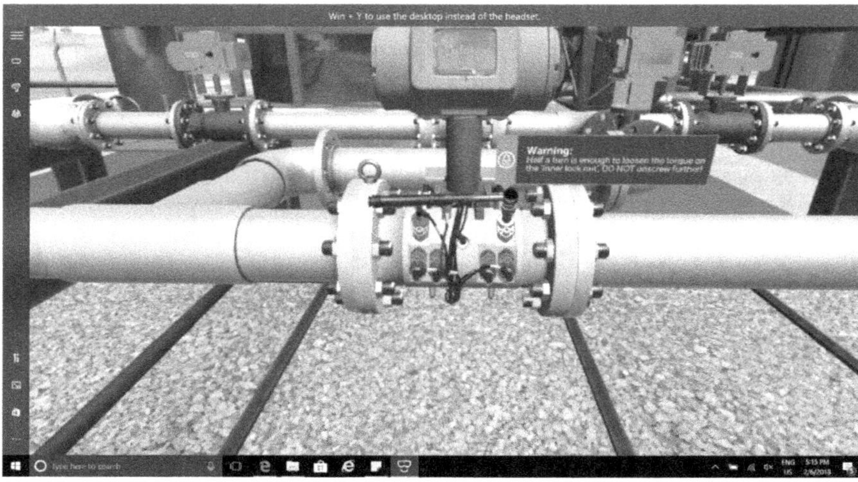

FIGURE 16.1. Virtual reality-based training for servicing an ultrasonic flow meter.

in the servicing of an ultrasonic flow meter. In this case, the meter being serviced is backed by a simulation of its operation that allows the trainee to diagnose faults with the meter, replace components if necessary, and check that it is operating correctly afterward.

There are several advantages to this approach. VR-based training can be more effective than learning practical skills from books or videos with faster learning and greater retention (e.g., Borsci et al., 2016). It avoids the need for physical equipment, and, as a result, training can be held in any location, such as an on-site facility rather than at an equipment vendor's location. Furthermore, it is relatively straightforward to construct virtual representations of plant and equipment that would never normally be available for training purposes, such as a large piece of equipment or an entire process unit. This allows field operators, for example, to benefit from field-based training in an environment that captures the essential elements of the plant they need to work in without the practical and safety issues of attempting to provide the same training in the actual plant.

While VR technology is a relatively mature technology and continues to develop, it does have several problems and limitations. One problem is that many users suffer from motion sickness when using VR due to the disorientation caused by the completely synthetic visual scene. Another problem is that this purely synthetic scene cuts the user off from the physical space they are in, making it difficult to interact with other people who are physically present. It is possible to remedy this second problem by creating virtual representations of the trainee and others using "avatars" that can interact within the virtual environment. This can facilitate collaboration among remote participants, including instructors to recreate a "face-to-face" learning experience even when the participants may be at different sites, cities, or countries. However, an alternative approach that avoids motion sickness and allows collaboration among co-located participants uses a variant of VR known as augmented or mixed reality (MR). MR differs from VR in that it provides an experience in which simulated virtual objects are located within a user's view of their physical environment using a see-through display in the form of a headset or glasses. These objects can be fixed in place with respect to the physical environment so that the user's experience includes a mixture of real and virtual objects.

Both VR and MR have their place in training delivery. MR is useful when it is appropriate to bring training-related equipment into the space used for training and collaboration with co-located workers is important. VR is useful when the training situation involves an environment that cannot be "brought indoors," such as the full-scale physical plant itself. Both technologies have many applications in an industrial digitalization effort, as will be seen in the following sections. In the context of competency development, these technologies also provide important opportunities for taking instruction out of the training facility and into the actual plant to help reduce reliance on "off-the-job" training, as will be seen in a following section.

The immersive experiences provided by VR and MR help deliver more effective training, but another aspect of effective training that can be enhanced through digitalization is the motivation of trainees to successfully achieve learning outcomes. A specific approach to this that applies across a range of training presentation techniques is the gamification of training, in which elements of game design and game play are introduced into the presentation of training material. Tasks to be learned or situations to be mastered have scores associated with them based on trainees' performance on the learning task. Trainees can compete against themselves or others in an effort to improve their performance. This can change learning into something that is exciting and fun as well as serious and competitive (Lahti et al., 2014). Again, simulation-based approaches are useful as they can be easily gamified. For example, handling a simulated process upset can easily become a game between console operators that compete based on their ability to quickly and appropriately respond to the upset. The effectiveness of gamification depends on a trainee's experience with game-based environments such as video games, but we might expect this to be less of an issue as more "digital natives" enter the workforce.

16.2.2 Knowledge Transfer

Implicit in any training scheme is the transfer of knowledge from knowledgeable people to less knowledgeable people. A formal training course captures the knowledge to be transferred in the form of a specific curriculum and associated training materials and experiences. Creating a training course typically involves significant time and effort that needs to be justified in terms of the significance of the learning outcomes, the number of people to be trained, and so on. There are, however, many cases where there is a need for knowledge transfer that does not warrant the cost or formality of a training course. For example, new field operators have numerous tasks they need to perform in a plant, many of which may require them to be shown what they need to do before they can do it themselves. This is often accomplished through one-on-one familiarization and instruction with an experienced operator. This needs to be repeated for each new operator, with an impact on productivity. In an environment of high staff turnover, this can be difficult due to the number of new operators needing training and the loss of experienced operators to provide the required guidance. In this case, there is a need to capture and transfer knowledge from experienced operators to less experienced operators without the overhead and cost of a formal training course. Instead of going to the effort of trying to capture all the knowledge needed to work in a plant in the form of training courses or procedures, another approach is to provide a more direct means of allowing an experienced worker to capture what they know and share it with others. There are several ways in which this might be achieved using some form of knowledge sharing platform. These platforms include a variety of mechanisms for capturing and sharing information, many of which include social media-style interactions that most people are familiar with from their use of technology in their day-to-day lives.

Knowledge sharing platforms aim to provide an easily accessible means of capturing and sharing knowledge that might otherwise be transitory or may disappear when someone leaves an organization. Examples from the process industries include miscellaneous undocumented field procedures, the exact location of a minor piece of equipment, or the quickest way to walk through part of the plant. This sort of knowledge might not warrant formal documentation or, even less likely, the preparation of a training course. However, with a knowledge sharing system in place, it is possible to capture knowledge like this so that it can be made available to others. However, to be effective, knowledge sharing needs to overcome three challenges. First, employees need to know that they have knowledge of value that should be shared. Second, they need to be motivated to share their knowledge. Third, there needs to be a mechanism for sharing knowledge that is easily accessible to both the sharer and the recipient of the knowledge. Layered on these challenges are various individual and organizational barriers that might compromise successful knowledge sharing, including trust, perception of job security, and recognition (Chong and Besharati, 2014).

The rise of Internet-based social media has provided platforms for knowledge sharing that are familiar to everyone. These platforms include text-based chat, photo sharing, and video sharing. As of 2020, YouTube, the most widely used video sharing platform, had 500 hr of video uploaded every minute and one billion hours of videos watched every day (Mansoor, 2020). A significant portion of the content uploaded is instructional content covering many aspects of life including fixing things, learning new skills, product reviews, travelogues, and so on. The ubiquity of the knowledge sharing and seeking activities facilitated by social media platforms has seen knowledge sharing platforms adopted in the enterprise setting, often weaving file sharing, Wiki-style content, text chat, and video sharing all into the one platform. An important aspect of social media-style knowledge sharing is the interaction between knowledge seekers and knowledge sharers. This interaction can include requests for information, feedback on contributions, etc. This interactivity can motivate people to share knowledge through feedback on their contributions and the satisfaction of contributing to a collective knowledge gathering effort.

The principles of social media-style knowledge sharing can be further adopted and adapted in an industrial setting by simplifying content creation and curation to make it very simple for someone to quickly capture content as they go about their work in a plant without having to necessarily plan ahead. Similarly, previously captured content can be made easy to find and navigate, removing barriers to accessing information. Figure 16.2 illustrates the use of video-based knowledge sharing that provides simple capturing and sharing of instructional videos in a mobile app designed specifically for the industrial environment.

While a social media approach can help overcome some of the barriers to effective informal knowledge sharing, such as the motivation to contribute, ease of access, and so on, the technology alone does not address all the individual and organizational issues that might need to be addressed in an effective knowledge sharing strategy. Many of these challenges apply more generally to the digitalization of the workforce and are explored in Section 16.4. Assuming these issues can be overcome, a knowledge sharing approach can provide an effective means to facilitate the capture and retention of knowledge that might otherwise not be available to all who might benefit from it or be lost to an organization altogether.

16.2.3 On-Demand Instruction

The value of knowledge sharing platforms is that they make it very easy to capture knowledge and then access that knowledge when needed. This helps to democratize the process of knowledge transfer to avoid the overhead and cost of developing more formal training courses. A further advantage of social media-style knowledge sharing is that content can readily be consumed on mobile devices, making it possible

FIGURE 16.2. Video-based procedure capture and sharing application.

to access information where and when it is needed. This approach to knowledge sharing enables a form of on-demand competency development where instructional material is delivered to the actual point of need.

This on-demand approach can also apply to the delivery of more formal training material. Electronic or video-based content prepared through a normal curriculum development process can also be delivered to the point where the knowledge is needed via mobile devices. An advantage of this approach is that instructional material can be made available to all workers as routine guidance that helps improve the productivity, consistency, and quality of work for experienced as well as inexperienced workers. This can be very effective, with significant demonstrated benefits (e.g., Robertson et al., 2018).

In the previous discussion of off-the-job training, it was noted that technologies such as VR and MR can provide more effective learning experiences. This approach extends to on-the-job situations where MR can be used to provide a form of direct, situated instruction in which step-by-step instructions can be rendered directly on the plant or equipment a worker is working on. For example, instructions for how to service a piece of equipment can be rendered on the equipment itself in an MR experience that steps the worker through the actions needed to complete the task. An example of this is shown in Figure 16.3.

In principle, it would be possible to avoid off-the-job training altogether with workers learning a new skill on the job when they first need it with additional reinforcement on subsequent occasions. However, this may not be desirable for safety or other practical reasons that mean that workers should have some experience with a procedure before they attempt to perform it in the field for the first time. Using technologies such as MR and VR, it is possible to provide both off- and on-the-job training experiences using essentially the same training material. In the off-the-job

FIGURE 16.3. MR-based instructions for servicing field equipment.

scenario, the training is delivered in MR or VR form that includes a virtual representation of the plant or equipment in question. In the on-the-job scenario, MR is used to render the guidance on the physical plant or equipment itself. Off-the-job training allows workers to get some experience in the required procedures before attempting them in the field. The on-the-job guidance ensures that they perform tasks correctly, completely, and in the correct order. Again, the on-the-job guidance can also be used by experienced workers to help improve the consistency and quality of their work.

16.2.4 Competency Management

Training delivery and knowledge transfer are key elements in addressing the need to enable the workforce with the skills they need to do the job, but a broader problem is managing the use of these approaches to ensure that the workforce is competent. Competency management is an approach to understanding current levels of workforce competency, the gap between that and what is required, and the competency development opportunities available to address that gap.

Competency management can be difficult as there may not be a systematic definition of what skills are needed by various roles, it might be hard to assess current levels of competency to identify gaps, and it may be difficult to marry learning opportunities with gaps where they can be identified. A useful concept in addressing these problems is the competency model. Most organizations have job descriptions for various roles that define the responsibilities of that role. However, this is usually not enough to properly understand what people need to know to do their jobs. A competency model takes this further by defining the specific behaviors required of people in a role. These behaviors are specified in a way that allows performance to be evaluated and learning opportunities for gaps to be identified.

Digitalization can help implement a competency model-based approach in an industrial setting by linking competencies with measurable outcomes derived from worker interaction with digital tools. For example, skills gaps in control systems engineering might be inferred from the performance of the control system, or in records of activity in engineering tools used to configure the control system, field operator competency might be inferred from the time taken to perform field tasks or any rework required, and activity on an operator console correlated with process performance might provide insight into the competency of a console operator. This enables continuous on-the-job assessment of competency that allows skills gaps to be identified more quickly, reducing the risk of negative impact of skills gaps.

Linking competency gaps with relevant learning opportunities helps to close the loop on competency management so that it becomes a continuous cycle of evaluation and competency development. With appropriate application of digitalization, competency management can move more toward *competency control*, in which the workforce continuously develops its competency to meet the changing demands of the organization.

Digitalization provides significant opportunities to greatly improve the way that an industrial organization ensures that its workforce is enabled with the competencies needed to run an efficient and safe operation, both in terms of overall competency management and the specific ways in which people learn the required skills and share knowledge.

16.3 EMPOWERING THE WORKFORCE

Digitalization has a large impact on many roles in an industrial enterprise. The following sections examine several key roles and the sorts of challenges they typically face. This is followed by an exploration of the opportunities for empowering these roles through digitalization in a number of key areas.

16.3.1 Illustrative Roles

Other chapters in this book deal with digitalization tools that help address the concerns of engineers, planners and schedulers, and so on. This section focuses on several key operational roles that need to work together to achieve and maintain efficient and safe production—console operators, field operators, and shift supervisors.

Console operators manage the real-time operation of the production process through interaction with the process via the automation system. Field operators perform a variety of tasks such as routine inspections, manual control of the process when needed, troubleshooting, data collection, and so on. Field operators typically work in close collaboration with console operators and often under their direction. Shift supervisors oversee day-to-day operation of a plant and help ensure that the operation of the process achieves production goals set by operations management through planning and coordination of the work done by console and field operators. The basic goal of this team is to operate the process consistently and efficiently within specified operating limits while avoiding unplanned losses in production capacity. An overarching goal is to run the process safely, minimizing the risk to personnel, the environment, and the plant itself. These roles and their key concerns are illustrated in Figure 16.4.

16.3.2 Console Operator

The primary task of a console operator is to manage the safe, real-time operation of the production process consistently within the specified operating limits for their scope of responsibility, typically a process unit. This will involve maintaining correct feed rates for the unit, together with maintaining correct operating conditions such as temperatures, pressures, and so on. Additional goals will be to minimize energy consumption, the use of raw materials and catalysts, and wear and tear on the unit. In a highly automated process, most or all this will be done by the automation system, and the console operator's task is to monitor the operation of the process and the automation system. An operator will take a more active role in managing the process when the production process needs to move to another

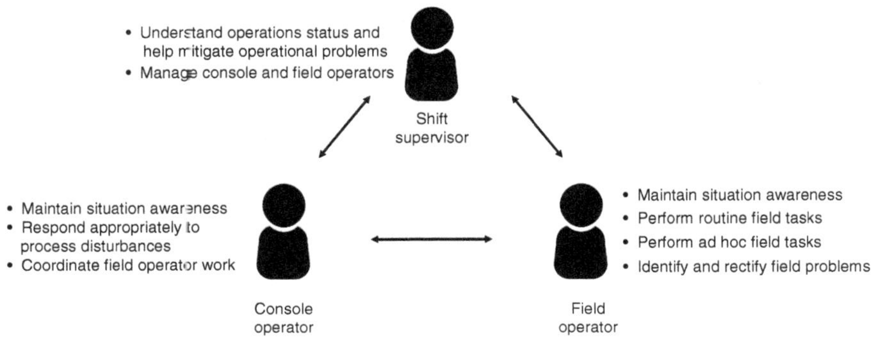

FIGURE 16.4. Key concerns for operations supervisors, console operators, and field operator.

operating mode, such as a change in the grade of product being produced, or when something goes wrong, such as an equipment failure, adverse weather event, or other process disturbance. A console operator's job is a difficult one with long periods of vigilance interrupted by abnormal situations in which an operator's actions can have a direct impact on the safety and profitability of the process. A console operator works closely with field operators, often relying on them to perform manual process operations, report field observations, and so on. They also have a key role to play in ensuring the safety of field operators and others working in the plant. Console operators work closely with shift supervisors to understand current production goals and any planned activities may affect production, such as changes in feedstock, maintenance activities, etc. Operating the process consistently within the prescribed limits requires an operator to maintain situation awareness, respond appropriately to process disturbances, and coordinate the activity of field operators.

Situation awareness allows an operator to understand what the production process is doing now and what it is going to do in the near future and what contingencies may affect the process and how. With sufficient situation awareness, an operator can identify issues as, or preferably, before they arise so they can take appropriate action to minimize the risk to production capacity and safety. For example, without adequate situation awareness, an operator may find it very difficult to assume control of a unit if a model-based predictive controller is unable to control the unit due to an equipment failure or significant process disturbance. Providing an operator with sufficient situation awareness requires careful management and presentation of information and needs to take current operating modes, equipment status, and planned activities into account. As production processes become more complex and the level of automation increases, the problem of obtaining and maintaining situation awareness gets more difficult (Endsley, 1995).

Responding appropriately to process disturbances requires an operator to make the correct process interventions (automation system parameter changes or direct interventions such as opening and closing valves) to return the process to desired operating range or to reduce or stop production if the situation requires. Deciding on the best course of action may be difficult due to complex process dynamics or other contingencies that may affect the future state of the process (e.g., equipment availability). Difficulty in deciding on the best course of action may, naturally, lead to conservative management of the process where a reduction in production or a shutdown is preferred to avoid risk to personnel or equipment.

Another critical aspect of a console operator's job is coordinating field operator activity. Field operators typically have routine, scheduled work to do, such as regular inspection rounds, but they also perform more ad hoc work in support of process operations. This can include manual process operations (e.g., opening a manually operated valve) and collection of data and observations from the field. This work is typically coordinated by the console operator over a two-way radio. These interactions can be error prone due to the nature of the medium and the nature of human voice communication more generally. Field operators may not clearly understand what they are being asked to do, and reports from the field may not be properly heard or understood. Even if a console operator understands reports from the field, they may not correctly transcribe those reports. For example, a report of a tank level from a field operator may be incorrectly entered into the automation system, leading to incorrect operation of the process. Another aspect of coordinating field activity is knowing where people are and what they are doing. This is critical to managing the safety of field workers and optimizing the allocation of work to operators (e.g., asking an operator who is closest to an issue to attend to it). This is an extension of a console operator's situation awareness, but is difficult to establish and maintain when information on what is happening in the field is limited to exchanges over the radio.

While there are many other aspects to a console operator's job, these considerations are central to their ability to perform their job successfully and provide good examples of how digitalization can help make them more effective.

16.3.3 Field Operator

The life of field operator can be a challenging one. They have a range of tasks to perform in difficult and often hazardous conditions while needing to coordinate with a range of other roles, such as console operators, shift supervisors, maintenance crews, and so on. Field operators perform a variety of tasks such as routine inspections, troubleshooting, manual control of the process when needed, data collection, and so on. Field operators respond to issues identified by others, and they also identify problems in the field themselves. Their work is complicated by the need to wear suitable personal protective equipment (PPE) that may impede their ability to move, see, hear, and handle things. Core to the field operator's role are acquiring and maintaining situation awareness of the plant they are working in, performing routine tasks, performing ad hoc tasks under direction of console operators, and identifying and rectifying field issues.

Situation awareness is a critical attribute for console operators. It is also critical for field operators, especially as their personal safety often depends on it. They need to be aware of process conditions around them, what other work is being done, the existence of safety hazards, and so on. This can be difficult as full situation awareness in the field can encompass a rich set of information, yet field operators are typically very information poor. They are typically limited to what they can see, hear, and feel, as well as what they can learn from the console operator over a two-way radio. Based on the limited information they have, it can be difficult for them to perform tasks as safely as possible—is the pressure in a process line below a safe threshold before breaking the line?—or they may not spot potential process or equipment problems because they have a limited view of what is happening around them. Situation awareness for a field operator overlaps with a console operator's situation awareness. Full situation awareness is the union of a console operator's situation awareness and the situation awareness of the field operators they work with. With limited opportunities for sharing information, situation awareness tends to exist in isolated islands, making it difficult for anyone to properly understand the bigger picture.

Performing routine, scheduled tasks such as inspection rounds correctly and efficiently is important for ongoing safety, equipment health, and performance of the process. This can be difficult when inspection rounds are paper based with no verification that the correct inspection tasks are performed at the right locations. Capturing the results of inspection rounds reliably can also be difficult, being subject to inaccuracies in recording observations in the field, made more likely by having to wear gloves and other PPE, and inaccuracies in transcribing and managing that information once the inspection round is finished.

As noted in the previous section, field operators also need to perform other tasks as directed by console operators. Performing the right task at the right place and at the right time can be difficult in a complex process unit. Ensuring these are done correctly has a large bearing on the reliability, efficiency, and safety of a field operator. The first problem is finding reliable ways of allocating work to field operators so that they unambiguously know what they are being asked to do. Field tasks and procedures can be complex with many steps to perform across many pieces of equipment. If a task is not carried out correctly or if the results of a task are not properly recorded, the task may need to be repeated. The consequences may be a lot worse if incorrect performance of a task presents a production or safety risk. A common error is for field operators to perform a required task on the wrong piece of equipment. The timing of field operations can also be important. It might be important to perform tasks as quickly as possible, such as opening or closing manual valves for a material movement lineup, where a delay in performing the tasks delays production. In other cases, activities of multiple field personnel may need to be coordinated so that things happen in the correct order. All of this can be difficult when tasks and procedures are paper based, positive identification of equipment is difficult (e.g., when there are banks of similar equipment such as valves, pumps, heaters, etc.), and the only means of coordinating field activity is a console operator communicating with field operators over a two-way radio.

Finally, field operators represent an opportunity to both identify and rectify issues in the field before they become a risk to safety and production. However, this can be difficult because they are typically reliant on others for relevant information and assistance. With sufficient information and knowledge, they would be in a better position to identify potential issues, decide what to do, and go ahead and do it, with obvious improvements in the efficiency with which issues could be dealt with.

16.3.4 Shift Supervisor

Shift supervisors oversee day-to-day operation of a process unit and help ensure that the operation of the unit achieves production goals set by planners and schedulers. They are the link between the console and field operators and operations management, helping to translate production goals into operational objectives for console and field operators. One of their primary goals is to maintain continuity of production across work shifts by making sure that each shift is suitably staffed and procedures are in place for an orderly handover from one shift to another. They play a key role in coordinating major process events such as unit start-up or shutdown and provide advice and assistance to operators in response to process disturbances, equipment failure, etc. They also help plan process operations to minimize the impact of scheduled maintenance activities. The health and well-being of console and field operators is another important concern for shift supervisors. To perform their job effectively, a shift supervisor needs to understand the current status of operations, planned activities that may affect operations, and any problems that threaten safety or production goals and manage the work of console and field operators.

Staying across the status of operations is key to a shift supervisor's role, although this does not constitute real-time situation awareness in same way it does for console and field operators. Rather, a supervisor needs to synthesize information on current operations together with information from many other sources such as production plans and schedules, maintenance plans and activities, human resources information, and so on. Staying up to date on current operations typically requires explicit effort by the supervisor (e.g., visiting to the control room, monitoring radio traffic, monitoring the automation system, etc.). With other things a supervisor needs to do, it can be easy for them to miss important events. Synthesizing information can be very difficult, typically requiring them to consult disparate information systems to collate and analyze the information they need. This is complicated by the fact that information in each of these system is organized differently requiring a lot of effort to locate the required information. Shift supervisors also need access to information across office, plant, and off-site environments.

Managing the work of console and field operators entails a number of challenges, but two worth pointing out here are ensuring continuity of operations across shifts and managing the well-being of operators. Continuous manufacturing operations typically involve two or three shifts of workers to cover a 24-hr period. Maintaining production performance across shifts requires an orderly handover from one shift to another. The difficulty with this is ensuring that there is enough information exchanged between the outgoing and incoming shifts so that the incoming shift can acquire situation awareness and an understanding of what is planned and expected on their shift. Various forms of shift logs are used for this, but the recording and transfer of information are often incomplete and error prone. This problem also applies to facilitating continuity across operators on the same shift. For example, rectifying an equipment problem may require coordination and information sharing between the supervisors, console operators, and field operators, which is difficult when the primary means of communication is via radio or phone calls.

The well-being of operators is a key concern for shift supervisors. Whether or not a console operator or field operator is fit for the task they are being asked to perform or whether they are being exposed to unacceptable environmental hazards requires monitoring of current worker health and their activity over time. Fatigue and stress levels may make it difficult for a worker to perform well. Cumulative exposure to environmental hazards such as noise may present long-term health risks. An even more basic question is whether a worker is wearing adequate PPE for the environment they are working in. This information can be difficult to monitor and collate to gain an accurate picture of worker well-being.

16.3.5 Impact of Digitalization

Many of the needs of the roles described above fall into three key areas—access to information, communication, and decision making. Advances in digital technology mean that digitalization of the workforce provides many opportunities for addressing needs in these areas in ways that greatly improve the ability of workers to do their jobs safely and efficiently.

16.3.5.1 Access to Information Good ways of accessing information are especially important in an industrial setting where there are vast amounts of information available, driven in large part, by digitalization. More information makes problems of accessing the right information at the right time more difficult. However, as well as adding to the amount of information available, digitalization also provides a range of advances in accessing and presenting information that have much to offer to console operators, field operators, and shift supervisors.

Advances in the presentation of information help provide a console operator with situation awareness and efficient ways of intervening in the process. The console operator's console is their primary source of information. Consoles using large, ultra-high-definition screens, touch panels, ambient information display, etc. provide a platform for presenting large amounts of process information organized in an easy-to-understand structure. Careful use of color and visualization techniques help direct an operator's attention to aspects of a unit's operation that may need their attention, such as alarm conditions. Orchestrated navigation is used to call up a volley of content across the entire console to display the information needed for a particular situation. Ambient alarm indication allows operators to step away from the console without losing sight of the alarm state of their unit. Touch panels combine the display of information with the means of interacting with the process so that an operator can manage the process without splitting their attention between different devices. An example of a modern operator console is shown in Figure 16.5.

A console operator's situation awareness can be further enhanced by providing information on field operator location through the use of real-time location services. This allows an operator to keep track of where people are without relying on them to report their location over the radio. Additional instrumentation of field operators can provide console operators and shift supervisors with information on a field worker's physical well-being and exposure to environmental hazards. Man-down and fall detection can alert console operators to accidents in real time.

A field operator's situation awareness can be greatly enhanced by providing them directly with the information they need without having to rely on others for it. Mobile devices and applications make it possible to give field operators access to the same process information available to console operators. With direct access to information, they are more able to understand process conditions around them and adjust their work accordingly. Additional information can include the existence of hazards, the presence of other workers, and current maintenance activities. The form factor used to deliver information is also important for workers wearing PPE and working with their hands. Wearable, voice-controlled mobile devices allow an operator to access information without interrupting their work. An example of such a device is shown in Figure 16.6.

A further enhancement to providing information to a field operator is to filter information based on the operator's context. For example, using a real-time location service, information can be filtered and presented to a field operator based on their specific location. A field operator could be automatically presented with

FIGURE 16.5. A state-of-the-art operator console.

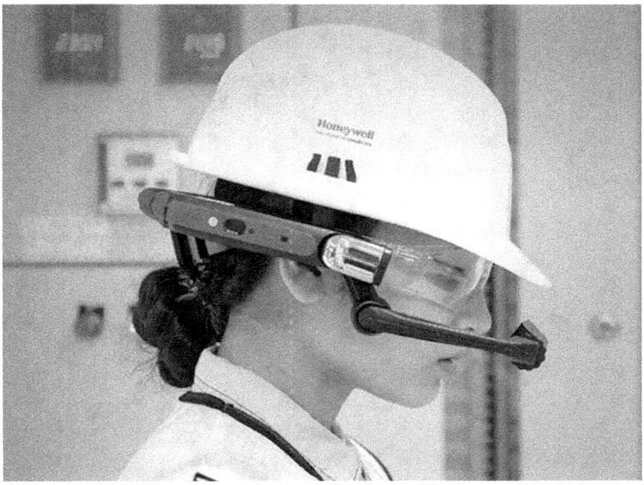

FIGURE 16.6. Wearable, near-to-eye voice-controlled mobile computer.

process conditions and alarms for the process in their immediate vicinity or service manuals for the type of equipment they are working on, avoiding the need for them to manually navigate an information system to find the information they need. This could be further enhanced by using MR to present information directly on the plant and equipment to which it relates. In effect, the physical plant itself provides the organizing structure for plant-related information, and all a field operator needs to do to access the relevant information is to look around the plant. Figure 16.7 illustrates the use of MR to provide a field operator with such a "situated HMI."

Making sure a field operator knows exactly what to do and that they do it either for routine or ad hoc work improves reliability, efficiency, and safety. Electronic procedures delivered via mobile applications can ensure that operators perform the correct steps in the correct order. Data and observations from a procedure can be collected directly in the app, including photos and videos. This data can be automatically sent on to the relevant systems, including the automation system, avoiding delays and errors in transcription. Electronic procedures can be further enhanced by making supporting documentation, such as specifications and reference manuals, accessible directly from the procedure on the mobile device. Positive identification of equipment can also be built into electronic procedures based on location, electronic tags, and visual object recognition to ensure a procedure is carried out on the correct piece of equipment.

Shift supervisors also benefit from mobility, allowing them to access information across all the locations in which they work. In addition to being able to navigate and view information from the relevant systems, mobile applications also allow a supervisor to receive notifications based on conditions relevant to their current concerns. Notifications may result from critical automation system alarms, underperforming performance indicators, completion of critical work activities, etc. With appropriately

FIGURE 16.7. Mobile application providing a mixed reality view of process data superimposed on the plant.

configured notifications, a supervisor can stay abreast of key aspect of process operations without having to rely on contact with console or field operators. Of course, they will make direct contact with those personnel if needed.

Another important impact of digitalization on access to information is being able to navigate and synthesize information from disparate sources more easily. This is achieved through the aggregation of information into data platforms and "data lakes" that allow the data to be queried and analyzed uniformly. The addition of semantic models and knowledge graphs to data platforms allows data to be organized and queried in ways that makes sense to the end user. For example, these mechanisms allow a shift supervisor to find all the information related to a process parameter, such as time series history for the parameter, lab samples, control strategies, plant equipment, maintenance activities, and so on. This effectively unlocks the data and makes it easy for a user to find what they are looking for.

16.3.5.2 Communication Communications between plant personnel is fundamental to effective operations under normal, abnormal, and emergency situations. It is often taken for granted that people know what, when, and how to communicate effectively. In fact, as plant conditions evolve, effective communication often means the difference between success and failure in managing the process.

As discussed above, a lot of communication that takes place in a plant serves information access needs, such as requests for process information from field operators and requests for information on the location of field workers from console operators. Enabling more direct and convenient access to information helps to free up communication channels like two-way radio for more important interaction between team members, such as troubleshooting collaboration, identification and discussion of safety hazards, and so on. However, two-way radio is not always the best medium for these interactions either. Something as simple as text messaging with photos can drastically reduce the time it takes to identify a problem and decide on a course of action. For example, if a field operator finds a leak in a process line, they can take a photo and send it directly to the shift supervisor. The shift supervisor can see the location and size of the leak and decide what should be done. Without this simple mechanism, the shift supervisor may have to visit the field to make the same decision. Richer forms of communication can further improve the effectiveness of interactions between team members with further gains in efficiency. Video calls make it even easier for shift supervisors, console operators, or other specialists to work with field operators, especially when field operators use wearable hands-free devices with built-in cameras that allow a remote colleague to share the field worker's view of the plant. Real-time remote support like this provides several benefits. The interaction is richer, allowing the remote specialists to better understand the situation and provide the right advice and spend less time doing it. Furthermore, the remote specialist can be located anywhere in the world, which means that they can leverage their expertise across geographies.

Once again, technologies such as MR provide additional opportunities for improving the efficiency of work. In the case of remote support, a field operator can be provided with an MR view of the plant that can be annotated by the remote supervisor or specialist to help walk the worker through a maintenance or troubleshooting procedure. These annotations can remain fixed in place relative to the physical environment, providing a very powerful way to provide remote assistance.

Mechanisms such as video calls, perhaps including MR capabilities, help improve the effectiveness of communication across space. As well as improving the effectiveness of communication across space, there is need to improve communication across space *and time*. The problem of shift handover serves to illustrate the problem. Handover between shifts requires communication between people separated in time rather than space. With real-time interaction, communication needs to occur at the boundary between shifts when both shifts are briefly together, making this short interaction critical to a successful handover. It would be better if communication between shifts could occur across time, without relying on the short time available at shift change. The problem of needing to communicate across time is very common in an industrial plant. Most activities in a plant occur over a period of time—hours, days, and weeks. A process problem may take hours to fix, equipment maintenance may take days, and a unit turnaround may take weeks. Around each of these activities, there is communication and collaboration among many people occurring at different times. Working effectively in these situations can be difficult when communication between people occurs over real-time channels such as radios and phones. People not party to a conversation miss out on what was discussed. One solution to these problems is to make communication persistent. Various forms of logbook or notetaking traditionally attempt to do this, but digitalization provides mechanisms for doing this much more effectively.

Email and text messages provide persistent communication that allows people to interact over time and refer back to the conversation at some later stage. Threaded conversations such as those used in Internet forums, bulletin boards, and newsgroups provide persistent conversations open to a larger number of people. Persistent conversations of this form are also common in social media platforms, together with mechanisms such as "likes" and "mentions" that help attract people to a conversation. The use of threaded conversations in an industrial plant is a powerful way to facilitate communication across space and time, especially when the conversations are directly associated with the things they refer to, such as notifications, electronic procedures, shift logs, and so on. For example, a notification of an operating limit violation may spark conversation among console operators, shift supervisors, process engineers, etc. If this conversation is considered part of the notification, then the notification provides the means of accessing the conversation and the context for the conversation. Figure 16.8 illustrates a conversation linked to a notification. This idea can be taken a step further by including the plant itself in the conversation. In this case a notification of a limit violation on a unit might be "posted" or "tweeted" by the automation system on the unit to which operations and engineering staff can respond with contributions to the conversation about the unit and the limit violation. These contributions may include photos and video supporting the conversation and "mentions" of others who should be aware of the conversation. Again, linking the conversation with the unit to which it refers helps to contextualize the conversation, making it easier to find, understand, and later analyze the conversation.

16.3.5.3 Decision Making Being able to make faster, better quality decisions has obvious value in the running of an industrial plant, where slow or poor-quality decisions can result in unplanned capacity losses and potential safety risks. Digitalization can help improve the speed at which decisions are made and acted upon, as well as the quality of those decisions by closing the loop on the decision-making process from detection of problems to resolution and by decentralizing decision making. The quality of decisions can be improved using analytics and simulations that allow the likely consequences of decisions to be more effectively assessed.

A shift supervisor working with operations and maintenance management can make and enact decisions much more quickly with analytics-based detection of problems, decision-making analytics, and automatic generation of instructions and work orders to enact a decision. For example, analytics monitoring data from rotating equipment may detect an emerging bearing problem. The operations team can be automatically alerted to the problem, after which they can use a variety of tools to decide what courses of action will lead to the least disruption to production, for example, fix the problem now or wait and fix it later. If the decision is to fix the problem now, work orders for the maintenance can be raised automatically. Delays in detecting the problem, deciding what to do, and doing it can be minimized. Decision support analytics can also be invoked automatically so that an alert comes to the attention of the operations team along with recommendations on what to do

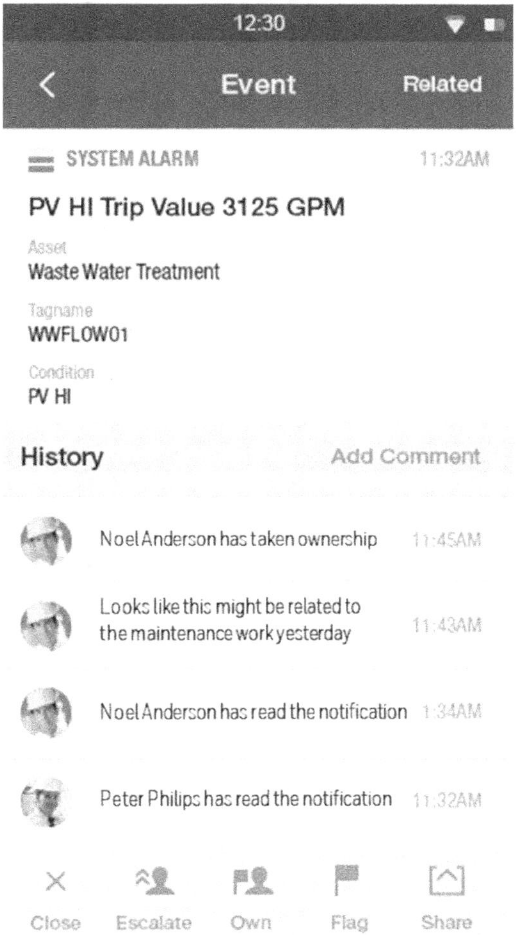

FIGURE 16.8. Notification with associated conversation thread.

about it, further reducing the time to deal with an issue. Figure 16.9 illustrates a solution that provides analytics-based real-time recommendations for maintaining production performance.

An important effect of digitalization on decision making is that it can allow decision making to move away from central engineering and support groups to those in the best position to identify problems and fix them. With better access to information and more effective communication, engineering and operational roles are more able to diagnose and fix issues within their scope of responsibility with less reliance on others for information and expertise. Field operators can benefit from better insight into the operation of the plant around them that they can correlate with their own observations, backed up with simple, rich communication with

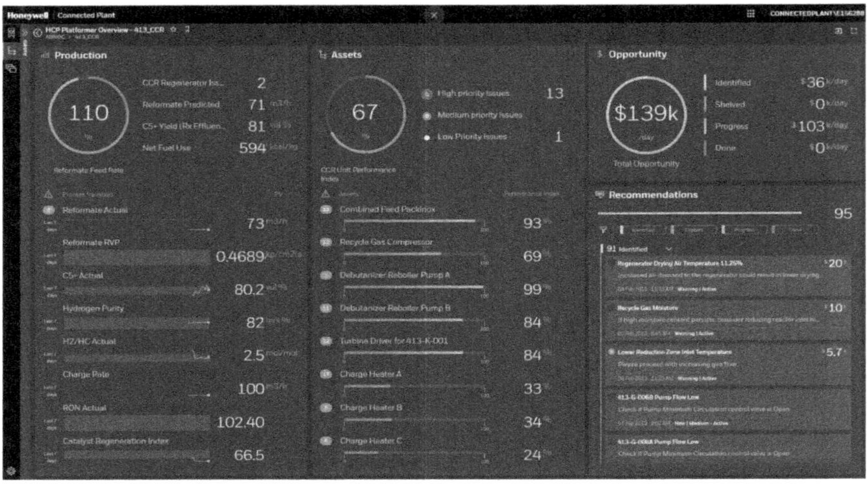

FIGURE 16.9. Real-time production status, analytics, and recommendations solution.

others that may allow decisions on remedial actions to be made on the spot, avoiding potential impact on production.

Models and simulations can provide console operators and shift supervisors with powerful decision-making tools. An essential part of a console operator's situation awareness is understanding not only the current state of the process, but the future state of the process as well. This typically relies on an operator's own understanding of process dynamics, which will vary from operator to operator. This can be made easier by tools such as model-based controllers that can provide a prediction of the process trajectory over the controller's prediction horizon. This provides an indication of what the process will do under the current operating conditions, but it does not help an operator decide what actions they should take to change the trajectory of the process from what it would be without their intervention. This is where dynamic simulations of the process can help. By simulating the behavior of the process with initial conditions derived from the current state of the process, an operator can explore the effects of different interventions on the future state of the process. This form of simulation-based "what-if" analysis can help an operator decide among possible courses of action. Similar comments apply to shift supervisors and other members of the operations team who can use models and simulations to explore and decide between problem-solving strategies.

16.4 DIGITALIZATION CHALLENGES

So far, this chapter has focused largely on the workforce benefits and transformation that digitalization can deliver. However, there are several challenges that need to be addressed for digitalization of the workforce to be successful and sustainable. These include churn created by the rapid development and obsolescence of the technology used in digitalizing the workforce, issues in the relationship between humans and automation systems, and human impacts of digitalization.

16.4.1 Technology Churn

The digitalization that has been described in this chapter is enabled by the rapid development of technology over the past decades. This poses challenges of its own, as new technology displaces older technology in a rapid cycle of development and obsolescence. How can a return on technology-related investments be realized when the lifetime of that technology can be very short? What strategies can be used to protect the investment made in the adoption of these technologies, given that rapid change in this technology is inevitable? Without a way of meeting this challenge, digitalization will quickly fall out of favor with those providing the necessary funding and resources.

It is important to acknowledge that many of the technologies used to digitalize the workforce are subject to rapid change and that this is as much an opportunity as a challenge. After all, the rapid development of mobile and wearable technology has provided, and will continue to provide, new opportunities for improving the effectiveness of the workforce. What is needed is a strategy that accommodates rapid development and adoption of technology while minimizing the impact of technology churn by separating the changing technology from more enduring investments in supporting infrastructure and changes to workforce behavior.

Much of the enabling technology such as mobile and wearable devices and related human–computer interaction technology depends on infrastructure such as wireless connectivity, back-end system integration, security services, location services, and so on. Investment in this sort of infrastructure will provide a platform for adopting and exploiting new mobile and wearable technology as it becomes available. This is a particular concern for industrial enterprises since infrastructure such as wireless connectivity and location services cannot be taken for granted, as is increasingly the case in the commercial domain. For example, comprehensive Wi-Fi coverage can be difficult to provide given interference and black spots caused by complex steel structures in a plant. Similar problems exist for 4G or 5G cell services and for GPS-based location services. As a result, careful planning and engineering is required to provide these enabling services using technology designed for industrial application, such as industrial wireless technology and locally deployed location services.

Longevity of infrastructure investments can be significantly enhanced through strategic adoption of industry standards. For example, adoption of infrastructure based on well-supported system interoperability standards will allow new technology supporting those standards to be more readily adopted and integrated with that infrastructure. Other areas where standards adoption is particularly important are in wireless communication, where NFC, Bluetooth, Wi-Fi, and ISA 100 are relevant, and security, where standard protocols such as OAuth and OpenID Connect facilitate secure, user-friendly integration of technology.

Another strategy for mitigating the effects of technology churn is to focus on fostering the changes in workforce behavior enabled by the introduction of technology. Changes such as more active knowledge seeking and sharing, increased worker autonomy that results from more localized decision making, and more frequent and richer communication between workers are behaviors that are likely to be further enhanced through the development of new technology. By fostering these behaviors as the outcome of digitalization, the focus can shift from technology as the solution to technology as an enabler of longer-term behavioral change. The investment is in behavioral change, which will outlive any one enabling technology.

16.4.2 Trust and the Ironies of Automation

Another area that deserves special consideration when applying digitalization is the degree to which humans trust the technology they are using. Digitalization increasingly creates situations where humans and technology work together with certain responsibilities that may have previously fallen to humans now falling to digital tools, such as model predictive control (MPC) and machine learning algorithms. For humans to work effectively with these tools, they need to have an appropriate level of trust in those tools. Too little trust and the digital tools will not be allowed to carry out their responsibilities as effectively as they might. Too much trust and tools may provide unchallenged erroneous results that negate the value they might otherwise deliver. This issue is well documented in relation to autonomous control systems but also applies to analytics, machine learning, and artificial intelligence-based tools in general.

Trust in an automated system can be described as the "attitude that an agent will help achieve an individual's goal in a situation characterized by uncertainty and vulnerability" (Lee and See, 2004). Trust in automation parallels trust in human colleagues in that it is necessary to know something about the motivation, expected behavior, and reasoning of an automated system before it can be assigned an appropriate level of trust. A key element in establishing an appropriate level of trust is a degree of transparency in the operation of an algorithm that allows an assessment of whether the algorithm is behaving reasonably in a given circumstance. With sufficient transparency, a human can decide whether an algorithm is behaving appropriately and treat the results provided by the algorithm accordingly. Insufficient trust may lead to humans actively subverting the behavior of an algorithm by disabling or constraining its behavior. Too much trust and it may be allowed to operate outside its design limits providing misleading decision-making support leading to suboptimal control of the process.

Providing sufficient transparency for algorithms such as MPC can be challenging without assuming a level of expertise in the operation of MPC controllers. Operators may not understand why a controller is manipulating the process in the way it is, with the result that they may disable the controller because they are uncomfortable with its behavior. Nonetheless, it is possible to provide a degree of transparency for MPC algorithms that helps build trust with an operator through the use of appropriate visualization and interaction techniques. The situation is more difficult with machine learning and deep learning algorithms, due to the opaque, nonlinear nature of the mechanisms used. This is an open research challenge with approaches such as visual analytics, embedded inference mechanisms, and anomaly detection currently being applied to help make the operation of machine learning techniques more transparent (e.g., Choo and Liu, 2018). As techniques for improving the transparency of these techniques become more widely available, it will become easier to successfully deploy these techniques in an industrial setting. Another approach to building an appropriate level of trust is to provide results together with probability distributions associated with the results. This provides additional information that can be used to assess the trustworthiness of the results.

In addition to trust issues, there is another aspect of automation that needs to be considered, particularly as the scope and sophistication of industrial automation increase with digitalization. It is tempting to think that increasing industrial automation will reduce the level of skill needed in human operators of the plant. However, the fact that most industrial plants face disturbances and failures that mean that automated control inevitably falls back to direct management by human operators from time to time can put those operators in a very difficult position. Operators face two challenges should they need to take over management of the process from the automation system. The first is that they may be taking over a process that is outside their "comfort zone," making it difficult for them to effectively manage the process. This is exacerbated by the fact that something unusual is happening, which is why the automation system has yielded management of the process. The result of this is that human operators need to be more skilled in managing the process than they would in the absence of automation. This is known as the "irony of automation" (Bainbridge, 1982). A related phenomenon is the so-called automation conundrum, in which the presence of automation shields an operator from the current state and dynamics of a process (Endsley, 2017). The resulting out-of-the-loop loss of situation awareness makes it hard for operators to take over management of the process, meaning that at the time they most need to intervene in a process, they are least able to do so.

One emerging approach to the issues outlined in this section, known as human autonomy teaming, builds on notions of transparency and trust to elevate the role of the automation system to that of a team member working alongside the human operator (Shively et al., 2017). Transparency, communication, and trust between the human operator and the automation system can lead to a working relationship in which there is a shared understanding of what each other is doing and why. This helps to make sure work is appropriately allocated between human and machine and ensures that responsibility can transition from one to the other without adverse impacts on production.

16.4.3 Human Impact of Digitalization

Another important area to consider in applying digitalization to the workforce is the impact that technology has on people beyond the productivity and safety benefits outlined in this chapter. The use of technology both within the workplace and outside may have unintended consequences for worker well-being, productivity, and motivation that may detract from the overall benefits promised by digitalization if not carefully considered. These include technology-induced stress, digital distractions, and privacy issues.

Digitalization has a profound effect on how workers go about their tasks. Indeed, this is the whole point of digitalization—to make operations for people safer and more efficient. However, the transformation wrought by digitalization represents major change in the lives of workers, and workplace change can be a major source of stress for workers. Workplace stress, in general, is a widely studied phenomenon that can lead to workers suffering physical symptoms of anxiety, feelings of being overwhelmed and frustrated, and a concomitant loss of productivity and risk of ongoing mental health problems. "Technostress" can contribute to the general causes of workplace stress—overwork, uncertainty, and lack of control—in the so-called areas of techno-overload, techno-invasion, techno-insecurity, techno-complexity, and techno-uncertainty (Tarafdar et al., 2007). Techno-overload refers to the degree of increase in the amount of work, change in working habits, and demand for improved task performance. Techno-invasion refers to the intrusion of technology into an individual's private life. Being "always connected" to work 24 hr a day is a common example of this. Techno-insecurity refers to situations in which users feel threatened about losing their jobs either to new technology or to other people who have a better understanding of the technology. Techno-complexity refers to the complexity of information systems and the effort required to learn how to use them. Finally, techno-uncertainty refers to the churn in technology discussed in Section 16.4.1. A full exploration of the effects of these sources of stress is beyond the scope of this chapter, but one effect is particularly pertinent to digitalization efforts. The level of discomfort in new technology resulting from technostress may lead workers to avoid using technology or, worse, subverting the technology. Obviously, workers failing to use a technology prevents the benefits of that technology being realized. Subversion of technology can have significant negative impacts. For example, technostress has been implicated in failure of employees to comply with organizational security protocols and procedures (Hwang and Cha, 2018). One strategy that can be used to help mitigate some of the issues associated with introducing new technology is to include workers in the design and rollout of the technology. A good example of this is the design of new consoles for console operators. End-user acceptance of a new console is critical for the success of the console in facilitating improvements in operator effectiveness. By involving operators in the design process for a new console, they get to steer the design based on their experience, gain a sense of ownership of the console, and are familiar with how it will work long before it arrives in their control room.

Enabling the workforce through digitalization leverages, in large part, the broader technological milieu that underpins all facets of modern life. In digitalization, we find valuable applications of this technology in industrial enterprises. However, the issue of workplace distractions caused by the ubiquitous connectivity we enjoy often raises its head. The source of these distractions may be work related or personal. Digitalization itself can introduce a new range of notifications, news feeds, and so on that help move information around a workforce. In many cases, workers will also have their personal smartphones or other devices with them in the workplace with the attendant risk that they will spend time checking social media feeds, responding to notifications, and so on, which have nothing to do with their jobs. How much of a problem is this? General anecdotal observations of people always being glued to their phones, as are common in the media, suggest that digital distractions at work could be a significant problem. However, the problem may not be as bad as it seems, and where it is a problem, there may be underlying causes that would lead to an impact on productivity regardless of the outlet for distraction. Many workers themselves believe that using social media-style platforms hurts productivity, but people tend to significantly overestimate the amount of time they spend on these platforms (Junco, 2013), so the problem may not actually be as bad as it seems. More importantly, distractions caused by access to social media and the like may be symptomatic of a more general lack of engagement that makes workers susceptible to distraction. Many workers simply do not have time to be distracted. However, in other cases, the nature of a job may provide an opportunity for distraction for which work and personal social media may become an outlet. For example, the vigilance component of a console operator's job provides ample opportunity for distraction, and in this case, the impact on job performance should be carefully considered as multitasking introduced by any form of distraction can be detrimental. These issues are, by no means, unique to industrial workplaces, but careful consideration should be given to them in an industrial digitalization effort so that workers are empowered and engaged in their work to reduce the opportunity for counterproductive distraction, especially where that may impact production efficiency or safety.

Privacy is another area of concern when empowering the workforce through digitalization. While digitalization provides many opportunities for improving the way workers go about their tasks using digital tools, those same tools also provide many opportunities for monitoring the behavior and performance of workers. In many cases, this may be one of the motivations for deploying tools such as mobile workflow tools. Being able to record where the workers are and what they are doing can be important for optimizing work procedures or monitoring the safety of workers. However, the active monitoring and recording of worker activity may raise privacy issues for workers when this monitoring goes beyond what is considered acceptable. In some cases, this may lead to worker dissatisfaction and rejection of the tools enabling the monitoring—a manifestation of techno-invasion—with consequences for the efficacy of the digitalization effort. In some jurisdictions there are laws governing worker privacy, while in others, issues such as these may be taken up by unions representing the workforce. Legislation, where it exists, such as the European General Data Protection Regulation (GDPR) can provide concrete guidance on what is acceptable. In other cases, it might be prudent to adopt of position based on respect for workers' attitudes to privacy and, where possible, provide workers with some degree of control over privacy-related issues. For example, smart PPE can provide clear benefits in the management of worker safety. In this case there is a strong argument to be made that the ability to monitor the location of a worker outweighs the workers' concerns for their privacy. Even so, workers may be permitted to opt out of location tracking when not working in certain parts of the plant. Once again, getting involvement and buy-in from workers during a digitalization effort can help workers understand the types of monitoring that are proposed, understand the benefits, and guide the design and implementation of systems to help address concerns they may have.

While this section has described a range of challenges that a digitalization effort may need to address, it is important to remember that these can be mitigated when they do arise and doing so opens the door to realizing the very significant benefits that digitalizing the workforce can deliver.

16.5 SUMMARY

This chapter has surveyed the impact of digitalization of the workforce in two broad areas—enabling the workforce though competency development and knowledge sharing and empowering the workforce through the application of digital tools to transform work practices. Through a well-planned digitalization strategy, it is possible to implement a range of solutions that allow workers to do more, do it more efficiently, and do it more safely. In many cases, these solutions are becoming essential to the ongoing viability of industrial enterprises as they ride out demographic change, changing market expectations, and a changing regulatory environment. However, while digitalization can deliver compelling benefits, there are a range of challenges that may need to be addressed in a successful digitalization effort. By carefully considering the application of digitalization to areas of greatest value along with strategies for mitigating the challenges that can arise, it is possible to make a significant step change in how the workforce is equipped to do the job and how they go about doing it.

REFERENCES

Bainbridge L (1982), Ironies of automation. In: Johannsen G, Rijnsdorp JE (Eds.), Analysis, Design and Evaluation of Man–Machine Systems. Pergamon Press, pp. 129–135.

Borsci S, Lawson G, Jha B, Burges M, Salanitri D (2016), Effectiveness of a multidevice 3D virtual environment application to train car service maintenance procedures, Virtual Reality **20** (1), 41–55.

Chong CW, Besharati J (2014), Challenges of knowledge sharing in the petrochemical industry, Knowl. Manag. E-Learn. Int. J. **6** (2), 171–187.

Choo J, Liu S (2018), Visual analytics for explainable deep learning, IEEE Comput. Graph. Appl. **38** (4), 84–92.

Endsley MR (1995), Toward a theory of situation awareness in dynamic systems. Hum. Factors 37(1), 32–64.

Endsley MR (2017), From here to autonomy: lessons learned from human–automation research, Hum. Factors **59** (1), 5–27.

Hwang I, Cha O (2018), Examining technostress creators and role stress as potential threats to employees' information security compliance, Comput. Hum. Behav. **81**, 282–293.

Junco R (2013), Comparing actual and self-reported measures of Facebook use, Comput. Hum. Behav. **29** (3), 626–631.

Lahti I, Rosin T, Qvist P, Vuorela V, Luimula M, Smed J (2014), IndustrySim: Finding the fun in industrial simulations. In: 2014 6th International Conference on Games and Virtual Worlds for Serious Applications (VS-GAMES), Msida, Malta. IEEE, pp. 1–4.

Lee JD, See KA (2004), Trust in automation: designing for appropriate reliance, Hum. Factors **46** (1), 50–80.

Mansoor I. (2020), YouTube Revenue and Usage Statistics. https://www.businessofapps.com/data/youtube-statistics/. accessed April 24, 2020.

Robertson T, Bischof J, Geyman M, Lise E (2018), Reducing maintenance error with wearable technology. In: 2018 Annual Reliability and Maintainability Symposium (RAMS). IEEE, pp. 1–6.

Shively RJ, Lachter J, Brandt SL, Matessa M, Battiste V, Johnson WW (2017), Why human-autonomy teaming? In: International Conference on Applied Human Factors and Ergonomics. Springer, Cham, pp. 3–11.

Tarafdar M, Tu Q, Ragu-Nathan BS, Ragu-Nathan TS (2007), The impact of technostress on role stress and productivity, J. Manag. Inf. Syst. **24** (1), 301–328.

PART 4

DIGITAL SOLUTIONS FOR SMART OPERATIONS

Part 4 includes four chapters. Chapter 17 explains the digital platform for Connected Plant, and Honeywell Forge is used as an example for illustration although other digital platforms can be used. Furthermore, the digitalization readiness assessment will be described in Chapter 18. The goal of digital readiness assessment is to determine the digital maturity of the organization and help the organization to understand the organization readiness for digital transformation given the current IT infrastructure and OT/ET capabilities.

In addition, the approach for managing digital projects will be illustrated in Chapter 19. It will explain how to build a business case for digital projects and what takes to develop and implement digital projects. What care must be taken when developing and implementing digital projects.

Finally, several meaningful case studies for digital projects are provided in Chapter 20 that demonstrate that the Connected Plant is more than just data collection and software tools. It is about applying intelligent models to reveal insights. It is about integrating new digital solutions with existing data infrastructure. It is about better management of asset, workflows, and knowledge via digital solutions.

As the result of the increased connectivity, more influx of data is generated from new contexts and the amount of data becomes overwhelming. That is why digitalization comes in and it is about how to make the data more intelligent in order to derive new insights and make more informed actions and decisions. The best outcome for digitalization is achieved by the proper integration of IT (data centric), OT (Operation Technology; control centric), and ET (Engineering Technology; model centric), which is designed to monitor current events and predict new events, detect early warning signs of failure, and make changes to process operations and enterprise business for better profit, efficiency, reliability, and safety.

Digitalization and Analytics for Smart Plant Performance: Theory and Applications, First Edition. Frank (Xin X.) Zhu.

17

HONEYWELL FORGE: THE PLATFORM FOR CONNECTED PLANT

MATT BURD[1] AND FRANK (XIN X.) ZHU[2]

[1] *Honeywell Connected Enterprise, Canada*
[2] *Honeywell UOP, USA*

17.1 HONEYWELL FORGE: A DIGITAL PLATFORM FOR CONNECTED PLANT

Honeywell Forge for Industrial is an example of an enterprise performance management solution that uses real-time accurate models and advanced analytics to deliver intelligent actionable recommendations for sustained peak performance. We use this example to illustrate what an industrial digital platform looks like and how it works.

Built with the Honeywell Forge platform, it is a powerful analytics software solution that provides real-time data and visual intelligence. Connectivity is based on an extensible platform (Figure 17.1) that provides an enterprise-wide, top-to-bottom view, displaying the status of process, assets, people, and safety.

17.2 IIOT FOR DATA INFRASTRUCTURE

Given the importance of data and information, it is important to treat and develop data as a central business asset. This means data is actively managed at all stages of the data life cycle (i.e., collected, stored, analyzed, shared, and archived) through defined data practices, standards, and policies. Commercial software offerings have

Digitalization and Analytics for Smart Plant Performance: Theory and Applications, First Edition. Frank (Xin X.) Zhu.

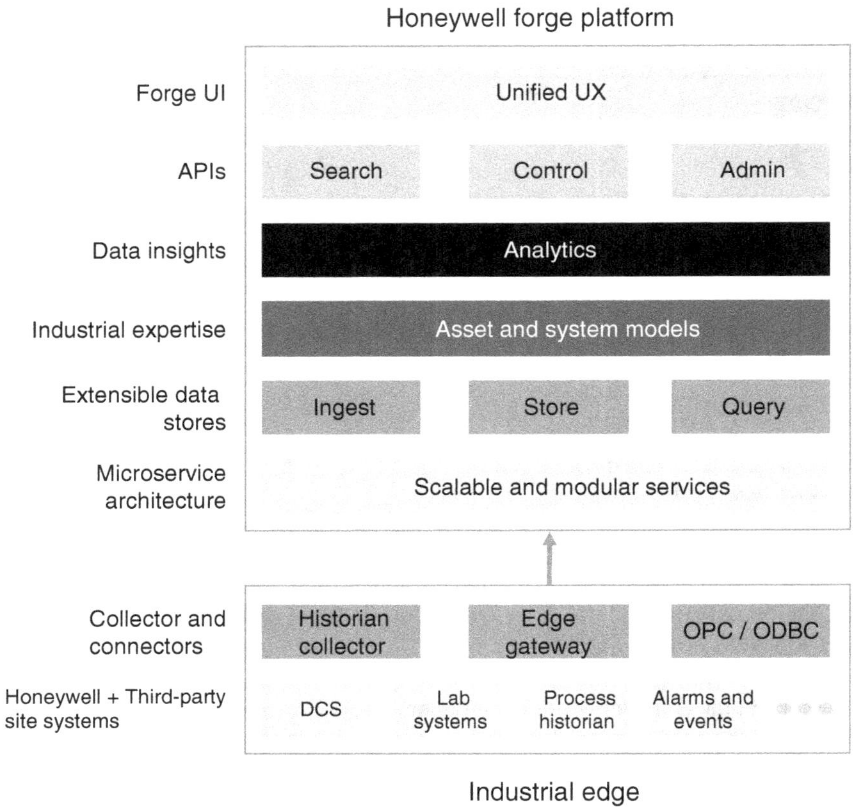

FIGURE 17.1. Honeywell Forge platform.

additional considerations including data governance, protection of personal information, and regional regulatory requirements related to data.

Figure 17.2 shows the data life cycle for the Connected Plant services. The process and lab data are collected directly from the site historian (see bottom left) and then enters the Honeywell Forge cloud, which connects different sources of data. Cloud offers several advantages including fast computation, large storage, and transparent data overview. Four main steps are performed in the Honeywell cloud:

(1) Data cleansing: Advanced data cleansing routines are executed, including data correction, mass balance reconciliation, data conditioning, component balancing, and simulation to ensure the best information is used as a basis for further processing.
(2) Advanced and fast computation: Cleansed data is run through the solution specific digital twins, for example, Reliability Advisor—this step would contain the process models to determine the current operation and the fault models to trigger any early detection and the appropriate resolution. For the

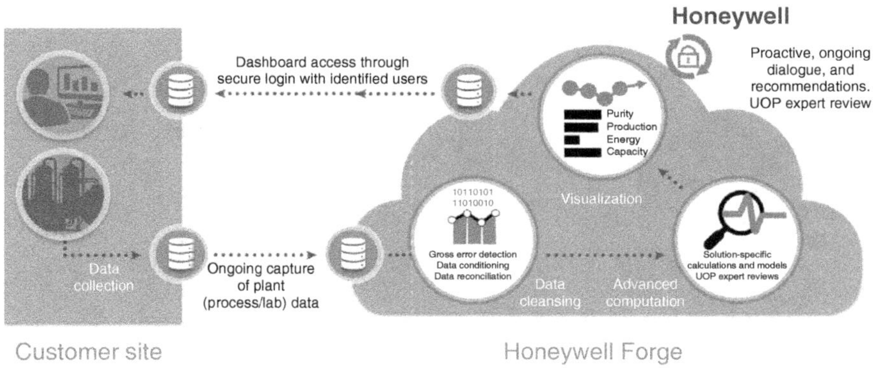

FIGURE 17.2. Data flow manufacturing supply chain.

Optimization Advisor digital twin where real-time economic data is integrated with real-time process data, optimization of the digital twin selects the right feed for a process and determines optimal process conditions and product yields.

(3) Friendly visualization: The results from the solution engine are presented to users in dashboards to drive situational awareness and improve understanding of constraints and opportunities; these dashboards are then accessible by the customer through the cloud.

(4) Turning data into outcomes: The last step needs to close the loop from the digital sphere back to the "real world" by translating conclusions from the data analyses into recommendations and—ultimately—actions. Many decision-making processes still require human involvement, while data analyses are often already automated and performed in real time.

In the following, an excellent example of the digital platform for Connected Plant is discussed in detail.

17.3 HOW IT WORKS?

Honeywell Forge supports the end-to-end capability to execute digital twins against process data and translate their output into actionable insights (Figure 17.3). Without this overall framework, even the best digital twins and analytics cannot be effective.

There are six parts to the digital twin framework:

(1) Edge connectivity for accessing actual plant process and quality data;
(2) Enterprise data for processing, storing, and contextualizing data;
(3) Digital twins for determining current and ideal performance based on actual plant conditions;

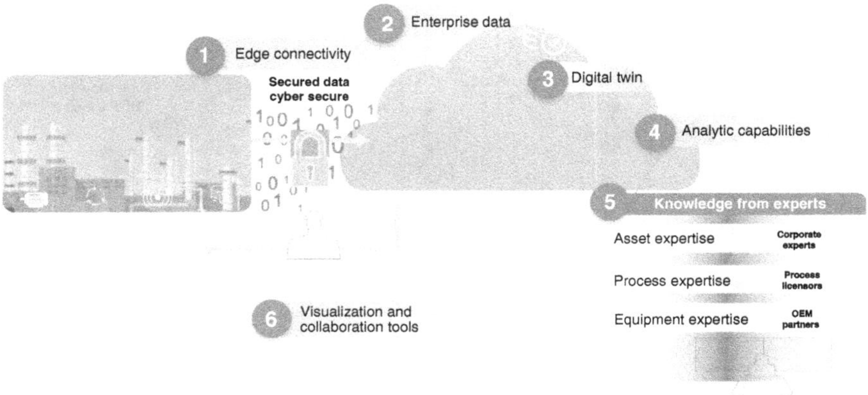

FIGURE 17.3. Forge digital twin framework.

(4) Analytics for evaluating and predicting issues and potential causes;
(5) Knowledge base for recommending follow-up actions;
(6) Visualization and collaboration tools for engaging users.

17.3.1 Edge Connectivity

Data is ingested into Forge from a variety of sources, typically process historians and laboratory information management systems. One or more Forge Connect data collectors are installed at side ("the edge") and send data securely to the Forge platform running in the cloud (Figure 17.4).

Forge Connect ensures robust, secure communication of data between site and cloud. A secure onboarding process ensures that only authorized data is sent to the cloud and that the cloud only accepts data from authorized collectors. After any network or system outage, data transfer will resume once communication is reestablished, and any data missed during the outage will be backfilled from the source system or Forge Connect's own cache.

Forge Connect supports accessing time series, alarm and event, and transactional data via variety of protocols including OPC, ODBC, and files. Once received in the cloud, raw data is stored as time series "tags" or events in warm storage to support interactive queries and to cold storage for archive purposes. Data can be further sent to a data lake for offline analytics development.

17.3.2 Enterprise Data

Honeywell Forge is a model-driven architecture. Site data is contextualized using an extensible object model ("asset model") where the equipment and process of the plant or facility are modeled (Figure 17.5).

Asset templates are used to facilitate configuration of asset instances with common structure. Honeywell has extensive libraries of asset templates for different

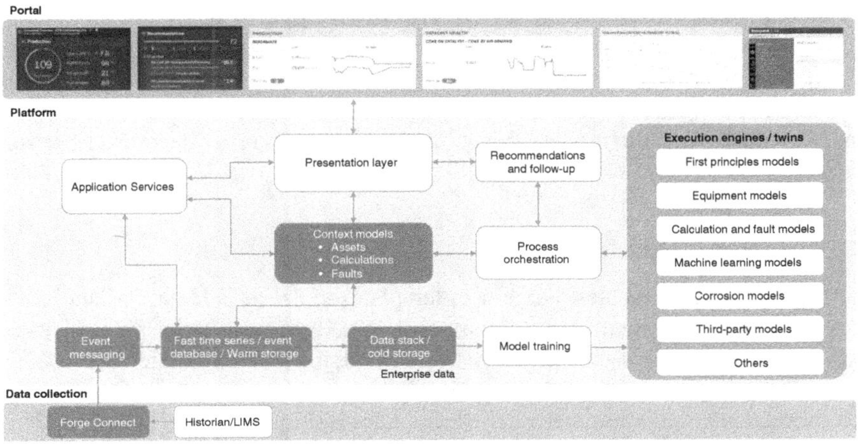

FIGURE 17.4. Edge connectivity in Honeywell Forge.

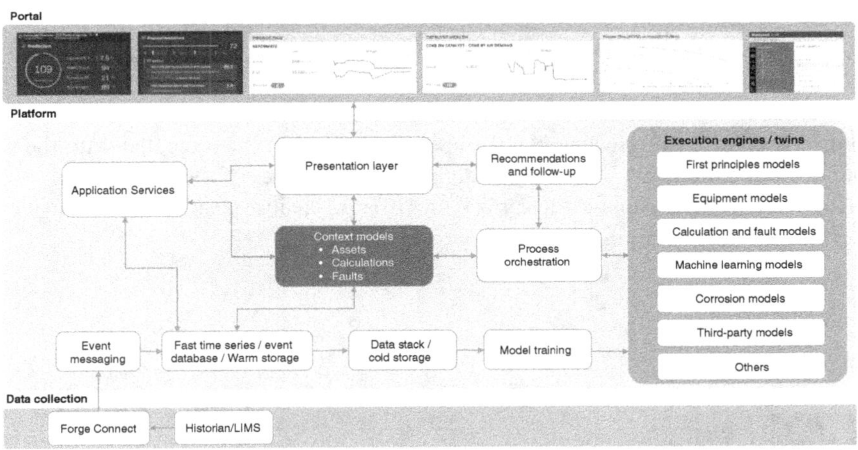

FIGURE 17.5. Context modeling in Honeywell Forge.

processes and equipment types. An asset template defines the typical properties for assets of that type, for example, a pump has inlet and outlet pressures, speed, flow, etc. Hierarchical or derived types can accommodate variations of a base type, for example, a reciprocating pump is a specialization of a base pump type and adds additional properties to the template.

Asset instances are configured to match the actual plant or facility using templates to define the expected attributes of the asset. Each attribute is then configured either as static value (e.g., capacity = 1,000 BPH) or with a reference to a time series tag that provides the value. Where possible this tag to attribute mapping is done automatically based on naming conventions, parsing and matching tag and attribute descriptions, or even comparing the behavior of the time series data with expected behavior.

This data contextualization is a key part of the digital twin solution. It maps data from the real-world process to a static digital representation of the process that digital twin analytics can understand and abstracts differences in naming conventions and other variations between plants. This allows standard calculations, simulations, fault models, visualizations, and more to be reused across similar plants.

17.3.3 Analytics Models

In addition to asset templates, Forge uses templates to define calculations and analytics and how they relate to assets. For example, the Honeywell Forge for Industrial Asset Performance Management (APM) solution defines pump efficiency calculations for the pump template, such that every time a pump is configured, the standard efficiency calculation is automatically executed for it (Figure 17.6). The calculation model defines the various types of calculations, which type of engine should run them, what the input and output parameters are, what the preprocessing requirements and prerequisites, what is the schedule, etc. The actual calculation or analytic logic may be defined in the template, or it may just be referenced. In this way the calculation model can be used to describe and control the execution of a variety of different process models.

Calculation templates can be linked with asset templates such that when as asset instance is create, any associated calculation instances are also created with their input and out parameters linked to the appropriate attributes of the asset.

Honeywell Forge supports a variety of different analytics models, which are described in more detail in Section 17.3:

- First principle models;
- Empirical models;

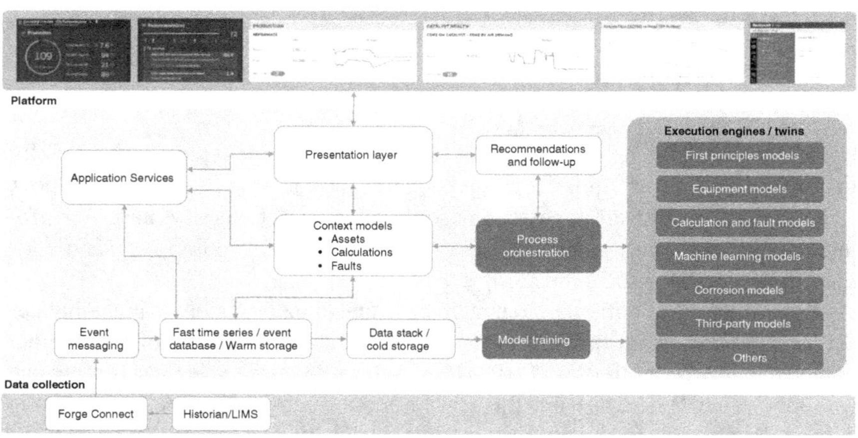

FIGURE 17.6. Analytics modeling in Honeywell Forge.

- Engineered models;
- User-defined models;
- Machine learning models;
- Built-in functions.

Use of the above models, based on current plant data, can determine current plant performance as well as what performance should be, could be, or will be. The models themselves make no judgment as to whether the difference between current and predicted performance represents an issue, an opportunity, or just normal variation. That is determined in the next step of the process.

Fault models are used to compare current and predicted plant performance to identify issues or opportunities and the potential causes or drivers for them. Honeywell Forge includes rich hierarchical symptom-fault models to identify abnormal conditions and their potential consequences. This supports drilling down from a high-level condition to understand the contributing factors as well as determining the potential impact a lower-level condition might have.

There can be multiple fault models for a given plant looking at different aspects such as process, equipment, control, and operations. Each of these can identify issues and opportunities in their domain, often targeted at specific persona in the plant, but may just be looking at the same core problem from a different perspective. An overall fault model layered on top can synthesize these different perspectives into an overall assessment of the situation and point to the true root cause.

Where possible the problem or opportunity identified by the fault model is translated to an economic impact. For example, reduced throughput due to equipment or other constraints can have a direct and measurable impact on plant margins.

17.3.4 Knowledge Base

Once a fault or opportunity has been identified, Honeywell Forge can make recommendations about the best corrective actions to take. This is initially based on expert knowledge that has been pre-seeded into the system by process and equipment experts. A recommendation service (Figure 17.7) presents this information in a consistent way regardless of source and supports workflows to track, close out, and document the recommendation follow-up.

Recommendation follow-up can be used to improve the overall knowledge based over time, as existing recommendations are validated (or not) or new cause-and-effect relationships are learned by users or analytics.

17.3.5 Visualization and Collaboration

Honeywell Forge provide rich tools for visualizing plant performance and guiding users to the key issues that require attention (Figure 17.8).

Dashboards (Figure 17.9) provide a high-level overview with drill-downs to support deeper investigations. Recommendation summaries give users prioritized

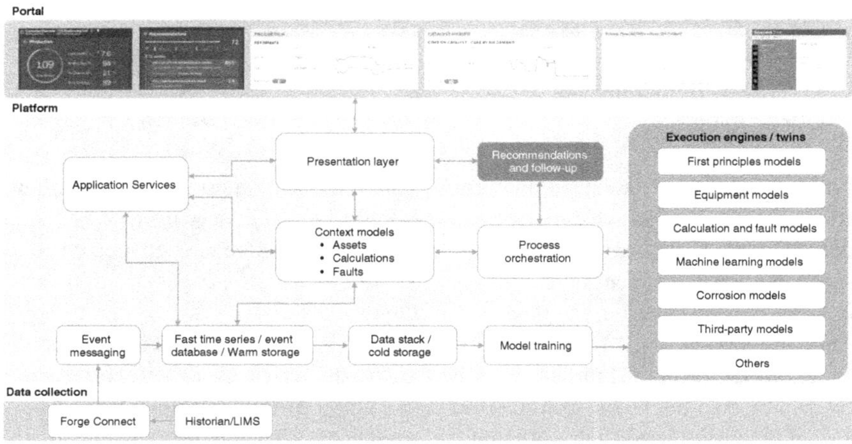

FIGURE 17.7. Recommendation service in Honeywell Forge.

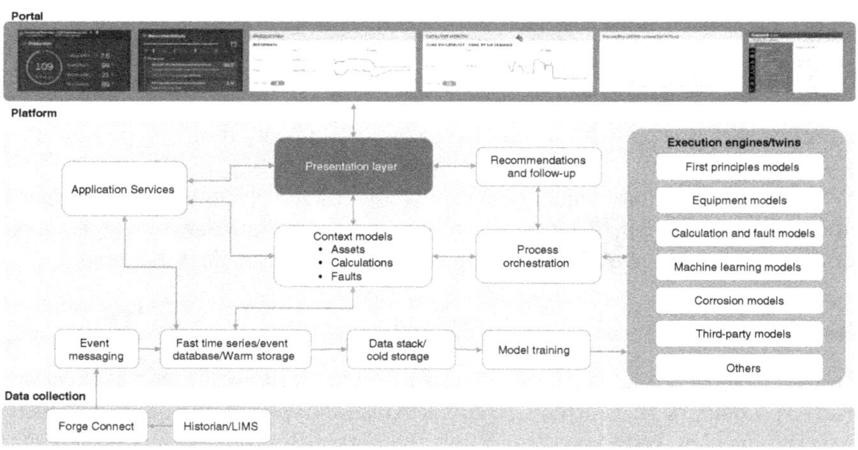

FIGURE 17.8. Visualization in Honeywell Forge.

actions to address current or potential issues and opportunities. Data analysis tools support ad hoc data exploration to assist in troubleshooting and process improvement.

17.3.6 Deployment

Honeywell Forge is based on a modern cloud architecture that supports rapid adoption of new features and the ability to scale to meet the needs of large and small customers. It is not only offered as a Honeywell-managed software as a service but can also be deployed in a customer's cloud or data center if needed.

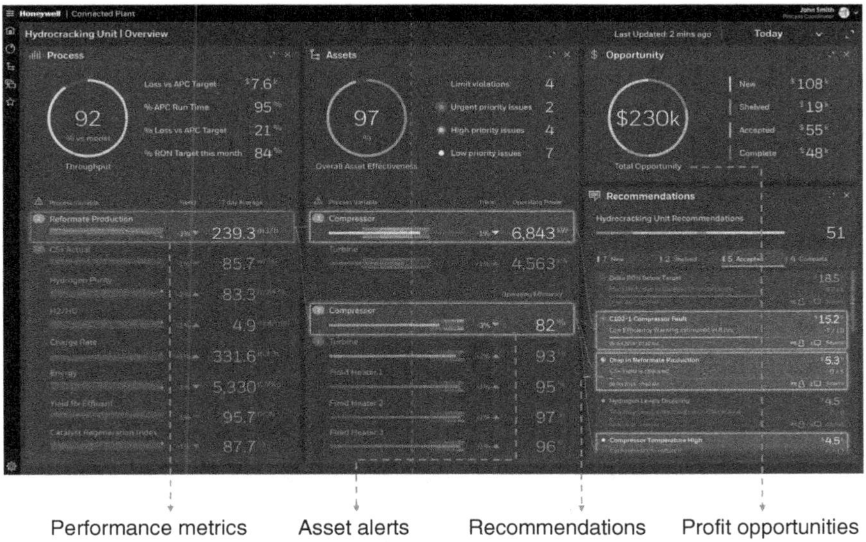

FIGURE 17.9. Visual display of Honeywell Forge for industrial.

17.4 INTELLIGENT MODELS BEHIND DIGITAL TWINS IN HONEYWELL FORGE

The cornerstone for a digital twin is a set of prediction model with high accuracy. Good models can help us to categorize, explain, and most importantly predict. Indeed, experience and information can be good teachers, but many times in asset management, we simply cannot afford to learn on the job. One cannot go through several incidents to learn how to protect critical equipment. Our experience shows that failure forecasting can be inaccurate up to half the time. Suboptimal service providers are unable to offer high-fidelity prediction models and thus cannot provide coordinated, integrated, and timely maintenance.

This is why models become so valuable in digital twin methodology: accurately predicting what will happen before it occurs and interpret the status of the installed base, thus enabling operators to quickly initiate maintenance measures when irregularities occur. As an example, Figure 17.10 provides an overview of modeling techniques used in digital twin for APM:

- **Rigorous models**: These are converted from process design simulation that consists of mass and energy balances, proprietary rigorous reaction models, product fractionation and impurity separation (Zhu et al., 2019), rotating equipment, fired heaters, heat exchangers, operating limits, and many more. In this manner, process design is integrated with feed conditions and production requirement. Process changes and technology improvement often provide

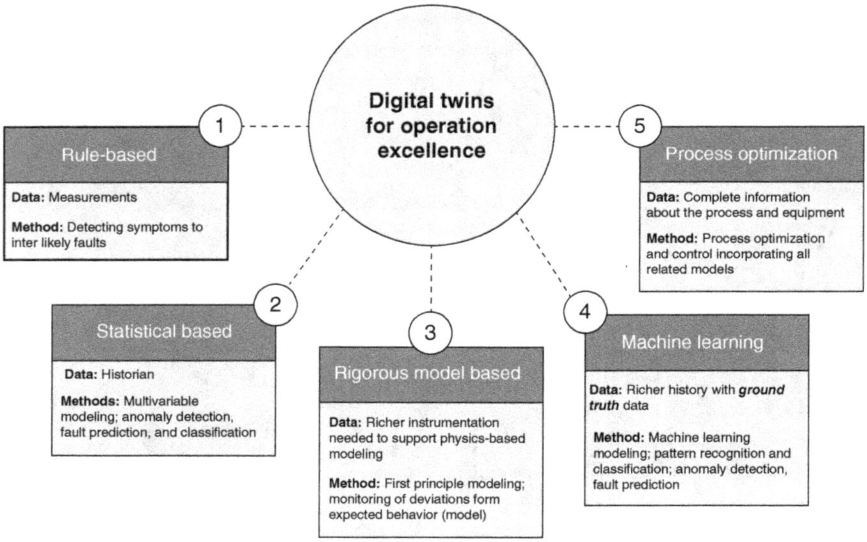

FIGURE 17.10. Model types in an APM digital twin.

business opportunities that enable more effective maintenance schedule and deployment of resources in the context of production needs.

- **Fault detection and diagnostics (FDD)** (what is wrong): FDD is the first natural opportunity for applying data analytics. Due to relatively small number of sensors and limited availability of historical data for reliability, it has been recognized that an integrated model and rule-based approach is the most easily deployable. Generalized rule sets were specified based on industry expertise and domain knowledge and can be easily incorporated and used working together with equipment models.
- **Descriptive models** (what may go wrong): Descriptive models identify a problem, and then predictive models can determine possible damage levels and maintenance options. The descriptive models can be those models for defining the operating windows for fired heaters, distillation columns, compressors, pumps, heat exchangers, etc.
- **Predictive maintenance** (what is likely to go wrong and what will be the consequences): The FDD approach is usually reactive—i.e., signs of faults are detected (what's wrong). Then the next step is moving toward predictive maintenance. The predictive analytics models can be developed based on rigorous models and statistic models such as principal component analysis (PCA) and partial least squares (PLS). In situations when there is enough ground truth data about various equipment faults, it should be possible to apply machine learning methods to train models for fault prediction. Predictive maintenance can leverage FDD-based algorithms to continuously monitor individual control and

equipment performance. Predictive modeling is then applied to a selected condition indicator that deteriorates in time.

- **Prescriptive maintenance** (what is the right solution): This is an emerging analytic technology that started to be embraced by oil and gas companies. Prescriptive analytics help to determine what is the best maintenance option and when it should be performed based on actual conditions rather than a time-based maintenance schedule. The prescriptive analytics moves beyond descriptive and predictive analysis. Once a problem is identified via descriptive analytics, then possible damage, failure, or maintenance options can be predicted by predictive analytics, followed by prescriptive analytics to select the right solution based on the company's capital, operational, and other requirements. The objective for prescriptive maintenance is to enable effective parts inventory and delivery, expert allocation with the ability to "fix it on the first visit." High-fidelity of failure forecasting models is critical to perform maintenance in a timely fashion, extend the useful life of critical machinery, keep maintenance costs down, develop optimal service centers, remotely fix troubled equipment, and maximize the cost efficiency.
- **Process optimization** (what are the optimal conditions?) This is about determining optimal conditions via adjusting setpoints and schedules. This approach can provide cost savings in the operation of equipment and reduce performance drifts. The optimized setpoints and schedules can be communicated directly to the underlying controllers, which allow an automated closing of the loop from analytics to control.

17.4.1 Digital Twin as Process Reliability Advisor

Leveraging Honeywell UOP process models with our extensive process experience, the Reliability Advisor evaluates a unit's performance to operate closer to constraints and diagnose and mitigate issues (Figure 17.11). With this level of insight, it can improve process reliability—which means better production yields and equipment health, and other parts of the facility can operate with the greatest efficiency.

Plant process and lab measurements are uploaded to Honeywell Forge where they are fed into first principles and other models to determine how well the process is performing versus design. Deviations are presented with a hierarchical fault model that leads user to potential causes and recommended actions.

17.4.2 Digital Twin as Process Optimization Advisor

The Optimization Advisor rigorously simulates an entire complex of process units to determine the maximum profitability of the complex, given current constraints and market conditions. To reach the optimum performance, operational recommendations backed by Honeywell UOP expert analysis are provided to help you reach optimal performance and asset utilization (Figure 17.12).

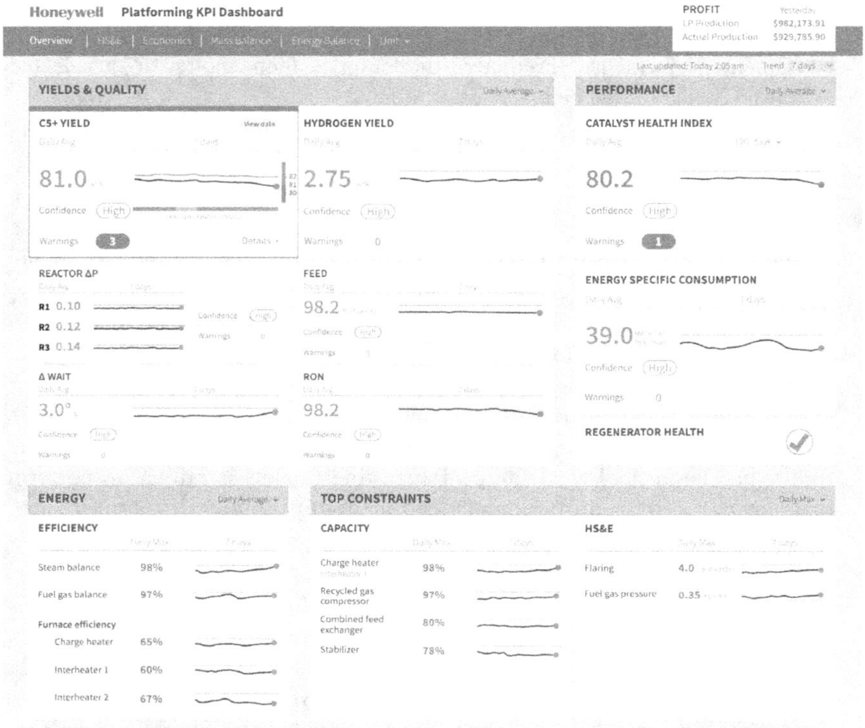

FIGURE 17.11. Screenshot for Process Reliability Advisor.

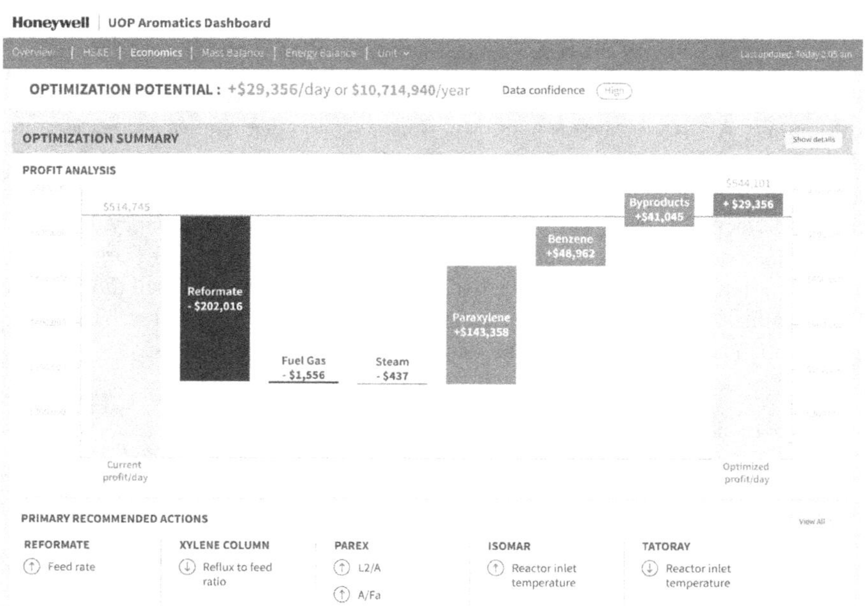

FIGURE 17.12. Screenshot for process optimization advisor.

17.4.3 Digital Twin as Profit Performance Monitor

Cloud-based performance monitoring (Figure 17.13) enables collaboration to understand what is limiting APC from driving the unit to its best potential. APC models are run as digital twins to evaluate opportunities to relax constraints and other actions that can improve performance. Providing visibility into the economic consequences of plant process performance as cost of lost opportunity ensures sustainable benefits over varying operating conditions, visualizing real-time unit profitability.

17.4.4 Digital Twin as Asset Performance Management

Cloud-based digital twins drive real-time equipment performance modeling and predictive analytics. Integrated process and asset models reveal root cause of bottlenecks, identify new levels of untapped productivity, and uncover potential problems.

Honeywell Forge APM goes beyond traditional machine monitoring and data gathering. By merging together decades of machine modeling experience with modern cloud analytics (Figure 17.14), Honeywell digital twins predict machinery availability, drill to the root cause of inefficient machine operation, and bring order to reliability and maintenance planning. APM deploys rapidly through a wealth of standard asset libraries and tools to integrate data sources quickly. On average, installation costs are recouped in 6 months or less.

17.4.5 Benefits of Digital Twins in Honeywell Forge

Honeywell Connected Plant's family of digital twins can anticipate operational complications, offering real-time solutions to overcome them, and can

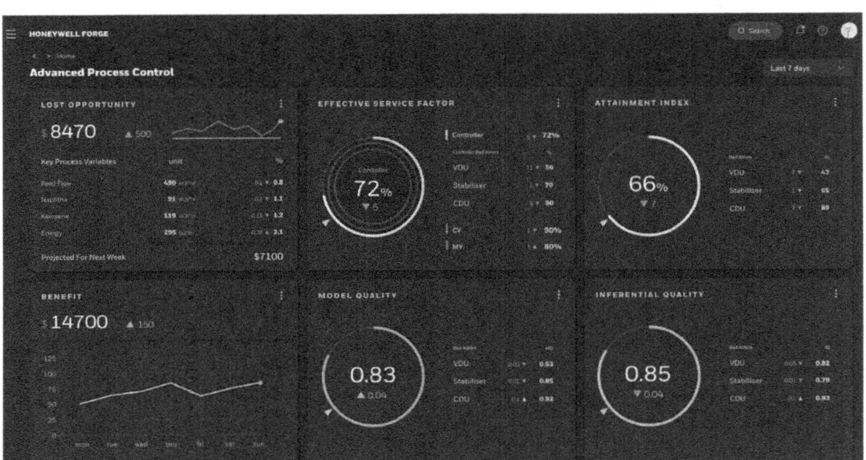

FIGURE 17.13. Screenshot for profit performance monitor.

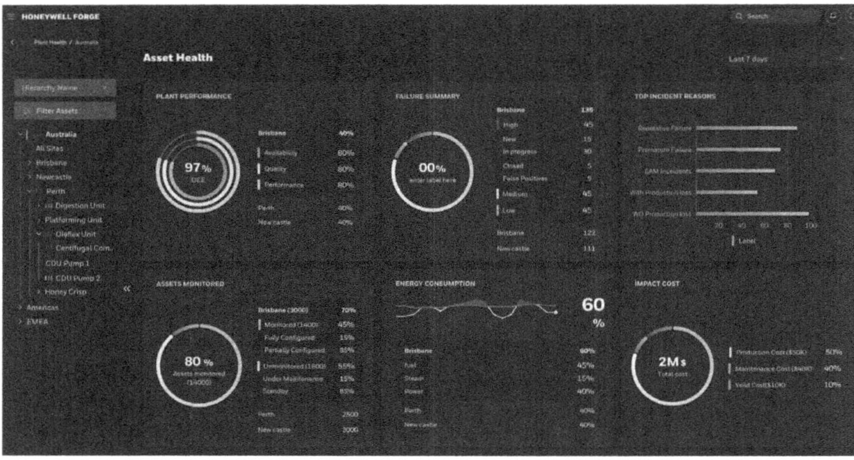

FIGURE 17.14. Screenshot for asset performance management.

reduce unplanned downtime of your plant. Our customers see improvements in plant performance and increased profitability by over \$0.30–0.50/bbl in refining and \$10–20/MT (Metric Ton) in petrochemicals. Example benefits include:

- \$7 million annually from optimization recommendations;
- \$3 million annually through energy savings;
- \$1.2 million annually from process efficiencies.

Honeywell Forge for Industrial provides actionable recommendations, highlighted with potential economic impact for intelligent business and operations. The portable and extensible software solution uses process and asset digital twins, role-based, and comprehensive analytics to enable top-to-bottom enterprise-wide visibility improving decision making and sustaining overall performance of the process, assets, people, and safety.

17.5 CYBERSECURITY

By embracing the digital age backed by over 50 years of distributed control system expertise, Honeywell worldwide cybersecurity centers of excellence, know-how, and field-proven technologies help customers embrace the digital age while protecting their industrial people,

Honeywell Forge Cybersecurity Platform (Figure 17.15) helps secure remote field assets from a single security and operations center. This field-proven technology automates the deployment and enforcement of system-wide security policies while focusing on cybersecurity essentials such as inventory visibility, patching, log collection, remote access, and compliance.

Complying with the NIST Cybersecurity Framework (Figure 17.16), the NIST SP 800-82 guidelines and leading international standards (including NERC CIP, ENISA, PERA, ISO 27001, and ISA/IEC-62443), Honeywell Forge Cybersecurity Platform improves the cybersecurity compliance posture across industrial organizations.

Deployed at thousands of sites worldwide in diverse sectors ranging from oil and gas, utility, chemical, mining, and manufacturing, Honeywell Forge delivers unrivaled visibility, reliability, and compliance for plant operations.

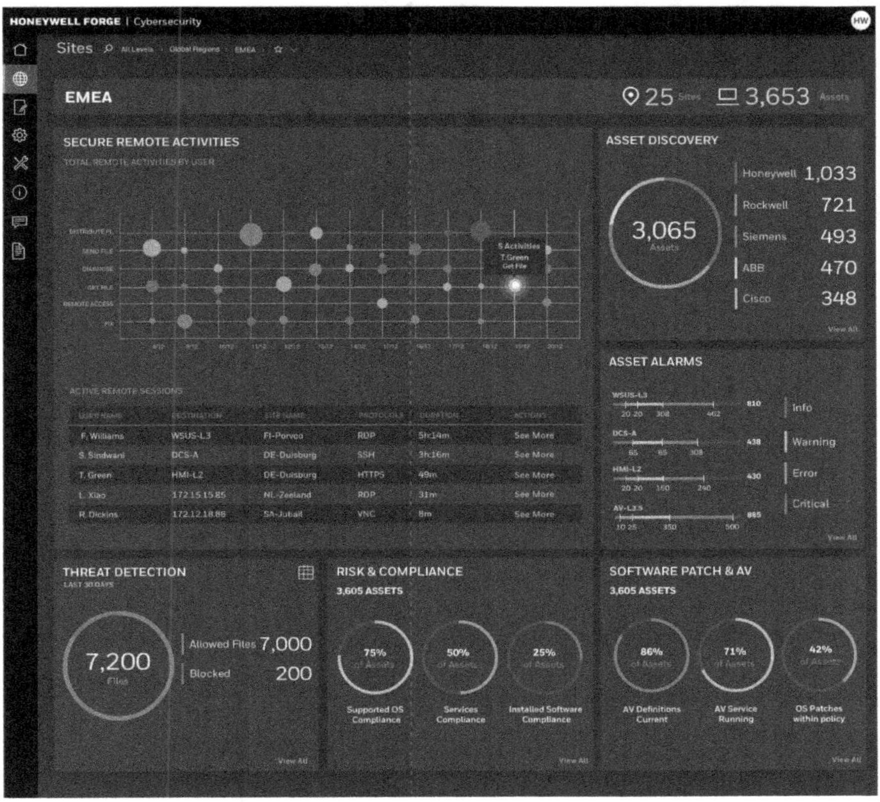

FIGURE 17.15. Cybersecurity in Honeywell Forge.

FIGURE 17.16. Cybersecurity dashboard in Honeywell Forge.

REFERENCE

Zhu FXX, Johnson JA, Ablin DW, Ernst GA (2019), Efficient Electrochemical Processes: Technology, Design and Operation, John & Wiley Sons, Inc., Hoboken, NJ.

18

DIGITAL REEDINESS ASSESSMENT AND SIX-STEP DIGITALIZATION JOURNEY

18.1 INTRODUCTION

Every manufacturing plant has been optimizing operations using engineering technology (ET) and operation technology (OT) successfully for many years. ET mainly consists of various models for simulation and optimization for business planning, operation scheduling as well as process models for monitoring and simulation, troubleshooting, equipment failure protection, environmental measure, and more. OT includes instruments, sensors, distributed control system, and advanced control. Furthermore, plants have been using information technology (IT) for data management to support OT and ET.

IT includes using computers, software, data storage, data analytics, networking devices, and other devices to generate, process, store, secure, and exchange all forms of electronic data.

Now there are new IT breakthroughs that include the Internet, cloud, wireless sensing, distributed computing, and big data science technology empowered by data analytics, artificial intelligence (AI), and machine learning (ML). These IT breakthroughs set the foundation for the Industrial 4.0 or Digitalization.

Digitalization is more than just data collection and software tools. It is about using cloud computing, Industrial Internet of Things (IIoT), data analytics, models, and automation to address end-to-end enterprise touchpoints.

Digitalization and Analytics for Smart Plant Performance: Theory and Applications,
First Edition. Frank (Xin X.) Zhu.

The best outcome for digitalization is achieved by the proper integration of IT (data centric), ET (model centric) and OT (control centric). The digitalization system is designed to monitor current events and predict new events, detect early warning signs of failure, and make changes to process operations and enterprise business for better profit, efficiency, reliability, and safety.

With the new IT technologies, companies strive to improve connectivity among people, assets, tools, and workflows to make the whole company/organization more collectively intelligent, agile, and efficient in business, technical, and operation management. It is natural for companies to consider digitalization to take advantage of the new IT capability for the good of the entire organization and gain competitive advantages.

The Internet makes big data available and use of data analytics can convert data to insights. As the result of the increased connectivity, more influx of data is generated from new contexts, and the amount of data becomes overwhelming. That is why digitalization comes in, and it is about how to make the data more intelligent in order to derive new insights and make more informed actions and decisions.

Companies are using digitalization through real-time supply chain optimization in planning, scheduling, and control via real-time data and optimization to determine when to process what feeds and how integration across supply, manufacturing, distribution, and retail can be obtained to maximize profit and efficiency and reduce inventory and working capital.

Companies can also use proactive intelligence that discovers impending abnormal events (price changes, supply and demand changes, asset failure, weather, etc.) and finds ways to mitigate their impact. Process engineering and reliability teams can use predictive analytics for process and asset reliability and maintenance improvement through effective real-time monitoring, anomaly prediction, diagnostics, and prescriptive analytics.

The goal of digital readiness assessment is to determine the digital maturity of the organization in order to know where a company stands in the digital maturity and identify what actions to take to satisfy the requirements and where to start for an effective digital journey. The readiness assessment helps the organization to understand your organization readiness for digital transformation given your current IT infrastructure and OT/ET capabilities. The readiness assessment will provide actionable recommendations for key stakeholders in the organization including technology adoption, strategy planning and execution, ecosystem management, key capability development and operations, talent development, cybersecurity, and risk and financial management. The end goal of the assessment is to identify strengths and weaknesses in the company's digital readiness and provide a plan for short- and long-term improvement.

18.2 DIGITAL READINESS ASSESSMENT

The digitalization readiness assessment plays an important role in building a long-term strategy for Industry 4.0. Readiness is defined as "the ability to capitalize on future production opportunities, mitigate risks, and be resilient and agile in responding to unknown future shocks" (World Economic Forum, 2018). There are different approaches to such an assessment that use different quantitative and qualitative key indicators. Some of the most significant assessments are derived from the approach developed by the World Economic Forum (2018), the Danish Institute Industry 4.0 (Faarup and Faarup, 2017), the Roland Berger readiness index (Berger, 2014), and Singapore Smart Industry Readiness Index (SIRI) (2017). The Singapore SIRI approach is developed by Singapore Economic Development Board (EDB) and applied for manufacturing companies. The SIRI approach consists of three building blocks and eight main pillars, shown in Figure 18.1. Each pillar comprises several dimensions, which are 16 in total as shown in Figure 18.2.

The following discussions provide a brief overview, and details can be seen in Singapore EDB (2017).

18.2.1 Technology Pillars

Technological advancement has been the cornerstone of the last three major industrial revolutions. Technology remains critical under Industry 4.0. New digital technologies such as cloud computing, ML, and the Internet of Things (IoT) are

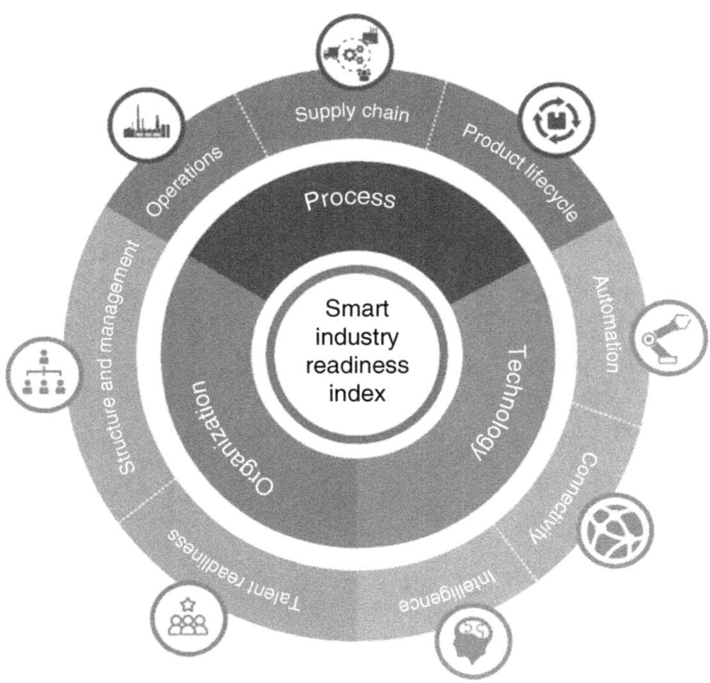

FIGURE 18.1. Three building blocks and eight pillars for digital readiness (Singapore EDB, 2017).

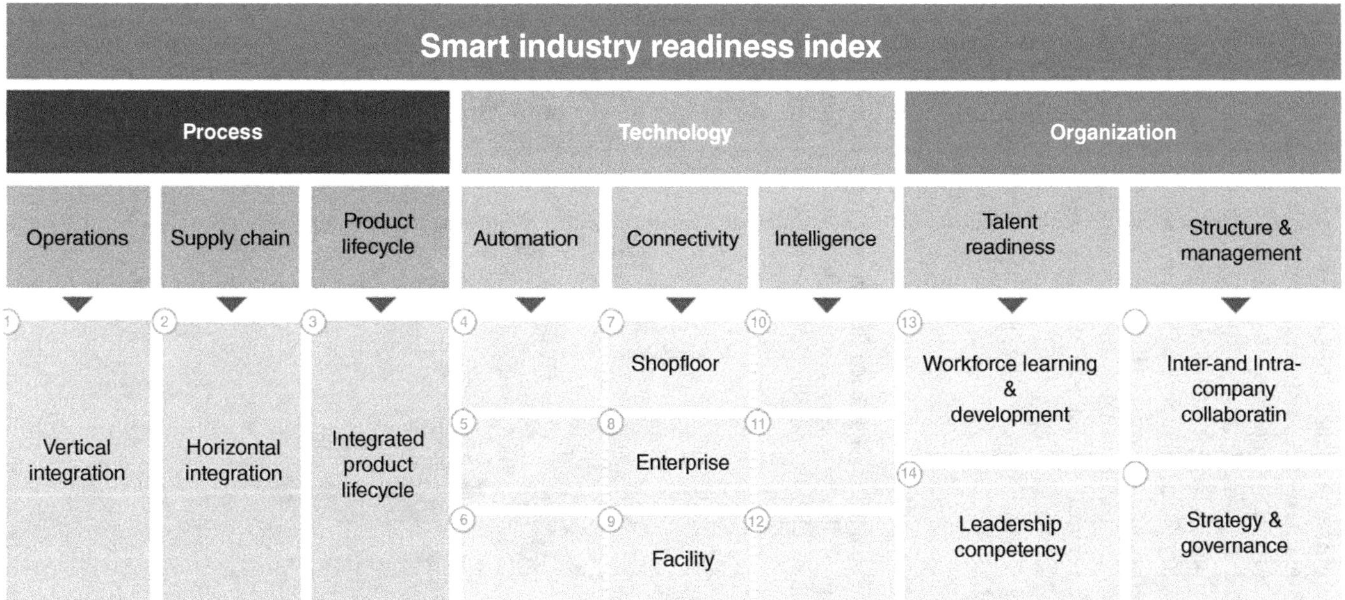

FIGURE 18.2. Three building blocks, eight pillars, and 16 dimensions for digital readiness (Singapore EDB, 2017).

creating a hyper-connected industrial landscape where physical assets and equipment are integrated with enterprise systems to enable the constant and dynamic exchange and analysis of data. These cyber–physical systems in turn make companies more agile and nimble.

For companies to realize their Industry 4.0 ambitions, a high degree of automation, ubiquitous connectivity, and intelligent systems are all necessary. To reflect this, the technology building block has been segmented into the three pillars of automation, connectivity, and intelligence:

(a) **Automation**: The application of technology to monitor, control, and execute the production and delivery of products and services. It not only frees workers from mundane and repetitive tasks but also enhances speed, quality, and consistency of execution. While automation has been and will continue to be a key enabler for companies, the role of automation is changing. To cope with rising demand for smaller batches and on-demand production, it is no longer sufficient to simply maximize efficiency. To adapt quickly to changing market needs, automation needs to be flexible instead of fixed. As automation systems become flexible, they will generate a wider range of products in smaller batches, without needing to invest in significant capital or time to overhaul or redesign processes. This allows companies to manufacture a greater variety of products with shorter turnaround times. At its most advanced stage, automation systems across all three layers will converge and interact dynamically with one another as a single integrated whole. This puts manufacturers in a more competitive position, helping them to pursue a large variety of global business opportunities and adapt to rapidly changing customer needs.

(b) **Connectivity**: Connectivity measures the state of interconnectedness between equipment, machines, and computer-based systems to enable communication and data exchange across assets. Like automation, the concept of connectivity has taken on new meaning. Every day, more and more devices and systems are being converted from wired and analogue formats to wireless and digital ones. Such IoT-enabled devices are also increasing in both quality and quantity, generating enormous amounts of data as a result. Technological advancements in smart devices, wireless infrastructure, and cloud computing also make it possible to integrate and analyze data from various data sources and different layers of complex operation and make informed decisions by seeing the whole picture. Likewise, systems that were once isolated can now be integrated, unifying the various facility, and enterprise systems through connected organization-wide networks. Interoperability—the ability to access data with ease across assets and systems—is key to achieving this. Once formal connections have been established across assets and systems, the higher bands measure the interoperability, security, speed, and agility of the network as a whole. These qualities allow interconnected systems to communicate with one another seamlessly and allow them to be reconfigured dynamically in response to changing needs. Companies need to standardize or make use of complementary communication technologies and protocols to establish more open, inclusive, and transparent communications networks.

(c) **Intelligence**: While automation provides the muscle for Industry 4.0 and connectivity acts as its central nervous system, intelligence is the brain powering this new age. Automation and connectivity focus on establishing linkages between equipment, machines, and computer-based systems for the collection and integration of data. Intelligence, conversely, is about the processing and analysis of that data. This is important as modern manufacturing is no longer just about finding ways to operate faster while reducing expenses; it is also about doing so in a data-driven and intelligent way. The benefits to be derived from the intelligence pillar are significant and far reaching. With technologies such as cloud and data analytics, the vast quantities of data generated can be processed and translated into actionable insights to diagnose problems and identify opportunities for improvement. It is believed that proactive data analytics will be the cornerstone of Industrial 4.0 and are central for process operations and decision making. With ML, highly intelligent systems can assist the workforce in predicting equipment failures and changes in demand patterns. At their best, these intelligent systems can also autonomously make decisions and respond to changing internal and external business needs.

As more advanced algorithms and models are introduced, computer systems will be able to detect deviations, predict potential failures ahead of time, identify likely causes, and recommend corrective actions. Ultimately, IT and OT/ET systems will autonomously learn and adapt to new needs while making decisions on their own to optimize processes, assets, and resources.

Proper integration of IT and OT/ET makes cyber–physical systems possible, arising from the convergence of the digital and physical worlds. This new age, commonly known as Industry 4.0, represents a paradigm shift for manufacturing on multiple fronts. Once solely focused on the execution of preprogrammed logic with control and operation functions as the focus of design, machines and devices are now part of intelligent, autonomous networks capable of communicating and interacting with one another. Processes are no longer static; instead, they are adaptive, self-corrective, and capable of responding to demands in real time. Rigid, centralized plant control systems also give way to decentralized intelligence and decision making, reshaping the basis of competition from scale to flexible production.

18.2.2 Process Pillars

Therefore, to maximize value, technology must always be applied in tandem with effective, well-designed *processes*. After all, using technology to digitize a poorly designed process will only result in a poorly designed digital process. Conversely, applying technology to a well-developed process will enhance its efficiency and enable the creation of new value.

From the beginnings of modern manufacturing, companies have used process improvements to lower costs and shorten their time to market. Previously, companies centered their efforts on improving the efficiency of individual processes. Under Industry 4.0, the concept of process improvements has expanded to focus on the integration of processes within a firm's operations, supply chain, and product life cycle:

(a) **Operations**: The first pillar is about maximizing profit margin from converting raw materials and labor into goods and services at the lowest cost. Operating companies are challenged to maximize the utilization of all their assets efficiently and safely while consistently hitting production targets, pushing assets longer and harder between turnarounds. There is a vast amount of data, but effectively identifying what is important, deriving actionable insights, and making informed decision remain a challenge. Digital transformation enables companies to unleash business potential, improve overall equipment effectiveness (OEE), and maximize values through unified process and asset digital twin, advanced industrial analytics, and predictive maintenance.

(b) **Supply chain management**: The second pillar will digitalize the planning and management of raw materials and inventory of a company's goods and services, all the way from the point of origin to the point of consumption through a central data hub and analytics engine. The digitalization of supply chains will also allow decisions about cost, inventory, and operations to be made from an end-to-end perspective rather than in isolation. This evolution benefits all players across the value chain, with greater speed due to reduced lead times, greater flexibility through real-time optimization for changing needs, greater personalization, greater efficiency, and greater transparency, both internally and for partners.

(c) **Product life cycle**: The third pillar refers to the sequence of stages that every product goes through from its initial conceptualization to its eventual removal from the market. These stages range from design, engineering, and manufacturing to customer use, service, and disposal. A robust product life cycle management framework has always been integral to manufacturing operations; however, shorter product cycles and a growing demand for personalization have accentuated the need for greater integration and digitalization across the different product life cycle stages.

18.2.3 Organization Pillars

Organization is the third building block of Industry 4.0. Often under-regarded, organization plays an equally important role alongside technology and processes. To remain relevant in the face of increasing competition under digitalization, companies must adapt their organizational structures and processes to allow their workforce to keep pace. Two major pillars are defined under organization, namely, people and intuitional systems:

(a) **People**: This component makes up the organization—the entire workforce from the top management down to the operational teams.
(b) **Institutional systems**: This component governs how the company functions.

Both components must be taken into account in order to fully reap the benefits of Industry 4.0. For instance, even a competent leadership team and workforce will be demotivated by rigid structures, inconsistent practices, and siloed processes. Likewise, open channels for collaboration and innovation will not be effective unless employees are informed and incentivized to use them. As such, the necessary enhancements must be made to people, represented by talent readiness, represented by structure and management, before a company can implement Industry 4.0 strategies effectively.

18.2.4 Benchmarking Digital Readiness

Most companies recognize the importance of digital transformation, but many of them do not know where and how to start and what to focus. Although the index method discussed above provided the common framework and guiding principles for all manufacturing companies to participate in, it requires a benchmarking assessment to reveal the gaps and opportunities to define the starting point and develop a roadmap for the digital journey. With a starting point assessment, focus areas, and a strategic roadmap, companies could feel great confidence in embarking on the digital transformation journey and achieving expected business outcome.

To provide aid for this purpose, Singapore EDB developed the second tool, namely, Assessment Prioritization Matrix, as an application tool, on top of the SIRI index theory.

The Prioritization Matrix is a management planning tool that helps companies identify high-priority SIRI dimensions that would deliver greatest impact to their organization. This tool was developed with the support of knowledge partners including McKinsey & Company, Siemens, SAP, and TÜV SÜD and was launched on April 1, 2019.

18.2.4.1 Prioritization Matrix **Prioritization Matrix** consists of several key components under each SIRI dimension, and it is a benchmarking tool for evaluating companies with its maturity based on 16 dimensions. The goal is to identify major gaps in terms of high-priority SIRI dimensions that would deliver greatest impact to their organization. Furthermore, the tool enables companies to formulate an effective Industry 4.0 roadmap with focused areas that will generate the greatest value. Having clearly identified focus areas drives both informed decision making and effective resource allocation. As the result, it creates synergy between the business's digital technologies and its business goals.

18.2.4.2 Assessment Matrix Score The Assessment Matrix helps companies identify the current Industry 4.0 maturity levels (ranging from Band 0 to Band 5) of their manufacturing facilities across 16 dimensions. This is referred to as the Assessment Matrix Score. The score at the beginning of the digital journey serves as the baseline for companies to measure the impact of potential changes and track the progress of their transformation. Using the Assessment Matrix Score also provides companies with a common language for goal setting in their digital transformation roadmap.

This approach is developed based on so-called the TIER framework (Figures 18.3 and 18.4), which provides a conceptual structure that underscores four key principles that companies must consider for a holistic prioritization. Companies can use the Assessment Prioritization Matrix to evaluate their current digital maturity, benchmark themselves against the peers, and identify potential gaps in their digital transformation.

18.2.5 Takeaways from the Digital Readiness Assessment

When conducting digital readiness assessment, companies often identify different starting points and focus areas for the digital journey due to their specific technology profile, organization situations, work process, and culture. Regardless of what differences may occur, the assessment must provide clear, actionable insights that can improve how you use, adopt, and integrate digital technology at your business. The following discussions intend to draw attentions to some key points for the assessment.

18.2.5.1 Integration of IT and OT/ET Proper integration of IT and OT/ET make cyber–physical systems possible, arising from the convergence of the digital and physical worlds. This new age, commonly known as Industry 4.0, represents a

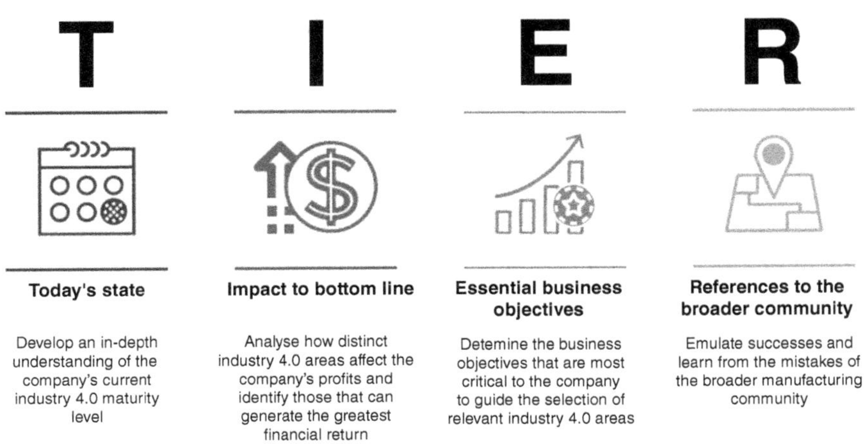

FIGURE 18.3. Definition of TIER (Singapore EDB, 2017).

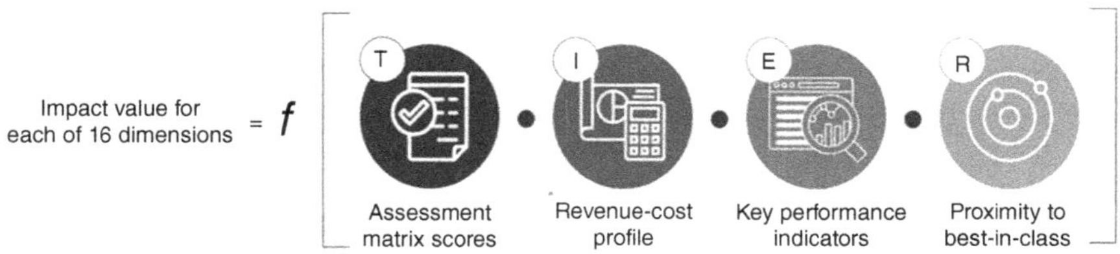

FIGURE 18.4. The prioritization matrix formula (Singapore EDB, 2017).

paradigm shift for manufacturing on multiple fronts. Once solely focused on the execution of preprogrammed logic, machines and devices are now part of intelligent, autonomous networks capable of communicating and interacting with one another. Processes are no longer static; instead, they are adaptive, self-corrective, and capable of responding to demands in real time. Rigid, centralized plant control systems also give way to decentralized intelligence and decision making, reshaping the basis of competition from scale to flexible production. As more advanced algorithms and models are introduced, computer systems supported by ML techniques will be able to detect deviations, identify likely causes, and even predict potential failures ahead of time. Ultimately, integrated IT and OT/ET systems will autonomously learn and adapt to new needs while making decisions on their own to optimize processes, assets, and resources.

Digital twin is an effective example of IT and OT/ET integration because it is a virtual digital copy of a physical asset. Thus, there are multiple different digital twins that cover various aspects of the asset life cycle and value chain. Most of them serve distinctly different purposes and rely on limited data sources and computing speeds. While individual point solution digital twins exist today, a future digital nirvana has one multipurpose digital twin. It is unrealistic to achieve the future state in one step. More likely, it will be achieved by the connectivity of valuable, high-performing individual elements. However, over time, these digital twins will become increasingly holistic and connected for multipurpose/dimensional deployment and will run off increasingly ubiquitous data sources. In this future world, an open architecture accommodating an enlarged digital twin "platform" allowing different digital twin tools switch-in/switch-out will be crucial (McIntee and Finnan, 2020).

One example of integrating both IT and OT/ET is shown in Figure 18.5.

18.2.5.2 Easy access to digital insights and actions Digital transformation is about the spread of information—access to it and act on it. Today, we are replacing hardware with software and physical items with intelligence. Digital twin insights streamed in real time to wearable accessories to frontline staff will better equip operators to make more holistic decisions. Sensory augmentation overlay of information on top of what is already observed and perceived by the operator will deliver additional measured information for operators to act on. Furthermore, inferred information derived through the application of complex physics- and chemistry-based model and AI/ML algorithms to measured data

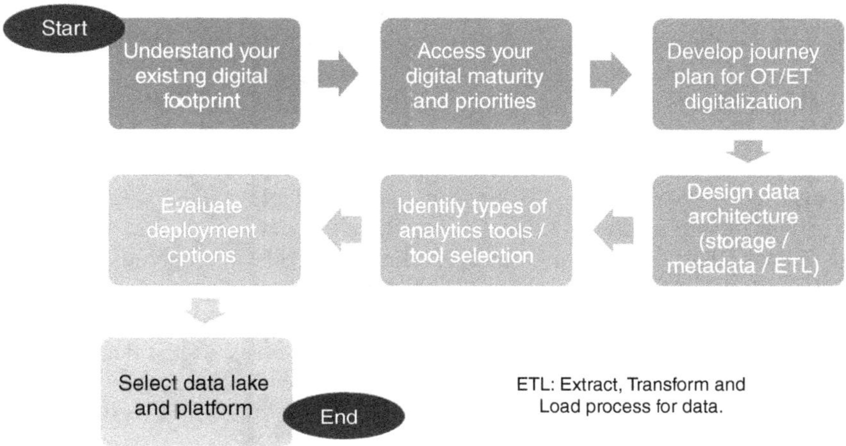

FIGURE 18.5. Illustrative example of integrating IT and OT/ET.

will mean that new insights will become available for parts of the asset where traditional measurements are not feasible or viable. The outcome is an increase in instrumentation and equipment productivity. Enhanced data readily available in the field will help minimize breakdowns, enable preventative maintenance, and reduce operating costs. The prediction of field device health and knowing the process interface conditions allow frontline operators to make better calls on shutdowns or failures. Within the field, key operator actions are controlled, and coordinated in real time through monitoring and control of work processes. This will minimize the learning curve for new operators, support change management, and enable vastly improved scenario validation through operator training simulation and an ISA106-compliant modular procedure automation solution. This will help ensure that field employees have the information they need to safely understand the bottom-line impact of their day-to-day choices. The new insights generated via integrated data and modeling improve the decision-making process at all levels of the enterprise by getting the right data to the right people at the right time so that right decision can be made. The result is increased profitability, and the assurance of knowing your plant is operating reliably and at the peak of its capabilities.

18.2.5.3 Technology alone is not sufficient The promise of digital transformation is to evolve into a more agile and enduring organization by taking advantage of advances in IoT, AI, cloud computing, data sharing, and increased connectivity. But the statistics are sobering: only 30% of major corporations are succeeding in their digital transformation efforts. Many projects failed to show value and ROI.

The million-dollar question is: why so many of these projects fail? Some people may think it is due to lack of technology such as AI or other intelligent models. The truth is: digital transformation projects do not fail because there is a shortage of AI. The answer is: the success is in combination of technology, people, and process. To succeed with digital transformation, the organization needs to evolve in all three areas, which is the reason why these areas are recognized as three building blocks for Industry 4.0 for manufacturing industry.

18.2.5.4 Integration of digital strategy and work processes People may rush to try new technologies without considering the broader organizational strategy for building effective work process. When many tools are deployed, individual departments are trying out the flavor of the day with no alignment to each other or the rest of the business. This creates a mess, and this might sound like a technology problem, but ultimately it is a process problem. The solution for avoiding this mess is that digital processes must be put in place for requesting and supporting new technology. Likewise, you will find it important to put processes around data science and the use of ML models. The digital strategy and processes will guide the companies to understand, govern, and manage business logic that increasingly reveals itself through data processing.

18.2.5.5 The risk of relying on analytics slavishly As analytics adoption increases, data scientists churn out more models of greater complexity using a variety of tools and programming languages. Getting these models into frontline operations should not require time-consuming and error-prone recoding steps to integrate the models into an operational system. At the same time, models that are built on softer features, such as customer behavior, can decay more quickly than those built on more stable features, such as customer age. And organizations increasingly view algorithms and ML models as an additional source of risk, for example, reputational risk from a potentially biased model. Model accuracy depends on how good or relevant the data are. Due to time-varying nature of industrial operation, such as catalyst aging, corrosion, fouling, and process reconditioning, the virtuous cycle of AI/ML data-driven model needs to be understood by both SME's and end users. The virtuous cycle consists of data collection, model update, and model redeployment. The key for AI/ML risk mitigation is manageable and cost-effective model update. An effective AI/ML model needs to work compatible with first principle model and other existing process knowledge, deal effectively with uncertainties that are time varying, and generate interpretable solutions. It is also necessary to consider trade-off between model accuracy and model complexity/maintainability. Part of your digital transformation strategies should be putting a process in place for operationalizing models.

18.2.5.6 It is all about value creation The measure of digital projects is crucial to transform big data into big value. Today, analytics is about more than just an algorithm. It is about integrating data-driven decisions. Who cares about the clever features of an algorithm if it never sees the light of day and never makes a difference for the organization?

The sheer speed of innovation means that there will undoubtedly be other tools and technologies that will become superior to others in one way or another. Companies will not be able to sit on the fence and wait for them to come; business must continue. Companies must put in place their best technologies and then plan for high levels of obsolescence. A large determinant of the future value of individual tools and technologies will become how compatible they are within a wider ecosystem/technology stack, along with the cumulative benefits of this integrated ecosystem/technology stack. A strong risk exists in applying technology for the sake of technology, especially with the pressure to deliver something with one of the latest buzzwords—Industry 4.0, IIoT, cloud, edge, big data, analytics, etc. Engineers and frontline managers must work backward from business goals and constraints, defining the problems and technologies to deliver outcomes, rather than starting with new technologies and looking for places to apply them. Being focused on outcomes and values as principal concerns will ensure that companies do not fall into the "inputs" trap. Examples of valuable outcomes include efficient delivery of value, discovery and capture of new value, sustainment of value achieved, and competitive advantage.

18.3 THE SIX-STEP DIGITALIZATION JOURNEY

To decide what digital applications and capabilities are required, one must identify the problems or pain points for operational improvements. Once these problems are known and understood, the next task is to develop value propositions or business cases for solutions to overcome these problems and prioritize them and define use cases. Then the team needs to develop a holistic digitalization approach for achieving operational excellence that addresses the selected use cases simultaneously. Not only will this eliminate duplicate or conflicting effort, but it also assures an optimum outcome with scalable implementation and measurable results. Furthermore, the team must determine the digitalization readiness as to where the company stands in the journey and what takes to achieve digitalization fulfillment for achieving operation excellence and sustaining the benefits in the journey. A typical digitalization journey takes six steps as shown in Figure 18.6 (Bonner, 2019):

- Step 1: Identify opportunities and define value proposition for the digital solutions.
- Step 2: Assess the overall readiness to determine focused area and the digital roadmap.
- Step 3: Define the use cases according to available data, domain knowledge, and expected impact.
- Step 4: Design data architecture, data sources, and analytics approach.
- Step 5: Implement the pilot program (proof of concept [POC]) for the use cases.
- Step 6: Based on the success of the pilot program, implement analytics solution at scale and disseminate to the other use cases. Identify high-potential operational, strategic, and emerging digital technologies and the roles they may play in transforming business models and processes.

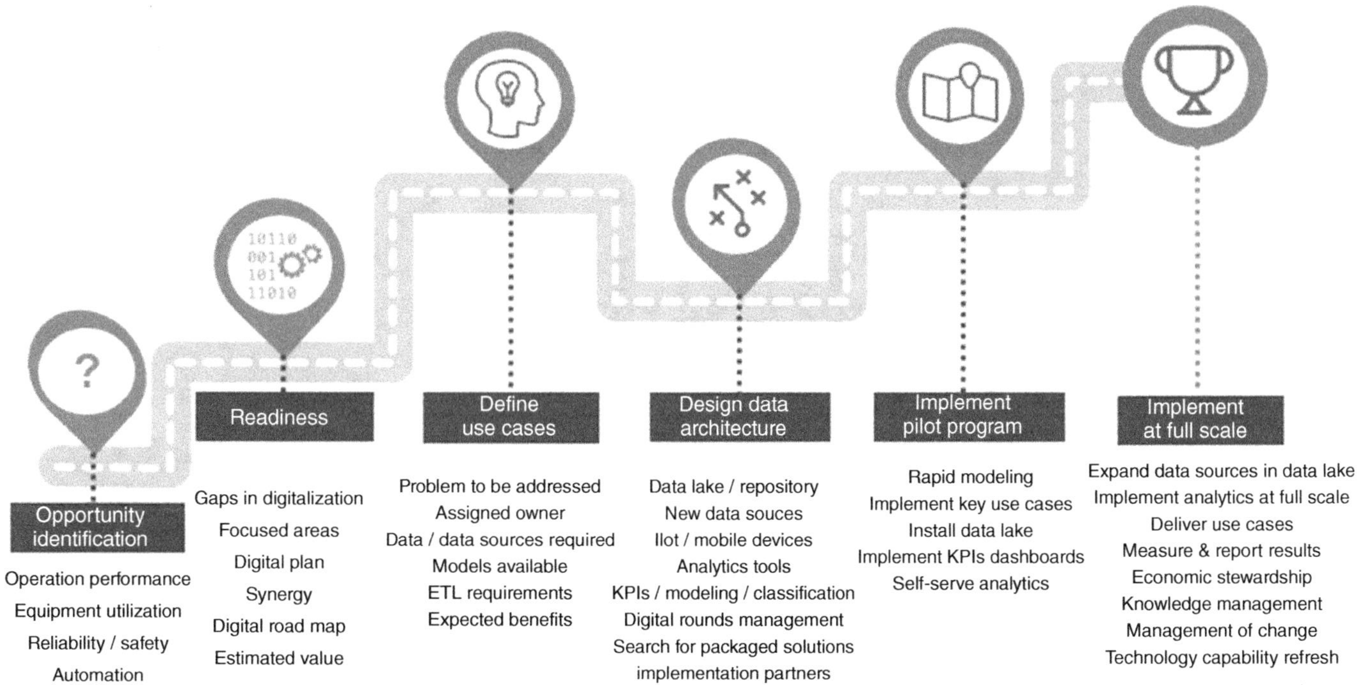

FIGURE 18.6. Six-step digital journey toward collective intelligence and value creation.

The detailed discussions for the six steps are given below. It is recommended to invite a specialized vendor to help in conducting the digital readiness survey and developing a comprehensive digital road map with the focus on the outcome and value creation.

18.3.1 Step 1: Opportunity Assessment

The first step in the digital journey is to know what are the problems and opportunities in the organization and determine the values if they can be captured via digitalization. This will become the motivation and reward for taking the digital journey. This may require an opportunity study to define the exiting problems, describe the challenges, and make the connections between the problems and digital solutions, eventually determining the value of solving the problems.

In this step, identify high-impact opportunities to determine where digital transformation is most valuable. The opportunities can be listed below but not limited to:

- Lower than planned yield;
- Utilization lower than industrial peers;
- Excess inventory;
- Excess waste material/high CPQ;
- Process and equipment reliability;
- Poor workflow management;
- Challenges in supply chain connectivity and planning;
- Lack of experienced workforce; Etc.

18.3.2 Step 2: Readiness Assessment

The lost opportunities were determined in step 1, which creates the incentive for considering digital transformation. In this step 2, the readiness index method will be applied to assess technology (IT/OT/ET), process (KPI/supply chain/product cycle), and organization (structure/talents) with the intention to answer the following questions: How can digital transformation benefit my company? Where should I start? What are the key gaps today and where are the opportunities tomorrow?

One example is that a major company is a leading process manufacturer with a diverse portfolio of consumer products generating an annual revenue of over US $30 billion. It has a 20-year-old plant in Southeast Asia. To ensure that the manufacturing facility continues to be cost competitive, the general manager decided to tap on the Smart Industry Readiness Index to help kick-start the plant digital transformation journey. Having completed the Assessment Matrix to understand the plant digital maturity level, the company applied the Prioritization Matrix to focus areas. Taking into consideration the company's current state, P&L components, KPIs, and proximity to best in class, the Prioritization Matrix has identified the following three Index Dimensions as the high-impact areas to focus its resources and attention on:

(a) Shop floor connectivity;
(b) Shop floor intelligence;
(c) Strategy and governance.

Improvements to (c) will enable the company to develop a more structured program within the organization so that it can design an action plan that will help it identify opportunities for improvement. Enhancing the maturity level of its "shop floor connectivity" and "shop floor intelligence" will allow the company to be more effective in its planning and scheduling abilities, which will in turn help optimize the management of its raw materials and consumables.

18.3.3 Step 3: Define the Use Cases

In this step, the most promising use cases for digital solution will be determined. This is accomplished by prioritizing the targeted focus areas according to available data, domain knowledge/support, and expected impact. These use cases will be used as showcases for the organization to see what takes for achieving the digital vision and what is the value of doing it and decide the next step moving forward.

When it comes to selecting use cases in the beginning of the digital journey, the advice is to start things small, valuable, feasible, and potentially scalable; pick a pilot project that can show value quickly; and communicate the success of the project broadly to build up confidence and credit as well as establish awareness and momentum.

For technology vendors, the advice is to start with consultative engagement approach to develop trust and become a partner in the digitization journey. This puts people at the center of the digital transformation project. Who are your customers in the organization, current and new? Do you know what they want? If you put those you serve at the center of everything, everything else will follow.

Once you understand who to serve and engage with digital transformation, your reason—the why—for transforming will be clearer. Consider some of the uses cases:

- Switch from offline periodic performance monitoring to real-time monitoring and anomaly prediction.
- Switch from reactive or scheduled maintenance to proactive data-driven predictive maintenance.
- Replace manual workflow management with digital connectivity and automated workflows; one example is that multiple scheduling spreadsheets with manual data inputs are replaced with a single software with automated data entry.

18.3.4 Step 4: Design Data Architecture

After defining the business case, expected outcome, and actionable insights from data, the next step is to work backward to find out what data are needed to achieve that and the suitable architecture for data collection and model deployment.

Process historian is still the primary data sources, which is supplemented by lab data and other data sources, such as digital rounds management/shift log, engine health management (EHM) data, intelligent edge devices, acoustic data, image, and video. Enterprise historian connects data from all sites in the enterprise and enables centralized data and subject matter experts (SMEs) and data scientists in the organization. Data architecture may have extensible API to enable integration with third-party apps and support real-time data streaming for online monitoring and decision making.

Data lake combines process data with other data sources in a standard data lake and contains structured and unstructured relational data. Data lake should be designed after the data and analytics requirements are fully understood.

The options for solution deployment include the following, depending on data governance and regulatory requirements:

- On-premise plant host (single site);
- Corporate or private cloud host (single or multisite);
- Public cloud-single tenant host (single or multisite);
- Public cloud-multitenant host (single or multisite).

18.3.5 Step 5: Implement the Pilot Program

The use cases selected will be implemented in digital platforms either in close loops on premises (e.g., for control applications) or open loops (e.g., for key indicator monitoring). These use cases serve for PoC purpose.

To make sure these use cases well defined, the following activities may be considered for each use case:

- Workshop to define business cases;
- Workshop to understand data and context;
- Data collection and review;
- Preliminary analysis and modeling;
- Model tuning and validation with more data;
- Review meetings after implementation.

Caution is required for the complexity of the models implemented in digital twin applications. As analytics adoption increases, data scientists churn out more models of greater complexity using a variety of tools and programming languages. Getting these models into frontline operations should not require time-consuming and error-prone recoding steps to integrate the models into an operational system. Organizations increasingly view complex models as an additional source of risk.

Clever use of analytics is crucial to transforming big data into big value. To do so, the digital team should consider putting a process in place involving SMEs to verify the opportunities revealed by digital twins and develop actions to capture them. AI/ML should leverage existing first principles and domain knowledge and provide new insights, not reinvent what is already known. Who cares about the clever features of a model if it never sees the light of day and never makes a difference for the organization?

18.3.6 Step 6: Implement the Digital Platform at Full Scale

To start this step, it is crucial to demonstrate how big data is transformed into big value. With the success in step 5 with selected use cases, the company may decide to disseminate the success for wide applications and high-impact digital projects. The team will develop an executable and supported project plan with considerations of investing in high-potential operational, strategic, and emerging digital technologies and the roles they may play in transforming business models and processes. Part of the plan should be putting a process in place for operationalizing models.

Advancements in digital technologies have made it easier than ever before to bring together data, processes, business systems, and people to create a single unified information backbone that can be managed digitally. For example, digital twin is one of such tools. It is a virtual representation of the physical assets, processes, and systems involved throughout a product life cycle. A digital twin offers two key benefits. Firstly, the information generated at each stage can be shared seamlessly, facilitating better decision making and enabling processes to be dynamically optimized in other stages. This allows companies to shorten their design and engineering cycles and respond to customer demands more quickly. Secondly, a digital twin removes the limitations of working with physical prototypes. By working off the digital twin, multiple prototypes can be created and tested virtually at speed, at scale, and at a much lower cost.

In this step, the team will provide a detailed digital transformation report, including business insights, value captured, and actions for further improvements.

18.4 RECOMMENDATIONS: A DIGITAL TRANSFORMATION MANAGEMENT SYSTEM

For most industrial enterprises, big data will manifest through the IIoT and will require a different approach to analytics. It will be difficult to extract maximum value from high volumes of fast-moving data with traditional data architecture. The following recommendations for embarking on the digital journey are provided by White (2016):

- **Start small and focus on a real business problem**. Many devises can be connected to the IIoT in a short period. However, it is best to only connect the things that are associated with business problems. The fastest way to impede a big data project is to expend resources with no measurable end result.

- **Pursue a project that promises quick value capture**. Do not go out of way to find a particularly nasty problem to solve. Your first big data project will be challenging, so pick an easy project. Use the data your already have, instead of the data that require new sensors. The project should have a relatively short timeframe (months, not years).
- **Assemble a multidisciplinary team**. Any successful IT project requires a blend of technical expertise and domain experts. Treating an IIoT project as purely as a technical problem will yield technical correctness but little business value. The IT team may need to learn and implement new technologies but will require operational and business insights to ensure there is value in implementation.
- **Measure return on investment on digital projects**. Before you start your first project, ensure that you have a process in place to measure the return on investment. If you cannot demonstrate tangible value for your first project, there likely will not be a second project. Make sure you understand the objectives as agreed upon with business leadership and document progress toward those objectives.

18.5 ESTABLISHING A DIGITAL TRANSFORMATION MANAGEMENT SYSTEM

It is necessary to have an effective digital management system in place to drive measures for the effectiveness of the digital projects and the value created. It is not enough to just have good intentions from management and motivation from personnel. Some years ago, in a chemical plant, a digital program kicked off with enthusiasm, slogans, revival meetings, and pledges. An audit showed visible improvements from month to month. Management expected the path of improvement to continue. Instead, improvement ground to a halt. The trend leveled off after a few months and gradually turned toward south.

What had happened? The rapid improvements seen at first came from correcting simple matters such as leaks, steam traps, insulation, and so on, which were detected by common sense. But after the obvious improvements were accomplished, the improvement curve leveled off and become stable at an unacceptable level. After a while, despondence kicked in and the program arrived at the dead end since there was no preparation to keep digging deeper for hidden opportunities and thus no resources were assigned.

The lesson learned is that intention and motivation can produce good results only if good opportunities are identified, sufficient resources are given, and enablers are provided. Therefore, it is necessary to put into place a digital management system with clearly defined roles and accountability, metrics and measurement, and review and actions. Simply put, a digital management system is about accountability, work process, metrics, audit, and communications. For sustainability, a digital management system must be integrated with existing technical management structure for continuous improvement, which is very important. Let us walk through these components one by one.

Accountability

The key element in the digital management system for a plant with multiple process units is to find a person as a digital manager who has the passion for driving operation excellence with good technical know-how, and communication skills. This person should come from within the plant so that he knows the people in the organization, processes, and existing management system. The role of the digital manager is to manage the digital work process, facilitate implementation of projects, handle interunit and off-site issues, and drive a digitalization culture in the organization.

Work Processes

There are two major fronts for a digital management system, which are capital projects and operational improvement projects. The work processes for these two fronts are different due to the nature of the work. For example, capital projects should follow techno-economic review process: define (process changes)–evaluate (cost/benefit assessment)–select (option prioritization)–execute (option implementation). In contrast, for operation improvements, the work process should be based on the operation process: define (key indicators, targets, and limits)–monitoring–gap analysis (current vs. targets)–actions/control.

The work process will define schedule, milestones, deliverables, and responsibilities. A well-developed work process can avoid cracks under which good ideas could fall through while bad ideas are selected. The work process should be put on test in trial for revision before it is integrated with the existing technical management structure.

With well-defined work processes for digital management, it is like setting the direction for the long-term cure. Often we observe situations where digital projects are driven by some strong personality without well-defined work process in place. When this person leaves, the campaign suffers as this would cause a major disruption to the digital program.

Scorecard and Tracking

Success in operating a plant depends on awareness and alignment of objectives. The digital system must be developed as part of overall business and operating plan. This starts with sharing annual target for cost reduction including milestones and benchmarks. Tracking must be in place for reviewing progress toward goals that could include monthly review with plant management and intranet site reporting scorecard of process performance.

Using a scorecard to show operation performance toward goals enhances alignment and awareness, while tracking progress promotes motivation. Drilling the performance scorecard from plant level to process area level until equipment level is the key for empowerment. This is similar to sports where the team will keep doing what they are doing well when winning. But when losing, the team will change the game plan. The essence of the scorecard is to tell if the team is winning or losing.

Documentation

Beyond the scorecard and tracking, it is critical to have a corporate-wide integrated documentation system for all management systems across the organization. When a new subsystem within a management system is created, it gets implemented throughout the corporation. The integrated document system, combined with the company intranet, can allow the company to integrate all management processes throughout the company.

Technical Target System

Key indicators define the basis for continuous monitoring. Technical targets for key indicators must be developed and applied across the organization. Many industrial facilities have the potential to increase the efficiency of their systems but have difficulty doing so because there are no metrics available to measure efficiency improvement. The absence of operation performance metrics makes it difficult to establish value structure for the improvements. Practically, review of key indicators and their targets should be part of process performance review, which is a key part of integrating digital management system with existing management structure. Since development of key indicators and targets is a big ticket for digital management system, it requires management support to acquire resources and technical capability.

REFERENCES

Berger R (2014), Industry 4.0 the new industrial revolution: How Europe will succeed, Think Act Industry 4.0, March.

Bonner P (2019), Application of big data analytics AI/ML/cloud computing on operations in refining & petrochemicals. In: AIChE Spring Meeting, New Orleans, LA, April.

Faarup J, Faarup A (2017), Global industry 4.0 readiness report 2016, Danish Institute of Industry 4.0 (DII 4.0).

McIntee A, Finnan K (2020), Five key innovation concepts to impact frontline engineers in 2020, Hydrocarbon Processing, February.

Singapore EDB (2017), The Singapore Smart Industry Readiness Index: Catalyzing the Transformation of Manufacturing, The Singapore Economic Development Board (Singapore EDB), December.

White D (2016), Big data—what is it? March special issue, CEP, AIChE Journal.

World Economic Forum (2018), Readiness for the Future of Production Report 2018, Insight Report in collaboration with A.T. Kearney, http://wef.ch/fopreadiness18.

19

DIGITAL PROJECT EVALUATION AND DEVELOPMENT

19.1 INTRODUCTION

For getting the job done, it might start with identifying a problem that can be solved via digital solutions. Regardless of whether it is a small project or launching a large-scale initiative, the steps are essentially the same. First, the team must prove that it can add value to the organization and worth the investment for kicking off the project. This phase is called business case evaluation. If the project is survived from the evaluation, the team then enters the design phase and starts to develop a digital solution. This chapter focuses on both value evaluation and product design phases with detailed discussions.

19.2 BUSINESS CASE EVALUATION

19.2.1 Business Case Generation

A business case provides the reasoning why a project should be initiated. A business case should describe the business needs, cost and value, as well as risk involved with the project such that the decision makers can determine that the expected business value will exceed the cost of performing the project.

Digitalization and Analytics for Smart Plant Performance: Theory and Applications, First Edition. Frank (Xin X.) Zhu.

The business case must contain sufficient information that justifies the investment in the project. Typical information that can form a part of the business case includes:

- Business need: identifies the problem that currently exists and the need for solving the problem and hence related investment.
- Evaluation methods: identifies what evaluation methods should be used for evaluation, such as a benefit-to-cost ratio, cash flow considerations, strategic options, opportunity costs, return on investment (ROI), and net present value (NPV).
- Benefit implementation plan: determines the value/benefits (rather than deliverables) that can be obtained.
- Resources required: the human and nonhuman resources needed.
- Timing: the major milestones for the project.
- Legal requirements: identifies any legal requirements that must be followed.
- Project complexity: This identifies how complex the project might be, perhaps even from a risk perspective, if the organization can manage the complexity and if it can be done with existing technology.
- Exit strategies: This identifies the criteria to be used to cancel the project if necessary.
- Project risks: This helps the decision makers evaluate the project by listing briefly the business, legal, technical, etc., as well as the risks of the project.

The above information is used not only to approve the project but also to be able to prioritize with other projects under considerations for investment.

19.2.2 Bubble Chart for Project Selection

Project managers should be actively involved in the preparation of the business case rather than taking it for granted that someone else will get it done correctly. Getting your project approved comes down to not only navigating the business case influencers and supporters but also how your project is viewed compared to others along the selection criteria previously discussed. The bottom line is project cost and risk versus reward.

A typical plot of value versus risk is so-called bubble chart, which is used for comparing different projects. Several forms are used and one example is shown in Figure 19.1. In this example, the size of the bubble represents the overall cost of the project. Along the "y" axis is the NPV, and on the "x" axis is the probability of project success. The probability of project success is calculated from the overall project risk. Using this plot, the selection committee wishes to choose a project that has the highest NPV, highest probability of success, and the lowest overall project cost. In the example in Figure 19.1, No. 2 project may have the best chance to be selected as it has the highest NPV, a low cost, and a probability of success that is about 57%. By balancing all of these factors, this project may bring the highest reward relative to the cost with a good chance of successfully meeting its objectives.

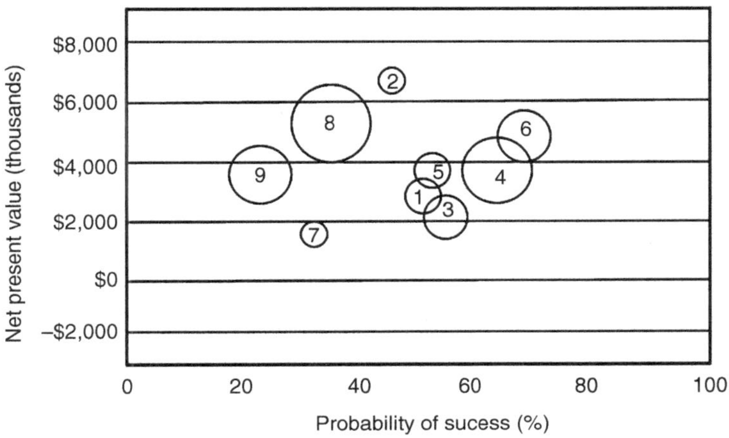

FIGURE 19.1. The project risk vs reward plot for project selection (bubble size based on total cost).

Other projects including No. 5 and 6 may be selected over this one based on the actual criteria used.

A comparison among projects is made along with the overall considerations as documented in the business case. Many criteria are used, and many business case influencers are factored in for selecting a project for approval. The fact still remains, however, that the project that provides the best NPV and fastest ROI with the high success rate will win almost every time, coupled with the best strategy for achieving the project business objectives.

19.3 DIGITAL PROJECT DEVELOPMENT STEPS

Regardless of whether you're kicking off a small project or launching a large-scale initiative, the steps are essentially the same. The nine-step development process is discussed below in details.

19.3.1 Step 1: Identify the Problem and Define Use Cases

The first step is to identify problems and opportunities that predictive analytics can help to solve. The problem can be low profit margin, poor asset utilization, low yields, excess inventory, poor process and equipment reliability, and inefficient supply chain management. There are two aspects of industrial analytics: prediction and interpretation. Some analytics solutions are for online monitoring and decision making. For example, asset performance management and predictive analytics softwares can be utilized to monitor the asset and unit performance and predict the impending machinery health issues and predict time to failure, which enable transformation from reactive to predictive maintenance. Some analytics solution can be used as a tool to identify key impactful variables and their effect on product quality for diagnosis, troubleshooting, and continuous improvement.

Analytics solution is not a replacement for fundamental models or the need for process knowledge and engineering experience. Instead, analytics solutions can leverage knowledge of chemistry, physics, and mechanics of the process and equipment to solve problems that are difficult to be handled by applying applying first principle model alone.

Before thinking about the data and analytics modeling, talk to people in the organization to gain good understanding on the business of the organization, the context of the problem, the use cases, and the work process for connecting the tool with end users to solve the problem. Simply put it, the team needs to identify a clear objective of what the tool must be developed for, which is to set the purpose for the project.

19.3.2 Step 2: Architecture Design

Based on the problem definition and use case specifications, we need to come up with the deployment architecture. An architecture provides visibility into the various layers, components, and technologies used in the solution; it provides a perspective on customer expectations; and most importantly, it gives us a flexible framework to plan for future expansion and adjustments. Enterprise historian connects data from all sites and enable subject matter experts (SMEs) and data scientist in the organization working together for developing big data analytics. High frequency data, such as vibration data, can be analyzed at the edge and be compressed before storing in the cloud. Data lakes contain massive and ever-growing amounts of structured and unstructured relational data. Data lake should be designed after the data and analytics requirements are fully understood. Reliability of data connectivity cannot be emphasized enough for achieving expected business outcomes. Vendor selection can be based on these insights.

What should this product look and feel like? At this stage, it is also good to discuss user interface. The difference between good and great user interface depends on small details. The user needs to play with different versions of designs to get firsthand experience before selecting and approving them.

19.3.3 Step 3: Get Data Right

After the problem is well defined and the objective for the project is figured out, it is time to prepare the relevant data. To develop predictive analytics models, a large amount of data with good quality is needed, which describe the process and represent the variation in equipment conditions, operating modes, and performance. The data sources consist of process design data, simulation data, and operation data and historical data, lab data, digital operator rounds data, etc. New data sources such as image, acoustic, and video data may also be included depending on the requirement of specific use cases.

After selecting the database and data set, you'll need to properly aggregate, transform, and clean your data. Data can be inaccessible if it resides in a database that's incompatible with the chosen analytics tool. It can also be inaccurate: statistically, up to 20% of the industrial data has been recorded incorrectly. This process of making data useful for analysis is called data wrangling. This process is often a very manual and labor intensive, especially with multiple silos of data. A significant amount of effort in data analytics projects is spent on wrangling data.

It is almost never a simple task in selecting the right data, cleaning the data to make sure the data for the project is clean and consistent, and then training your model. Finally, one critical element of data preparation is to make sure the data is compliant with data privacy regulations.

19.3.4 Step 4: Get Analytics Right

You've chosen a problem to solve and gathered the data needed, and now it's time to choose a predictive analytics modeling technique, build models with various algorithms, and select the most suitable one for your applications. This step is when the real fun starts as analytics models can help to get insights from data and predict future trends. This is also an iterative process; we learn as we go, not before we go.

The analytics models must be tested, verified, and approved by SMEs. Model interpretability is critical for industrial operation, since operators and decision makers must establish a level of trust in the analytics solution. Application of the models is vital for the organization monetize the investment for the digital projects.

19.3.5 Step 5: Get Visualization Right

With high-quality data and analytics models developed from the steps above, it is the time to perform visual analytics by building visualization capability including graphs. Visualization helps to quickly explore and understand the data, the trends and relationships to communicate the insights. One can get graphs to show some trends, get insights, and even search predefined patterns in historical data to test hypothesis of domain experts. There are some quite sophisticated visualization softwares available commercially that can aid in data visualization and visual analytics. Some are general-purpose tools and some are domain specific, such as specific to process manufacturing time series data. Before jumping directly to complicated AI/ML models, significant insight and value can be gained by visual analytics.

19.3.6 Step 6: Get Prototype for Proofs of Concept (PoC)

The purpose of this step is to build a prototype in a timely manner. The main goal is to prove its effectiveness as fast as possible to justify the investment. Building a porotype is to get the first results and demonstrate the value for the digital project.

Start with something simple for rapid proof of concept (PoC), and the prototype can help the end users and stakeholders to get a taste for what they are going to get. These will be the first users of the product, and their feedback will shape the direction of the predictive analytics tool. Incorporate their suggestions and expand the beta test to others in the organization.

Acceptance tests are typically performed in test benches where endpoints and other operational parameters can be simulated and easily scaled to detect any defects.

19.3.7 Step 7: Deployment

This step is to full rollout the final product. Here, the developed product is transitioned from its development environment to the live cloud or edge connecting with real-time data, where it can be accessed by the masses. This phase can be completed with relative ease; however, a number of situations can arise when moving digital hosting, and each deployment should factor in time to allow for bug fixes and code adjustments. For example, model accuracy depends on how good the data are. Because of time-varying nature of industrial operations, such as normal aging and fouling of equipment, reconditioning, and new opportunity feed, data analytics application will start to decay over time as the data that is being used to make predictions is different from what the model was trained on. Hence, data analytics application requires model performance monitoring and model update with new data if needed.

In the absence of product- or deployment-specific security certifications, solution acceptance testing of the deployment is critical. A security acceptance criterion can set the expectations that a deployment must conform to, in terms of functionality, safety, security, availability, and compliance.

19.3.8 Step 8: Optimization

This step is not a typical phase of the digital product development process. This is because optimization should take place throughout the project and takes a number of different forms. In the initial discovery phase, optimization takes the form of user surveys and interviews to test the assumptions that the team makes. In the prototype step, it takes the form of prototype user testing, allowing users to interact with wireframes and mock-ups to test concepts and workflow ideas. In the deployment step, optimization takes the form of quality assurance (QA) testing, making sure to quash bugs as they are discovered. After deployment, it takes the form of model maintenance and improvement. Finally, optimization takes the form of a project postmortem, analyzing the overall project to see how elements of the process can be improved to provide solid digital solutions in the future.

19.3.9 Step 9: Project Lookbacks

The lookbacks should be conducted for all projects to see if they deliver the promised results and address what went well and what did not and why. Care should be taken in order to avoid interdepartmental disputes. The emphasis should be on learning, avoiding mistakes, and doing better the next time. The review should answer the following questions:

- What benefits has each project delivered?
- Are projects delivering the economic return promised at the time of approval?
- Are projects completed on schedule?
- How does the cost estimate compare with the actual cost for each project?

- Is the deviation in the project performance caused due to differences in economic assumptions used (feed costs, product values, capacity required), technical performance of the project, cost, or schedule?
- For projects with prediction errors, what are the root causes and how to fix them?

19.3.10 Go/No Go Decision

In the nine-step approach, go/no go decision point should be included at some of the steps by the appropriate management or stakeholders at each development stage gate review. Assessment criteria should be clearly defined considering required accuracy, interpretability, and complexity. Analytics may or may not work depending on data availability and quality as well as mechanisms for proper data mapping. When facing obstacles, it is suggested to extend the data pipeline to add more data, try new features in the model, or change the objective function or relax certain constraints. To be sure, good algorithm choices could lead to incremental improvements. But there is usually far more to gain from other activities like integrating new data sources because every bit of explanatory power inherent to the data might be exhausted by the models already. The best improvement strategy requires a full context holistic solution from different functions.

19.4 REMARKS ON DIGITAL PROJECT DEVELOPMENT

Several comments must be made for attention during product development.

19.4.1 Fit into the Organization Workflows

Digital tools can create value only if they are integrated and operated with organizational workflows. One must approach software solutions from the perspective of business outcomes rather than just a technology solution. Digital tools must be integrated with workflow in an automated manner that are quick to implement and flexible enough to accommodate changes over time.

Many organizations have unique processes that require approvals, which standard off-the-shelf systems are not designed for. It is important to digitize these processes, integrate them with the current systems, and enable you to build and extend these flows to give you complete flexibility over how work is managed, without investing in complex and expensive system customizations.

Traditionally, businesses seeking integration have either built direct system-to-system integrations, which are cheap but prone to breaking, or invested in large integration or middleware projects, which are more robust but relatively expensive, slow to deliver, and poorly suited to integrating with external entities. Additionally, many consultancies that offer integration services are technology (rather than business) focused and lack the functional knowledge to understand the end-to-end process and what an organization is trying to achieve. This can result in integration designs and workflows that do not adequately anticipate the future needs of the business and need to be reworked and tested as the business changes.

19.4.2 Include Subject Matter Experts (SMEs)

The team must include SMEs as early as possible. Industrial data are highly correlated and complicated by physics and control system, not to mention safety, reliability, and accuracy requirement. To drive data analytics innovation in process industries, the following aspects have been proven to be effective in building a high-performance data science team.

(1) Data analytics needs to be specialized or verticalized.
(2) The team should include SMEs and educate them in data science.
(3) Data scientists should explore the various ideas generated by SMEs, develop different algorithms required, and iterate through the stream of hypothesis generated by SMEs.
(4) The team needs to be equipped with the latest technology.

Automated ML software should be applied to replicate the tasks that have been manually performed by data science process. The software will allow users to implement ML solutions without having to write code or preselect algorithms, which can perform both simple and moderately sophisticated analytical tasks. Automated ML creates a new class of "citizen data scientists," which enable SMEs to bring their own expertise and unique skills to the analytical process. Data scientists are more productive for repetitive steps for the model building process to be automated, allowing them to use their unique expertise for selecting and fine-tuning models.

19.4.3 Empower End Users

For a predictive analytics project, you need to think hard about not just delivering the information but also empowering the end users to act on it. How can you make that information useful? First, you must get it to the right people. Then, you should suggest next steps based on the information and ideally let users take action without leaving the application.

When you're developing the predictive analytics solution, consider who needs to know about it and what they may want to do about it. The key is to make it easy for the end users to see insights and take actions within the application.

Don't wait too long to involve the stakeholders and other departments in the organization. Start with executives or team leaders who will need to buy into the project. As stakeholders become advocates, they can help promote the initiative to others, making sure you get the cross-functional data you need and setting your project up for long-term success.

Now is also the time to consider who will be the end users of your predictive analytics application. Try to gather a mix of people from different roles and departments so you get well-rounded feedback on the solution. Don't just pick the users who will agree with you. Also, talk to the people who will be maintaining the solution and integrating it with the rest of the IT operations.

19.4.4 Get the Project Basis and Data Right in the Very Beginning

In most cases, a digital project is conducted under a tight timeline. When the plant management approves a digital project, they want to see the benefit to be captured as quickly as possible. However, it happens from time to time that a digital project cannot generate correct results on time or is unable to address the key concerns sufficiently and hence turns out to be unsuccessful. In majority of such cases, it can be traced back into one source: failure to define the basis correctly.

It is true that getting the basis and data right could take a lot of time and effort in an early stage. People may have concern on the project expenditure and schedule as the digital project usually faces a tight timeline and budget. However, the time is worth to spend to identify what are the objectives and requirements first before finding ways to achieve it. Getting the basis right is like "look before leap" if simply putting it. The benefit is avoidance of rework down the road of the project.

Getting project design basis right implies that we need to determine and agree on:

- Business objectives: There could be different objectives such as throughput increase, technology upgrade, changes in feeds and/or products, energy reduction, and environmental regulations.
- The key indicators for measuring the success of the digital project.
- The scope, deliverables, and schedule.

Due to the importance of getting the design basis and data right for a complex digital project, following guidelines could be used:

- Assign a dedicated coordinator who can ask right questions and extract required information from plant management, marketing engineer, commercial planners, process engineer, maintenance engineer, utility engineer, and energy engineer for issues pertinent to their responsibilities.
- Engineers with different responsibilities will provide their input of basis and data: The process engineer will need to generate test run data if needed as well as explain process steps and postulate functional descriptions. The commercial engineer will provide input from business planning reality and help develop the project objectives and product specifications. The utility engineer will provide data for utility and infrastructure and bring the local plant situation into focus as it relates to the newly postulated facilities.
- Assign a process engineer with good experience for the process unit of interest who can provide input and feedback in developing simulation for the process and prepare tables and graphs that can summarize the design basis effectively.

- The instrumentation engineer who is assigned to collect basic data should be dedicated to the task of gathering data and allowed to be free from other responsibilities for the period of data extraction and reconciliation. This engineer should proceed to involve in the feasibility assessment even if it is not a full-time involvement. However, with design basis defined correctly, other people will be brought "up to speed" quickly.
- Bring out all possible surprises (good or bad) as early as possible. Do not bring surprises to management without proper answers (reasons and solutions).
- Good communications and review meetings are the key to get the basis and data right.

19.4.5 Root Cause Analysis When Analytics Does Not Work

All of us could dread when analytics does not work and ask, "why does the analytics model does not work?" The only thing worse than that question is not having any answers! There are many possibilities and often you want to understand exactly what is driving a given model prediction error—especially if it is unexpected.

Understanding the underlying causes of model errors is known as root cause analysis. There are two general types of root causes. One type comes from modeling aspect. Chemical engineers have always prided themselves on their modeling capabilities, but, in this new era, modeling would go beyond unit-operation modeling techniques. On the other hand, data scientists try to tackle any problems using data-driven modeling such as stochastic models, ML models, and so forth. There are pros and cons for these two approaches. An effective digital twin software should include model selection to determine what models are used for what kind of problems and related data (see Chapter 17 for details).

The other type of root causes comes from data quality. Model prediction accuracy also depends on the quality of incoming data, that is, error-prone, unreliable data often means error-prone, unreliable prediction. Data errors may come from different sources as they may include process transience or measurement error due to noise, instrument accuracy, instrument installation or calibration. Steady-state detection analyzes a data set for statistically similar values in key operating conditions over a certain period of operating time, from which selected data is used as the basis for process simulation and optimization. For example, steady-state data must be used for parameter estimation to minimize differences between model predictions and measured variables. However, the presence of random and system errors results in gross errors (or outliers) in the process data that must be identified in the phase of data reconciliation and gross error detection and removed from the treated data set. As such, data cleansing techniques such as steady-state detection, data reconciliation, and gross error detection are essential to isolate and correct measurement errors and ensure more accurate and reliable data-based analyses. An effective data cleansing method is developed (Raftery et al., 2019) and implemented in Honeywell Forge, a digital twin platform. This method is integrated into the data workflow to make sure the treated data will be correct for fit-for-purpose applications for process optimization, control, and monitoring of equipment reliability.

19.5 S-CURVE FOR PROJECT REVIEW AND MANAGEMENT

S-curve analysis is an important project review and management tool (Figure 19.2). It allows the progress of a project to be tracked visually over time and forms a historical record of what has happened to date. Analyses of S-curves allow project managers to quickly identify project growth, slippage, and potential problems that could adversely impact the project if no remedial action is taken.

The S shape indicates that projects are typically slower when the project is just getting started (a flatter part of the S) and then pick up speed afterward reaching the steepest rate during the middle period of the project when there is a hive of activity and progress (the steeper part of the S), then flattering again toward the end of the project when the digital tools developed are tested and defects are being rectified.

However, the S-curve won't always be a perfect S in every case. The shape of the S-curve will differ from project to project—but the theory behind these S-curve examples remains the same.

19.5.1 Determining the Project Growth

With comparison of the baseline and target, S-curve can quickly reveal if the project has grown well (the target S-curve finishes above the baseline S-curve) or contracted (target S-curve finishes below baseline S-curve) in scope. Figure 19.3 shows the

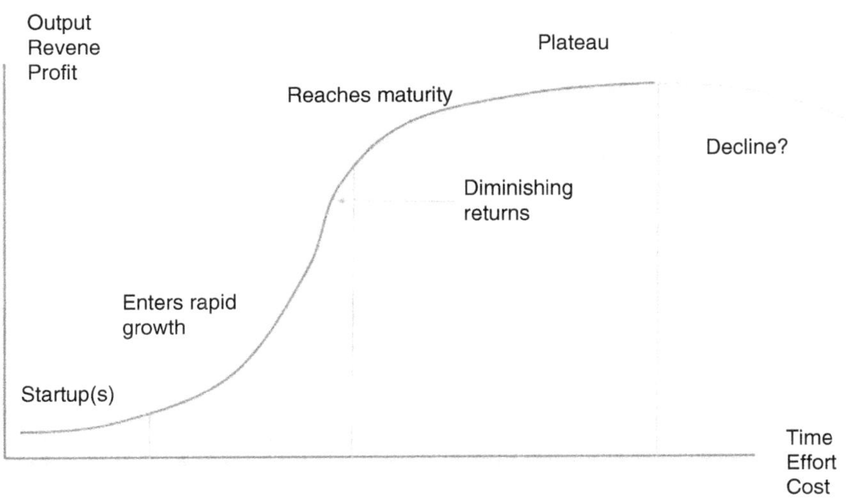

FIGURE 19.2. A typical S-curve.

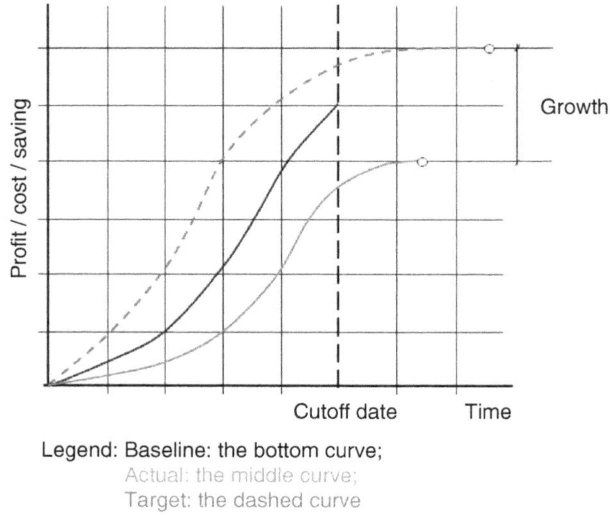

FIGURE 19.3. Calculating project growth using S-curves.

growth case. A change in the project's scopes implies a reallocation of resources (increase or decrease) and the very possible requirement to raise contract variations. If the resources are fixed, then the duration of the project will increase (finish later) or decrease (finish earlier), possibly leading to the need to submit an extension of time.

19.5.2 Determining Progress

Comparison of the target S-curve and actual S-curve reveals the progress of the project over time (Figure 19.4). In most cases, the actual S-curve will sit below the target S-curve for the majority of the project (due to many factors, including delays in updating the production schedule). Only toward the end of the project will the curves converge and finally meet. The actual S-curve can never finish above the target S-curve. If the actual S-curve sits above the target S-curve at the cutoff date, the production schedule should be examined to determine if the project is truly ahead of schedule or if the production schedule contains unrealistic percentage complete values for ongoing tasks.

19.5.3 Determining Slippage

Slippage is defined as the amount of time that a task has been delayed from its original baseline schedule. The slippage is the difference between the scheduled start or finish date for a task and the baseline start or finish date as indicated in Figure 19.5. Slippage can occur when a baseline plan is set and the actual dates

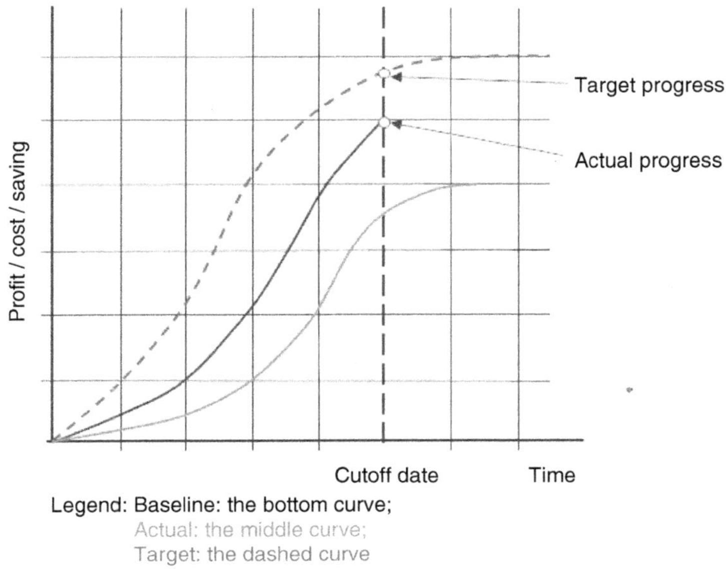

FIGURE 19.4. Calculating project progress using S-curves.

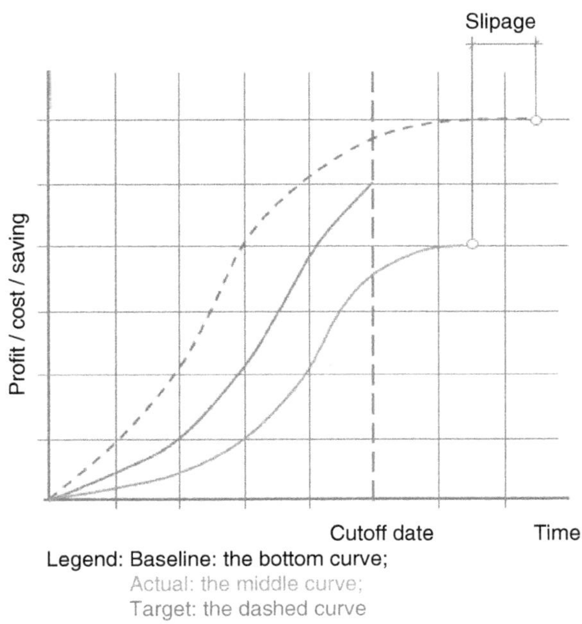

FIGURE 19.5. Calculating project slippage using S-curves.

subsequently entered for tasks are later than the baseline dates or the actual durations are longer than the baseline schedule durations.

Comparison of the baseline S-curve and target S-curve quickly reveals any project slippage. Additional resources will need to be allocated or additional hours required in order to eliminate (or reduce) the slippage. An extension of time may need to be submitted if the slippage cannot be eliminated or reduced to an acceptable level.

19.5.4 Create and Update S-Curves

The S-curves are designed to give a quick overview of the life cycle of a project development in different phase and how close the actual project performance versus target. S-curves is a powerful tool to help a project team staying on top of the project.

S-curve software enables a project manager to set their baselines in the software and then collect data in real time as the project progress on site. This has been made possible by mobile technology, whereby workers on-site can enter production and cost information into their mobile devices that is instantly synced to the cloud.

19.6 BASICS OF ECONOMIC ANALYSIS

Economic analysis at early stages in a digital project is primarily useful for project selection. More detailed economic analysis can be completed when the project is more clearly defined.

Different economic approaches are used for evaluating investment benefits, which include ROI and NPV and internal rate of return (IRR).

- **ROI:** The return of a project will make at a certain investment over several years specified—the higher the ratio between the return and investment, the better.
- **NPV**: The return a project will make at a specified discount rate. Ideally this should be a high/positive value.
- **IRR**: The yearly return % of the investment—the higher, the better.

19.6.1 Return on Investment (ROI) Approach

The basic ROI calculation is to divide the net return (B) by the cost of the investment (I) and to express this as a percentage. ROI, while a simple and popular metric, may be easily modified for different situations. The ROI formula is

$$\text{ROI\%} = \frac{B}{I} \times 100 \tag{19.1}$$

A project is more likely to proceed if its ROI is higher—the higher the better. For example, a 200% ROI over 2 years indicates a return of double the project investment over a 2-year period. Financially, it makes sense to choose projects with the highest ROI first and then those with lower ROIs.

Issues with Basic ROI Calculations

ROI calculation may be erroneous and may appear to point to the wrong conclusions. Issues to be concerned about include the following:

- ROI calculations can be manipulated if you are not careful.
- Project savings, income, and expenditures should be measurable and realistic. But sometimes these measures are not easily measurable and their realism is questionable.
- Project benefits may be attributable to more than one improvement—so care needs to be taken to avoid double counting.
- When forecasting costs and benefits, it is not always possible to obtain a high degree of certainty with respect to the accuracy of project costs and benefits.

Calculation Considerations

Relevant factors to consider in ROI calculations include:

- **Timeframe:** The timeframe for calculating ROI for IT projects may vary. Three years is common for hardware projects, as technology is often obsolete after 3 years. However, 5 or more years is often used for new software systems. Changing the timeframe can make significant differences in ROI calculations. Try to be consistent from project to project.
- **Consistency:** ROI calculations should be consistently applied across all IT system projects. Consistency also applies to the assumptions behind the ROI calculations. For example, treatment of inflation and taxation (corporate and VAT/sales taxes).

19.6.2 Net Present Value (NPV) Approach

The NPV approach involves assigning a rate of return (r) that is specific to the project and then computing the present value of the expected stream of revenues (R_i). It provides indication of what the present value of the investment is considering the investment cost while realizing the future revenue from the investment at the present time. The NPV is, simply, the present value of future cash flows minus the investment price, taking inflation and returns into account. If the investment is initially expended, it is counted as negative revenue. A NPV formula is expressed as

$$\text{NPV} = -I + \frac{R_1}{(1+r)} + \frac{R_2}{(1+r)^2} + \frac{R_3}{(1+r)^3} + \cdots + \frac{R_N}{(1+r)^N} \qquad (19.2)$$

where R_1 is the first year revenues, R_2 the second year revenues, and so on. N is the last year to account for revenue (R_N) from the investment (I). The analysis uses a minimum acceptable return, r. This NPV is defined as the difference between the present value of the expected returns (r) and the initial investment required to generate the returns. If the NPV is positive, the NPV will add value to the business. If it is negative, the investment will subtract value from the business.

If assume a constant revenue over time, Equation 19.2 will reduce to

$$\mathrm{NPV} = -I + R \times \sum_{i=1}^{N} \frac{1}{(1+r)^i} = n \times R - I \tag{19.3}$$

where n is the factor converting revenue across N number of years into present value as

$$n = \sum_{i=1}^{N} \frac{1}{(1+r)^i} = \frac{1}{r}\left(1 - \frac{1}{(1+r)^N}\right) \tag{19.4}$$

Typically n takes a value within the range of 5–10. For example, if $N = 10$ years and $r = 10\%$, $n = 6.14$. If $N = 20$ years and $r = 10\%$, $n = 8.51$. If $N = 10$ years and $r = 15\%$, $n = 5.02$.

19.6.3 Internal Rate of Return (IRR)

The IRR approach is similar to NPV. The difference is that rather than assume an acceptable return r, the IRR approach solves the equation NPV = 0 in Equation 19.2 to determine an effective interest rate or IRR. In other words, IRR is the interest rate at which the NPV of all the cash flows (both positive and negative) from an investment equal zero:

$$0 = -I + \frac{R_1}{(1+\mathrm{IRR})} + \frac{R_2}{(1+\mathrm{IRR})^2} + \frac{R_3}{(1+\mathrm{IRR})^3} + \cdots + \frac{R_N}{(1+\mathrm{IRR})^N} \tag{19.5}$$

If the IRR determined from Equation 19.5 exceeds a company's required rate of return, that project is desirable. Otherwise, if IRR falls below the required rate of return, the project should be rejected.

When the question is whether to undertake a project or not, NPV is a stronger metric than ROI and IRR for making investment decisions. This is because NPV determines the present value of making the investment and not just the amount of time needed to realize the investment. For this reason, NPV has become the most common approach to investment decisions. The IRR approach gets the profit-maximizing answer only if it agrees with NPV. For revamp projects, IRR is more widely used.

The ROI approach indicates how quick a project can provide retune against the investment. Thus, the ROI approach would select the project with the faster payback—it may make a small amount of money very quickly; but it is not obvious that this is a good choice. However, the choice is clear with NPV—select the project with the higher NPV because it has the highest value and produces the strongest benefit for the company.

REFERENCE

Raftery J, Keels J, Bowman D, Ulas-Acikgoz S, Zhu FXX (2019), Developing Data Cleansing Methodology for Honeywell Connected Plant Solutions, Vol 11, Issue 2, Exponent, Honeywell.

20

APPLICATION CASE STUDIES

20.1 INTRODUCTION

The Connected Plant methodology aims to connect **people**, **processes**, and **assets** within the organization much better than what has been doing with traditional digital solutions. The Connected Plant wants to achieve the desired outcome for customers.

The case studies provided in this chapter demonstrate that the Connected Plant is more than just data collection and software tool. It is about applying intelligent models to reveal insights. It is about integrating new digital solutions with existing data infrastructure. It is about better management of asset, workflows, and knowledge via digital solutions.

20.2 APPLICATION CASES FROM DIGITAL TWINS

A digital twin can identify reliability and optimization opportunities. The following cases demonstrate the capabilities of digital twin for assisting companies to achieve their operation excellence (Gupta, 2020; Honeywellprocess.com, 2019).

20.2.1 Case 1: APM Drives Capacity Improvement

The challenges The plant at Orange had incurred several costly unplanned shutdowns, and the penalty for unplanned downtime has been rising rapidly. There is a need to

Digitalization and Analytics for Smart Plant Performance: Theory and Applications,
First Edition. Frank (Xin X.) Zhu.

reduce the degree of unplanned downtime by transforming from schedule-based or highly reactive maintenance to predictive maintenance. At the same time, the plant management wanted a way to extend service intervals without increasing the risk of an unplanned event. Costly unplanned shutdowns combined with operational challenges, long-lived assets, and a worn plant infrastructure made it an ideal target for the Honeywell APM digital twin solution, which is selected by the plant management. The objective of using the APM digital twin is quickly to uncover hidden inefficiencies, within existing plant resources and without interrupting operations in order to ensure on-time delivery given that the plant has been running at full capacity. The plant sought to increase uptime yet needed a digital tool to enable to understand existing plant data and derive actionable insights.

What has been done

- Implementation included APM training for reliability, maintenance, and process engineering personnel, and operational procedures were updated to include roles and responsibilities of the system's users. A documented chain of command was created for APM alarm recognition and an approval workflow established for implementing corrective actions.
- Of the selected assets, four assets were recognized as opportunities to reduce downtime, prolong maintenance intervals, and capture untapped throughput potential—cumulatively reducing both planned and unplanned downtime of plant machinery and increasing overall plant efficiency.
- The most significant finding involved three critical machines where engineers ran descriptive analytics and found unique "shift" patterns in the machines bearing vibration and temperature signatures. Interestingly, the shifts revealed the ability to predict failure of a nearby packing gland (i.e., seal) rather than the bearing itself. Previously the glands were a preventative maintenance item and replaced early, but with the new insight, the plant realized they could confidently prolong the change-out interval.

Results

- The plant avoided 3 days of unnecessary downtime. Instead, the plant was able to run uninterrupted through the end of the year and plan the packing replacement in conjunction with the plant's planned 2019 turnaround.
- By the closing of its fiscal year, the plant facility had captured additional revenue as a direct result of the APM implementation. Income on the incremental revenue is setting a recovery rate of project costs to within one year.

Remarks

- The most significant contributors to the derived benefits were found directly from descriptive analytics. The project team created a unified historical data set from the data of machines and their associated processes. Using the unified data set as the basis for applying modern analytics, the engineers developed enhanced predictive models that both recognized impending machine and process deterioration and provided accurate estimates for prolonging time to required maintenance.

- This case demonstrates the value that today's cloud analytics can identify hidden asset degradation and magnifies how assets interact with the processes they serve. Leveraging the capabilities of Honeywell APM, insight was derived from asset monitoring that revealed hidden improvement opportunities in the plant's production process—improvement opportunity that would otherwise have gone unnoticed.

20.2.2 Case 2: Resolving Operation Gaps in a FCC Unit

The challenges A refinery in North America was exploring digital solutions to improve operation excellence in the FCC unit.

What was done After initial review of the refinery's operating practices, Honeywell implemented Process Reliability Advisor that enabled:

- Early event detection and resolution with embedded root cause analysis;
- Operations to run closer to constraints by leveraging a digital twin of the process.

The results

- Identified pathway to increase overall throughput by +10% (US MM$ 5/year) through optimization of the Stripper and gas concentration unit to mitigate H_2S contamination that was impacting alkylation.
- Optimization of wet gas compressor and main column heat balance for increased capacity.
- Higher conversion.
- Proactive identification of issues for unit stability.

20.2.3 Case 3: Plant Reliability and Optimization

The challenges A refinery operating in Asia wanted to improve operating margin, energy management, and increase overall profitability and reliability from their naphtha hydrotreating unit, CCR Platforming™ unit, and corresponding fractionation columns.

What was done Honeywell deployed the Process Reliability Advisor and Process Optimization Advisor that have found the following opportunities to improve the performance of the unit:

- The solution recommended adjustments to their CCR Platforming unit (including reactor inlet temperatures, H_2/HC ratio, and reactor pressure), which allowed them to operate closer to plant constraints. This enabled the customer to increase unit throughput.

- Using the tuned process model, a unit performance evaluation revealed an opportunity to improve yields and product quality through reducing the CCR Platforming unit's separator pressure while operating within the equipment constraints.

The results Process Reliability Advisor enabled the customer to maintain the reliability of the plant while maximizing profitability through optimization of the complex. They have achieved a net margin improvement of US $13,200/day amounting to US MM$ 4.5/year.

20.2.4 Case 4: Aromatics Complex Optimization

The challenges A petrochemical plant operating a reforming unit and aromatics complex observed that their naphtha feed quality fluctuates unexpectedly, which makes it difficult to optimize their aromatics complex operation.

What was done Honeywell implemented Process Optimization Advisor that leveraged the tuned process models to maximize profitability of the units through key operational changes. Major operating changes were recommended and implemented, which are summarized as follows:

- Naphtha prefractionation unit—Increased feed rate to the reforming unit by optimizing debutanizer column and naphtha splitter column.
- Reforming unit—Improved aromatics yield in reformate by 5 wt% by optimizing WAIT and saved utility costs through adjusting debutanizer column and depropanizer column reflux.
- Aromatics complex—Improved aromatics yield by debottlenecking reformate splitter, reduced ring loss in Isomar, and increased xylene production in Tatoray.

The results After implementing recommendations from the Process Optimization Advisor, the customer achieved a profit of US MM$ 5/year through improved aromatics yields in paraxylene production.

20.2.5 Case 5: Product Quality Prediction for Polymerization Process

The challenges A North American chemical company needed to reduce off-spec material production and increase working capital and profitability. Final product samples are taken at the end of processes to measure key product qualities in the lab. The long lag time from process and lab analysis means that there is a high potential for long periods of subprime material production before getting back on-spec.

What was done Honeywell Connected Industrial (HCI) developed and implemented machine learning-based predictive model to provide real-time prediction of final product quality:

- The prediction model was developed using process data, laboratory data, and advanced data analytics coupled with deep process understanding.
- Proper data aggregation was conducted by leveraging plant operational knowledge.
- The final selected model is not only a good mathematical fit to historical data but also consistent with polymerization reaction mechanism.

The results Predictive analytic models provide real-time prediction of final product quality before lab results are available, enable process adjustment without delay, and minimize off-spec product production.

20.3 APPLICATIONS FROM OTHER DIGITAL PROJECTS

In some cases, digital twins could identify opportunities, but these opportunities require further investigations and evaluation by SMEs to come up with an implementation plan until the value is captured and the benefit can be sustained via continuous improvements. The following discussions explain the procedure for implementing such opportunities to capture the value.

20.3.1 Case 6: Heat Exchanger Fouling Mitigation

The crude unit has a charge rate of 200,000 bbl/day, and crude density is 35.6API. There is a crude preheat train consisting of 18 heat exchangers with 10 before and 8 after the desalting system. Crude oil from storage tanks is fed to the heat exchangers of the crude preheat train, initially at ambient temperature. The crude oil is then heated through the preheat train to 480 °F (250 °C) at entry to the feed furnace.

Out of the furnace, the crude is fed to a distillation column where valuable product streams, such as gases, kerosene, gasoline, and diesel, are distilled and collected. After the process unit is operated for less than 6 months, severe fouling occurs in the heat exchangers and leads to a decline in furnace inlet temperature by as much as 65 °F, which results in burning of significant amount of extra fuel in the furnace to make up the temperature necessary for effective distillation.

20.3.1.1 Know the price for fouling mitigation The average heat recovery from the preheat train is 378 MMBtu/hr versus the maximum heat recovery of 511 MMBtu/hr if every exchanger is in relative clean condition. This implies that 133 MMBtu/hr is lost due to fouling. The plant engineers determined that if 50% of this lost heat could be captured by better fouling control, about 70 MMBtu/hr of process heat can be recovered and thus 82 MMBtu/hr of fuel could be saved in the feed heater. In monetary terms, it could save approximately US MM$ 4/year with natural gas prices at $6/MMBtu. This savings does not consider additional benefits of larger throughput achievable as the furnace capacity is limiting for the unit. Thus, the plant management decided to implement a fouling control project to capture the benefit.

20.3.1.2 Rating assessment The rating assessment was conducted for heat exchangers in the preheat train, and the assessment results indicated that fouling occurred, in a varying degree, to most heat exchangers, but the severe fouling is concentrated only in few heat exchangers under high temperature located near the terminal of the preheat train. These heat exchangers mainly involve services provided by reduced crude (RC), heavy vacuum gas oil (HVGO), and vacuum residue (VR). There are no downstream heat exchangers available to offset the loss of heat recovery in the terminal of a preheat train network.

The severe fouling in these exchangers caused higher pressure drop, which results in vaporization of crude by more than 3% before entering the feed heaters and thus flow path control valves wide open. Crude flashing made it impossible to control feed flows with eight paths. This could be a safety hazard if some furnace tubes filled up with vapor and formed a stratified flow pattern. The worst case could happen if the furnace was allowed to continue to operate in this unsafe mode. This is because under this flow pattern, these tube walls directly exposed to combustion flames could become too hot. This could eventually lead to tube rupture. The finding of this reliability risk made the fouling mitigation project even more urgent.

20.3.1.3 Root cause analysis Root cause analysis was conducted that revealed the following:

- **Crude properties**: During offline cleaning, the bundle from one of the very fouled heat exchangers was pulled out for inspection, and chemical analysis of fouling materials was performed (Table 20.1). It was found that asphaltene deposits was dominant (Figure 20.1), which is caused by polymerization of asphaltenes. This explains the reason why the most severe fouling occurs under high temperature. In comparison, the crude corrosive properties (e.g., reactive sulfur and naphthenic acids) were found to be much less responsible for depositional fouling.
- **Low velocity**: Low velocity was another primary cause for promoting fouling that occurs in some of the heat exchangers. The velocities in these exchangers were found to be 3–4 ft/s versus 6 ft/s recommended for these services.

TABLE 20.1. Fouling Deposit Compositions from Lab Test

Materials	wt%
Iron sulfides/oxides	15
Asphaltic material (soluble in benzene)	20
Asphaltic material (insoluble in benzene)	50
Inorganic salts (CO_3; SO_4 Cl)	15

FIGURE 20.1. Asphaltenic deposits found in the heat exchangers. *Source*: Trevor S. Bird.

- **Poor desalter operation**: It was found that salts were not removed sufficiently from the desalter; it makes fouling worse.
- **Exchanger geometry**: The heat exchangers located in the terminal of the preheat train suffered the most severe fouling. Use of helical baffle bundles could help in mitigating fouling much better than shell-and-tube bundles currently used.

20.3.1.4 Explore fouling mitigation options Based on the fouling mechanisms and compositions found and fouling mitigation methods available, a fouling mitigation strategy was developed as below:

(a) Offline cleaning: Proper cleaning of crude–HVGO exchanger (heavy gas oil) restored the performance of this exchanger.

(b) Operation change: Improve desalting operation to minimize salt-related fouling.

(c) Design change: Replace existing vertical baffle bundle with helical baffle for crude–residue exchanger. This design change was straightforward to implement, which prolonged the operation for this exchanger.

(d) Flow path change in maintenance: Increase flow velocity from 3 to 5 ft/s for two exchangers.

Let us go through each of these fouling mitigation solutions in details.

20.3.1.5 For offline cleaning of crude–HVGO exchanger Since this exchanger suffered severe fouling, it was difficult to clean using water jetting alone. To have effective cleaning, a combination of different cleaning methods was employed for two shells of this exchanger:

- Clean method 1: Use hot light cycle oil (LCO) to clean two bundles for E1007-A/B.
- Clean method 2: Then hydroblasting E1007-B using water jet.
- Clean method 3: Combination of water jetting and use of abrasive balls for cleaning E1007-A.

The performance trend in UA is plotted in Figure 20.2, which shows significant improvements at different levels (1–3) for different cleaning methods. As the result, the bare metal performance for E1007 was established (Figure 20.3).

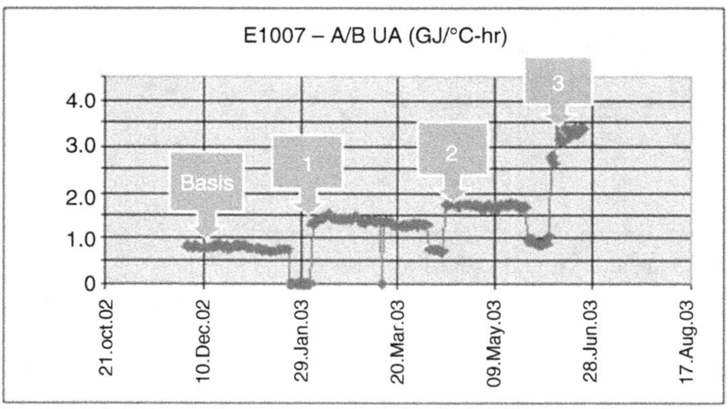

FIGURE 20.2. UA improvements for different cleaning methods.

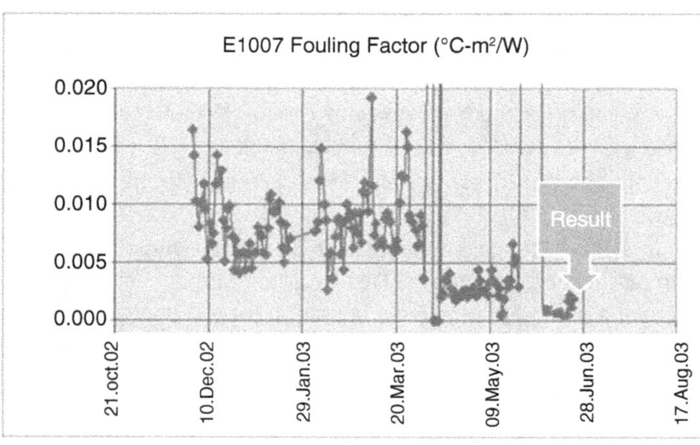

FIGURE 20.3. Bare metal performance established after cleaning.

20.3.1.6 Improve desalting performance Crude requires desalting to minimize fouling and corrosion caused by salt deposition on heat exchanger surfaces and acids formed by decomposition of chloride salts. This crude unit has one-stage desalting design that can reduce the salt content in crude by 95%. The desalting and operating objectives are to remove salt, solids, and metal while minimizing the loss of oil in water phase due to oil entrainment. The key operating parameters such as desalting temperature, water injection, and pH control were optimized to achieve the target desalting performance as shown in Table 20.2.

20.3.1.7 Design change: Change bundle type A major change was made to E1011 and E1012, the crude–residue heat exchangers arranged in parallel, which are the last two exchangers in the preheat train before fired heaters. The crude was placed on shell side in design with the purpose of reducing pressure drop for the crude. The asphaltene deposit quickly built up from the dead flow zones (Figure 20.4) and then filled up the flow channels on the shell side where crude was placed.

The design change was implemented to replace existing two tube bundles with single helical baffle bundle (Figure 20.5). The helical flow eliminated dead flow zones as well as enhanced heat transfer on the shell side due to the countercurrent flow. Not only heat transfer duty was augmented greatly, but pressure drop was reduced at the same time.

20.3.1.8 Flow path change in maintenance After implementing bundle replacement for the two crude–residue exchangers, plant engineers focused on other two crude–residue exchangers, namely, E1009 and E1010, located upstream of E1011 and E1012. High pressure drop from measurement signaled that both E1009 and E1010 were also experiencing severe fouling. The rating assessment indicated the velocity of the crude was too low only at 3.9 ft/s by design versus required velocity of 6–7 ft/s. The solution to this problem was to increase the tube passes from four to six, and thus velocity increased to 5.8 ft/s at the expense of higher pressure drop than four-tube passes. The extra pressure drop came from

TABLE 20.2. Benchmark Desalting Performance for the Single-Stage Desalting

	Unit	Benchmark
Salt in desalted crude	ppm	0.6–6
Salt removal	%	90–98
Solids removal	%	50–80
Solids in desalted crude	%	0.1–1.0
Na removal	%	95+
Oil in effluent water	%	0–2
Iron removal	% inorganic iron	50–75
Slops reprocessing	% of crude charge	0.1–1.0

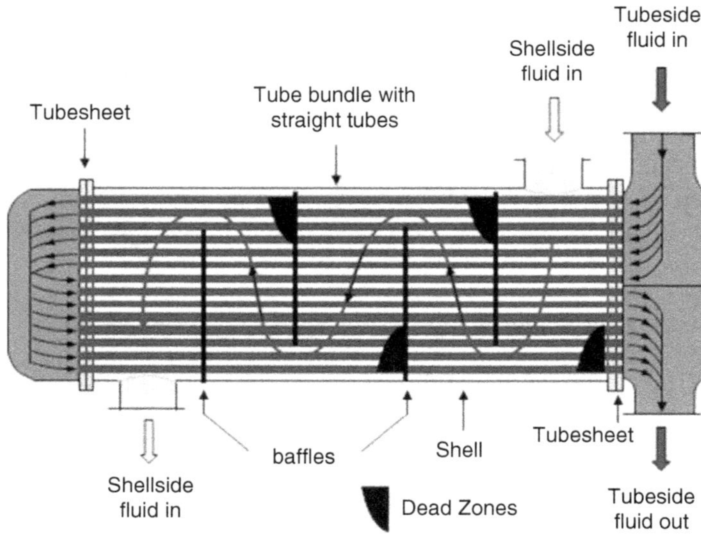

FIGURE 20.4. Dead flow zones in straight baffle shell-and-tube exchanger (one-shell pass and two-tube passes).

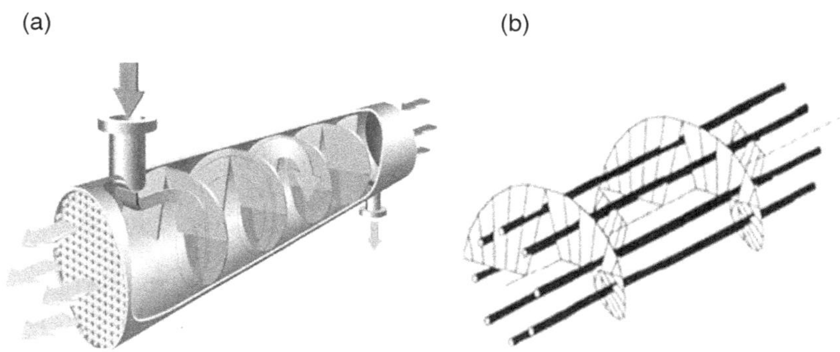

FIGURE 20.5. Helix shell-and-tube exchanger: (a) outside view and (b) inside view.

replacement of the existing bundles with helical baffle bundles for E1011 and E1012 as explained above.

20.3.1.9 Measuring and sustaining benefits The overall energy saving benefit from the above improvements was about 80 MMBtu/hr worth about US MM$ 3/year on a sustainable basis. Importantly, the unsafe operation with feed flashing before heaters was resolved. The implementations required cleaning of one heat exchanger, replacement of four tube bundles for two exchangers, and change of tube passes for two exchangers as well as improvement for desalting operation. As a measure to sustain the benefits, monitoring system and clean cycles were developed for these heat exchangers. The details for heat exchanger fouling assessment can be seen in Zhu (2014).

20.3.2 Case 7: Distillation Optimization

Operation of separation columns involves trade-off between energy use and product recovery and purity. When energy usage is lower than the requirement, product recovery and purity suffer. On the other hand, when energy is more than the requirement, product purity and recovery are improved. The optimal trade-off determines the operating target. The operating parameters involved in the trade-off include reflux ratio, feed temperature, column temperature, reboiling duty, column pressure, and so on. The questions are: how to obtain the energy target for a fractionation column and how to achieve the most economic operation of the column?

Let us look at an example that consists of a debutanizer column (Figure 20.6) for which White (2012) gave excellent explanations. This example gives perspective and guidelines in principles for optimizing a separation column, and this example is reproduced here with permission from AIChE.

The feed and product specifications and prices for this example are listed in Table 20.3. Both products have tiered prices: on-specification product is priced

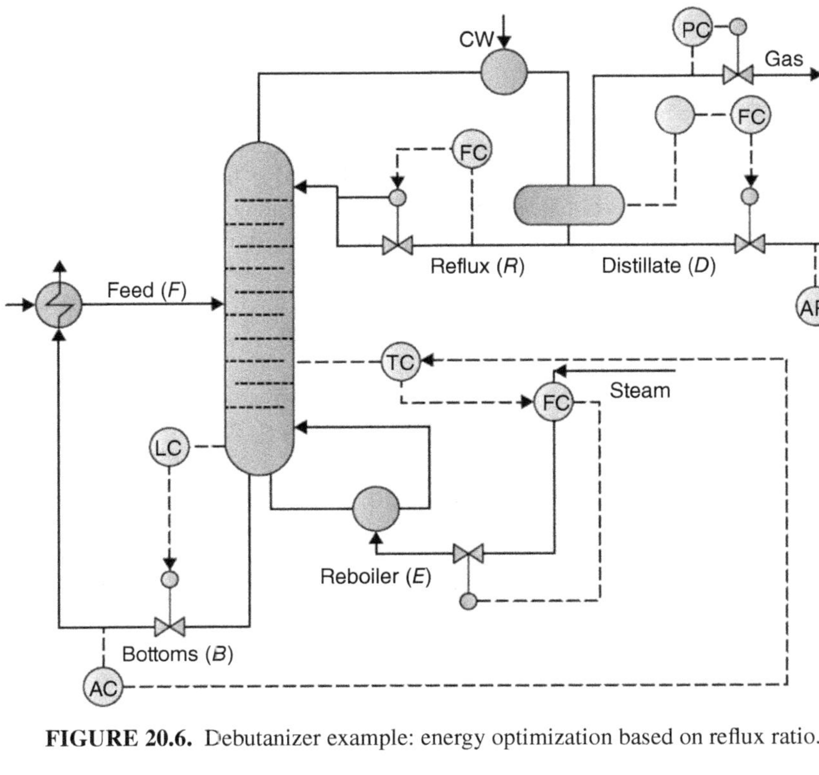

FIGURE 20.6. Debutanizer example: energy optimization based on reflux ratio.

TABLE 20.3. Product Specifications and Prices

Stream	Composition/Specification	Value
Feed, 20,000 bbl/day	25% C_3 25% nC_4 25% nC_5 25% nC_6	$60/bbl
Bottoms product = C_5	≤5% C_4	$80/bbl
	>5% C_4	$60/bbl
Top product = C_4	≤3% C_5	$60/bbl
	>3% C_5	$40/bbl
Steam		$15/MBtu

Source: From White DC (2012). Reprinted with permissions of American Institute of Chemical Engineers.

much higher than the one that does not achieve specification. If the top product, butane, achieves the specification, that is, less than 3% C_5, it is sent to downstream unit for further processing leading to eventual sales. Off-spec butane will be used as fuel that has a lower value than selling as a product. Similarly, the bottoms product, pentane, if achieving specification, is used for making high-value product. Otherwise, it will be sent to tank for reprocessing. In operation, the operator changes column temperature and reflux to feed ratio to achieve the product specification.

20.3.2.1 Develop economic value function for distillation optimization The objective function (economic value) representing the economic operating margin for the column is shown in Figure 20.7, which is defined as the difference between the value of products (top butane product and bottom pentane product) and costs of feed and energy. The value function features two discontinuities. The first, which occurs when the composition of the bottoms product is about 1% butane, corresponds to a change in the top product from off-spec to on-spec. The second discontinuity occurs when the bottoms product becomes off-spec at 5% butane.

20.3.2.2 Setting operating targets with column bottom temperature To choose the bottom temperature target, first assume that the reflux rate is fixed and that the bottoms product is on-spec but the top product off-spec because of its high pentane content. This would correspond to a very hot bottom temperature. When the bottom temperature is slowly reduced, the amount of bottoms product increases, but the percentage of butane in the bottom also increases simultaneously. As the amount of pentane product increases, the total product value improves. The middle line in Figure 20.7 represents the normal operation.

Usually one would select a temperature target such that the bottom composition is as close to the specification limit as possible. There will always be some variability in the control performance due to external disturbances and limitations on loop control action. If composition control is poor and has a high variance, the observed composition probability distribution could look like a normal distribution in relation of operating margin as shown in Figure 20.8.

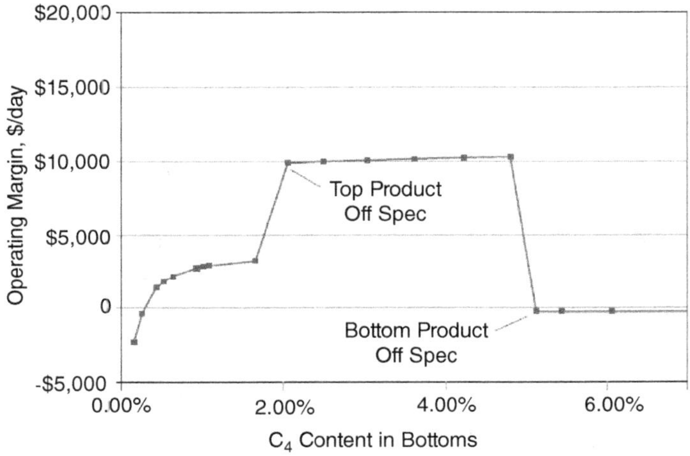

FIGURE 20.7. Operating margin as a function of the bottom composition. *Source*: From White DC (2012). Reprinted with permissions of American Institute of Chemical Engineers.

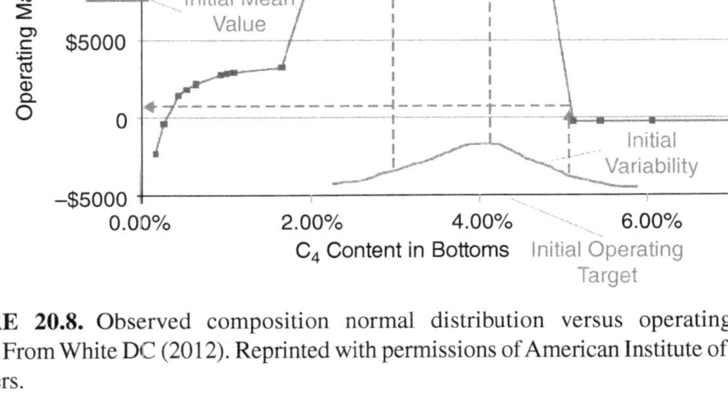

FIGURE 20.8. Observed composition normal distribution versus operating margin. *Source*: From White DC (2012). Reprinted with permissions of American Institute of Chemical Engineers.

The product composition target is the mean value of observed composition distribution as shown in Figure 20.8. The mean value of the operating margin in Figure 20.8 is calculated based on the weighted average composition of the observed distribution—that is, the percentage at each composition is multiplied by the margin value at that composition to determine the overall value.

Figure 20.8 shows that part of the column operation is the bottoms product being off-spec. The mean product value does not correspond to the value at the mean of product composition, which is also the operating target. This is because of the nonsymmetrical nature of the operating margin and low value of off-spec products.

After reducing the variability through improved control-valve performance and reduced measurement error, the new mean value of the operating margin increases at the same operating target or bottom composition target (Figure 20.9). It can be seen that reduced variability results in increase in the mean value of operating margin.

20.3.2.3 Setting operating targets with column reflux ratio The above discussions involved constant reflux ratio. Next consider the situation where the reflux rate is varied and bottom temperature is constant. Fundamentally, reflux rate provides internal reflux needed for separation in the tower, and it is generated by either feed heater or reboiler. Thus lower reflux rate saves energy, but too low reflux rate could negatively affect product quality. On the other hand, high reflux rate could improve production of more valued product.

In tower design, the reflux ratio is determined based on the trade-off between operating cost in reboiler and capital cost for the tower. In other words, use of more separation stages requires less reflux rate and thus less reboiling energy but at the expense of additional capital cost. The minimum reflux ratio is calculated based on Underwood (1948). A tower requires an infinite number of stages to achieve the minimum reflux ratio. To make separation feasible and at the same time a tower affordable, a reflux ratio larger than the minimum is used. Typical design reflux ratios are 1.1–1.3 of the minimum reflux ratio. With a high reflux ratio, the number of theoretical stages is lower, resulting in lower capital cost for a tower but at the expense of higher reboiler duty; and vice versa. The optimal reflux ratio in tower design is determined based on the minimum total cost.

However, in operation, optimal reflux ratio is determined based on the trade-off between product value and energy cost. When the reflux ratio increases, the separation improves at the expense of increased reboiling duty (Figure 20.10). As the result, top product rate decreases, while the bottoms product rate increases. As shown in Figure 20.10, the cost of reboiling duty presents a linear relationship with reflux ratio, but the product quality is nonlinear and presents a different trend as reboiling duty.

Figure 20.11 shows the operating margin for different energy prices, assuming constant product prices. The optimum reflux rate depends on the price of energy. At a high energy price, the optimum reflux rate is at the minimum value, which allows the column to maintain the top product in specification. At the lower energy prices, the optimum reflux rate increases.

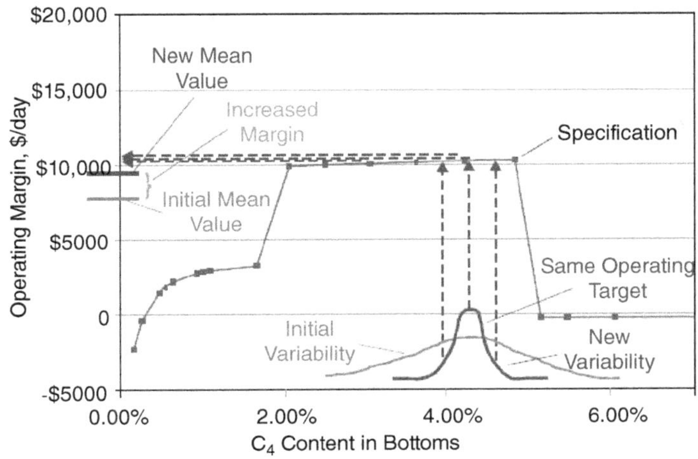

FIGURE 20.9. Improved composition normal distribution versus operating margin. *Source*: From White DC (2012). Reprinted with permissions of American Institute of Chemical Engineers.

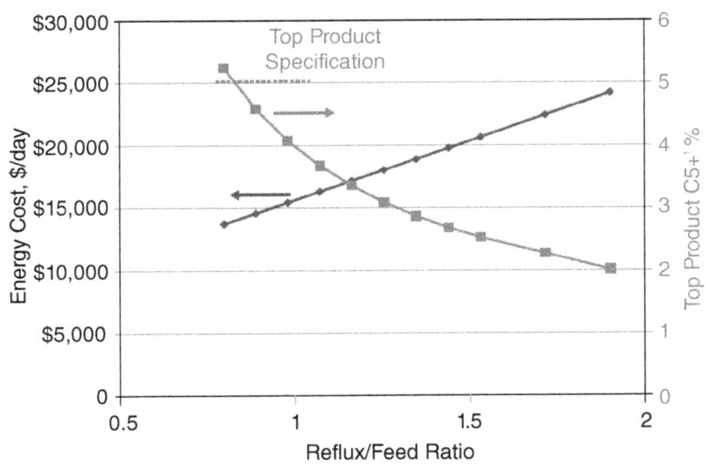

FIGURE 20.10. Energy-separation trade-off: energy cost increases linearly as reflux rate, while the top product quality improves. *Source*: From White DC (2012). Reprinted with permissions of American Institute of Chemical Engineers.

The conclusion is that operating targets should be a function of energy costs rather than a fixed number even with fixed composition limits. It is common to observe separation columns operating at reflux rate that are 50% higher than the optimum. For the debutanizer column operation discussed here, such an operation could cost operating margin in excess of $500,000/annum.

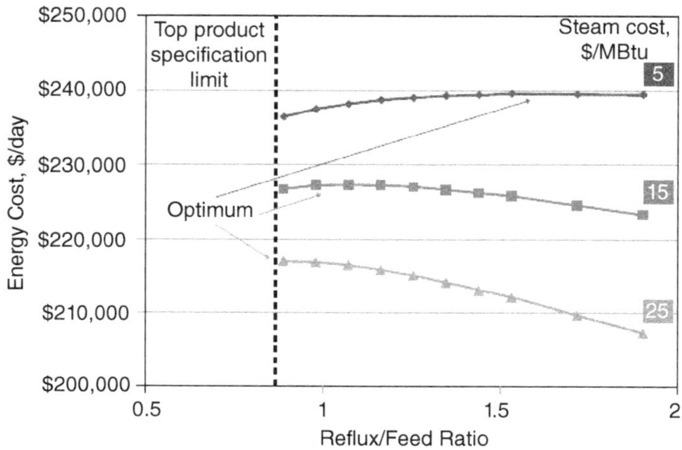

FIGURE 20.11. Optimum reflux rate depends on energy price. *Source*: From White DC (2012). Reprinted with permissions of American Institute of Chemical Engineers.

20.3.2.4 Setting operating pressure It is generally known that reducing the operating pressure of separation columns reduces energy consumption. This is because the lower the tower pressure, the less heat required for liquid to vaporize (thus less energy required) and the easier for vapor to penetrate into liquid on the tray deck (thus better separation). Yet many columns are operated well above their potential minimum pressure. One may ask: if benefit of reducing pressure is well known, why is it not widely implemented? There appears to be three primary reasons for this.

First, changing column pressure requires simultaneously changing the bottom temperature setpoint to hold the product composition at their targets. This is difficult to do manually—advance composition control is required.

Second, changes in column pressure have other impacts such as changes in the off-gas rate, the amount of reboiler duty, and hydraulic profile of the plant. In the case of partial condensation, pressure control can interact with the overhead receiver level. While these effects are real, their magnitude is sometimes exaggerated and cited as reasons for not making any changes.

Finally, plant personnel frequently do not agree on the amount of operating margin required to handle major disturbances. For instance, questions often arise about the dynamic response of an air-cooled condenser to a rainstorm and the ability of the overall control system to handle such conditions. A well-designed overall control system for the column can compensate for such disturbances.

The condenser pressure controls the tower pressure and thus the feed tray pressure. There is a pressure valve in the overhead that can be used to control tower pressure. The lower limit of the tower pressure depends on column overhead condensing

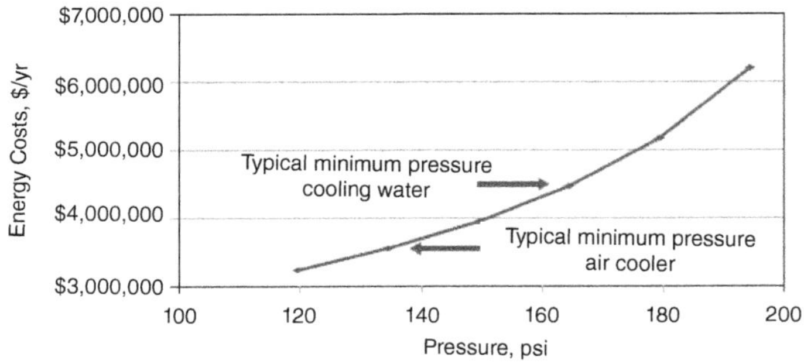

FIGURE 20.12. Pressure has significant effect on energy cost. *Source*: From White DC (2012). Reprinted with permissions of American Institute of Chemical Engineers.

duty, net gas compressor capacity, and column flood condition. Many of the new advanced process control (APC) systems are using pressure control to save energy. During extended turndown periods, reducing pressure up against an equipment limit can improve efficiency (Figure 20.12).

20.3.3 Case 8: Integrated Planning and APC

A North American oil refinery contracted Honeywell and Piccolo Control Services to perform a benefit study of this new MPC cascade technology (see Chapter 13). The study (Lu, 2014) is composed of two steps:

- Step 1—Estimate the amount of potential benefits unrealized in the form of "giveaways."
- Step 2—Assess how much of the potential benefits can be obtained by applying the new multiscale MPC cascade technology.

Step 1 is a qualitative analysis that sets up the directions for step 2. Step 2 is a quantitative analysis via rigorous simulations using 12 contiguous months of actual operating data. The simulation conditions and scenarios are determined from the qualitative analysis of step 1.

The distillate production of the refinery, is selected as the initial scope of the study for three reasons. Firstly, distillate products command a premium price over gasoline products in North America, and improving the production efficiency in that part of the refinery can have a significant impact on the profit margin. Secondly, the distillate production involves fewer units than its gasoline counterpart. Therefore, it takes less effort and costs less for the customer. And thirdly, the margin improvement percentage can be extended to the gasoline production since both sides have a similar planning-control gap. Note that the scope includes the FCC unit to ensure that the commitment of other product deliveries (such as gasoline) can be fulfilled.

During the study period, many distillate products were delivered—Seasonal Diesel, No. 1 Diesel, Marine Diesel, and Furnace Fuel Oil, just to name a few. Among them, the Seasonal Diesel accounted for a significant majority of total product delivered.

For each distillate product, 13 product qualities (aka properties in oil refining) must be lab-tested and certified before the product can be shipped. Five of the 13 qualities are more difficult to meet than the rest and thus are tightly controlled in each of the five distillate units with MPCs. They are cloud point, sulfur content, distillation 90% point, flash point, and pour point.

Figures 20.13 through 20.17 show the important product quality specifications for Seasonal Diesel and the actual product properties. Of all the properties analyzed, only one product property (cloud point) tracked its quality specification changes over several seasons in the year. The other four product properties had a significant amount of giveaways (meaning the refiner delivered overqualified products for the given grade, often at a higher production cost).

Figure 20.13 shows the cloud point of Seasonal Diesel in Celsius. For most of the study year, it tracked its seasonal spec changes, though there was some room available for improvement in the middle of the study year.

The pour point is almost always parallel to the cloud point, and thus it is not explicitly controlled in the refinery. Once the cloud point is close to its spec limit, it becomes an inactive dependent property and thus will not be further analyzed here.

Figure 20.14 shows the sulfur content of Seasonal Diesel in ppm. For most of the study year, sulfur was overtreated at the expense of excess consumptions of hydrogen and hydrotreating catalysts.

Figure 20.15 shows the distillation 90% point of Seasonal Diesel. For an entire study year, the distillation 90% point was far below the required diesel cut. In fact the final product was closer to the kerosene cut than to the diesel cut. One of the causes was that the distillation 90% point and the cloud point were correlated under

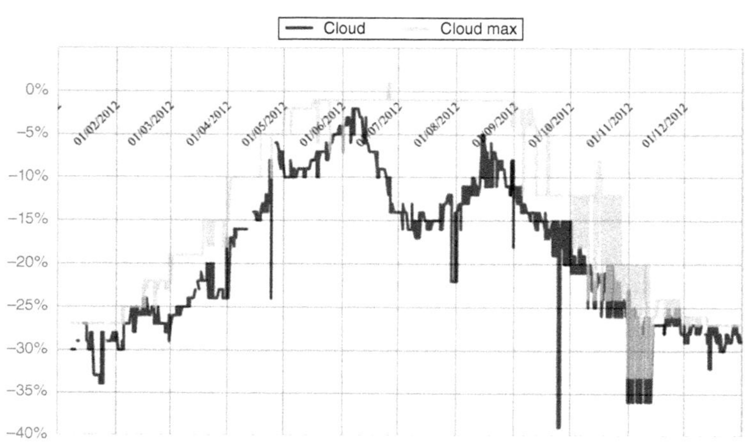

FIGURE 20.13. Cloud Point of Seasonal Diesel (°C).

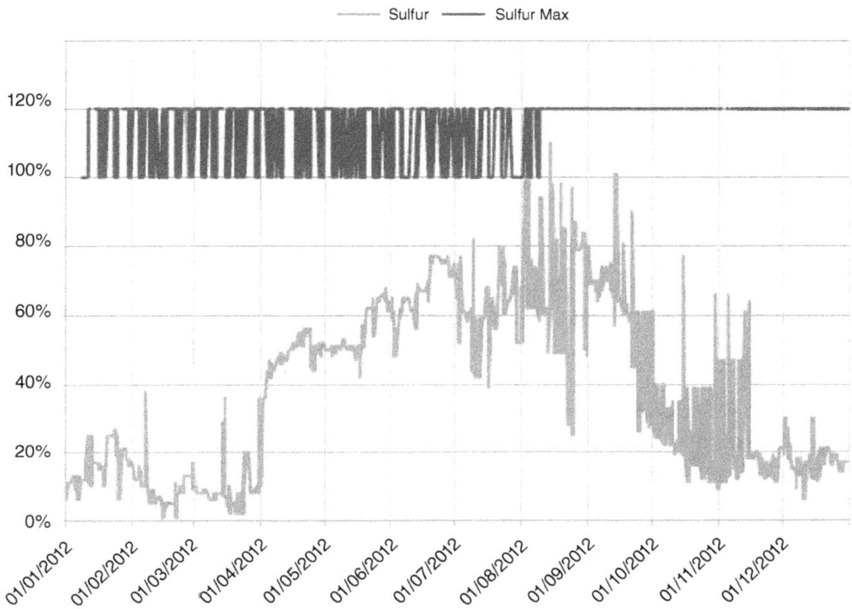

FIGURE 20.14. Sulfur Content (ppm).

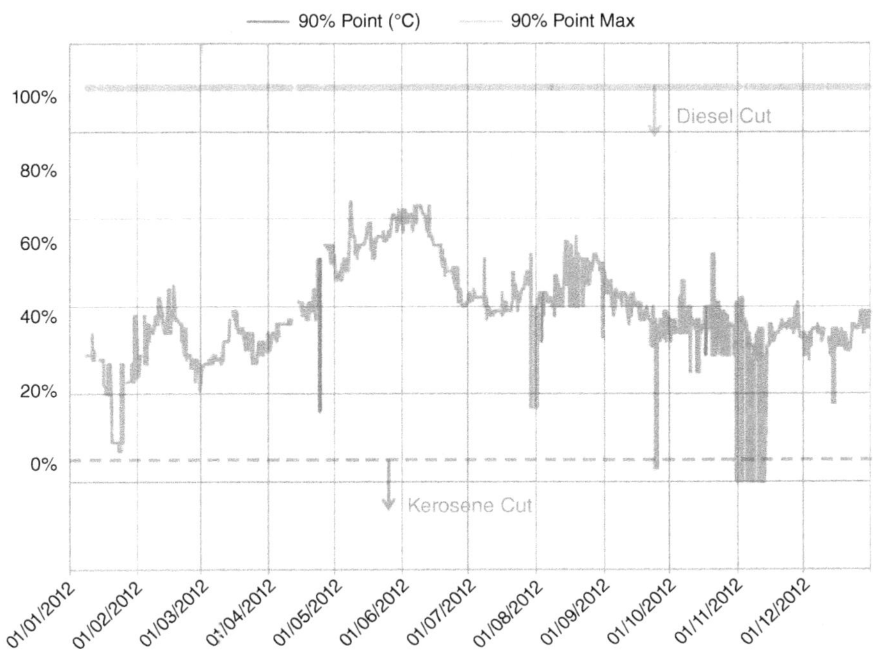

FIGURE 20.15. Distillation 90% Point (°C).

certain operating conditions, and a simple correlation analysis indicates that they were more correlated in the summer than in the winter season. Another possible cause was a lack of guidance from planning, and thus the unit-level MPCs were not directed to push these two properties apart and minimize the giveaways of both.

Given the fact that there was only a 3-month period in the summer where the cloud point was close to its high limit, this suggests that for the remaining 9 months, the cheaper heavier crude feed could be increased to crude unit C, whenever the capacity is available, to offset the more expensive lighter crude that was used in crude units A and B. If the new MPC cascade technology is applied, the amount of crude swap would be maximized by the MPC cascade until either distillation 90% point, cloud point, or both ride on their spec limits.

A revealing fact is that distillation 90% point (or its correlated CV, such as cloud point) is controlled at every product unit by MPC controllers to ±1~2 °C. APC engineers spend a lot of time and painstaking effort to step-test, design, tune, and finally achieve good control performance. But in the end, 50 °C of distillation 90% point is given away despite the control efforts. This suggests that either distillation 90% point should not be controlled within ±1~2 °C in every unit or the amount of giveaway should be reduced to a value that is comparable with the control precisions at each production unit. This disconnect shows how big of a gap currently exists between planning and control.

Figure 20.16 shows the flash point of Seasonal Diesel. For an entire study year, the flash point was far above the required diesel spec limit of 45 °C. In fact the product was one grade better (overqualified) and was even above Jet A fuel at 51.1 °C (Jet A typically commands a higher price). This means that an increased amount of less expensive naphthas could have been routed into the diesel pool to generate more valuable diesel products while still meeting the flash spec. If the new MPC cascade technology is applied, a maximum amount of naphthas can be controlled and redirected into the distillate pool until the flash point rides on its spec limit of 45 °C.

Another revealing fact is that there was a lack of guidance from the planning model to APC controllers on how flash point should be controlled or optimized. As a result, flash point was not controlled tightly at any unit. In the end, 10–25 °C of flash point was given away. If the planning tool had modeled flash point more sufficiently and provided guidance to the APC controllers, the flash point giveaways could have been significantly reduced. This reiterates how big the gap is that currently exists between planning and control.

Figure 20.17 shows the Cetane Index of Seasonal Diesel. For an entire study year, the Cetane Index was far above the required diesel spec limit of 40. In fact, the product was again one grade better (overqualified) and was even above the EU Diesel Plus (its spec limit is 46). Cetane Index was not controlled directly at any distillate unit, and only the planning tool monitors the Cetane Index in open loop. An offline planning solution cannot by itself effectively control any properties without a significant giveaway. If the new MPC cascade technology is applied, the Cetane Index can then be controlled in the master MPC in closed loop, and the amount of giveaway can be significantly reduced. Note that for this specific customer, the Cetane Index does not carry much value in its specific market. The savings from the Cetane Index will not be included in the benefits.

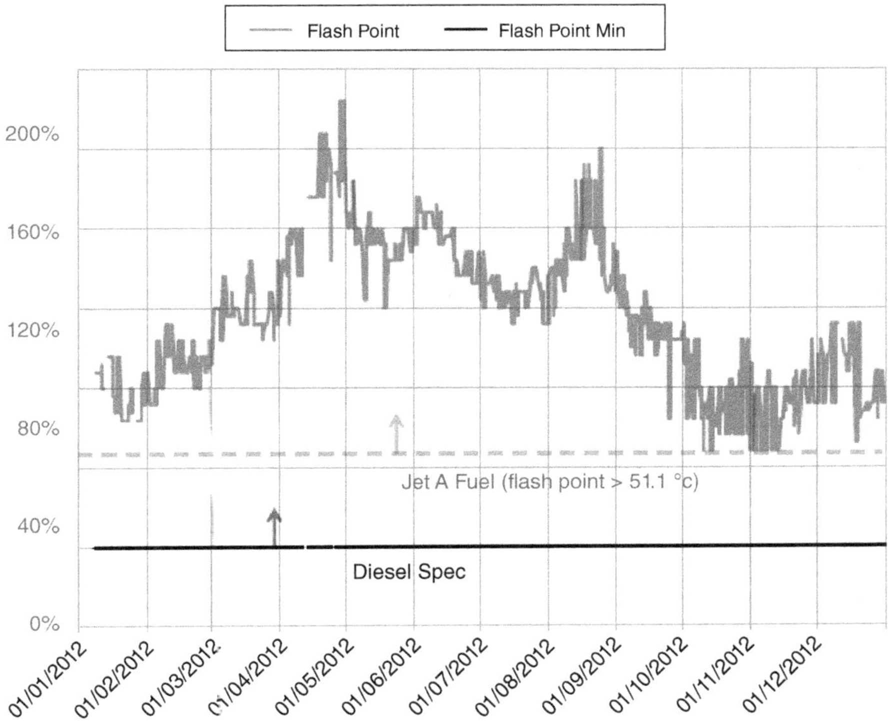

FIGURE 20.16. Flash Point (°C).

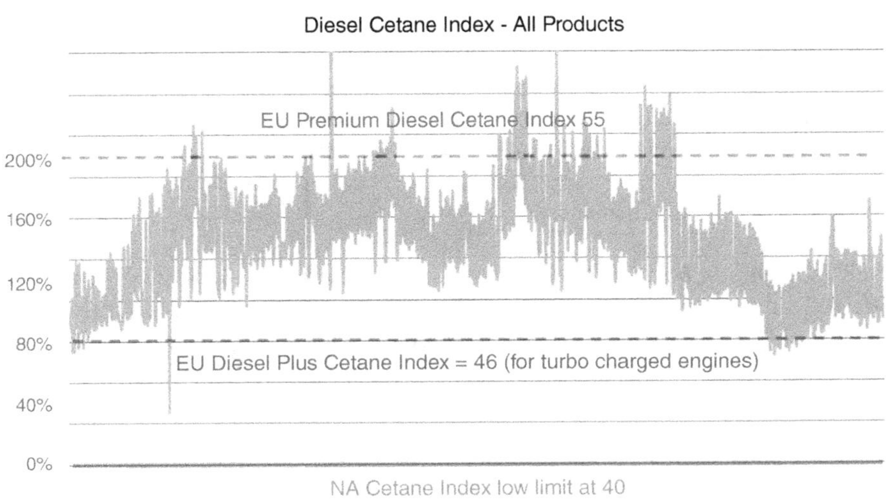

FIGURE 20.17. Cetane Index of Seasonal Diesel.

With the above qualitative analysis, it is conservatively estimated that at least US MM$ 10/year additional profit can be captured with the new multiscale MPC cascade technology. A majority of the US MM$ 10 profit margin improvement will come from feed cost reduction, yield improvements (from the unit itself and, more often, from routing an optimal amount of suitable intermediates to each unit for the known product deliveries), better trade-offs between distillate and naphthas, reduction in property giveaways, more accurate real-time planning results, better coordination between distillate production and blending, and the savings of catalyst and hydrogen at three reactive units. Only a very small portion will come from crude feed increases because the production rates of the three crude units are already near their maximums. At the time of writing this paper, a set of rigorous simulations are being devised, using the actual operating data from the plant, to compute a more accurate estimate of the profit margin improvement derivable from applying this new multiscale MPC technology.

With respect to the gasoline pool, it is also estimated that the obtainable profit improvement will be a bit less than twice that of the distillate pool. The overall profit improvement for a mid-sized (200 MBPD) refinery could be as high as US MM$ 20/year. As of July 2014, the aforementioned rigorous simulation study of the MPC cascade technology, using the real refinery operating data, has been performed. The benefit results, which have been reviewed by the paying customer, confirm that greater than US MM$ 24/year can be achieved in its diesel pool of the refinery without any production increase.

20.3.4 Case 9: Overall Process Optimization

This example comes from Loe and Pults (2001) and is reproduced, with permission from AIChE, for the purpose of explaining how a tower could be optimized. This example provides a case history of how operational improvements for a single deisopentanizer (DIP) fractionation tower are identified, implemented, and sustained. To do so, the current operation is simulated and assessed. Then improvement opportunities are identified, and the limiting factors are determined. Optimal solutions are obtained by optimizing tower ISBL conditions (Inside System Battery Limits) and OSBL conditions (Outside System Battery Limits). The improvements on this single tower have generated over $500,000 as compared with historical operation over the first 6 months.

20.3.4.1 Design basis The DIP tower, shown in Figure 20.18, processes light straight-run naphtha from two sources. The LPG fractionation unit debutanizer bottoms consists of primarily isopentane and normal pentane (iC_5 and nC_5), with small fractions of butane and C_{6+} components. The overhead from the naphtha fractionator tower contains mostly C_5 and C_6 paraffin compounds, with some benzene and C_6 naphthenes and a small amount of butane. The combined feed to the DIP is typically in the range of 9,000–15,000 bbl/day.

The DIP overhead product is normally rich in isopentane and is routed to gasoline blending along with other high-octane, low Reid vapor pressure (RVP)

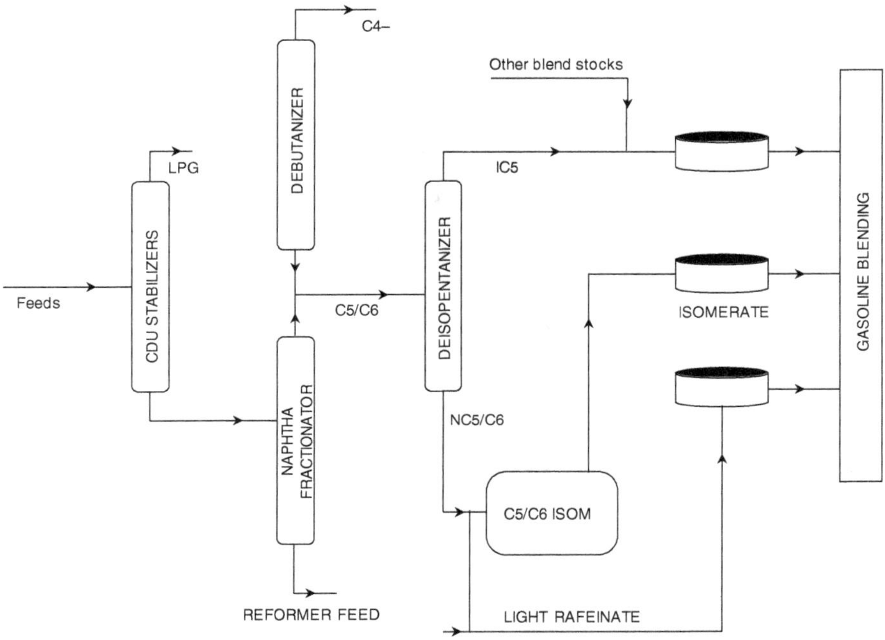

FIGURE 20.18. Deisopentanizer flowscheme. *Source*: From Loe B, Pults J (2001). Reprinted with permissions of American Institute of Chemical Engineers.

gasoline components. The DIP bottom, which is rich in nC_5 and C_6 paraffins, is routed to the light naphtha isomerization unit, along with light raffinate from the aromatics extraction unit.

The DIP tower has 50 trays in comparison with 70+ trays used in a typical DIP. As a consequence, the DIP tower often has a difficult time making a good split between isopentane and normal pentane components.

20.3.4.2 Current operation assessment Historically, the tower had been operated with a target of 10% nC_5 in the overhead and 20–30% iC_5 in the bottoms product. The tower was reported to be limited by reboiler or condenser duty. One of the two steam reboilers had been out of service for some time, and 20–30% of the condenser fin fan motors were not operating and in need of repair. The DIP equipment had not been a maintenance priority, in part because no economic penalty had been calculated for having a reboiler or condenser out of service. There was also a concern that the tower could flood if both reboilers were placed in service.

The DIP process control was accomplished with a DCS system equipped with an APC algorithm. The controller was set to target 10% nC_5 in the overhead and 10% iC_5 in the product and would increase reboiler steam and reflux rate until

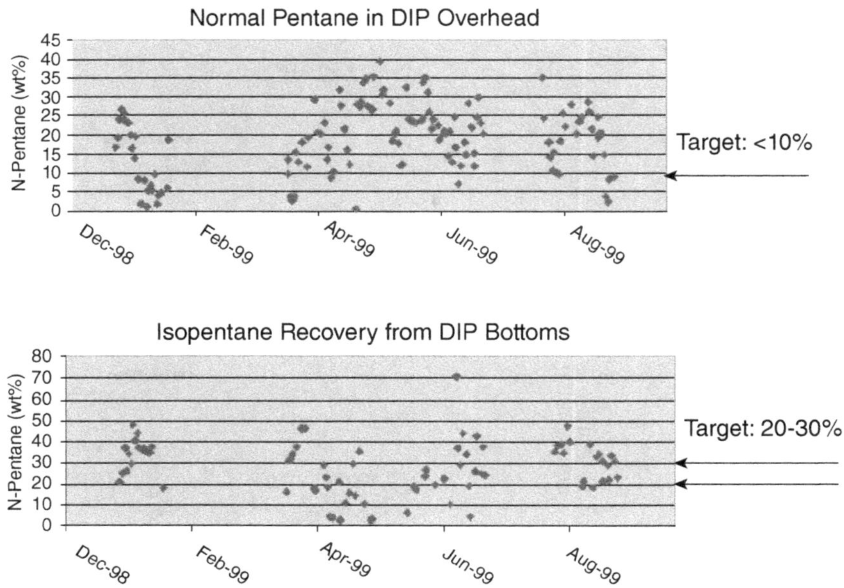

FIGURE 20.19. Variation of DIP performance. *Source:* From Loe B, Pults J (2001). Reprinted with permissions of American Institute of Chemical Engineers.

reaching the maximum limits for these flows. The tower pressure was also controlled within a specified range by the APC, and this could indirectly limit the reboiler duty as well, if the tower pressure increased beyond its set maximum. Inferential estimates for product iC_5 and nC_5 qualities were calculated based on tower temperatures and pressures, and a bias for these values was continually updated based on daily lab data.

To establish the historic performance, the operating and laboratory data were collected from the past year, eliminating periods of known equipment failure or poor unit volume balance. The data showed an average of 19% nC_5 in the DIP overhead and 27% iC_5 in the bottoms product, with a wide variation in the product quality, as shown in Figure 20.19. An average of 18% of normal butane was observed in the overhead product, indicating poor debutanization in the upstream fractionation towers.

The process design for the tower showed an available reboiler duty of 51,000 lb/hr of steam, but the average for the data collected showed only 27,000 lb/hr. The data shows an erratic variation in reboiler duty.

20.3.4.3 Simulation Using the averaged process and lab data, a simulation model for the DIP was developed. The feed rate and composition to the tower were fixed, as well as the reflux rate, tower pressures, and overhead rate. The model results for reboiler duty, tower temperatures, and compositions compared favorably with the unit operating data, as shown in Table 20.4.

TABLE 20.4. Simulation Results Versus DIP Operating Data

Result	Data	Model
Reboiler duty (MMBtu/hr)	23.6	26.7
Top temperature (°F)	143	153
Bottom temperature (°F)	179	183
nC_5 in overhead (wt%)	18	16
iC_5 in bottoms (wt%)	31	30

Source: From Loe B, Pults J (2001). Reprinted with permissions of American Institute of Chemical Engineers.

TABLE 20.5. Simulation Results at Designed Reboiler Duty

Result	Design	Model
Reboiler duty (MMBtu/hr)	48.8	48.8
Condenser duty (MMBtu/hr)	45.0	48.1
Max jet flood (%)	85	74
Max downcomer backup (%)	50	37
nC_5 in overhead (wt%)	NA	5
iC_5 in bottoms (wt%)	NA	18

Source: From Loe B, Pults J (2001). Reprinted with permissions of American Institute of Chemical Engineers.

The calibrated model was used to simulate DIP performance at the design reboiler duty to determine if available condenser duty and tower tray capacity would be adequate for this operation. As shown in Table 20.5, the predicted condenser duty for this operation was only slightly above the design value, and tower tray parameters indicated that flooding was unlikely. Also the separation of iC_5 and nC_5 improved dramatically versus the historical operation as would be expected with the increased tower traffic.

20.3.4.4 Define the objective function The profitability of the DIP column was determined based on the value of separating iC_5 for direct blending to gasoline and nC_5 to be used as feed for the C_5/C_6 isomerization unit, less the utility and downstream unit opportunity costs incurred to do so. Lighter feed components, such as *n*-butane, were assumed to always be fractionated into the DIP overhead, and components heavier than nC_5 were assumed to always be found in the DIP bottoms stream. Thus only the disposition of iC_5 and nC_5 components was considered in the profitability calculation.

Therefore, the objective function for optimizing the DIP tower is defined as below:

$$\begin{aligned}\text{DIP upgrade value} = &\ \text{overhead value} + \text{bottoms value} \\ &- \text{feed value} - \text{reboiler steam cost} \\ &- \text{Isom operating cost} - \text{Isom capacity penalty}\end{aligned}$$

The DIP feed and overhead values are calculated as the gasoline blending value of the iC_5 and nC_5 in this stream, with corrections for road octane and RVP of these components versus those of conventional regular gasoline. The DIP bottoms stream is normally processed at the isomerization unit, where 75% of the exiting C_5's are assumed to be iC_5. After this equilibrium conversion, the value of the resulting iC_5/nC_5 stream is calculated at gasoline blending value as described for feed and overhead above. The reboiler steam cost is calculated assuming a 70% generation efficiency from refinery fuel gas, and the isomerization unit operating cost (for fuel, power, and catalyst) was taken to be the same value per barrel as used in the refinery planning model.

During some periods, the refinery isomerization unit has more feed available than can be processed. If additional DIP bottoms is produced, less capacity is available to process light raffinate from the aromatics extraction unit. The Isom capacity penalty, or the opportunity cost for processing additional DIP tower bottoms at this unit, was therefore estimated by evaluating the octane upgrade of light raffinate.

20.3.4.5 Offline optimization results Once the economic evaluation criteria were determined, it was implemented in the simulation model. Numerous case studies were conducted via process simulation to determine the optimum operating point for the DIP tower under different scenarios. It quickly became apparent that in nearly all economic and operating situations, maximizing the DIP reboiler duty up to the maximum limit gave the highest profitability.

For subsequent case studies, the simulation was completed with maximum reboiler duty, and tower pressure and nC_5 content of the overhead product were also fixed. These constraints completely specified the tower operating conditions.

The DIP profitability was first examined for scenarios where the isomerization unit has available capacity. The DIP feed rate and overhead nC_5 content were varied, and profitability calculated, as shown in Figure 20.20. This analysis showed that the optimum target was around 5% of nC_5 in the overhead, regardless of the tower feed rate.

Profitability was then examined assuming the isomerization unit was at its maximum charge rate, and additional production of DIP bottoms would result in bypassing of light raffinate around the Isom, direct to gasoline blending. The cost of losing the light raffinate octane upgrade can vary between \$2 and 5/bbl, and so simulation cases were completed for both of these scenarios as shown in Figure 20.21. In these cases, the optimum nC_5 in DIP overhead target is dependent on the charge rate to the tower. At low charge rates, the available reboiler duty is sufficient to obtain good separation between iC_5 and nC_5 components so that minimal iC_5 is lost into the DIP bottoms when targeting 5% nC_5 in the overhead. At higher charge rates, more iC_5 is lost to the bottoms stream, and it is more profitable to increase the overhead nC_5 target, reducing the DIP bottoms rate to the Isom unit and allowing additional raffinate upgrading. Thus the optimal nC_5 in overhead target varies between 5 and 20%, depending on DIP charge rate and the value of light raffinate upgrading.

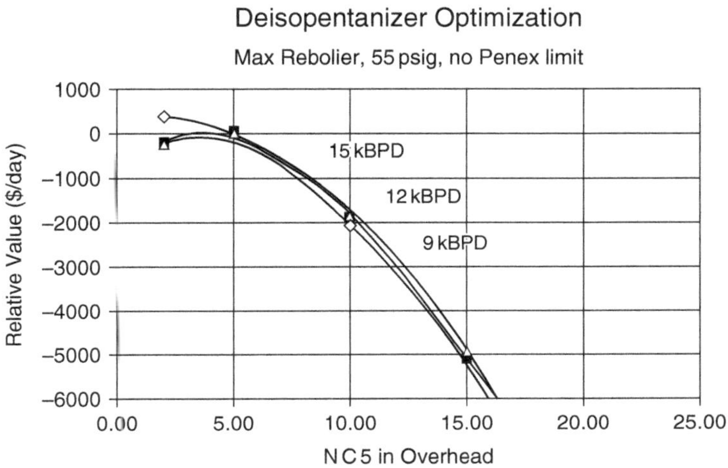

FIGURE 20.20. Optimization without Isom capacity constraint. *Source:* From Loe B, Pults J (2001). Reprinted with permissions of American Institute of Chemical Engineers.

Based on a comparison of the optimal tower operation as determined above and the historical performance, an incentive of around US MM$ 1.5/year was identified to improve DIP fractionation.

20.3.4.6 Optimization implementation In order to realize the benefit indicated by the optimization results above, several unit hardware, process control, and operating philosophy changes were needed. Firstly, it was clear that both reboilers would be required, so the spare bundle was leak-tested and returned to service by operations. Several fin fan motors were also quickly repaired to ensure that design condenser duty was available and overpressurization of the tower would not limit the reboiler duty that could be applied.

Secondly, the optimum target for the DIP overhead nC_5 was implemented into APC system. The APC controller on the DIP DCS system was reconfigured to operate at this nC_5 target while maximizing the reboiler duty as limited by the high limit on tower pressure. This APC system allowed the DIP operation to be maintained at an economic optimum, accounting for the isomerization unit capacity and the economics of the day.

Thirdly, communication of the new operating philosophy was also critical in improving DIP performance. Operators and unit supervisors were trained on the importance of always maximizing the reboiler duty and setting the nC_5 in overhead target. The iC_5 content of the bottoms stream was still measured by lab and inferred analysis, but this was no longer a tower control variable. The reboiler duty and overhead nC_5 were tracked on a daily basis, and performance for these key performance indicators (KPIs) was discussed at weekly operations and planning meetings.

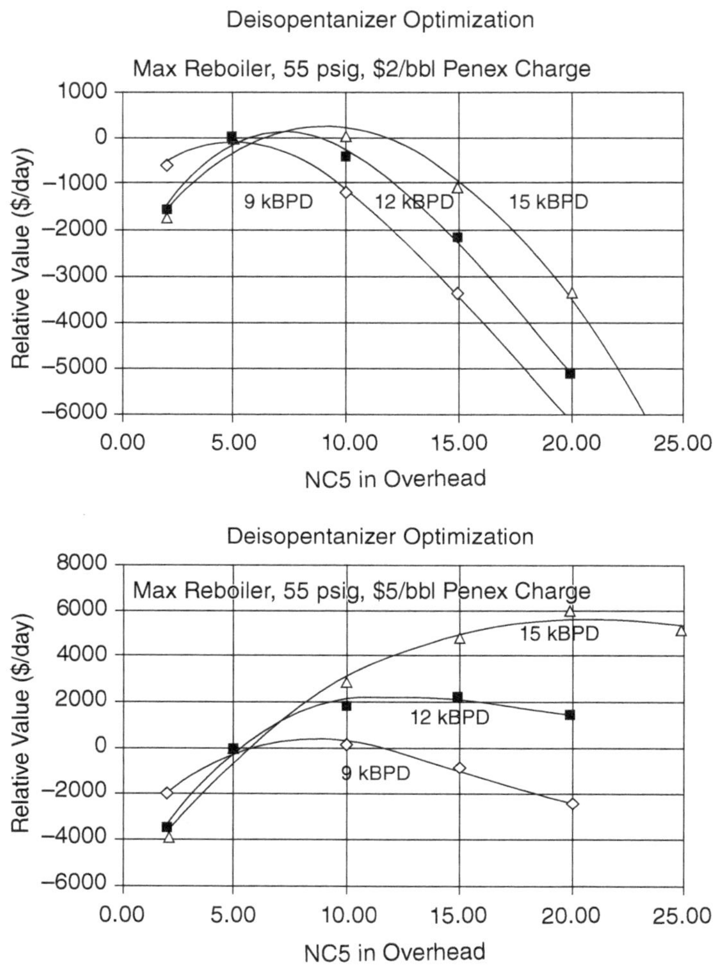

FIGURE 20.21. Optimization with Isom capacity constraint. *Source*: From Loe B, Pults J (2001). Reprinted with permissions of American Institute of Chemical Engineers.

20.3.4.7 Online optimization results The economic benefit of improving DIP fractionation was tracked on a monthly average basis shortly after implementation of the new optimization strategy. Economic performance versus the baseline operation is shown in Figure 20.22 using actual monthly averaged economics and unit operating and lab data. Monthly benefits of over $100,000 were achieved in several cases during the summer months, when octane values were at their highest level. As octane values dipped during spring and fall months, benefits from the improved DIP fractionation dropped off as well.

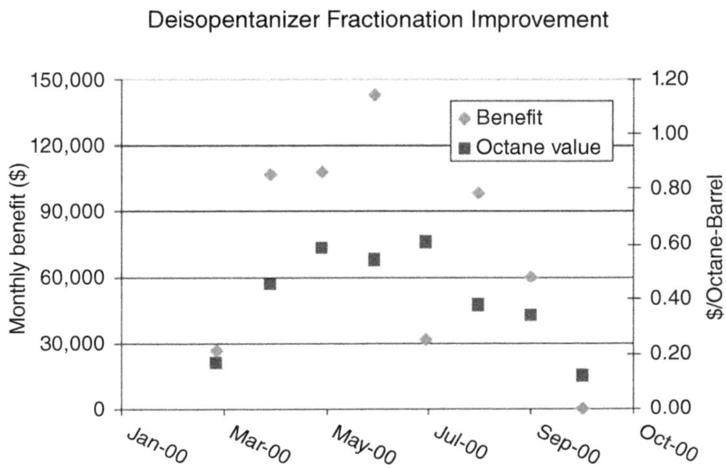

FIGURE 20.22. DIP economic improvements. *Source*: From Loe B, Pults J (2001). Reprinted with permissions of American Institute of Chemical Engineers.

A significant drop in benefits can be seen in Figure 20.22 for July. This was due to poor operation of the DIP tower, caused by a high butane content in the tower feed from the crude unit stabilizers. The C_4's caused the DIP tower pressure to increase up to its safe operating limit, and the reboiler duty was cut back to avoid overpressuring the tower. This resulted in a reduction in fractionation efficiency and profitability during part of July and represented one of the challenges encountered in sustaining the improved DIP performance.

20.3.4.8 Sustaining benefits Without ongoing attention, optimization improvements from initiatives such as that on the DIP tower tend to fade over time, for a multitude of reasons. Challenges to the new level of performance must be tackled as they arise, whether they result from hardware or control problems, misunderstandings, or operating changes in other parts of the refinery. Tracking the economic benefits of the initiative is a critical element of sustaining the change, since knowing the lost profits associated with a loss in performance helps to set work priorities within the refinery. Several problems occurred during the first 6 months of improved DIP operation, which reduced the profit derived from this tower, and these issues were quickly addressed to sustain the improvement.

During the summer months, a change in routing of a portion of the refinery condensate resulted in a hydraulic constraint on the amount of condensate that could be removed from the DIP reboilers into the condensate header. This caused reboiler duty reduction, leading to fractionation efficiency drop. The economic calculations clearly showed that the benefit derived from the additional reboiler duty was much higher than the value of recovering the condensate. For this reason, condensate was safely spilled into the sewer until normal condensate header operation was restored, and DIP fractionation was maintained.

A second challenge occurred when it was noticed that the DIP reflux drum temperature had increased above 140 °F, which was higher than the recommended rundown limit to tankage. Initially reboiler duty was decreased to ensure safe operation. However, in meeting with the tank farm operators about the problem, it was found that the DIP overhead mixed with several other much larger streams before entering a gasoline blending tank. Calculations showed that the effect of the higher DIP rundown temperature on the tank temperature was minimal and DIP reboiler duty was again increased.

Another problem in maintaining reboiler duty was identified when the DIP tower pressure increased due to butanes in the feed, as mentioned above. It was found, however, that the maximum tower pressure was set well below the vessel design pressure, and so the safe operating limit of the tower could be increased after an appropriate management of change (MOC) review. Simulation modeling showed that operation at the higher pressure did not significantly impact the tower profitability if the reboiler duty could be maintained near the maximum. The pressure limit was increased, and performance of the tower again improved.

A fourth reduction in DIP profitability was noticed when the APC controller began cutting reboiler duty, for no apparent reason. Further investigation showed that the inferred nC_5 content of the overhead stream had been deviating from the lab value for a few days because of a computer glitch. This was quickly corrected and DIP operation returned to normal.

Although all of these obstacles reduced profitability for a short period, timely identification and resolution of the constraints averted potentially long periods of underperformance.

REFERENCES

Gupta A (2020), Digital transformation in refining & petrochemicals: A digital twin approach. In: 24th Refining & Petrochemicals Technology Meet, Bengaluru, India, January.

Honeywellprocess.com (2019), Honeywell Cloud APM Helps Avoid Machinery Downtime within its PMT Division—Case Study, May.

Loe B, Pults J (2001), Implementing and sustaining process optimization improvements on a de-isopentanizer tower. In: AIChE Spring Meeting, Houston, April.

Lu JZ (2014), Bridging the gap between planning and control: A cascaded MPC approach. In: The 19th World Congress of the International Federation of Automatic Control, Cape Town, South Africa, August 24–29.

Underwood AJV (1948), Fractional distillation of multi-component mixtures, Chem. Eng. Prog. **44**, 603.

White DC (2012), Optimizing Energy Use in Distillation, Chemical Engineering Process (CEP), AIChE, March.

Zhu XX (2014), Energy and Process Optimization for Process Industries, Wiley/AIChE.

INDEX

absorbers, 142
adsorption, 143
advanced pattern recognition model, 52
advanced process control (APC), 194, 233, 317, 493
 for ammonia plant, 232
 benefits, 232
 for copper smelter fired heaters, 231
 distillation column application, 231
 dynamic operation modeling, 226–229
 features, 221
 for fired heaters, 231
 hydrocracker reaction, 231
 implementation challenges, 232
 model predictive control, 225–226
 MPC-based optimization approach, 230
 multivariable control, 222–225
 operator acceptance, 232
 for pulp digesters, 231
advanced regulatory control, PID controller modeling
 cascade control, 217–218
 control parameter tuning, 220–221
 disturbance input changes, 220
 feedforward control, 218, 219
 quantitative control performance measures, 219–221
 setpoint input change, 219–220
alternative hypothesis, 75
American Petroleum Institute's (API) RP 584, 280
analyzer data, 38
APC *see* advanced process control (APC)
APM *see* asset performance management (APM)
aromatics complex optimization, 482
Arrhenius activation energy, 111, 112
Arrhenius equation, 111
artificial intelligence (AI), 13, 437
 characteristics, 40–41
 general, 50–51
 narrow, 50
artificial neural networks (ANNs), 51
 description, 236
 perceptron, 237–242
assessment matrix score, 445–446

Digitalization and Analytics for Smart Plant Performance: Theory and Applications, First Edition. Frank (Xin X.) Zhu.

assessment prioritization matrix, 444
asset management
  digitalized
    digital twin for asset performance management, 285
    intelligent models, 286–290
  gaps in current practice
    ineffectiveness in communication and execution, 284
    knowledge and skill gap, 284
    no sufficient consideration of process conditions, 283–284
  predictive maintenance, 277
  preventive maintenance, 276
  proactive
    advanced pattern recognition, 278
    condition indicator (CI) method, 279
    fault-injection method, 278–279
    operating windows, 280–283
  for process safety (*see* process safety management)
asset performance management (APM), 10, 426
  Connected Plant functions, 26–27
  digital twin for, 285
    descriptive models, 58
    fault detection and diagnostics, 58
    predictive maintenance, 58–59
    prescriptive maintenance, 59
    process optimization, 59
    rigorous models, 58
  Honeywell PMT Specialty Material plant, 299–300, 477–479
asset template, 425
autocorrelation, 101–103

backpropagation, 242
batch reactors, 127
Bayes' theorem, 73–75
best-in-operation energy performance (OEP), 338–339
big data, 65
  definition, 35–36
black-box models, 39
bubble chart for project selection, 460–461
business case evaluation, 459

catalyst deactivation, 135–136
catalyst effectiveness factor, 117–120
catalyst poisoning, 135
catalytic reforming process, 175
  optimization results for, 174, 176
  reaction network for, 124
causality *vs.* correlation, 96–97
central limit theorem, 86
Cetane Index of Seasonal Diesel, 496–497
chemical efficiency, 7
χ^2-test, 87
cloud-based digital twins, 26
cloud-based performance monitoring, 432–433
collective intelligence, 15–16
combined feed exchanger (CFE), 341
competency control, 397
competency development, 389–390
competency management, 397–398
computerized maintenance management systems (CMMS), 297
concave functions, 163–165
conditional probability, 72–73
condition-based monitoring (CBM)
  and meter diagnostics, 257–258
  and meter uncertainty, 260–267
    CBM packaged solution, 261
    with near-real-time uncertainty analysis, 262, 263
    phase 1 and phase 2 implementations, 260–261
    technically auditable measurement, 263, 265
connected metering, 254, 258–260, 273
  future of, 270
    within a digital mesh, 272
    future connected 4D ecosystem, 271
  IIoT connected architecture for, 262, 264
Connected Plant, 477
  description, 12, 19, 23
  digitalization
    artificial intelligence, 13
    collective intelligence, 15–16
    components of digital system, 12–13
    computing power, 13
    digital twin, 13
    human's role, 16–18
    Internet of Things, 13
    objectives, 13–14

digital twin software features
cybersecurity, 32
IIoT as data infrastructure, 29
integrate simulation and analytics, 30–31
leverage domain knowledge, 31
ML algorithm, 30
visualization and collaboration, 31–32
digital twin system
benefits of, 28–29
description, 27
foundation for, 28
functions of
asset performance management, 26–27
integrated energy, 25–26
Process Optimization Advisor, 25–26
profit performance optimization for total site, 27
reliability enhancement, 24–25
workforce competency management, 27
human–digital interactions, 32
key activities and data, 23
metering ecosystem, connectivity within, 267, 268
for operation excellence, 32
outcome for customers, 11–12
Connected Plant Performance (CPP), 19
consistent operation patterns, 7, 8
console operators, 398–400
continuous flow reactors, 127–128
continuous flow stirred tank reactor (CSTR), 127–128
process model for, 43–44
reactor volume, 130
control chart
multivariate, 71–72
stable and unstable operation, 70–71
control limits
normal distribution, 70–72
Six Sigma, 88, 89
control parameter estimation, 206–207
control variables (CVs), 193
convex functions, 163–165
convexity, 163–164
convex optimization problem, 165
convex space, 165
convolution neural network (CNNs), 247
corporate energy audit, 364
corporate-wide integrated documentation system, 457
correlation models
linear, 94–100
nonlinear, 100–101
cost-based cleaning strategy, 179
critical exponents, 94
crystallization, 143
cybersecurity, 434–436

Danish Institute Industry 4.0, 439
data analytics, 437
need for, 38–39
process, 39–40
data-driven models, 36, 39
data flow manufacturing supply chain, 422–423
data lake, 453
data science
vs. conventional data analytics techniques, 36
description, 36
data technology (DT), 4, 13
data wrangling, 463
debutanizer column, 487
decoupling, 223
deep learning, 51, 237, 239
applications, 52–53, 247
deep reinforcement learning, 235
for model sustainability, 57
deep neural networks, 247
deep reinforcement applications, 53–54
deisopentanizer (DIP) fractionation tower, 501
operational improvements for
current operation assessment, 181–183
defining objective function, 183–184
defining project basis, 180–181
DIP economic improvements, 187
operation monitoring, 187
optimization implementation, 185–187
simulation model, 183
simulation results at designed reboiler duty, 183
simulation results *vs.* operating data, 183
sustaining benefits, 187–188
verification and what-if analysis, 184–185

delta yield method, 153
design data architecture, 453
design efficiency, 7
digital economy, positive-unexpected consequences of, 267–269, 274
digitalization, 3–4, 20
  artificial intelligence, 13
  collective intelligence, 15–16
  components of digital system, 12–13
  computing power, 13
  digital twin, 13
  human's role, 16–18
  Internet of Things, 13
  objectives, 13–14
  readiness assessment
    benchmarking digital readiness, 444–445
    building blocks and pillars, 439–440
    definition, 438
    organization pillars, 443–444
    process pillars, 442–443
    takeaways from the, 445–449
    technology pillars, 439, 441–442
digitalized asset management
  digital twin, asset performance management, 285
  intelligent models, digital twin cornerstone
    advantages of, 289–290
    first principle models, 287
    machine learning models, 287–288
    optimization models, 288–289
    process operation simulation, 286
    rule-based models, 287
    statistic models, 287
digital operation efficiency tools, 8
digital project evaluation and development
  basics of economic analysis
    internal rate of return, 474–475
    net present value approach, 473–474
    return on investment approach, 472–473
  remarks
    empower end users, 466–467
    fit into the organization workflows, 465–466
    include subject matter experts, 466
    project basis and data, 467–468
    root cause analysis, 468–469
  S-curve for project review and management, 469
    create and update, 472
    progress, 470–471
    project growth, 469–470
    slippage, 470–472
  steps
    analytics, 463
    architecture design, 462
    deployment, 464
    getting data, 463
    go/no go decision, 465
    identify the problem and define use cases, 461–462
    optimization, 465
    project lookbacks, 464–465
    prototype for proofs of concept, 464
    visualization, 463
  value and risk evaluation
    bubble chart for project selection, 460–461
    business case generation, 459–460
digital projects
  distillation optimization
    debutanizer column, 487
    economic value function, 488–489
    operating pressure, 492–493
    operating targets with column bottom temperature, 488–481
    operating targets with column reflux ratio, 490–492
    product specifications and prices, 487–488
  heat exchanger fouling mitigation
    change bundle type, 485–486
    desalting performance, 485
    flow path change in maintenance, 485–486
    fouling mitigation options, 483
    measuring and sustaining benefits, 486–487
    offline cleaning of crude–HVGO exchanger, 484
    price for fouling mitigation, 481–482
    rating assessment, 482
    root cause analysis, 482–483
  integrated planning and APC, 493–499
  overall process optimization

current operation assessment, 499–500
design basis, 498–499
objective function, 501–502
offline optimization, 502–503
online optimization, 504–505
optimization implementation, 503–504
simulation, 500–501
sustaining benefits, 505–506
digital transformation, 3
management system, 454–455
accountability, 456
documentation, 457
scorecard and tracking, 456–457
sustainability, 455
technical target system, 457
work processes, 456
digital twin, 13, 446
application cases
APM drives capacity improvement, 477–479
aromatics complex optimization, 480
plant reliability and optimization, 479–480
product quality prediction for polymerization process, 480–481
resolving operation gaps in a FCC unit, 479
for asset performance management, 285
descriptive models, 58
fault detection and diagnostics, 58
predictive maintenance, 58–59
prescriptive maintenance, 59
process optimization, 59
rigorous models, 58
within automation-of-maintenance agenda, 266
Connected Plant
benefits of, 28–29
cybersecurity, 32
description, 27
foundation for, 28
IIoT as data infrastructure, 29
integrate simulation and analytics, 30–31
leverage domain knowledge, 31
ML algorithm, 30
software features, 29–32
visualization and collaboration, 31–32
for energy management, 360–361
Honeywell Forge
asset performance management, 433–434
benefits of, 434
descriptive models, 430
fault detection and diagnostics, 430
predictive maintenance, 430–431
prescriptive maintenance, 431
process optimization, 431
process optimization advisor, 431–432
process reliability advisor, 431–432
profit performance monitor, 432–434
rigorous models, 429–430
integration of multiple digital twins, 16
diminish return model, 92
direction, linear regression model, 96
distillation, 142
modeling, 136–138
optimization
debutanizer column, 487
economic value function, 488–489
operating pressure, 492–493
operating targets with column bottom temperature, 488–491
operating targets with column reflux ratio, 490–492
product specifications and prices, 487–488
distribution models
exponential decay probability distribution, 90–92
exponent *vs.* power law, 94
normal, 86–90
power law distribution, 92–94
dynamic modeling approach, for process control, 194
based on first principles, 195–197
based on step testing, 204–206
conservation balance equations, 195
control parameter estimation, 206–207
Laplace transformation and transfer function, 201–204
linearization about the operating point, 197–204
state space model, 199–200

Economic Development Board (EDB), 439
edge connectivity, 424–425
effective diffusivity, 118

elementary reaction, 111
rate law, 113
empirical models, 109–110
energy audit, 364
energy benchmarking
EPI, 338
indications, 337–338
OEP, 338–339
PEP, 339
TEP, 340
energy intensity
calculate process, 331–335
definition, 328–329
fuel equivalent, 329–331
remarks for, 337
for total site, 333, 336–337
energy management
accountability, 362
corporate energy audit, 364
digital twin for, 360–361
documentation, 363
scorecard and tracking, 363
technical target system, 363–364
work processes, 362–363
energy performance index (EPI), 338
engineering data, 38
engineering technology (ET), 4, 13, 437
engine health management (EHM), 453
enterprise data, 424–426
enterprise-wide cybersecurity solutions, 32
equation-oriented (EO) simulator, 141, 142, 178
equilibrium-stage based distillation modeling approach, 137–138
European General Data Protection Regulation (GDPR), 415
exponential decay probability distribution, 90
diminish return model, 92
half-time prediction, 91–92
exponent *vs.* power law, 94
extractive distillation, 142
extreme value theory, 93

"face-to-face" learning, 392
fault detection and diagnostics (FDD), 58, 430
fault detection, PCA and PLS for, 106
feedforward neural networks, 239 *see also* multilayer perceptron (MLP)
field operator, 400–402
first-order-plus-dead-time (FODT) model, 206
first principle models
and characteristics, 42–45
description, 109
differential and algebraic equations, 39–43
fixed bed reactors, 128–129
fixed time-based cleaning cycle strategy, 179
flow meter
fundamental requirement, 253
history, 254–255
fluid catalytic cracking (FCC) unit, 304, 358
fluidized bed reactors, 129
Forge Connect, 424
F-test statistic, 82, 87
fuel equivalent
definition, 329
energy intensity based on, 331
for fuel, 330
for power, 330
for steam, 330
fundamental models *see* first principle models
Furnace Fuel Oil, 494
future connected 4D ecosystem, 271

gas chromatography (GC), 120
general artificial intelligence, 50–51
General Data Protection Regulation (GDPR), 415
generalized linear mixed model (GLMM), 103
goodness of fit, linear regression model, 96

half-time prediction, 91–92
hazard and operability (HAZOP), 291, 294
Hazard and Operability Study, 294
heat exchanger clean cycle optimization, using Excel, 180
heat exchanger fouling mitigation
change bundle type, 503–504
desalting performance, 485

flow path change in maintenance, 485–486
fouling mitigation options, 483
measuring and sustaining benefits, 486–487
offline cleaning of crude–HVGO exchanger, 484
price for fouling mitigation, 481–482
rating assessment, 482
root cause analysis, 482–483
heavy vacuum gas oil (HVGO), 482
helix shell-and-tube exchanger, 485–486
heterogeneous reactors, 128
homogeneous reactors, 128
Honeywell and Piccolo Control Services, 493
Honeywell Connected Industrial (HCI), 481
Honeywell Forge
cybersecurity, 434–436
data infrastructure, 421–423
digital twins
asset performance management, 433–434
benefits of, 434
descriptive models, 430
fault detection and diagnostics, 430
predictive maintenance, 430–431
prescriptive maintenance, 431
process optimization, 431
process optimization advisor, 431–432
process reliability advisor, 431–432
profit performance monitor, 432–434
rigorous models, 429–430
schematic, 421–422
visual display of, 31
working of, 423–424
analytics models, 426–427
deployment, 428
edge connectivity, 424–425
enterprise data, 424–426
knowledge base, 427–428
visualization and collaboration, 427–429
Honeywell PMT Specialty Material plant, 477–479
background, 299
benefits, 300
challenges, 299
results, 300
solution, 299–300
Honeywell's Energy Dashboard, 363
human impact of digitalization, 413–416
human neuron activity, 236–237
human's role, in digitalization, 16
missing-in-the-middle components
explainer, 17–18
sustainer, 18
trainer, 17
hybrid models, 41
hydrocracking unit (HCU), 321

inconsistent operation patterns, 7, 8
independent protection layers (IPLs), 291
industrial digital system, 13, 14
Industrial Internet of Things (IIoT), 254, 259, 437
industries
challenges
asset management, 9–10
competency management, 10–11
operation variability, 7–8
optimization, 8, 9
process efficiency improvement, 7–8
rigorous model-based control, 8–9
safety management, 10
supply chain management, 6, 7
workflows integration and automation, 11
Industry 4.0, 442, 445–446
inflection point, 167, 168
information technology (IT), 4, 13, 275, 437
initiating event frequency (IEF), 291
integrated energy, Connected Plant functions, 25–26
integrated kinetics and reactor modeling, 131–134
integrated production planning and process control
challenge, 303
decomposition strategy
demonstration, 310
intermediate flows, 310–313
master model and submodels, 309
with recycles, 313
MPC-based integration approach
features of, 318–322
1 to n site-wide optimization model formulation, 314–318

integrated production planning and process control (*cont'd*)
  rigorous model-based integration approach, 323–324
  site-wide optimization
    current practice in, 304
    example, 307–308
    NLP model, 304, 305
    planning model and MPC models, 304, 306
integrity operating windows (IOW)
  critical limit, 280
  dynamic IOW limits, 281–282
  informational limit, 281
  integration of, 283
  standard limit, 280
Intelligence, 442
internal rate of return (IRR), 472
Internet-based social media, 394
Internet of Things, 13
Interoperability, 441
irony of automation, 413
ISA106-compliant modular procedure automation solution, 447
isothermal reactor, process model for, 43–44

Karush–Kuhn–Tucker (KKT) approach
  necessary conditions, 166
  second derivative (Hessian), significance of, 167
  sufficient conditions, 166–167
key energy indicators (KEI)
  application of, 340
  definition, 341–342
  deheptanizer column, xylene isomerization unit
    problem, 346–347
    reboiling duty and toluene, 347–350
  digital twin for, 360–361
  economic evaluation for, 350–354
  product fractionation section
    effective measure for the energy needs, 345
    energy needs development, 345–346
    understand process characteristics, 344–345
    understand the energy needs, 345
  reaction section
    effective measures for the energy needs, 343–344
    energy needs development, 344
    understanding the energy needs, 342–343
    understanding the process, 342
  remarks for, 346
  simplify, 342
key indicators, 457
knowledge base, 427–428
knowledge transfer, 391, 393–395

Laplace transfer function, 201–204
Laplace transformation, 201–204
layer of protection analysis (LOPA), 291–294
least squares regression, 95
left-tail *p*-test, 79, 80
level indicating transmitter (LIT), 292
level-to-flow cascade control, 218
linear correlation models
  multivariable, 96–98
  single-variable, 95–96
linearization about the operating point
  Laplace transformation and transfer function, 201–204
  nonlinear function, 197, 198
  state space model, 199–200
linear programming (LP)
  optimization, 148
    assumption, 151
    delta yield vector and base vectors, 154–156
    incorporating the base–delta yield model, 156–158
    linearization about current operating point, 152
    linear objective function, 149
    nonlinear function, 152–153
    refinery optimization problem, 149–152
  planning model, 158–159
liquid–liquid extraction, 143
local optimum, 162–163
lumping kinetic modeling (empirical based), 120–122

machine learning (ML), 40, 41, 51, 55, 437, 465 *see also* artificial neural networks (ANNs)
  deep learning applications, 52–53, 247
  deep reinforcement applications, 53–54
  and first principle models, integration of, 55–56
  reinforcement learning, 243–246
  robust low-rank matrix approximation, 56
  straightforward applications, 246–247
  supervised learning, 51–52, 242–243
  unsupervised learning, 52, 243
magnitude, linear regression model, 96
manipulated variables (MVs), 193
Marine Diesel, 494
mathematical programming, 41
mean, 67
  and variance, 93–94
mechanical integrity (MI) SMEs, 283
mechanism-driven models, 36
mechanistic models *see* first principle models
median, 67
membranes, 143–144
merchant refinery model, 369
meter diagnostics and condition-based monitoring, 257–258
metering devices
  condition-based monitoring
    and meter diagnostics, 257–258
    and meter uncertainty, 260–267
  flow meter, 253–255
  supervisory systems, 255–257
metering supervisory computer (MSC), 256–257
micro-kinetic modeling (MKM), 124–125
mixed integer linear programming (MILP), 171
mixed integer nonlinear programming (MINLP), 171
mixed models, 103
mixed reality (MR), 393
model calibration, 323
model predictive control (MPC), 304, 305
model predictive control (MPC)-based integration approach
  features of
    dynamic path from current to optimal point, 319
    feedback loop for solution consistency, 319–320
    integration of planning optimization via multiscale modeling, 318–319
    proxy limits, 320–322
  vs. rigorous model-based approaches
    dynamic path, 325
    model maintenance, 325
    process nonlinearity, 324–325
  three-tiered structure of, 314, 315
  1 to n site-wide optimization model formulation
    tier 1 formulation, 315–317
    tier 2 formulation, 316, 317
    tier 3 formulation, 317–318
model predictive range control (MPRC), 314
modular-based simulator, 178
molecular-type homologous series (MTHS) matrix, 122
  for gasoline stream, 123
moving bed reactors, 129
multi-effects models, 103
multifunction reactors, 128
multilayer perceptron (MLP)
  deep learning, 239
  for predictive control of chemical reactors, 240
  for recognizing handwritten digits, 240
  training of, 241–242
multilevel models, 104
multiphase-phase reactors, 128
multi-plant product distribution, 49
multivariable linear correlation models, 96–98
multivariate control chart, 71–72
Murphree tray efficiency, 138

naphtha reforming process modeling
  calibration, 140–141
  feed composition data, 139
  four-heater reformer components, 139, 140
  objective, 138
  simulation methods, 141–142
  simulation software, 140

naphtha reforming reactions, 131
  dehydrogenation, 121
  hydrocracking
    of naphthalene, 122
    of paraffin, 121
  PNA lumps for, 121
  ring closure, 121
narrow artificial intelligence, 50
net energy, 329
net present value (NPV) approach, 477
neural networks, 237
NIST Cybersecurity Framework, 435–436
No.1 Diesel, 494
nonconvexity, 163–164
nonconvex optimization problem, 166
nonelementary reaction, 111
nonelementary reaction rate law, 114
nonlinear correlation models, 100–101
nonlinear optimization problem
  goal of, 161
  naphtha reforming reaction, 167
    mixed integer optimization, 171
    overall optimization model, 170–172
    reaction kinetic modeling, 168–169
nonlinear programming (NLP) problem, 161
non-regenerated adsorbers, 143
normal distribution, 65
  central limit theorem, 66–67
  characterization, 66
  control limits, 70–72
  covariance, 68–69
  mean and median, 67
  probability density function, 66
  standard deviation and variance, 67, 68
normal distribution model
  based on the same average, standard deviation, and intervals, 87
  for Six Sigma, 88–90
  online data distribution, 86
null hypothesis, 75
numerical data
  flow measurements, 37
  level measurements, 38
  pressure transmitters, 37
  temperature measurement devices, 37

off-the-job training, 391, 393
one-tailed test, 82
online analytical processing (OLAP), 384
operating windows
  dynamic IOW limits, 281–282
  integration of IOW Limits, 283
  IOW critical limit, 280
  IOW informational limits, 281
  IOW standard limit, 280
operational excellence, 14
operation efficiency, 7
operation technology (OT), 4, 13, 437
operation variability, 90
operator console, 404
optimality conditions *see* Karush–Kuhn–Tucker (KKT) approach
optimization modeling, 41
optimization models
  building process, 190
  components, 48
  description, 47, 147
  determining optimal values for decision variables, 189
  distillation reflux rate, 49
  effective sales tools, 190
  equality functions, 48
  inequality constraints, 48
  linear programming problems, 148
  model-based control, 190
  nonlinear programming problems, 148
  objective function, 48
  performance prediction, 189
  purposes, 148
  quadratic programming problems, 148
  sensitivity analysis, 189
order of reaction, 111
organization pillars, 443–444
outliers, 100
overall equipment effectiveness (OEE), 443

packed catalytic reactor, reaction volume, 130
partial least squares (PLS) regression, 105–106
pathway kinetic modeling (semiempirical based), 122–124
Pearson correlation coefficient, 98–100
peer energy performance (PEP), 339
perceptron, 237

multilayer perceptron
deep learning, 239
for predictive control of chemical reactors, 240
for recognizing handwritten digits, 240
training of, 241–242
single perceptron, training of, 238–239
petroleum industry, workflow integration
current supply chain management exist, 382–383
integrated work process for supply chain management, 383–385
manufacturing supply chain, 367–368
supply chain digital twin
integration of engineering models with, 388–389
workflow integration and automation, 385–388
supply chain elements
crude oil evaluation, 372–373
linear program (LP) tool, 369
planning, 373–375, 381
price forecasting, 372
process operation and control, 380, 382
product demand, 371–372
refinery model, 369
refinery scheduling, 375–382
strategic planning, 370–371
supply and distribution (S&D), 369
PID block
comparator, 210, 211
controlled variable, 209
manipulated variable, 209–210
PID control attributes, 210, 211
PID control modes, 210, 211
PID controller modeling
advanced regulatory control
cascade control, 217–218
control parameter tuning, 220–221
disturbance input changes, 220
feedforward control, 218, 219
quantitative control performance measures, 219–221
setpoint input change, 219–220
derivative control mode, 212, 214–215
integral control mode, 211, 214
noninteracting form, 215–216
proportional mode, 211
proportional-plus-bias equation, 212–213
PIDFF algorithm, 218, 219
plant reliability and optimization, 479–480
plug flow reactor (PFR), reactor volume, 130
polynomial regression, 101
power distribution model, 86
power law
distribution, 92–94
vs. exponent law, 94
power recovery turbines (PRT), 357, 358
predictive analytic models, 37
predictive analytics, 42
predictive maintenance, 58–59, 288, 430–431
prescriptive analytics, 59
prescriptive maintenance, 59, 288, 431
principal component analysis (PCA), 38–39, 42, 56, 105
Prioritization Matrix, 444
proactive asset management
advanced pattern recognition, 278
condition indicator (CI) method, 279
fault-injection method, 278–279
proactive intelligence, 438
probability of failure on demand (PFD), 291
process control, 193
advanced process control, 194, 221–233
dynamic modeling approach, 194
based on first principles, 195–197
based on step testing, 204–206
conservation balance equations, 195
control parameter estimation, 206–207
Laplace transformation and transfer function, 201–204
linearization about the operating point, 197–204
state space model, 199–200
field devices, 194
production control, 194
regulatory control, 194
process data
analyzer data, 38
characteristics and attributes, 37–39
engineering data, 38
numerical data, 37–38
process data analytics, attributes of, 39–40
process energy, 329

process optimization advisor, 25–26, 431–432
process pillars, 442–443
process plant operation activities, 4–5
process reliability advisor, 431–432
process safety management
  safety digital twin advantages
    life cycle management tools, 296
    specialists, 297
    valuable for plant operators, 297–298
  traditional safety strategy
    HAZOP, 291
    LOPA, 291–293
    safety integrity level analysis, 293–294
product life cycle, 443
product quality prediction for polymerization process, 480–481
profit performance monitor, 432–434
P-test, 77, 79–81

Q-learning, 53
quality assurance (QA) testing, 464
quantile–quantile (Q-Q) plot, 87, 88

range control algorithm (RCA), 226
rate-based distillation modeling approach, 136–137
reaction catalyst, 111, 113
reaction constant, 110–111
reaction kinetic model, 44–45
reaction kinetic modeling
  Arrhenius activation energy, 111, 112
  elementary reaction rate law, 113
  for solid catalyst, 116–120
  lumping kinetic modeling (empirical based), 120–122
  micro-kinetic modeling, 124–125
  nonelementary reaction rate law, 114
  order of reaction, 111
  pathway kinetic modeling (semiempirical based), 122–124
  reaction catalyst, 111, 113
  reaction constant, 110–111
  reaction mechanism, 114–116
  reaction rate model, 110
  reversible reaction, 113–114
  steady-state approximation, 114
reaction mechanism, 114–116
reaction rate, 126
reaction rate model, 110
reactor design, 126
  equations and plots for single reactor, 130, 131
  modeling, 129–131
reactor types, 126
  batch reactors, 127
  continuous flow reactors, 127–128
  fixed bed reactors, 128–129
  fluidized bed reactors, 129
  heterogeneous reactors, 128
  homogeneous reactors, 128
  moving bed reactors, 129
  multifunction reactors, 128
  multiphase-phase reactors, 128
  semi-batch reactors, 128
  single-phase reactors, 128
  slurry reactors, 129
  trickle bed reactors, 129
  two-phase reactors, 128
real-time safety management (RSM) approach, 10
recursion techniques, in linear programming, 159–160
reduced crude (RC), 482
refinery optimization problem, 149–152
refinery scheduling
  crude oil scheduling, 375, 377
  process unit scheduling, 375, 377–378
  product blending and lifting schedule, 375, 378–380
regenerated adsorbers, 143
regression coefficient, 95
regulatory control
  feedback control, 207–208
  feedforward control, 208
  PID loop
    PID block (*see* PID block)
    PV block, 209, 210
    valve block, 209, 210
reinforcement learning (RL), 53–54, 57
  Q-function, 243, 245–246
  Q-learning, 245–246
  with reward/penalty function, 243, 244
Reliability Advisor, for fluid catalytic cracking (FCC) unit, 25
reliability-based cleaning strategy, 179
reliability enhancement, Connected Plant functions, 24–25

return on investment (ROI) approach, 472–473
reversible reaction, 113–114
right-tail p-test, 80, 81
rigorous model-based control, 8–9
rigorous model-based integration approach
  vs. MPC approach
    dynamic path, 325
    model maintenance, 325
    process nonlinearity, 324–325
  site-wide and process systems
    consistency check, 324
    model calibration, 323
    optimization, 324
    state estimation, 323
rigorous process models, 109
Roland Berger readiness index, 439
root cause analysis, 469–470
80–20 rule, 93

safety instrumented systems (SIS), 293
safety integrity level (SIL), 290
scorecard and tracking, 456–457
S-curve for project review and management, 469
  create and update, 472
  progress, 470–471
  project growth, 469–470
  slippage, 470–472
Seasonal Diesel, 494–498
semi-batch reactors, 128
semiempirical models, 110
separation technology
  adsorption, 143
  crystallization, 143
  distillation, 142
  liquid–liquid extraction, 143
  membranes, 143–144
  technology maturity, 144
sequential linear programming (SLP) techniques, 309
sequential quadratic programming (SQP) approach, 162
sequential simulation approach, 141, 142
shift supervisors, 402–403
shop floor connectivity, 452
shop floor intelligence, 452
significant variables, 97
simulation-based optimization
  in Microsoft Excel, 178–180
  with simulation software, 178
Singapore Smart Industry Readiness Index (SIRI), 439
single perceptron, training of, 238–239
single-phase reactors, 128
single-variable linear correlations, 95–96
single-variable regression model, 96
site-wide energy management strategy
  effective energy use via optimizing operation, 355–357
  high capital investment via process changes, 357–358
  low capital investment via increased heat recovery, 357
  minimize waste and losses via diligence, 355
  potential saving, 360
  utility optimization for reduced energy generation cost, 358–359
site-wide optimization, 304
Six Sigma
  control limits, 88, 89
  definition of, 88, 89
  operation variability, 90
six-step digitalization journey, 449–451
  design data architecture, 453
  digital platform implementation at full scale, 454
  opportunity assessment, 451
  pilot program, 453–454
  readiness assessment, 451–452
  use cases, 452
slippage, 470–472
slurry reactors, 129
Smart Industry Readiness Index, 451
solid catalyst, rate kinetic modeling for, 116–120
Solow growth model, 86
standard deviation (SD), 65
standard deviation and variance, 67, 68
state estimation, 323
state space model, 199–200
statistical models
  advanced modeling techniques
    autocorrelation, 101–103
    generalized linear mixed model, 103
    multi-effects/mixed models, 103
    multilevel models, 104

statistical models (*cont'd*)
partial least squares regression, 105–106
principal component analysis, 105
correlation models, 94–101
description, 85
distribution models, 86–94
objective, 85
statistics, 40, 41, 65
Bayes' theorem, 73–75
conditional probability, 72–73
control chart, for minimizing operation variability, 46
correlation, 47
data set applications, 45
normal distribution, 65–72
role in data science, 45
statistic test
alternative hypothesis, 75
for a given confidence level, 76–78
F-test statistic, 82
methods, selection of, 81–82
null hypothesis, 75
P-test, 77, 79–81
statistical significance, 75
t-statistic, 82
type I error, 82–83
type II error, 83
steady-state approximation, 114
step function, 238
stochastic optimization methods
application in process industries, 171, 177
description, 177
Monte Carlo simulations, 177
randomization principle, 177–178
stochastic programming, 171
straight baffle shell-and-tube exchanger, 485–486
straightforward machine learning applications, 246–247
strippers, 142
structure-oriented lumping (SOL) approach, 122
Student's *t*-test, 87
subject matter experts (SMEs), 281, 453, 464
successive linear programming (SLP), 158–160, 324
successive quadratic programming (SQP) methods, 160
supervised learning, 51–52, 242–243
supervised monitoring and diagnosis, 56–57
supervisory control and data acquisition (SCADA) systems, 255–257
supervisory systems, 255–257
supply chain management, 443

technically auditable measurement, 263, 265
technical target system, 457
techno-complexity, 414
techno-insecurity, 414
techno-invasion, 414
technology energy performance (TEP), 340
technology pillars, 439, 441–442
techno-overload, 414
technostress, 414
techno-uncertainty, 414
temperature control loop
control actuators, 208
controller, 208
measurement device, 208
piping and instrumentation (P&I) diagram, 208, 209
theoretical models *see* first principle models
Thiele modulus, 117
TIER framework, 445–446
time series measurement data, 38
time series model, 52
traditional safety strategy
HAZOP, 291
LOPA, 291–293
safety integrity level analysis, 293–294
transfer function
for FODT model, 206
trickle bed reactors, 129
t-statistic, 82
two-phase reactors, 128
two-tailed test, 82

universality, 94
unsupervised learning, 52, 243
UOP Isomar process, 341

vacuum residue, 482
value and risk evaluation

bubble chart for project selection, 460–461
business case generation, 459–460
video-based knowledge sharing, 395
virtual meter, 267, 271
virtual reality (VR), 391–392

wearable, near-to-eye voice-controlled mobile computer, 404–405
white models, 39, 41, 55
workforce competency management, 27
workforce digitalization
communication, 390
competency development, 389–390
decision making, 390
digitalization challenges, 410–411
human impact of digitalization, 413–416
technology churn, 411–412
trust and the ironies of automation, 412–413
empowerment
console operator, 398–400
field operator, 400–402
illustrative roles, 398–399
impact of digitalization, 403–410
shift supervisors, 402–403
enabling the workforce, 390–391
competency management, 397–398
knowledge transfer, 393–395
on-demand instruction, 395–397
training delivery using VR and MR, 391–393
on-the-job performance of workers, 389
World Economic Forum (2018), 439

xylene isomerization unit, 341

Z-statistic, 76
Z-test, 76, 79, 82

www.ingramcontent.com/pod-product-compliance
Lightning Source LLC
LaVergne TN
LVHW011533200926
848484LV00001B/249
9781119634034